The Michigan Historical Reprint Series

Reprints from the collection of the University of Michigan University Library

The Scholarly Publishing Office
The University of Michigan
University Library

PART II.

CHAPTER XIV.

THE Xth CHAPTER OF GENESIS.

"Consilium igitur fuit tractatui de Paradiso pro appendice subnectere breué expositionem decimi capitis Geneseos de humani generis propagatione ex stirpe Noæ. Ex quâ non veteres modo sed et *nouitios interpretes horum ignoratione à sacri Scriptoris scopo sæpe aberasse pateret*. . . . Itaque hoc restat vnicum, vt ad sacram anchoram hoc est ad Scripturam confugiamus: Quæ non solum in genere docet *omnes homines ex vnô semine esse editos*, nempe ex Adamo in creatione, et post diluuium ex Noâ et tribus filiis, sed et recenset nepotes Noæ, et qui populi ex singulis ortum duxerint."

(PHALEG seu DE DISPERSIONE Gentium et Terrarum divisione facta in *ædificatione turris Babel* — auctore SAMVELE BOCHARTO: 1651.) 567

Preliminary Remarks.

Two centuries intervene, as well as many thousand miles of land and water, between the completion of BOCHART's unsurpassable labors and the seemingly-audacious resumption of his inquiries in the present volume. The author of *Geographia Sacra* would smile, with more complacency perhaps than some of our readers, did he know that the edifice raised by his enormous erudition, in old scholastic Belgium, had been taken to pieces stone by stone; and, after a scrutinizing, but frugal, rejection of time-rotted superfluities, has been reverentially rebuilt, in the piny-woods of Alabama, on the rough, though beauteous, shore of Mobile Bay.

It is with some regret that, in order to compress their work into a portable *tome*, the authors lop away unsparingly the evidences of studies to which many months were conjointly and exclusively devoted: but, at present, they must content themselves with the briefest synopsis of *results*. Their references indicate the sources of all emendations proposed — by far the greater bulk of which (with the sole exception of MICHÆLIS's criticisms of seventy years ago)[568] arise from discoveries made by living Egyptologists, Hebraists, Cuneatic-students, and similar masters of Oriental lore. These references will establish, that, in the conscientious application of enlightened learning to the *Hebrew Text of Xth Genesis*, commentaries of the genuine English evangelical school have ever played an insignificant part. Where the latter sometimes happen to be right, their facts are taken—generally at second-hand, and mostly without acknowledgment—from Bochart; and wherever, more frequently, they are wrong, they have either ignored his text or the very-accessible criticism of Continental archæologists. Of trivial value in themselves, such popular commentaries possess less weight in science; and, having wasted their own time in hunting through dozens of them for a new fact or an original observation, the authors will spare the reader's by leaving them unmentioned.

> "*Priscorum mendax commenta est fabula vatum,*
> *Sincerumque nihil, nil sine labe fuit.*
> *Sordibus ex istis densa et caligine lucem*
> *Eruere, humanæ non fuit artis opus.*
> *Desperata aliis unus tentare* BOCHARTVS
> *Ausus, et ignotas primus inire vias.*"

"The ethnographic chart[569] contained in the tenth chapter of Genesis, presents," says Dr. Eadie, "a broad and interesting field of investigation. It carries us back to a dim and remote era—when colonization was rapid and extensive, and the princes of successive bands of emigrants gave their names to the countries which they seized, occupied, and divided among their followers. This ancient record has not the aspect of a legend which has arisen, no one can tell how, and received amplification and adornment in the course of ages. It is neither a confused nor an unintelligible statement. Its sobriety vouches for its accuracy. As its genealogy is free from extravagance, and as it presents facts without the music and fiction of poetry, it must not be confounded with Grecian and Oriental mythe, which is so shadowy, contradictory and baseless—a region of grotesque and cloudy phantoms, where Phylarchs are exalted into demigods, born of Nymph or Nereid, and claiming some Stream or River for their sire. The founders of nations appear, in such fables, as giants of superhuman form—or, wandering and reckless outcasts and adventurers, exhibiting in their nature a confused mixture of divine and human attributes; and the very names of Ouranos, Okeanos, Kronos, and Gaea, the occupants of this illusory cloud-land, prove their legendary character. In this chapter there is, on the other hand, nothing that lifts itself above vulgar humanity, nothing that might, nothing that did not happen in those distant and primitive epochs. The world must have been peopled by tribes that gave themselves and their respective regions those several names which they have borne for so many ages; and what certainly did thus occur, may have taken place in the method sketched in these Mosaic annals. No other account is more likely, or presents fewer difficulties; and, if we credit the inspiration of the writer of it, we shall not only receive it as authentic, but be grateful for the information which it contains. Modern ethnology does not contradict it. Many of the proper names occurring on this roll remain unchanged, as the appellations of races and kingdoms. Others are found in the plural or dual number, proving that they bear a personal and national reference (*Gen.* x. 13); and a third class have that peculiar termination which, in Hebrew, signifies a sept or tribe (x. 17)."[570]

The above scholar-like definition of what Dr. Hales styles "that most venerable and valuable Geographical Chart, *the tenth chapter of Genesis*,[571] indicates the absolute impossibility of obtaining satisfactory glimpses of a large portion of humanity's earliest migrations without discussing, at the very threshold of inquiry, that antique *document.* Apart from this fundamental classification of some human primordial wanderings, bootless indeed would be attempts to follow the cobweb threads of our own ancestral creepings, backward from America to Europe, and thence to their primitive European or Asiatic starting-points. Every aboriginal tradition we Anglo-Saxons cherish, is but a ray of morning light, flitting though it be, projected from the Aurora of our Eastern homes.

"The *Orient,* with her immense recollections that touch the cradle of the world, as this itself touches the cradle of the sun, with her seas of sand, beneath which nations lie forgotten, endures still. She preserves, yet living in her bosom, the first enigma and the first traditions of the human race. In history as in poetry, in religious manifestations as in philosophical speculations, the East is ever the antecedent of the West. We must therefore seek to know her, in order to become well acquainted with ourselves." [572]

But, before the historical character of this *Ethnic map* can be appreciated — before our unhesitating acceptance of it as a witness demonstrably credible — its antiquity, its nature, and its authorship, are indispensable points of preliminary inquiry.

The authors of the present work, impressed with the necessity of using the *Xth chapter of Genesis* as a "ground-text" for a large section of their anthropological researches, coincided in the opinion that an "Archæological Introduction to its study" ought to preface their adoption of its data. In consequence, it was decided, that the labor involved in such undertaking should be allotted to that one of the writers whose Oriental specialities naturally indicated him as performer of the task. Too complex in nature, no less than too bulky in size, to serve for a chapter in the text of "Types of Mankind," this *Archæological Introduction* now becomes a SUPPLEMENT to the work itself; thereby preserving its own unity, at the same time that to the reader it is equally accessible, being bound up in the same volume.

The perusal, then, of the SUPPLEMENT is recommended to the reader previously to his further continuation of this work; because the paragraphs upon Xth *Genesis,* hereto immediately following, are projected under the impression that such will be the natural course.

Which taken for granted, we place before us CAHEN'S *Genèse,*[573] for the Hebrew text of Xth *Genesis,* and proceed to its critical dissection. The method we shall adopt, if at first sight novel, will be found strictly archæological. It would be unphilosophic to set forth with any theory as to age, authorship, or true place, of this *document,* in the arrangement of the canonical books. These points can resile solely through exegetical analysis of the document itself; which — written in the *square-letter* Hebrew character (not invented prior to the third century after c.); divided into words (a system of writing not introduced in the earliest Hebrew MSS.—tenth century after c.); punctuated by the "Masora" (commencing in the sixth, and closing about the ninth century after c.); and subdivided into verses (not begun before the thirteenth century after c.)—now presents itself to our contemplation.

Section A.—Analysis of the Hebrew Nomenclature.

Omitting, for the present, any comment upon *verse* 1: "Behold the generations of the children of Noah, Shem, Ham, and Japheth; they had children after the deluge"—our point of departure is *verse* 2. "The children of Japheth," *eldest* of the three brethren; whose descendants, upon grounds to be justified hereinafter, we denominate

Iapetidæ, *or White Races.*

[Before proceeding, let me mention that, after our *Genealogical Table* was in type, Prof. Agassiz favored me with the loan of by far the most important work I have ever met with on Japethic questions: viz., *Voyage autour du Caucase, chez les Tcherkesses et les Abkhases, en Colchide, en Géorgie, en Armenie, et en Crimée,*[574] *par* Frederic Dubois de Montpereux. Extreme was my satisfaction to perceive that *our results* not only had been anticipated, but that they were so accurate as to demand no alterations of the Table. Following the profound researches of Omalius de Halloy,[575] and of Count John Potocki,[576] the personal explorations of M. Dubois supersede everything printed on "Caucasian" subjects. I have made the freest use of his ethnological inquiries, as will be perceived under each Japethic name; but it is not in my power to convey to the reader adequate knowledge of the maps with which this magnificent folio *Atlas* is profusely adorned. On these, the successive displacements occasioned by the migrations, &c., of ancient "Caucasians" are so skilfully shown, that one's eye seizes instantaneously some 2500 years of history. To take GoMeR, or *Kimmerians*, as an example. Beginning in the

6th cent. b. c.—	Pl. VIII*a*.	gives	"Primitive Georgia before the invasion of the Scythians (Khazars)."
5th "	" IX.	"	"Scythia and Caucasus of Herodotus."
3d "	" X.	"	"Periplus of Scylax Caryandinian."
1st "	" XI*a*.	"	"Tauride, Caucasus, and Armenia of Strabo."
1st cent. a. c.	" XII.	"	"Tauride, Caucasus, and Armenia of Pliny.'
2d "	" XIII.	"	"Arrian's Periplus of the Black Sea."
6th "	" XIV.	"	"Wars of the Romans and Persians."
10th "	" XV*a*.	"	"Massoudi's description of Caucasus," &c.

Now, on such maps, the transplantations of these *Kimmerians* can be followed, almost station by station: so minutely, that one might infer that GoMeR-*ians* became known to the Hebrew geographer after they had abandoned the northern Tauride to the Scythians, b. c. 633, and had settled about Paphlagonia, on the south-eastern side of the Black Sea. And so on with all the *Iapetidæ* of Xth Genesis. It need hardly be said that, in common with Bochart and ourselves, Dubois perceives *nations* and *countries*, and not individuals, in the Hebrew chart.—G. R. G.]

בני יפת—BNI-IP*h*T*t*—"Affiliations of Japhet."—*Gen.* x. 2.

1. גמר—GMR—'Gomer.'

Essentially Indo-Germanic, this name, as well as all those of Japethites, is irresolvable into Semitish radicals; and its Hebrew lexicographic affinities, such as to '*complete, consume*,' &c., are rabbinical, spurious, and irrelevant.

(1 *Chron.* i. 5, 6)—"GOMER, and all his hordes—" (*Ezek.* xxxviii. 6). In Homer and in Diodorus, Κιμμεριοι; in Herodotus, Βοσπορος Κιμμέριος. In Josephus the *Galatæ* are called Γομαρεις; possibly also understood in the Scytho-Bactrian *Chomari, Comari,* of Ptolemy. These are, undoubtedly, the *Gomerians, Cimmerians, Crimeans,* who, under the various forms of *Cymr, Kymr, Kumero, Cimbri, Cambri,* and *Galatæ, Gael, Gauls, Kelts, Celts,* figure as a branch of *Celtic* migrations in later European history. If Celtic migrators be considered anterior to the age of Xth Genesis, we should not hesitate in adopting the Germanic *Sigambri, Sicambri*; or the *Gambrivii,* or the *Gamabriuni,* as memorials of 'Gomer.' Rawlinson evolves 'Tsimri' from the cuneatic legends of Khorsabad.

The name G*i*M*e*R*ian*, in endless forms, is scattered from Asia Minor to Scandinavia, for the following historical reason. About the year B. C. 633, the Scytho-Khazars expelled the Kimmerians from *Kimmericum.* One set of fugitives sought asylum in Western Europe; while the other skirted the eastern shores of the Black Sea; and, settling in and around Phrygia, became known to the writer of Xth Genesis. Bochart had happily remarked "Itaque omnibus expensis terra *Gomer* mihi videtur esse Phrygia, cujus portio est regio Κατακεκαυμένη." This word signifies the '*burnt*-district:' and Dubois thoroughly establishes that the volcanic nature of such Kimmerian localities explains all their mythic associations with the infernal waters, Styx, Phlegethon, Cocytus, Acheron, &c., which cluster around the naphtha-springs and mud-volcanoes of the present *Iénikalé.*

The Tauric Chersonesus, north of the Black Sea, would seem to have been the extremest geographical boundary assumed by the Hebrew writer; and by a simple transposition of letters, GMR (GRiMea) is still apparent in the name of this early Kimmerian halting-place, viz.: the *Crimea.*[577]

2 מגוג—MGUG—'MAGOG.'

Indo-Germanic, or Scythic; and, therefore, not the Hebrew "he who *covers* and *dissolves.*" (*Gen.* x. 2; *Chron.* i. 5; *Ezek.* xxxviii. 2; xxxix. 6).

MAGOG is not associated with GOG until the times of Ezekiel, during the Captivity, from about 'the 30th year' of Nabopolassar, 595 B. C. down to 572 B. C. (*Ezek.* i. 1; xxxix. 17). In the post-Christian but uncertain age of the writer of the Apocalypse (between A. D. 95 and the Council of Laodicea, which rejected it as apocryphal, 360–369, A. D.,) 'GOG and MAGOG' appear together as *nations* (*Rev.* xx. 20); whereas, seven to eight centuries previously, GOG, "the Prince of *Rhos,* Meshech and Tubal," would seem to have been understood as the proper name of a *king.* King James's version (*Ezek.* xxxviii. 2, 3, &c.), by "Chief *prince* of Meshech and Toubal," effaces RAS (*i. e.* RHOS; the river *Araxes,* and the nation *Rhox*-Alani, or Alains), and perpetuates an error detected by Bochart 200 years ago.

Arab tradition, under the appellatives YADJOOJ and MADJOOJ, prolongs the union down to the seventh century after Christ; with the commentary, that they are two nations descended from Japheth; GOG being attributed to the Turks, and MAGOG to the *Geelàn,* the Geli and Gelæ of Ptolemy and Strabo, and our *Alani.*

In ancient Greek and Latin, Γιγας, *Gygas,* read also *Gug*-as, signified *giant;* and oriental legend associated giants with Scythians in the north of Asia. MAGOG has been assimilated to the *Massagetæ* (perhaps *Massa*-Getæ, *Masian*-Getæ, of Mount *Masius*) who are to *Getæ* what MAGOG is to GOG; the prefixes of *ma* and *massa* being considered intensitives to indicate either the most honored branch of the nation, or the whole nation itself. Tacitus and Pliny mention the '*Chau*corum gentes,' and the *Chauci,* among powerful tribes in Germany at their day; and GOG may underlie these migrations.

EZEKIEL groups GOG with *Rhos, Toubal* and *Meshech;* and, inasmuch as Roxalani, Tibareni, and Moschii, no less than the transplanted Crimeans (GOMER), were geographically located in Asia Minor, between the Black Sea and the Caspian, the habitats of them all lay in that region. By Strabo, the country of *Gog-arene* (Gog-aïranian? *aïr* = man; '*man* of Cauc-asus'?) is placed near that of the *Moschi*. Josephus renders the name of Magog by *Scythians;* and Jerome, "Magog esse gentes Scythicas immanes et innumerabiles, quæ trans Caucasum montem et Mæotidem paludem, et prope Caspium mare ad Indiam usque tendantur."

But, ingenious as they are, such etymologies become henceforth superfluous through Dubois's excellent suggestions. The Hebrew word is M*a*-GUG. The first syllable refers to the *Maïotes, Mœtes, Mates, Meotes:* tribes of the Sarmates, *royal*-Medes, Sauro-Madaï, (*i. e.*, Tauric Medians, transplanted from the Taurus to the east of the Caspian,) of the Sea of Azof. The second syllable, GUG, is simply the Indo-Germanic word *Khogh*, 'mountain' (as in the celebrated diamond, *Kôh-en-noor*, '*mountain* of light'); which has been preserved in the Hellenized name *Kauk*-asos, or *Cauc*-asus, from the time of Herodotus, B. C. 430; as also in the "inscription de Périsades, premier archonte du Bosphore, en 349 avant J.-C." Having thus fixed GUG to a 'mountain,' *Cauc*-asos, the root of asos is instantly recognized in the national name of the *Osses, Osseth, Yases, Aas, Asi;* whence the continent of 'Asia' derives its European designation. These *Osses*, or *As*, are traceable in the ancient *Jaxamates*, or Yas-Meotes, as perfectly as in the modern *Jazigees, Yasyghes* (or *Yas*-Djiks), 'Jaz-Djiks'; who now call themselves *Tcherkesses*, by us corrupted into 'Circassians.' They have been likewise termed *Ovsni, Acias, Akas*, and even *Kergis*, by the old travellers; and while the first syllable of their ante-historical name yet floats over the Sea of AS*of* (Azof), and lives in the Abkh-*Ases*-mountaineers, it has been borne to *Asaland* (land of the Asa) no less than to *Asgard* (city of the Asa), in old Scandinavia. In this manner ably sums up Dubois, "As far back as history mounts, she finds within the angle circumscribed between the Cauc-asus, the Palus Méotis, and the Tanais, an *Asia-proper*, inhabited by a people, 'AS,' of Indo-Germanic race:" and we discover, in the *Ma*-ïotes of the 'mountain' *Cauc*-asus, the long-lost and mystified *nation*, Ma-GUG, of Xth *Genesis*.

Thus, this collective name of Magog designated one of many barbarous Caucasian hordes, roaming of yore between the Euxine and the Caspian, including, probably, *Gothic* amid Scythic families; and Gog has left, even to this day, besides the living *Osses*, a trail still visible in the very etymon of his ancient homestead, the CAUC-*Asian* mountains.[578]

3. מדי — MDI — 'Madai.'

Indo-Germanic, or Scythic. Not Hebrew, 'covering,' 'coat,' &c.

The LXX transcribe Μαδοι, in lieu of Μεδοι. The Persian word *madhya*, the 'middle,' its supposed derivation. Herodotus counted seven nations, and says their ancient name was *Arioi*, the 'braves'; that is, *Arii*, 'Arians.' It is probable, however, that the root *aïr*, which in Scythic tongues means 'man,' may have been assimilated to *Ari*, 'lion,' in the alien speech of Semitic nations. The name is spread over a vast area, from *Arhan*, 'Armenia,' through *Irân*, 'Persia,' to the conquering *Aryas, Ayras*, of Hindostan.

In primitive times, the *origines* of all nations were personified; and, according to Strabo, *Medus*, son of the mythological *Jason* and *Medea*, was the progenitor of the Medes. The name *Madah* occurs in the seventh century, written in Assyrian cuneiform, on sculptures from Khorsabad; and Rawlinson transcribes *Mâdiya* from the innumerable legends of Behistun and Persepolis, deciphered through his acumen.

Ragœ 'Media,' was called *Ruka* by the Egyptians of the XVIIIth dynasty; and perhaps *Matai* is Media itself.

The name *Mede* still survives in *Hamadan* (Ecbatana), just as that of *Arian* (Aria, Arii) in the HaRA of 1 *Chron.* v. 26.

They are the *Medes:* and further reference to Scriptural or to classical passages, in their case, is superfluous.[579]

4. יון — IUN — 'JAVAN.'

Indo-Germanic; and not from the Hebrew, 'mud,' or 'oppressor.'

In this instance, the *Masoretic points* (not added to the Text until after the fifth century of our era), and the modern Jewish reading of V for U, alone obscure a name whose literal meaning springs out at first glance.

"The barbarians called all Greeks by the name of *Ionians*," says the Scholiast on Aristophanes: and the Greeks revenged themselves by terming all other people *barbarians.*

The LXX correctly transcribe Ιωυαν; for Ιαονες is the older form in Homer; a name to be distinguished from the later Ιωνες, according to Pausanias. Herodotus recounts how the Athenians, previously called *Pelasgi*, received the name *Ionians*, from ION, son of *Xuthus;* the traditionary ancestor of the Ionian race.

In DANIEL xi. 2, where King James's version renders *Grecia*, the original has IUN; but the age of this document not ascending earlier than B. C. 175–160, in the reign of Antiochus Epiphanes, we go back to the 27th March, B. C. 196, date of the coronation of Ptolemy Epiphanes at Memphis, recorded on the *Rosetta Stone;* where the word Ελληνικοις, in *Greek*, is rendered, on the corresponding *demotic* and *hieroglyphic* texts, by IUNiN: a name given by Egyptians to the Greeks at every age, back to the earliest records we possess in which *Ionians* are mentioned — documents anterior to Xth *Genesis* by some centuries, because ascending to the XVIIIth dynasty.

Upon the Assyrian monuments of Khorsabad, the same name, JAOUNIN, is read by cuneiform scholars, as early as the eighth century B. C.; and upon the Persian sculptures of the Achæmenidan dynasty, in the sixth century B. C., the *Greeks*, as YUNA, or *Ionia*, frequently appear.

Javanas, or *Yavanas*, is the Hindoo appellative of the Greeks, in the "Laws of Manou," who therein are classed among the *Soudras*, or 'degenerates'; and, although the fabulous antiquity of these Sanscrit records has sunk far below the pretensions of the so-called *Mosaic*, their compilation certainly ascends to the fourth century of our era, if not beyond. While, finally, among the Arabs, ancient and modern, *Yoonân* is the generic name for Greeks in general, and *Ionians* in particular.

By IUN, or *Ionian*, the writer of Xth *Genesis* seems to class the Greeks collectively, as far as they were known to him; and *Ionia*, on the western coast of Asia Minor, is the approximate limit of its geographical application.[580]

5. תבל — TtBL — 'TUBAL.'

Indo-Germanic. Not the Hebrew, 'he who is *conducted*,' &c.

The LXX place before *Thubal* another son of Japheth, called *Elisa;* but ISAIAH, by exiling "those who escape" to "*Tubal* and Javan, the *states* afar off," shows that, in the idea of the writer of the second (or spurious) part of the oracles ascribed to this prophet, *Thubal* ranked among distant northern nations of the gentile world. Connected, in EZEKIEL, always with *Meshech*, by whom *Tubal* is immediately followed in Xth *Genesis*, these two nations of the "uncircumcised" must have lain close together in Hebrew geography.

Iberia, from the roots EBR, and υπερ, 'beyond,' or, so to say, 'the *yonderer*,' was the name of an Asiatic country east of Colchis, south of Caucasus, west of Albania, and north of Armenia; in short, corresponding to *Georgia* of the present day; classically denominated *Imeriti.* The substitution of B for M, at once changes the *Imeriti* into the *Iberiti:* to which prefixing the antique particle T, we obtain the *t-Ibarenes* of Herodotus and Strabo: a designation equivalent to *ultra*-Caucasians. The word *Iberian*, in the sense of 'yonderer,' was given to many remote nations by aliens to the formers' autochthonous traditions.

Identified as the Τιβαρηνοι of Strabo, who, by Herodotus, are located with the *Moschoi*, they seem to have been subject to *Gog*, CAUC-*Asus*, in the days of Ezekiel, and to have supplied slaves and brazen vessels to the bazaars of Tyre.

Through the common mutation of R for L, *Tubal* is fixed among the *Tibareni*, (about Pontus, on the south-east of the Black Sea, in the neighborhood of Colchis,) from ante-historical times down to the Christian era; and it is in vain, therefore, that Spanish orthodoxy, in efforts to affiliate its ancestry with some Genesiacal worthy, (confounding the Celto-*Iberes* with the *Iberians* of Asia,) should claim TUBAL as progenitor of *Spaniards*.

"The identity of Thobel, or Tubal, with the *Georgians*," holds Dubois, whilst substantiating Bochart, "is nowadays well recognized; because Flavius Josephus expressly says, that Tubal represented the Iberians of his time, the Iberians of Pliny, of Strabo, of Procopius, who are the Georgians of our day. The transition between *Tubal* and *Iberia* is the *Tibareni* of Herodotus. This name has never been, among the Georgians themselves, that of the nation; they give themselves the generic name of *Karthles*: but it has remained in their capital *Tbelissi*, our Tiflis." The root *υπερ*, *over*, 'ultra,' probably underlies *T-ibar-eni*, and its Hebraicized form of T*t*uBaL; as well in the Hispanian *Iberes*, as in the Caucasian *Iberians* — both being a "people *beyond*."[581]

6. משך — MSK — 'MESHECH.'

Indo-Germanic. Not from the Hebrew, 'drawn with force,' &c.

Erroneously substituted for the Shemite *Mash* (in 1 *Chron*. l. 17), and confounded with the Arabian *Meseq* (in *Psalm* cxx.), by the forty-seven translators of King James's version; mere analogy of sound has led some commentators to behold in MESHECH the parent of the *Muscovites*, incarnated founder of the city of *Moscow!* At the same time that the Arabic version transcribes *Khorassân!*

As above stated, "Tubal and Meshech" were deemed cognate nations by the writer of Xth *Genesis* and by Ezekiel; confirmed by Herodotus — Μοσχους μεν και Τιβαρηνους; and the concurrent testimony of Mela, Pliny, Stephanus, and Procopius, places the Μοσχοι, or Μεσχοι, on the *Moschian* range, adjacent to Iberia, (*Tubal*,) Armenia, and the Colchide, between the Black and Caspian seas; still called *Mesidji-dâgh*, or 'Meshech-mountains,' by the recent Turks. The *Misek* of Rawlinson's cuneatic inscriptions?

More ancient than classical, Hebraical, Assyrian, or other extant annals, is the name of MESHECH. Early as the age of Ramses II., in the fourteenth — fifteenth century B. C., or prior to the fugacious era of Moses, (even supposing the Xth chapter of *Genesis* to proceed from his individuality,) the *Maasu*, [Masii, Moschii,] whose cognomen is still preserved in "Mons *Masius*" of the Taurus chain, are chronicled on Egyptian papyri, inscribed in days contemporary with Ramses's reign.

'Meskhes' is the Georgian appellative for the people of Moskhike, or *Moschic*. They were a mixed population of primitive Phrygians (Thargamosians) and Medes, on the southern slope of Caucasus; who in classical geographies, as the *Mosunicoi*, *Mosynœci*, *Moschici*, are always neighbors of the Colchians, the Tibareni, the Khalybes, &c.; while Ezekiel, as above shown, groups together, in the land of *Gog* (*i. e.*, Caucasus), nations under the sway of the "Prince of Rhos, *Meshech*, and Tubal;" that is, the Araxians, the *Meskhes*, and the Iberians — inhabitants of that mountainous region.

MESHECH and *Moschi* are identified.[582]

7. תירס — T*t*IRS — 'TIRAS.'

Indo-Germanic. Not hebraically, 'demolisher,' &c.

Occurring but twice, no light can be gathered upon this appellative from other Biblical sources than the context of *Gen*. x., and its repetition in 1 *Chron*. i. 5.

The Armenian historian, Moses Chorenensis, remarks — "Our antiquities agree in regarding *Tiras* not as the son of Japheth, but as his grandson."[583]

Θραξ, 'Thracia,' is unanimously reputed to be the ethnological synonyme of *Thiras*.

and the river Τιρας, 'Tyras,' of Ptolemy, flowing into the Euxine, now called *Dniester,* to be its geographical, as *Thuras,* Mars, was its mythic, correspondent.

TIRoaS, and *Troas,* in western Mysia, so closely resembling each other, it is not impossible that the Troad is intended by the Hebrew writer; especially since the *Teucri* were perhaps of Thracian origin: but no reasonable objection can be raised to the usual attribution of *Tiras;* and Thrace, the *Thraces,* or *Thracians,* may be safely assumed as the "ultima Thule" of Hebrew knowledge, towards the north, in the time of the writer of Xth *Genesis;* whose dim horizon in that direction was doubtless similar to that of the Egyptians during the XVIIIth dynasty. *Sesostris* (in this narrative, Ramses II.) had pushed his conquests into Thrace, according to Herodotus and united classical tradition. *Thriksu,* 'Thracians,' are recorded in hieroglyphics at the ruined temple north of Esneh, among the conquests of Ptolemy Evergetes I.[583]

Gen. x. 3. — בני גמר — BeNI-GMR — 'Affiliations of the Crimea.'

8 אשכנז — ASKNZ — 'Ashkenaz.'

Indo-Germanic; and, although traced to a 'fire that distils,' so alien to Hebrew, that even Rabbinical philologers abandon it, as "obscure." In consequence, some perceive the parent of the *Germans!*

Oriental Jews call those of their co-religionists who are settled in Germany *Ashkenazìm,* which has been confounded with the ASKNZ of Xth *Genesis;* whereas the real source of this mistake lies in their intonation of the Indo-Germanic name, *Sassenach, Sascenak,* old form of our word *Saxon.*

ASKIN, ISQIN, in many dialectic varieties, is the national name of the *Basques;* and inasmuch as nobody seems to know whence they came to Biscayan neighborhoods, we pass on this suggestive similitude as cautiously as it was given to us.

Repeated in 1 *Chron.* i. 6, the "Kingdoms of Ararat, Minni, and *Ashchenaz,*" seem to have been limitrophic in the time of Jeremiah — 629 to 588 B. C. — and hence the province termed *Asikinsene* by Strabo has been looked upon as its equivalent.

The Phrygians appear to have been anciently called *Ascanians;* and footprints of this migratory name are traceable throughout Bithynian vicinities, in *Sinus-Ascanius, Ascanius-lacus* and *amnis;* and likewise in Lesser Phrygia — *Ascania,* and *Ascaniæ-Insulæ.* Ascanius, son of Æneas, bore the original patronyme from Troas to Latium. Bordering on the Black Sea, these *Ascanian* similarities receive natural explanation through Pliny, "Pontus *Euxinus,* quondam AXENUS;" and Ευξεινος, the *Euxine,* or Black Sea, preserves a mnemonic of *Ascanians* and *Ashkenaz.*

Rawlinson perceives analogies between *Askenaz* and the *Arzeskan* mentioned in cuneiform inscriptions of the Nimroud obelisk, the date of which is now assigned to about 860 B. C.

"Pontus," says Bochart, "olim *Ascenaz,* Græcè Αξενος, quasi inhospitalis dictus;" which wears very much the guise of an Hellenic play upon a foreign word. Potocki, followed by Dubois, "finds the *Askhanaz* (Rheginians of Flavius Josephus) in the Mysian-Askanians, who came from Great-Mysia, and established themselves in the Phrygia of Olympus: it was a Germanic colony." May not ASKN, as *Ascanian,* or as *Euxine* be an adjective to *a*Z, the *Asi?*

Suffice it for our purposes, to accept the southern coast of the *Euxine* as one of the pristine habitats of a people called Ashkenaz.[584]

9 ריפת — RIPT*t* — 'Riphath.'

Also Indo-Germanic; not 'medicine,' nor 'pardon.'

Owing to the slight distinction between the letters ר, *resh,* R, and ד, *daleth,* D, of the modern *square-letter* character in which the Hebrew text is written, some copyist has bequeathed to us a dilemma — whether the R*iphath* of *Gen.* x. 3, should be D*iphath*, or the D*iphath* of 1 *Chron.* i. 6, R*iphath!* Commentators agree, however, in preferring *Riphath;* and, while some, following the pseudo-Josephus, have identified the name with Great Britain, there are many claimants for *France!* The LXX read Ριφαθ, in Xth *Genesis.*

Josephus restricts the name to Paphlagonia; in which country Mela places the *Riphaces.*

Mons Niphates (snowy), in Armenia, through the substitution of N for R, has learned defenders. But the Ριπαια ορη, the *Riphæis montibus*, and the *Rhipæas* placed by Ptolemy where no *mountains* exist, near his imaginary sources of the Tanais, or Don, are the favorite localities chosen for *Riphath.*

To this view there are weighty objections. If the *Montes Rhipæi*, or *Hyperborei*, be the Ural chain, they were too remote even for the vision of geographers who wrote at least nine centuries later than the author of *Gen.* x. The mere accidental analogy of a proto-syllable — RIP-*ean* with RIP-*a*T*t* — when the second radically differs, (the only ground upon which the hypothesis rests,) cannot be allowed as negative proof against simpler reasons; especially when the geographical position of the Riphæan mountains, save as the tenebrous hyperborean limit of Greek geognosy, is utterly unknown.

The writer of Xth *Genesis* must have had some reason, more or less scientific, for the order in which he mapped out the nations he enumerates. In the present instance, among the "affiliations of the *Cimmerian*," or Crimea, he places *Riphath* between the *Euxine* (Ashkenaz) and *Armenia* (Togarma); confirmed by Latin writers who station the *Rhibii* east of the Euxine.

"*Riphath*," adds Dubois, from the authentic researches of Potocki, "is the veritable and most ancient name of the people Shlave. *Hénètes* and *Honoriates* are but translations of a Sclavonian word which signifies honored, distinguished." The Latins added a letter to *Enètes;* which, becoming *Venetes, Venedes, Vendes, Vinides*, and *Wends*, was the title of those *Wendo-Shlaves* from whom descended the ancient Prussians, together with the present Lithuanians, and whence *Venice* inherits her name.

Paphlagonia for the country, and *Riphaces* for its inhabitants, corroborated by the opinions of Josephus ond Mela, sufficiently define the position of RIPHATH.[585]

10. תגרמה — T*t*GRMH — 'TOGARMAH.'

Indo-Germanic, or Scythic; not, 'which is all bone'!

"They of the house of *Togarmah* traded," in the fairs of Tyre, "with horses, horsemen, and mules," in the time of *Ezekiel* xxvii. 14; and, based upon this text, Moses Chorenensis derives the Armenians, Georgians, &c., from THARGAMOS, grandson of Noah.

Its classical similitudes are visible in the *Trocmi, Trogmi*, about Pontus and Cappadocia; and, at the Council of Chalcedon, there was a bishop, τροχμαδων, of the *Trogmades.* Josephus makes *Aram, Minyas*, and *Khoul*, adjacent to *Togarmah.*

The name of Armenia now is *Arhan*, identical with IRAN, *Iriana*, original cradle of Persians.

The "History of Georgia," compiled in the reign of Vakhtang V., King of Karthli, in 1703–'21, is one of the rarest works. Dubois translates some curious extracts of its commencement: — "According to these traditions, the Armenians, the Georgians, the inhabitants of Rani (*Arran*), of Movakani (*Chaki, Chirvan*, and *Mougan*), of Hérèthi (*Cakheth*), the Lesgians, the Mingrelians, and the Caucasians, all descend from the same father, who was called THARGAMOS. This Thargamos was the son of *Tarchis*, son of *Avanan*, son of *Japhet*, son of *Noah*, and was a valiant man." Like Moses of Chorene, in the fifth century, Vakhtang wished to hitch his local traditions on to Biblical origins. The former historian metamorphosed the names *Zrouan, Didan*, and *Habe-*

dosth (which he found in an old Chaldæan volume), into "Shem, Ham, and Japheth;" and the race of Habedosth, *Merod*, *Sirath*, and *Thaklath*, became, in his pious hands, "Gomer, Thiras, and Thorgomus!" "It was thus that he reconciled the sacred with the profane, and that the *Haïk* of the ancient Chaldæan volume, son of *Thaklath*, was superimposed upon *Thorgomus*, as a descendant of Japheth." History abounds with similar fraudulent genealogies. Thus, skilfully observes Jardot, "Ràshid-ed-Deen, Vizir of the Emperor Gazan-Khàn, has left at the commencement of the fourteenth century, upon the origin of the *Mongols*, erroneous notions, which Arab, Turkish, and Persian historians have copied; and even Aboo 'l-Ghàzee, Governor of Kharizm, in 1654. Misguided by a false religious sentiment, Ràshid-ed-Deen attached the antique traditions of the nomad hordes of Asia to those of the Jews, as preserved in the Koràn: — *Japhet, son of Noah, transported himself to the East, and it is from him that descend the people of those countries, afterwards partitioned between two brothers, Tatar-Khàn and Mogoul-Khàn.* All this recital is fabulous, and does not correspond with any of the accounts furnished by the Chinese." Even in our day, the "Caucasian" missionary is stipended to instil into the ill-furnished crania of African Hottentots and Australian Papuas the fond hope that they are positively and lineally descended from *Ham!*

The *Turks* did not approach the Euphrates from their aboriginal hive on the confines of China until about 1000 A. D.; and consequently all ascriptions of the name *Togarmah* to them seem to be linguistically and historically fallacious. Whether in the appellative 'Turcoman' there be any demonstrable connexion, we will not aver or deny. But the Armenians, a primordial people upon their native mountains, call themselves "the house of *Thorgom;*" and there is no good reason to suppose that *Armenia* is not TOGARMAH.[586]

Gen. x. 4. — בני יון — BeNI-IUN — "Affiliations of IONIA."

11. אלישה — ALISH — 'ELISHAH.'

Indo-Germanic; not, 'God that gives help.'

Elisa, 'Elis,' on the coast of Peloponnesus, one of the earliest historical settlements of Greece, divides with *Hellas* the honor of being catalogued in Hebrew geography. The former, Ἐλις, or the *Elide*, would seem supported by EZEK. xxvii. 7 — "blue and purple from the isles of *Elishah;*" purple-bearing shells having been abundant, anciently, on the Laconian shore. The latter, Ἑλλας, whence Ἑλληνες became the national name for Greeks, does not appear to have possessed, in the times of Homer (whose disputed era cannot be much removed from that of the writer of Xth *Genesis*), the pan-Hellenic extension it had acquired about the fifth century B. C., when Herodotus and Thucydides flourished: having previously been restricted to a district and town of Thessaly. But, adds Grote, no sooner do we step beyond the "first Olympiad, 776 B. C., our earliest trustworthy mark of Grecian time," than the quicksands of mythical legend engulph the criteria by which the relationship of facts can alone be decided. Thus, to the Judaic compiler of Xth *Genesis*, IUN, *Ionia*, would seem to have been the parent of ELiSaH, *Elis*, or *Hellas*. On the contrary, Grecian tradition reverses the order; and *Ionia*, in Asia Minor, becomes an affiliation of Hellas, about 1050 years B C. There is no *Sh* in Greek alphabets, and consequently that articulation was foreign to the people. The author of Xth *Genesis* wrote A, L, I, S, H, in the unknown alphabet he used. ELISHAH, is not older than the Masora Rabbis. The LXX read Ἐλισά.

Either view, however, establishes a close affinity between *Ionians* and *Hellenes*, or *Eleans;* and Greeks in general, as well along the shores of the Morea as on the isles of the Archipelago, would adequately represent the geography of ALISH; but, in view of restricted knowledge (and no *Sh*), it seems more probable that *Æoles* and *Æolia*, in Asia Minor, were the nation and country intended by the writer of Xth *Genesis*.[587]

12. תרשיש — T*t*RSIS — 'Tarshish.'

Indo-Germanic (?), or Semitic (?); not, 'contemplation.'

Perhaps, in endeavoring to attain the exact point of view of the author of Xth *Genesis*, this is the most enigmatical problem left to modern solution; although commentators of the present day slide over its difficulties, and range themselves under one of two schools: the first of which claims *Tartessus* on the Spanish, the second, *Tarsus* on the Cilician coast, to be the true locality.

The question is so far important, that in it is involved the occidental limit of the geographical knowledge of the Hebrews at the time when Xth *Genesis* was compiled; and, as customary, modern orthodoxy, which discovers the *Chinese* in the SINIM of *Is.* xlix. 12 — the *Negroes* in K*h*aM, *Ham*, of *Gen.* x. 1! and the "ten lost tribes of Israel" in the *American* aborigines, contends for the widest interpretation.

Scriptural texts require the word Tarshish to be classed under three categories: —

A. — *Tarsus*, Ταρσος — now *Tarsous*, on the coast of Caramania — an ancient city on the river Cydnus: birth-place of Paul, and sepulchre of Julian. Between T*t*aRSIS of Xth *Genesis*, or other passages of the text, and T*a*RSoS, there is no difference, philologically, except a "mater lectionis," or vowel, which, in palæography, is vague. The *Masoretic points*, like the Greek tonic accents, are unauthoritative, beyond indicating the traditionary phonetism of post-Christian writers in either tongue: and the *Masora* commences only six centuries after Christ.

The amphibious adventure of Jonah, which, the Rev. Prof. Stuart says, "plainly savors of the miraculous," might possibly indicate the Spanish *Tartessus*, as the correspondent of *Tarshish* during the uncertain, but recent, age at which this prophetic book was composed — a treatise that must not be confounded with the scientific and more ancient document — Xth *Genesis*.

[The NaBI, 'Jonah,' rebelled against I*e*HO*ua*H's command, "go to Nineveh," and therefore encountered the fate from which Perseus delivered *Andromeda*, viz.: that of deglutition by "a great *fish*," or monstrous *cetus* — the *Whale:* which became a sempiternal emblem of icthyophagy, when, assuming the forms of *Cepheus* and *Cassiepea*, it ascended to the heavens, or, as *Glaucus*, descended to the sea. In 1850, a paragraph, started in the New York "Sunday Messenger" by Major Noah, went the rounds of the religious and profane newspapers throughout the Union. It asserted that the portrait of the Prophet Jonah had been found on the walls of *Nineveh!* Here he is (Fig. 355).

Fig. 355.[588]

Ουανες, *Oannes* (of Berosus) as IOAN*es;* and *Jonah*, 'Jonas,' as IONAS; both being *i*-ON-*es* = 'the sun' — were identified long ago with *Dagon*, DAG-ON; *i. e.* the "sun in *pisces*," incarnated in this Assyrian fish-god. The same mythe lies in *Atergatis*, or *Derceto*, and especially in those Christian forgeries called the "Sibylline verses," beneath the acrostical Ιχθύς.

I should not hesitate, but for the above præternaturalities, in reading the *Tarsus* of Cilicia as the destination of the ship whereupon Jonah took his passage, and "*paid* the fare," on an obedient voyage from Joppa to Nineveh, (as a convenient route anciently, before *steam*-navigation, as now "cæteris paribus"), for compliance with the "tetragrammaton's" behests: but he spitefully "rose up to flee unto *Tarshish*, from the presence of ADONAI"; and, in consequence, while Jonah was righteously punished for his obduracy, it seems that his intention was to escape through a western, in lieu of proceeding in an easterly, direction; and therefore *Tartessus* of Hispania, or elsewhere so long as *Jonah* could realize a contrary, would appear to have been the country for which the vessel cleared, and wherein dwelt her consignees. — G. R. G.]

B. — *Tartessus*, Ταρτησσος, probably a Phœnician emporium, whether among the *Tartessii* in the vicinity of the present Cadiz, or at some other point within the Mediterranean, lay unquestionably in Spain. Hither Solomon and Hiram dispatched their commercial navies (1 *Kings* x. 22; 2 *Chron.* ix. 21); and thence, about the time of the Babylonish captivity (*Ezekiel* xxvii. 12; *Jeremiah* x. 9), *silver*, *tin*, *iron*, and *lead*, were imported, through Tyre, into the Levant. The presence of *silver*, *tin*, and *lead*, upon Egyptian mummies of every age back to the XVIIIth dynasty, establishes, beyond dispute, epochas far earlier than those of any Hebrew writers, Moses inclusive, for relations of trade between the Nile and whatever western regions, probably Spain, whence those articles were introduced: so, no doubts on relative antiquity need arise upon Iberian *Tartessus*. It corresponds perfectly to *Tarshish* in later parts of Hebrew annals. But there is a third element in the discussion, unknown to Anglo-Saxon divinity, which it is due to our contemporary Michel-Angelo Lanci, Professor of Sacred Philology at the Vatican, not to overlook.

C. — *Tarsis* does not proceed from *Tur-ṣus;* but from the old Semitic root *rasas*, preserved in Arabic, meaning 'to wet,' 'to lave.' With the primeval feminine article T prefixed to it, *Tarshish* means 'land laved by the sea,' that is, *the sea-shore;* and, in consequence, "vessels of *Tarshish*" often signifies *coasters*, irrespectively of any geographical attribution. For example — we should read, "thou breakest the *coasting*-vessels" (not ships of a place called *Tarshish*,) "with an east-wind." (*Ps.* xlviii. 7.) Again, "The kings of maritime states (*Tarshish*) and of inland regions (*Iìm*) shall present offerings." (*Ps.* lxxii. 10.) And finally, not to digress here on that most prolific theme, the mistranslations consecrated in King James's Version, compare "Sheba and Dedan, and the merchants of Tarshish, with all the *young lions* (!) thereof" — (*Ezek.* xxxviii. 13) — with Lanci's lucid *Italian* rendering: "The inhabitants of the strong places of *terra-firma*, Saba and Dedan, and the maritime merchandizers and their colonists will say to thee" — (*Gli abitatori de' forti luoghi di terra ferma, Saba e Dedan, e i mercatanti marittimi e i loro coloni diranno a te.*)

This derivation of *Tarshish*, from T-*rasas*, bears upon the geographical inquiry so far as concerns the marine position of a territory to which the name is applied.

The following passages are note-worthy in our discussion: —

1st. — (2 *Chron.* xx. 36.) Jehoshaphat "joined himself with him (Ahaziah) to make ships to go to *Tarshish;* and they made the ships at *Etsion-gaber.*" Now, this arsenal lay near *Elath*, on the Elanitic arm of the Red Sea, not far from *Akaba;* and therefore, in those days, the Jews were not likely to have intended a circumnavigation of Africa to reach *Tartessus* in Spain! Nor is it probable that, after building galleys at enormous cost on the Red Sea, the Hebrews contemplated transportation backwards over the Isthmus to launch them again on the Mediterranean.

2d. — (1 *Kings* xxii. 48.) But we learn that "Jehoshaphat made ships of *Tarshish* to go to *Ophir* for gold: but they went not; for the ships were broken at Etsion-gaber." What other construction but "coasting voyages" will suit *Tarshish*, in the former passage? What other than "coasting vessels" could go by sea from Akaba to *Ophir* (on the Persian Gulf, as we shall see,) in the latter?

Here, then, without question, *Tarshish* refers to "coasters," or "maritime merchandizers," sailing down the Red Sea towards India, and not to Spain.

3d. — (2 *Chron.* ix. 21.) "For the king's (Solomon) ships went to *Tarshish* with the servants of Huram; every three years once came (back) the ships of *Tarshish*, bringing gold and silver, S*h*iN-HaBIM (*teeth*, of elephants?), KUP*h*IM (apes), and T*h*KIIM (peacocks?)." The parallel passage 1 *Kings* x. 22, enumerates the same articles, but has "fleet of *Tarshish.*" So, "coasting vessels," and not a locality, seems intended by both writers. This is confirmed by Gesenius, who says that "a *ship of Tarshish*" meant "any large merchant vessel in general."

All the articles named, with one exception, might have been imported equally well from the African coast of the Gates of Hercules, opposite to the Spanish *Tartessus*, as from Southern Arabia, *Ophir*, &c.; because *elephants* abounded in Barbary, even in Roman times; while "*Apes*-hill," at Gibraltar, even now corresponds to the opposite Atlantic range, where *apes* are as common as African *baboons* in Arabia; whence the latter are brought now-a-days to Cairo.

But the exception excludes *Spain*, and all Northern Africa. The singular T*t*K, pointed *Thuk*, like its homonyme *Taoðk*, and *Taoðs*, in Arabic, Turkish, &c., is considered to mean 'peacock.' If so — and there is no actual impossibility that such a "rara avis" should have been brought *via* Arabia by the coasting trade — *India* is the country of *peacocks;* and therefore these birds were not procurable at *Tartessus*, in Spain, 1000 years B. C.

Peacocks are not impossible; but a new reading is submitted, equally destructive of Spanish *Tartessii* in these texts.

It is certain that *cocks* and *hens* (the common fowl), as well as *geese*, are never mentioned in the canonical writings of the Hebrews. Nor *fowls* in authentic works of Homer; nor by Herodotus. The Pharaonic *Egyptians* knew not the common fowl; using *geese*, *ducks*, and these birds' *eggs*, instead. But one instance of possibly a "*cock's* head," and that a stuffed specimen, occurs on Nilotic monuments. It is in the "Grand Procession" of tributes to Thotmes III., as Pickering first indicated. Etruscan vases, being of later manufacture, are no exception to the rule that the common fowl had not reached Europe, or Asia west and north of the Euphrates, or Africa, before the conquests of the Achemænians, B. C. 540, downwards. It is also positive, that the centres of creation for this bird are Indo-Chinese and Australasian; and that, like *peacocks*, they had to be imported into Arabia from India. Now, in Arabic, a *cock* is called 'Dèyk,' D*i*K. Stripped of the modern Masora, the Hebrew word is T*t*K, or D*i*K. May not the common fowl, in lieu of peacock, be alluded to in the above passages? It is as probable as *pheasant*, proposed by others; and about the same ages (B. C. 1110) *white pheasants*, probably from *Caffraria*, were received at the court of *Tching-wang*, in China; according to Pauthier.

Bochart, following Eusebius's Θαρσεῖς ἐξ ὧν Ἴβηρες — the Iberians of Spain — and the generality of English commentators, fix upon *Tartessus* as the equivalent for *Tarshish* of Xth *Genesis*. Continental orientalists of our day lean towards the Cilician *Tharsis*, *Tarsus;* upon the earlier authority of Josephus, and of Jonathan, the Chaldee paraphrast. And, without dogmatizing in the least upon either view, the order in which Ionic affiliations succeed each other — *Æolia*, *Tarshish*, *Kittim* the Cyprians, and *Rhodanim* the Rhodians — coupled with the geographical proximity of Rhodes and Cyprus to *Tarsous*, on the Caramanian coast, seems confirmatory of those opinions which select *Tarsus*, in Cilicia, as the locality indicated by the writer of Xth *Genesis* for TARSHISH. There is no difficulty with regard to the antiquity of Cilician *Tarsous;* because Mr. Birch read, long ago, "This is the vile slave from *Tarsus* of the sea," inscribed in hieroglyphics, during the thirteenth century B. C., over a captive of Ramses III.[589]

13. כתים — KT*t*IM — 'KITTIM'; plural of KiT*t*.

Language uncertain. Not, 'they that bruise,' or *gold;* nor, 'hidden," &c.

Three Mediterranean countries have been supposed by commentators to be figured by the various etymons of this word: *Italy*, *Macedonia*, and *Cyprus;* besides many "islands." The first, resting solely upon the fanciful analogies of Κετια, in Latium, and Κετος, a river near Cumæ, although supported by the erudition of Bochart, may now be dismissed without ceremony.

Kittim, as Μακετια, after Alexander's conquests had made Macedonia renowned, is the acceptation in which it appears in two latest books of the Hebrews — Daniel (xi. 30) and 1 Maccabees (i. 1); equally canonical in archæology.

The books belonging mainly to the period between Alexander (B. C. 330) and the Babylonish captivity — say, from Hilkiah's high-priesthood, about B. C. 630, downwards — give to *Kittim* a wider extension than can well be deduced from Xth Genesis; for Jeremiah (ii. 10) and Ezekiel (xxvii. 6) speak of the states or "isles of *Kittim:*" the latter with reference to works in *ivory* thence imported. Greece was celebrated for chryselephantine manufactures, certainly in the 30th Olympiad, 660 B. C., and perhaps before.

In the Hebrew text of the doubtful parts of Isaiah (lxvi. 19), *Tarshish* (Tarsus), *Phul* (probably Pam-*phylia*), *Lud* (Lydia), *Thubal* (Paphlagonia), *Javan* (Ionia), and *Kittim*, are grouped together; hence their proximity is inferable.

Josephus adopts the Oriental form of personification when he relates that "*Kethimus* possessed the island of *Kethima*, which now is called Cyprus; and from this, by the Hebrews, all islands and maritime places are termed Kethim."

Hence, modern researches unite upon the island of *Cyprus* as the centre-point of probabilities — *Citium*, χιτιον πολις, of Ptolemy, a city in Cyprus, now *Kiti;* and the Phœnician *Citiaci*, applied by Cicero; justifying the adoption. Confirmed, moreover, by Boeckh's Greek inscriptions, wherein איש כתי, a 'man of KiT*t*,' is explained by Κιτιευς; a *Kitian*, or Cypriote.

But the true position of *Kitium*, as Cyprus, is now fixed by "coins of the anonymous kings of Cittium;" no less than by a cuneatic inscription of the time of the Assyrian king Sargon (recently found at Larnica, and conveyed to Berlin), which carries the name back to the eighth century B. C. Egyptian monuments, elucidated by Birch, enable us to behold it again in hieroglyphics of the thirteenth century B. C., where the "Chief of the *Khita*, as a living captive," surmounts one of the prisoners of Ramses III. Nor is this our earliest record; because the KeFa, portrayed in the "Grand Procession" of Thotmes III. [*supra*, p. 159, Fig. 82], are said to come "from the isles in the sea," *i. e. Cyprus;* and, again, "*Khefa* (Cyprus), *Khita* (Kettiæi)," stands registered in the sculptures of Amunoph III., at Soleb. So the people, and their island, are as old as the XVIIIth dynasty, or the sixteenth century B. C.

The inhabitants of *Cyprus* in particular, and of the adjacent coasts and islands in general, are undoubtedly the KiT*t*IM (*Cypriots*) of the later projector of Xth *Genesis*— a conclusion ratified by their propinquity to the nation immediately succeeding.[590]

14. דדנים — DDNIM — 'Dodanim'; plural of *Dodan*.

Between D*odanim* of Xth Genesis, and R*odanim* of 1 Chron. i. 7, a literal discordance, produced by the error of some unknown transcriber, leaves the decision for posterity (as Cardinal Wiseman declares in respect to 1 *Tim.* iii. 16) to "rest on what judgment it can form amid so many conflicting statements!" Who, from the text alone, can tell whether we must read R*odanim* in Xth Genesis, or D*odanim* in 1 Chronicles? In consequence, conjecture has had full scope; and Bochart's ingenious assimilation of the river *Rhodanus*, Rhone, has been seized upon by a standard Anglican divine (Bishop Patrick, to wit), who beholds in *France* the country of the Rodanim! "Our old chroniclers," says Champollion-Figeac, "equally robust etymologists as able critics, do they not found the realm of *France* by *Francus*, one of the sons of Hector, saved expressly from the sack of Troy!" The Hungarians caused Attila to descend from Nimrod in a straight line; the *Danes*, from the *Danai* issuing from *Dodona*, crossed the *Danube*, to which they gave their name, and finally settled in the country they named *Danemark!*

Dodanim possesses advocates; and of course *Dodona*, in Epirus, site of Græcia's most ancient oracle, at once suggests that the *Dodonæi* must be the people intended. Nor, except its remoteness from the neighborhood of other proper names whose geography is tolerably positive, can a negation be absolutely demonstrated.

However, the Samaritan Pentateuch, reading *Rhodians* where the LXX have Ῥόδιοι, affords a preponderating vote in favor of the R. And, other conditions being equal, this fixes attention on the isle of *Rhodes;* by excluding the possibilities of D. Its early Grecian occupancy; its location between *Cyprus* and *Æolia;* and their common affiliation from *Ionia;* support the view that Ῥοδος, the roseate island of the *Rhodians*, was the habitat of the Genesiacal RoDANIM.[591]

HAMIDÆ, *or Swarthy Races.*

בני חם — BNI-K*h*M — "Affiliations of HAM." — *Gen.* x. 6.

15. כוש — KUS — 'CUSH.'

By the LXX, and in the Vulgate, this word, whenever translated, is made to figure under the Greek form of Αιθιοπια, *Æthiopia.* Through Cruden's Concordance, it appears that *Cush* is transcribed in King James's Version as if in the primary Hebrew Text the name had occurred only *five* times: whereas, if we restore to its relative passages in the Text the original KUS, in every instance where in our version we find its supposed equivalents, '*Ethiopia*,' '*Ethiopian*,' '*Ethiopians*,' it will be perceived that *Cush* is repeated, (5+34=) *thirty-nine* times in the canonical Hebrew Scriptures.

It may occur to a simple believer in plenary inspiration to inquire, why, and upon what principle of logic or philology, the *translators* of our *authorized* version—"By Her Majesty's special command — appointed to be read in Churches" — took upon themselves the suppression of the Hebrew word KUS*h* thirty-four times, and its preservation only five? How happens it, that strict uniformity was not adopted; and that they did not either substitute *Ethiopia* all the way through, or preserve the original *Kush* in every instance; according to the consistent method of Cahen, in his much more accurate translation? To answer such queries is beyond human power, because the aforesaid translators did not know themselves: but some explanation may be found in the fact that, little versed in *Hebrew* literature, the fifty-four revisers, in 1603, followed the *versions*, and not the Text; as our Part III. thoroughly establishes.

Investigation must first be directed towards the Hebrew triliteral KUS. Its translation by the Greek word *Ethiopia* is a secondary inquiry. כוש, KUS, are its radicals; and must have been its components, at whatever time, and in whatever alphabet, anterior to the Hebrew *square-letter* (not invented until the third century after c.), the Xth chapter of Genesis was first written. The diacritical points, added by the Masoretes after the sixth century of our era, makė its sound KUS*h;* whilst, as regards its original Hebrew phonetism, the terminal *Sh* is (Chaldaically) likely, and we adopt it in the form KUS*h*.

What did KUS*h* signify, in the mind of the compiler of Xth Genesis? There is not one *per mil* of our contemporary divinity-students who will not glibly reply — "*Ethipia*, to be sure — *Africa*, above Egypt"!

[Five years have passed since the authors of the present volume denounced such answer to be simply ridiculous (J. C. N.: *Biblical and Physical History of Man*, 1849, pp. 138–146;—G. R. G.: *Otia Ægyptiaca*, 1849, pp. 16, 133–4). Between replies so diametrically opposed there can be no reconciliation. One of the two must be absolutely false. Among the many, however, who have felt themselves called upon to contravene our assertions, not having hitherto met with one person really acquainted with the Hebrew *alphabet*, we may be excused by Hebraists from recognizing as "Biblical authorities" those teachers who (even the articulations of א, ב, ג, being to them unknown) are yet ignorant of the A, B, C, of Scriptural language, meanings, and history.

It was the authors' intention, when projecting "Types of Mankind," to publish an investigation of *Ethiopian* questions, sufficiently copious and radical as to leave few deductions ungrounded; and their MSS. were prepared accordingly: but, so much extra space has been occupied by Part I., that "copy," to the extent of some 200 of these pages, must be suppressed for the present. The reader will, in consequence, be lenient enough to accept dry *references*, in lieu of logical argument. If "truth" be the object of his search, we feel confident that our bibliographical indices will at any rate place such reader on the easiest route of verification. — G. R. G.]

Bochart's words show that we were not the first, by more than 1000 years, to claim

"Arabia" for KUS*h*, instead of "Ethiopia." "Chus alii Æthiopiam, alii Arabiam explicant. Priorem interpretationem præter Hebræos fere quotquot sint, etiam Græci sequuntur, et vulgatus interpres, et Philo, et Josephus, et Eusebius, et Hieronymus, et Eustathius in Hexæmeron, et author Chronici Alexandrini, et chorus patrum vniuersus. Arabs etiam nuper editus qui hic habet אלחבש Abasenorum seu Abissinorum terram, id est Æthiopiam. Posteriorem è veteribus, quod sciam, *solus* Jonathan, in cujus paraphrasi *Gen.* x. 6, pro Hebræo *Chus* est ערביא *Arabia.* . . . Ex iis quæ hactenus à nobis disputata sunt, credo constare luce clarius Chusæos in iis locis habitasse quæ supra indicauimus, nimirum supra Ægyptum ad Rubri maris sinum intimum, in parte *Arabiæ Petrææ et Felicis.*"

Circumscribed within a few pages, our part limits itself to the production of such atoms of new data as have been attained since Bochart's day: beginning with the four rivers of *Eden.*

"The name of the second river, Gihon; that which encompasseth all the land of KUS*h*" (*Gen.* ii. 13) — part of the *Jehovistic*, and consequently later document — may be dismissed from the discussion; because, relating to ante-diluvian epochas, its geography is unknown. If there ever was an universal *Deluge*, all land-marks were necessarily obliterated. If there was not, as some geologists now maintain, the *Bereshith* (from *Gen.* i. 1 to *Gen.* vi. 9, rabbinical division) ceases to contain history; and, when not accepted in the allegorical sense maintained by learned Christian fathers, must be abandoned, by science, to thaumaturgical ingenuity; while the KUS*h* of *Gen.* ii. remains to be sought for "near the isle *Utopia* of Thomas Morus. Utopia! expressive name! — invented by the satirical Rabelais (Pantagruel), and afterwards applied by the great Chancellor of England (Sir Thomas More) to the beautiful land (Oceana) of which he dreamed—this Greek noun seems made expressly to indicate the sole degree of latitude under which the poetic marvels of the grand Atalantic island (and of the *four rivers* in Eden) could have ever been produced. It has been believed," continues Martin, the ablest critic upon Plato, "that it [the river *Gihon*] might be recognized in the New World. No: it belongs to *another world*, which exists not within the domain of space, but in that of fancy."

In the geographical nomenclature of Xth Genesis, KUS*h* is the "son of *Kham;*" a name applied to *Egypt* and her colonial affiliations: of which some are African, and others, such as *Canaanites*, indisputably Asiatic. To which continent did the Hebrews refer the name KUS*h?*

In 1657, Walton, the upright and most proficient compiler of *Biblia Polyglotta*, inveighed against the notion that KUS*h* could be the African "Æthiopia;" citing the best scholars of his day to the same effect. So, again, Beroaldus, Bochart, and Patrick, following the Targum of Jonathan, the Chaldee paraphrast — third to eighth century after Christ — render KUS*h* by *Arabia*, on the subjoined, among other grounds: —

1st. Moses' wife is termed a KUS*hean* (*Num.* xii. 13). Tsipora was a daughter of Jethro, the *Cohen* (priest) of Midian (*Exod.* ii. 16, 21; iii. 1); and Midianites being Arabians, here KUS*h* is Arabia. No other wife is given to Moses in the Pentateuch; nor can any supernaturalist so torture the plain words of its text as to prove, to a man of common sense, that Moses ever visited *Ethiopia* above Egypt. The Abbé Glaire, Doyen de la Sorbonne, whose two volumes — models of erudition and style that protestant divines would do well to imitate—lie before us, never resorts to such pitiful subterfuges.

2d. "I will make the land of *Mitzraim* a waste of wastes, from the tower of Syene even unto the frontier of KUS*h*" (*Ezek.* xxix. 10). Syene being *Assouân*, at the first cataract, on the border-line of (Ethiopia) Nubia and Egypt, the writer cannot mean "from Ethiopia to *Ethiopia*," but from Syene to KUS*h*, beyond the Isthmus of Suez, on the north-eastern frontier of Lower Egypt, and consequently here indicates *Arabia.*

Modern researches furnish more critical light. In the first place, Dr. Wells sustains, and, to a certain extent, demonstrates, that the word KUS*h* refers exclusively to the Asiatic "Ethiopia," and never to African localities; summing up his reasonings with, "the nation of Cush did first settle in *Arabia;* and the word is, generally, to be so understood in Scripture." In the second, believers in the *unity* of all mankind's descent from "Noah and his three *sons*," must concede that *Nimrod*, and many other affiliations of KUS*h*, settled in Assyrian vicinities; even if offshoots did afterwards cross through *Arabia* into Africa, and there, owing to "effects of climate," originate *Nigritian* races; beginning with the comparatively high-caste *Berber*, and descending down to the lowest grade of *Bosjesman*—always along a sliding scale of deterioration, from the valley of the Nile to the Cape of Good Hope—where, unfortunately, 200 years of occupancy have not yet transmuted Dutch *Boers* into animals different from those left behind them in Holland and Flanders.

The text most triumphantly quoted to prove the African hypothesis is *Jerem.* xiii. 23.—"Can the *Ethiopian* change his skin, or the leopard his spots?" A glance at the Hebrew shows that here, as in other instances, the fifty-four revisers of King James's version blindly copied the LXX, or the Vulgate; because "Can the KUS*hean* change his skin" leaves the question vague until the real application of KUS*h* be determined. The same proclivity leads many divines to cite another text, from the so-called "Song of Solomon," in behalf of their negrophile theories.—"I (am) *black*, but comely. . . . Look not upon me, because I (am) *black*, because the sun hath looked upon me: my mother's children were angry with me; they made me keeper of the vineyards; (but) mine own vineyard have I not kept." (*Cant.* i. 5, 6.) The absence of notes of interrogation in Hebrew palæography, coupled with the philological inanity of modern translators of this ancient erotic ballad, perpetuates a delusion, removeable by Lanci's rendering:—"I (am) *browned*, but comely. . . . Look not [disparagingly] upon me that I (am) *browned* ["fosca" = tawny, dark], because the sun has *tanned* me: the sons of my mother [*i. e.* my step-brothers] becoming free to dispose of me [according to Oriental usage], posted me (as) custodian of vines; my own vine, have I not guarded [taken care of] it?" Besides, as it has been remarked on the above interrogatory of Jeremiah,—"If Cush means a Negro, then we have revelation to prove that climate will not change a Negro into a white man; if it means an Arab (dark) Caucasian, then it will not change a white man into a Negro!"—Indeed, the ultra-high-church orthodoxy of a living English divine, and profound, whilst fantastic, Orientalist, unhesitatingly endorses this critical view.—"Among the great land-marks of national descent, none, it may safely be affirmed, are *surer*, or *more permanent*, than those physical varieties of form, countenance, and color, which distinguish from each other the various *races of mankind*. . . . In Arabia, one of the earliest seats of post-diluvian colonization; a country rarely violated, and never occupied, by a foreign conqueror; and peopled, in all ages, by the same primitive tribes, . . . peculiarity of form and feature may be justly received, in any specific or authentic example, as evidence of identity of origin, little, if at all, short of demonstration. This principle we are enabled, by Scripture, to apply as an index to the *Arab* tribes descended from Cush, and especially to the posterity of his first-born, Seba."

If we had penned the above paragraph ourselves, we could not have embodied more forcibly Morton's decisive opinions on those "primordial organic forms," which are perpetuated to this day, as the Rev. Charles Forster, B. D., justly remarks, among "the various *races* of mankind."

After the citation of "Can the *Cushite* change his skin?" the geographer of Arabia proceeds:—"This *indelible* characteristic of race would seem to identify with the families of Cush the inhabitants of the southern coast" of Arabia. "Now, since the Cushites generally were distinguished by the *darkness* of their skin, and the Sebaim (*Isa.* xlv. 14), particularly, were noted for the procerity of their stature, if we find, in Arabia or its vicinity, a race uniting both distinctive marks, the probability certainly is not a low one, that, in that race, we recover a portion of the family of Seba." In testimony whereof, the reverend author quotes Burckhardt's description of the Dowaser tribe of Arabs — "*very tall men*, and *almost black*" — as well as passages from Chesney, Niebuhr and Wellsted, corroborating the *dark* complexion observed by these authoritative travellers among Bèdawees of the Persian Gulf; to whom we could add multitudes, were they needed.

Having indicated to the reader sufficient sources to substantiate the existence at this day, in Southern Arabia, of tribes *dark enough* to justify Jeremiah's simile (xiii. 23), we might proceed at once to the identification of KUS*h* in its geographical affiliations. Inasmuch, however, as one of the objects of the present work is to bring the archæological and ethnographical facts contained in Hebrew literature from out of a deplorable mysticism into the domain of science, there are other scriptural passages that claim priority of analysis.

1st. *Isaiah* (xi. 11) — "from Assyria, and from Egypt, and from Pathros, and from KUS*h*, and from Elam, and from Shinar, and from Hamath, and from the islands of the sea." Circumscribed within the geographical limits to be established for the Hebrew writers, *Southern Arabia* is here the equivalent of KUS*h*, because, otherwise, an immense peninsula, very familiar to them, would be omitted.

2d. *Isaiah* (xviii. 1, 2) — the prophet in Palestine here apostrophises *Egypt*. We have given Rosellini's rendering in Part III., and need merely now remark that "The rivers of KUS*h*" have no relation to the Nile, nor to "Ethiopia" above Egypt, but are the *torrens Ægypti*, the "streamlets of Mizraim"—the *Besor*, *Corys*, now "Wàdee el-Arish;" the winter-brook, or *Seyl*, which divides Palestine from Egypt at Rhinocorura. Indeed, this is, and has ever been, the boundary-line; the extremest West; beyond which, towards Africa, the word KUS*h* never passes, in the geography of the earlier Hebrews: and, from that occidental line, it stretches backwards to the Euphrates and its lower territories south-east of Syria. The term "earlier" Hebrews is used advisedly, to distinguish those parts of their literature that belong to times preceding the Captivity, from others composed during and after, when KUS*h* may have possessed a less restricted sense.

The most formidable objection to the Asiatic restriction of KUS*h* would seem to originate from 2 Chronicles (xiv. 9, 12; xvi. 8), where the rout of "Zerah *the* KUS*hean*," with a *million* of combatants, by Asa, is described — events attributed to the year 941 B. C. But this has been ably overthrown by Wells, sustained by the later work of Forster; who shows that *Gerar*, whither Zerah *the* KUS*hean* fled, "lay on the border of the Amalekites and Ishmaelites, between the kingdom of Judah and the wildernesses of Shur and Paran;" and, consequently, the scene lies in Arabia, and Zerah was some marauding potentate, probably *Shèykh* of a powerful Arab horde, whose foray was repelled into the "land of KUS*h*," Southern Arabia, whence he came. Saracus, moreover, (the classical transcription of Zerak-*us*,) was a proper name among *Kushean* dynasties descended from *Nimrod*, and also in Arabian traditions. To the Egyptologist, in consequence, the now-preposterous identification of Zerah *the* KUS*hean* with OSORKON (as *o*S*o*RK*on*, or SRK), second king of the XXIId dynasty of Bubastites, has long ceased to be of interest, because this text has no relation to Egyptian, any more to "Ethiopian," events.

The narrow circle of geography comprehended by all ancient nations situate around the Mediterranean as late as the *Persian period*, in the sixth century B. C., to which the Hebrews form no exception, forbids any such deduction as Jewish acquaintance with Nigritia. That analogy and comparison of the *literal* texts do not require KUS*h* to be sought out of South-western Asia in general, and Arabia in particular, in any Scriptural passages, could be shown text by text, did space allow. The "onus probandi" of the contrary may now be left to "le théologien" — for, as Letronne philosophically observed, "ici le rôle de l'hagiographe commence; celui de l'archéologue finit." "Le théologien," neatly declares Cahen, "en traduisant, ne perd jamais de vue son église, son temple, sa synagogue; borné par cet horizon, il allonge, raccourci, taille, entretaille, contretaille, les pensées de son auteur, jusqu' à ce qu'elles aient la dimension voulue pour entrer dans l'enceinte sacrée. Tel est *le faire du théologien; nous ne le blâmons pas; mais ce n'est pas le nôtre.*"

The reader, who may be pleased to verify the exactitude of the following *results*, will be enabled to do so through the references appended to this condensation of a complete chapter of our work, which lack of room compels us to curtail.

In hieroglyphics coeval with the XIIth dynasty at least, or 2200 years B. C., an African nation, situate immediately south of Egypt, always bore the following designation, in one of many dialectic forms — as "KS*h*I, *barbarian country*"; or spelt KAS*h*, KeS*h*, KiS*h*, or KS*h*; with or without the terminal I.

FIG. 356.[592]

K

S*h*

I

— *country, barbarian.*

The human portraits, wherever accompanying this name on the monuments, are invariably *Africans*, but more generally of the dark mahogany-colored Nubian than of the jet-black Negro type.

We contend that this proper name, which, indigenous to African *Nubia*, was ascribed by the ancient Egyptians to *Nubians* alone, has no relation (except through fanciful resemblances, produced in modern times, through corrupt vocalizations of Rabbis on the one hand, and of Copts on the other,) to the Hebrew word KUS, conventionally pronounced *Kush*, which, to the Jews, meant "Southern *Arabia*," and no country or nation out of Asia.

To render this clear, one must commence with a query — When, and how, was the Old Testament translated into *Coptic?* Quatremère, sustained by the old Coptologists, claims, "que la Bible avait été traduite sur le *texte hébreu* en langue Égyptienne." De Wette and the Hebrew exegetists aver, that "the origin of these versions (*Memphitic* and *Sahidic*) is probably to be referred to the end of the third and the beginning of the fourth century; for at that time Christianity seems first to have been extended to the Egyptian provinces [it had not even then reached the temple of *Osiris* at Philæ]. Both follow the Alexandrian version, but it is doubtful which of the two is the oldest."

The question is somewhat important, inasmuch as upon it hinges whether the Copts followed the LXX's Greek mistranslation of Αιθιοπια, or the original Hebrew word KUS. There can be little doubt that such translators imitated the Alexandrian Version, and not the Text; and substituted *Ethaush* and *Koush* for "Æthiopia." Champollion gives P-KA-N-NGHOOSH, NEGOOSH, and ETHAUSH, from various Coptic topographical MSS., as synonymes for the Greek Αιθιοπια, the Arabic *el-Habesh* (Abyssinia), and the vulgar Ethiopia; while Lenormant states — "the Coptic books employ the same expression (*Kousch*) that is frequently met with in its altered form, *Ethosch.*" Peyron and Parthey establish the same fact; but Lanci's deeper philology traces *Ethaosh* into two Semitic radicals, *heet* = 'form,' and *abes* = 'to-be-black."

Champollion's *Grammaire, Dictionnaire,* and *Notices Descriptives,* prove that the great master, whose discoveries were made through *Coptic,* always transcribes the ancient hieroglyphical KS*h* by the modern Coptic form of *Kousch,* or *Khoosh.* Hence, it has been universally taken for granted that Champollion's Coptic transcript of the old hieroglyphical African name of KiS*h* is identical with the Hebrew Asiatic KUS — that both are comprehended under the Greek maltranslation of "Ethiopia" by the LXX — and thus Arabs and Nubians, the Arabian Peninsula and the Upper Nile, Hamitic and Semitic distinct roots, have become jumbled up into "confusion worse confounded!"

Now, it so happens that the old hieroglyphical KS*h* is never written with a medial '*u*,' which is a radical "mater lectionis" in the Hebrew KUs — a strong point of dissimilarity to begin with. On the former word, Birch had critically remarked — "The term Kash is a fluctuating and uncertain territorial appellation: it is *supposed* to be the Kush of Scripture, the *Thosh* or *Ethosh* of the Copts, which, after all, is merely 'the frontier.'" We have already [*supra*, pp. 256–9] furnished abundant extracts from Mr. Birch's more recent definitions of KS*h*'s localities above Egypt.

But, in addition to the perplexing difficulties of archaic Egyptian and Hebrew names, and the anachronisms of modern philologers, there is a third element of medley, on which it behooves us to say a few words: viz., *Ethiopia*, and *Ethiopians*. Indeed, it is the prevalence of misconceptions upon the latter which lies at the bottom of mistakes concerning the former.

Already in A. D. 1657, the scholarship of Walton protested against "Ethiopian" delusions, with a citation from Waser — "Græci *Æthiopiam* deducunt ab *αἴθω cremo, uro*, et ὤψ, ὠπός, *facies, aspectus*, quia a solis vicinitate ita uruntur et torrentur, ut atro sint colore." Hence it is immediately perceived that *Ethiopian*, meaning simply a 'sun-*burned-face*," possessed at one time a generic application to the *color* of the human skin, and not an attribution to one specific geographical locality. During Homeric ages, by Αἰθιόψ, the fair-skinned Hellenes merely meant a foreigner *darker* than themselves; and, by Αἰθιόπια (the existence even of true *Negro* races being then utterly unknown to the Greeks) early Grecian geographers understood (not our modern "Ethiopia" above Egypt) the countries of all *swarthy* Asiatic and Barbaresque nations — Persians, Assyrians, Syrians, Arabs, Phœnicians, Canaanites, Jews, Egyptians, Carthaginians, and Libyans — especially those situate along the coast of the Mediterranean from the Orontes to Joppa.

This fact has been established beyond all controversy by the vast erudition of a Letronne, a Raoul-Rochette, and a Lenormant.[593] Its etymological truth can be verified in any Greek lexicon; while it is adopted, although not with sufficient archæological rigor, in the popular cyclopædias of Anthon and Kitto.

Want of space alone compels us to suppress many pages of extracts from the three first-named savans; through which it would become demonstrated that Αἰθιόπες, in all writers down to the fifth century B. C., meant nothing more than "visages brulés"; that is, "sun-*burnt-faces*." By way of example, take *Memnon*, who by Hesiod is termed Αἰθιόπων βασιλῆα, and by Homer, the most beautiful of men. Pausanias, Strabo, Diodorus, Æschylus, and Herodotus, affirm that he was an Asiatic demigod, probably from *Shusan*, or *Chuzistan*, on the confines of Persia. Now, Hesiod never meant that modern interpreters should understand that *Memnon* was "king of the *Ethiopians*" — of our Ethiopia above Egypt! The poet wrote that Memnon was "king of the *burnt-faces*;" that is, his followers were a *dark-skinned* people, such as the *Cushite*-Arabians are on Persian confines to this day. It is the same in Homer's "Eastern and Western *Æthiopians*" — again the same in Herodotus's *Ethiopians*, enrolled in the Persian army of Xerxes; some of whom were Asiatics, and others Africans — and, not to enumerate instances by the dozen, it is the same in Ælian's Indians (Hindoos), whom he terms *Æthiopians* also. In all these cases, the writers meant "sun-*burned-faces*" of the so-called "Caucasian" type; and it is but the inanity of modern *littérateurs* which ascribes any of the above *Æthiopians* to countries south of Egypt.

However, the time came, (after the Persian conquest, B. C. 525, and hardly before Ptolemaic days,) that Greek geographers, having discovered that there was a race "nigro nigrior" whose habitat lay south of Egypt, began to restrict *Æthiopia* and *Æthiopians* to the mahogany-colored *Nubians* and to the jet-black Negroes; and it is in this, the later specific, not in the older generic, sense, that scientific geographers understand a name which, without such reservation, is as vague as *Indians* (East and West Indies, and American aborigines!); as *Scythian* (from the Himalaya to the Baltic!); or, as that wretched term "Caucasian."

Now, it was during the prevalence of such geographical misconceptions—when *Africa* meant little more than Carthaginian and Cyrenaic territories along the face of Barbary; when *Asia* signified Asia Minor — in the interval between Eratosthenes the first scientific geographer, and Strabo the second — whilst Hindostan was termed *Ethiopia*, or *vice-versa* — pending the notions that the Nile and the Indus were one and the same stream; and that a circumambient ocean surrounded what little of a *flat* and stationary earth was known to Alexandrian science: — during such, and hundreds of similar cosmographical views since proved to be false, it was, we repeat, that the *Jews* of Alexandria, (having forgotten not only their parental Hebrew, but even the Chaldee dialect subsequently acquired through the Captivity,) caused the books of the Old Testament to be translated into *Greek;* in the form preserved to us under the mystic *No.* LXX, and by us consecrated as the *Septuagint:* translations fluctuating in date between B. C. 260, and B. C. 130.

Books of different origins, translated at different epochas, and by different persons, necessarily teem with imperfections; nor can uniformity be expected from literary labors under those circumstances, and in such uncritical times. Geographical criticism was certainly not a paramount object with any of these "uninspired" translators. They never foresaw archæological discussions that occur now, 2000 years after their day, in a language not formed for 1500 years later, by a distinct people, (whose infantine traditions attain not their Alexandrine lifetimes,) and on a Continent (6000 miles from Alexandria) whose existence was still undreamed of, even sixteen centuries after the original *Septuagint* MSS. were completed. In consequence, some of the Hellenizing Jews, or Judaizing Hellenes, when they met with the Hebrew word KUS*h*, simply transcribed it into Greek characters as Κοὺς, ΚΩC, or ΚΩΣ: others translated KUS*h* by Αιθιοπια — a word at that time equally applicable, etymologically in the sense of 'sun-*burned faces*,' no less than geographically, to *India*, Persia, Arabia, and the *Nubias*, indifferently to its Asiatic or African association. And this explains why, after 2000 years, the imaginary sanctity of Hebrew and Greek *words*, accidentally preserved in recent MSS., or through Latin and other re-translations, and despite innumerable recensions, enables us yet to admire in King James's version the English transcript of *Cush* only five times, and its Alexandrian substitute, *Ethiopia*, some thirty-four [ubi supra]; at the same time that, in the far elder and original Hebrew Text (copies of which, only about 800 years old, have come down to us), Providence permits our counting the triliteral KUS*h* in about forty different places.

Under these circumstances (notoriously accessible to anybody who can read English), to quote the *Septuagint* authoritatively on doubtful relations of "Ethiopia," as if it had applied to Africa exclusively at the time when this Greek literary work was in progress, may be exceedingly praiseworthy on the part of professional hagiographers, but, archæologically, is "vox, et præterea nihil," leaving the radical issue untouched.

But there is yet one more rock of confusion to be indicated, upon which the adopters of Wilford's Puranic delusions, Faber's fantastic reconciliations, and Delafield's American extravaganzas, have always split. It occurs when, through disregard of philology and palæography, they prefix an S, or other sibilant, to the Hebrew KUS*h;* and, reading SKUCH, *Scuthi*, Σκυθαι, &c., make this patriarch the father of *Scythians*, *Sacæ*, Saxons, Scotchmen, and even of *American Indians!* One blushes to treat such absurdities seriously in A. D. 1853. Nevertheless, the disease is inveterate with many writers "à qui il ne manque rien que la critique;" and it behooves us to note our "caveat," because, as Bishop Taylor says, "it is impossible to make people understand their ignorance; for it requires *knowledge* to perceive it, and therefore he that can perceive it hath it not."

A dry recapitulation of the *results* of studies, that could not be presented in full under half this volume, together with references through which the reader may verify exactness, is all that the authors can now offer on the *hieroglyphical* KS*h*, the *Hebrew* KUS, and *Greek* Αἰθιόπια.

1st. That the KeS*h* were African aborigines — probably similar to the *Barâbera* of the present day; but were not NAHSI, *Negroes*.

2d. That their habitat, from the XVIIth dynasty downwards, was closer to Egypt than that of any other Africans — probably Lower Nubia, because the KeS*h* are the first people encountered in Egyptian expeditions above Philæ.

3d. That their name, still preserved at Tutzis in *Kish*, was never KuS*h*, but KeS*h*, *Kish*, or *Kash*.

[Lower Nubia, nearest to Egypt, would seem to have been the residence of the *Kish*, or KeS*h*, anciently; just as we find a similar people, the *Barâbera* (who present striking similarities), there now. A curious little fact comes in opportunely to support this position. The ruins of the ancient town of *Tutzis*, or Tusis, the military station "Dodecaschœni," are identified in the modern Gerf Hussèyn. A Coptic papyrus, found there in 1813, established that its former name was *Thosh;* and the similarity of this word with "Ethaush," the Coptic form of "Ethiopia," or *Koush* [ubi supra], was long ago pointed out by Wilkinson, who ascertained, moreover, that the present Nubian name of *Tutzis* is KISH.]

4th. That this appellative, *KeSh*, in hieroglyphics, refers to a special Nubian people, without the slightest relation, linguistically, geographically, or anthropologically, to *Tirhaka*, beyond the fact that, like his pharaonic predecessors, he conquered and ruled over them [*supra*, p. 264, Fig. 186.]

5th. That the African KeS*h* of the hieroglyphics are totally distinct from the Asiatic KUS*h* of the Hebrew writers, and are never implied by the latter in this term.

6th. That the confusion, still prevalent on this subject, proceeds from an insufficient examination of old Hebrew ethnic geography on the one hand, and of Egyptian records on the other, after starting with a fundamental error as to the Greek word "Æthiopia."

7th. That KUS*h* of Xth Genesis denotes *Arabia* in its widest sense, and Arabian tribes of *dark* complexion.

8th. That, except perhaps in two or three doubtful instances, in the later biblical books, where geographical precision is sacrificed to poetic license, the biblical word KUS*h* never crosses the Red Sea into Africa; and, even if it be sometimes coupled by a conjunction to *Phut*, and to *Lud*, it never embraces those races we term *Negro*— the context, in every case, being susceptible of more rational exegesis.

9th. That KUS*h* in Hebrew is radically distinct from the Nubian KeS*h* of hieroglyphics, as well as from the *Kish* of our present day.

10th. That KUS*h* is not Σκυθαι, *Skuth*, or *Scot!* does not include Scythic, Indo-Germanic, Tartar, Mongolian, or other races outlying the boundary of ancient Hebrew geography.

11th. That, excepting as regards its application to Asiatic tribes of *dark* complexion, KUS*h* cannot be rendered by Αιθιοπια, in the sense in which this Greek word was used during Ptolemaic times at Alexandria, and by ourselves, without leading to equivoque; but, if we restore to "Æthiopia" its old Homeric meaning of "sun-*burnt-faced*-people," there is no doubt that the KUS*h*, mentioned in parallel ages by Hebrew writers, were sometimes included among the *Eastern*, *i. e.* Asiatic, *Æthiopians* of Hesiod, Homer, and Herodotus.

12th. That, in archaic anthropology, *Æthiopian* is as vague an adjective (without specific warning, on the author's part, of the meaning he attaches to it) as *Scythian*, *Indian*, or *Caucasian*, and therefore had better be avoided by ethnographers.

13th. That the *Coptic* KHOUSH, and *Thaush*, or *Ethosh*, belong to post-Christian days, and represent "Ethiopia" in the corrupt sense in which the Hebrew name KUS*h* was already understood by the Hellenistic Jews called the LXX, and by Josephus. The former word, meaning *dark*, was naturally applied by Egyptian (Copts) *Jacobites* to African families and localities above the first cataract of the Nile; the latter, meaning "the *frontier*," and also (through dialectic mutations of K and T*h*), being a homonyme of KHOUS*h*, was a natural transcript of "Ethiopia;" a name which, from similarity of sound as much as from identity, in *Coptic* days, of association with Africa above Egypt, had been previously given to the *Nubias* by Alexandrian writers.

14th. Finally, that, unless *words* and *names* are restricted to the acceptation in which they were used by each writer in *his own age*, the natural history of humanity, greatly dependent as it is upon historical phenomena, can never rise to the level of a *positive* science; and that sublime sentence, "the proper study of mankind is *man*," mouthed by rote without perceptions of its lofty import, and still overlaid by theological clap-trap, will never reach practical realization.

To us, therefore, KUS*h* of Xth Genesis means *Asia* geographically, *Arabia* topographically, and the *dark Arabs* ethnologically. We pass on to classify KUS*hean* affiliations, in hopes that they will justify our *à priori* assumptions.[594]

KUS*h* as ARABIAN.

We have shown in the foregoing *résumé* that, amid geographical personifications of the Hebrews, KUS*h* was *Asiatic* generally, no less than *Assyrian* and *Arabian* espepecially. In consequence, it seems rational to seek for KUS*hean* origins among Arabic traditions, and Arab localities.

And here it is that the *Recherches Nouvelles* of Volney take precedence over all those made during the first quarter of the nineteenth century. Volney: "Un des hommes les plus pénétrants de ce siècle. . . . Si, parmi nous, Volney a profité des écrits de Richard Simon, ce n'est pas parceque Volney était imbu des principes de l'école matérialiste, mais à cause de l'instinct scientifique qu'il possédait profondément et qui, dans ses écrits, s'est souvent fait jour, en dépit même de ses préjugés philosophiques." Orthodoxy can find no fault with the words of Lenormant, whose views are eminently catholic, even in archæology. We gladly follow his example, when taking departure, in Arabian inquiries, from Volney. Nevertheless, since the peace of 1815, multitudes of scientific Europeans, profoundly versed in Arabic lore through arduous studies, or far more adventurous travels, have given to Arabian researches a propulsion similar to that received, since 1822, by Egyptian, and, since 1843, by Assyrian. *Primus inter pares* among the above, whether in the cabinet or on the road, ranks M. Fulgence Fresnel. Than his opinion French and German scholarship at this day recognizes none higher: because, in addition to a mind disciplined by thirty years of devotion to this speciality, no man, in Arabian investigations, has yet enjoyed M. Fresnel's facilities of actual observation. We select him, then, as our standard authority on KUS*h*, and *Cushites:* supporting it by the concurrence of distinguished Orientalists to whom his publications are familiar.

The arbitrary Ptolemaic repartition of the Peninsula into *Happy*, *Desert*, and *Petræan* Arabia, has long ago been abandoned by geographers. To the Arabs these foreign divisions were unknown. Into the varied districts designated by such alien names, old Arab tradition recognizes the introduction of three races, forming three distinct nationalities; whose several origins being lost in the night of time, Mohammedan writers have appropriated, through the Koràn, Hebrew genealogies in the absence of history; so that it is now impossible to separate much of the exotic from the autocthonous. These three divers stocks of primitive *Arabian* nations, *i. e.*, *â*RaB, *Western* men — according to Ebn-Dihhiyah, followed by Fresnel and Jomard — were,

1st. The ARBA, or ARIBAH, *Arabs* par excellence — subdivided into nine tribes, claiming descent from IRAM (*Aram* of *Gen.* x. 23), son of Shem: from whom the semi-Egyptian, semi-Hebrew, ISHMAEL is said to have learned Arabic!

2d. The MOUTA'ARIBA, *naturalized* and not pure Arabs; whose genealogies ascend to QAHTÀN (*Joktan* of *Gen.* x. 25), son of Heber, son of Salah, son of Arphaxad, son of Shem.

3d. The MOUSTAARIBA, still less pure Arabs; descendants of ISHMAEL, son of Abraham and Hagar.

These, in general, are reputed to be the *surviving* Arabs; in contradistinction to the lost tribes of AD, THAMOOD, &c. &c., destroyed for their impieties, between the times of "the prophet HOOD" (*Heber* of *Gen.* x. 24) and Abraham. "But the spirit of that entire table (*Gen.* x.), in which names of people, cities, and lands, are personified, leads us to conclude," says Gesenius, "that *Heber* was not an historical, but only a mythical personage, whose name was first formed from that of the people. This was, doubtless, the case with Ion, Dorus, and Æolus."

None of the above nations, however, attribute their descent to an *Hamitic* affiliation through KUS*h:* and Hyde sustains that the *Cushites* migrated from *Chusistân*, or Susiana, to the shores of the Euphrates and Persian Gulf; whence it is probable their offshoots spread over Southern Arabia, and eventually crossed the Red Sea, in common with Arabs of the Semitic stock, into Abyssinia and other Upper Nilotic provinces.

With the *Ishmaelitish* tribes of Arabia, as they are not included in Xth Genesis, our inquiries have little to do. Their distribution has been worked up, as completely as the subject admits, by Forster; although the attentive comparisons of Fresnel result in but nine or ten nominal identifications of Arab tribes mentioned in the Bible, while above forty biblical tribes are wanting in the lists of the Arabs. The purely Semitish families of Xth Genesis are allotted their own places in our Essay. To determine KUS*hite* occupation of Arabia is our object, now that, except as "sun-*burned-faces*," they had no relation to African "Ethiopia," at the remote age of our historical horizon.

No one will dispute that, in the idea of the writer of XIth Genesis, the affiliations of SHEM, HAM, and JAPHETH, catalogued in the Xth, assembled, when "*the whole earth* was of *one* language," on the plain of *Shinar* (*Gen.* xi. 1, 2), whence they were dispersed by miraculous interposition. Among the number was KUS*h*, the father of NIMROD; and consequently *Asia*, on the banks of the Euphrates, was the primitive starting-place of himself and children, viewed as *men*. Conceding to orthodoxy their departure thence towards Africa, *Arabia* was inevitably their road and halting-place. The only differences between debaters are questions of time: our view being that the KUS*heans* remained there for indefinite ages, and that their African emigrations were partial, as well as chronologically recent; to be demonstrated, anon, by the Arabian concentration of their several descendants.

The many scriptural citations of our preceding remarks establish that KUS*hites* were still in Arabia at a far later period: a notable instance being ZERAH *the Cushite*, in the time of Asa; to place whom in Africa, because the *Lubîm* and *Cushîm* are united in 2 *Chron.* xvi. 8, when the *Cushîm* alone are recorded in the historical narrative (2 *Chron.* xiv. 8–14), merely to accumulate proofs that no confidence can be given to either account at all, is, to say the least, incautious. The KUS*heans* were yet in Arabia, at the time of Jeremiah's (xiii. 23) interrogatory, "Can the *Cushean* change his skin?" which contrast, we have shown, applies to the *dark* Arabian tribes, abounding in Arabia then as now. But, lest our application should be considered dubious, this fact must be contemplated from a more philosophic point of view.

It is acknowledged by the highest ethnological students of our generation, Prichard, De Brotonne, Jacquinot, Bodichon, Pauthier, and others, that wherever in Austral-Asiatic latitudes, Hindostan for example, tradition yet pierces through the gloom of time, the *dark*, or black, families of mankind (specimens of whom also survive there to our day) have invariably preceded colonizations by the *Whites*, or higher castes. It is also claimed by Kenrick, Bunsen, De Brotonne, and Lenormant, that the great *Hamitic* migration westwards through Arabia antedates the Semitic: in other words, that KUS*hites* were settled in Southern Arabia prior to the arrival of *Djourhomidæ*, *Joktanidæ*, or *Abrahamidæ* — Semitish tribes, like the Hebrews, of fairer complexion. The new doctrines advanced in this volume [*supra*, Chapter VI.] relative to the improving gradations of *type*, in humanity's scale, when we consider each family of mankind, one by one, from the Cape of Good Hope to the Caucasian mountains, show how a *dark* group of men ought to present itself in Arabia, as the immediate Asiatic successors of the swarthy Egyptians: *Egypt*-proper, according to ancient opinions, now corroborated by zoological facts, being far more Asiatic than African in its natural history and phenomena. What group answers all these conditions but the one to which, from immemorial time, the name of KUS*h* has been appropriately referred? Even as late as the fifth century after Christ, Syrian authors, cited by Assemani, designated *Himyarite* Arabs by the name of KUS*hites*.

And this brings us to the point where Fresnel's discoveries establish the entity of a *fourth* group of "Arabs," distinct from Semitish families, dating in Southern Arabia from ante-historical ages to the present hour.

Carsten Niebuhr, in 1763, first announced to Europe the positive existence in Southern Arabia of inscriptions which old Arab authors had characterized as *Musnad*, 'propped up,' and had considered anterior in age to Islâm, no less than to the present *Neskee* and its parent the *Cuphic* writing of Mohammed's day. De Sacy, 1805, with his usual acumen, investigated the subject; Seetzen, 1810; Gesenius, 1819; Kopp, 1822; and Hupfeld, 1825; chiefly from *Ethiopic* (Abyssinian) data, advanced its study; until Wellsted, 1834, and Crittenden, (officers attached to the East India Company's surveys,) discovered inscriptions of the highest interest, cut in the old Himyaritic alphabet, at *Hisn Ghoràb*, &c.

The learned critique of our friend Prof. W. W. Turner would greatly simplify an expository task, could we herein digress upon these Himyaritic inscriptions, the earliest date of which falls far below the Christian era. To his scathing refusal of "one particle of sympathy for Mr. Forster" viewed as translator (!) of the *Himyaritic*, we beg leave to add ours in respect to this gentleman's more recent "*Sinaic* Inscriptions—Voice of Israel from the Rocks of Sinai"; and to apply Turner's just strictures to both of the Rev. Mr. Forster's fabrications. "His wholly false and inconclusive method of deciphering the inscriptions, the bombastic strain in which he dilates on his achievements, and above all the disingenuous artifices by which he seeks to disguise the hollowness of his pretensions, render his performance [whether Himyaritic, or Sinaic, or, worse than either, his last pseudo-*hieroglyphical!*] deserving of all the ridicule and censure it has met with." It is sufficient now to mention, that Hunt's refutation also lies before us; together with the *Recherches sur les Inscriptions Himyariques de San'â, Khariba, Mareb*, &c., through which Fresnel's claim to the resuscitation of ancient Himyar is universally acknowledged.

M. Fresnel's IVth and Vth Letters to the *Journal Asiatique*, "Djiddah, Jan. and Feb. 1838," give a sprightly account of his rencontre with a "piratical grammarian" yclept *Moukhsin;* through whose and other fortuitous aids, he constructed the vocabulary of a still living tongue, spoken at *Zhafâr* and *Mirbât*, in Southern Arabia; which speech, now unintelligible to Semitic Arabs, is called *Ehkìli* by native speakers, and *Mahri*, or *Ghràwi*, by surrounding tribes. This extraordinary language, whose existence was unsuspected until 1838 by modern philologers, possesses thirty-four to thirty-five *consonant* articulations, six pure *vowels*, and as many *nasal*—approximately, some forty-seven different sounds; among which three are utterly inexpressible in any European alphabet; and one is altogether too inhuman for any man but a true Zhafârite to enunciate! Of the twenty-eight articulations current during Mohammed's time in the Hedjàs, two have become superfluous in the vernacular Arabic (*Dàrig*) of Cairo; nevertheless the old Arabic alphabet of twenty-eight articulations is too poor, by nineteen phonetics, for tribes living at Mirbàt and Zhafâr!

[They completely destroy, Fresnel states, "la symétrie du visage." Even Moukhsin thought the facial contortion ridiculous; though he told M. A. d'Abbadie that none of his tribe pronounced three of those letters on the *left* side of the mouth. "Pour rendre le son du ث il faut chercher à prononcer un Z, en portant l'extremité de la langue sous les molaires supérieures du coté droit"—such is "*Himyaritic* euphony"! Having humbly endeavored, "in auld lang syne" at Cairo, to imitate my friend M. Fresnel's attempts to rival Moukhsin's mode of oral articulation, I was, and still am, at a loss to define the agonies of its intonation, otherwise than by reprinting how, "while (this letter) somewhat resembles the 'LL' of the Welsh, (it) *can* be articulated only on the *right side* of the mouth—being something between 'LLW,' a *whistle* and a SPIT!"—G. R. G.]

Gesenius had divided Semitish languages, classified as they are too vaguely, into three main branches: —

1st. The *Aramæan*, spoken in Syria, Mesopotamia, and Babylonia. This is again divided into East and West Aramæan; that is, the Chaldee and Syriac.

2d. The *Canaanitish*, or *Hebrew*, spoken in Palestine and Phœnicia. Of this the Punic is a descendant.

3d. The *Arabic*, of which the Æthiopic is a parallel branch. The Samaritan is a mixture of the Hebrew and Aramæan.

To the above, Fresnel's discoveries add a *fourth:* viz., the "Ehkèelee" of the inhabitants of Mirbàt and Zhafâr; one which he considers among the richest and most ancient in the world — allied to the Ethiopic, but more archaic; preserved in Arabia by a peculiar family (long cut off from the rest of mankind by wild Bédawees of the Semitic stock, with whom, it is said, the Zhafârites never intermarry) — descended probably from the *Homeritæ;* in whose name classical annalists have preserved to us the original word *Himyar* (Arabicè, *Ahmar*), 'the *red*-men,' as the distinguishing title of the once-great *Himyarites* of Saba and Mariaba.

"He who enters Zhafâr *Himyarizes*," is an ancient Arab proverb, which shows that the Zhafârites were different, in some striking peculiarities, from Semitish tribes, and that visitors were constrained "to speak the language of the country;" as unintelligible even now to Ishmaelite and Joktanide Arabs as the *Basque* is to Frenchmen or Spaniards. Now, this tongue and the tribes that speak it, are considered by M. Fresnel to be the true relics of KUS*h;* owing as much to the abundance of words foreign to Arabic contained in its dialects, as to the singular characteristics of the speakers themselves; whose antiquity at *Zhafâr* reaches beyond all history. The daring of Dr. Arnaud, (who, at Fresnel's instigation, penetrated where no European ever reached previously to 1844, and copied multitudes of Himyaritic inscriptions on the ruined edifices of Sana, Khariba, and Mareb,) has confirmed, in all important respects, the existence of these human vestiges of KUS*hites* in their earliest Arabian homestead "even unto this day": and the men, their language and monuments, having now been found, our results on Xth Genesis may be finally tabulated as follows: —

1st. That by KUS*h* the Hebrew chorographer meant *dark* tribes of Southern Arabia, who probably inhabited that section of the peninsula prior to immigrations of strictly Semitish Arabs. They are the *Homeritæ* of Greek and Roman writers; *Himyarites* of Arab history; remnants of whom, speaking *Ehkìli*, still residing at Mirbàt and Zhafâr, are living witnesses of the indelibility of primordial *types.*

2d. That other compilers of Scripture corroborate this view, and prove that in Hebrew geography the KUS*hìm* — bounded at the extreme west by the "rivers of *Cush*" on the Isthmus of Suez — spread across the peninsula to the banks of the Euphrates; perhaps eastwardly to *Chuzistàn* and *Susiana.* Their settlements, as Forster has shown with commendable felicity, lay dotted around the Arabian coasts of the Red Sea and Persian Gulf; separated originally from the intrusive *Joktanides,* (as the writer of Gen. X. accurately remarks, v. 30), by a line drawn from "*Mesha,* as thou goest unto *Sephar*" — the former being the *Zames Mons* in Central Arabia of Ptolemy the geographer; the latter, Mount *Sephar*, at the extreme south-west of the peninsula, where in Ptolemy's time dwelt the *Sapharitæ;* and where at *Zhafâr,* Fresnel's researches (unquoted by Forster) prove their *Ehkìli* descendants to live still.

3d. That before future hagiographers place KUS*h* in *Africa,* as the Hebrew name for Nigritian races (of whom Cush, scripturally and physically, is no more the father than Abraham himself), it might be well, perhaps, if they re-read their "Bibles" with a little attention; and not perversely close their eyes to the new lights that Oriental science is continually shedding upon an ancient code which, Lanci emphatically and truthfully observes, "is the more honored and revered as thought dives into it to illustrate and *comprehend* it."

As *Southern* Arabia, and as *dark* (*himyar*, 'red') Arabian tribes, KUS*h* takes his rightful position once more in Xth Genesis.[595]

16. מצרים — MTsRIM — 'MIZRAIM.'

Semitic; but certainly not the Hebrew 'tribulation,' &c.

As it stands, is the plural of MTsR. With the Masoretic points, added since the sixth century after Christ, it is a *dual*, MITSRAÏM, meaning the *two* MTsRs. In the singular, MTsUR, it is the name (by modern natives referred also to the city of *Cairo*,) through which Egypt is designated in the form MUSS'R, not merely by her present Arabicized people, but by all Oriental nations: and there being no dispute as to the application of MTsUR by Semitic races to the *land of Egypt*, from the present hour back to the remotest period for which we possess records, our genesiacal purposes would be served sufficiently on reading *Egypt* for MTsR*aim*, were it not for foolish rabbinical notions, vulgarly current, that, misunderstanding the principle of Oriental personifications, still treat of "*Mizraim*" in Xth Genesis as if *he* had been really a *man*, "*son* of Ham," another individual! One might as reasonably maintain that all the *Russias*, or the "*two* Russias," mean a human being actually resident in Muscovy! Pandering to no such historical falsehoods, we briefly set the reader on the "royal road" to their refutation.

The earliest personification of *Matzur*, the singular of MTsRIM, is not in the Bible, but in Sanconiathon; a very ancient Phœnician writer, who flourished (none will dispute) some time *before* Philo Byblius, about the second century after C., translated into Greek such fragments of his works as reach our day through Athenæus, Porphyry, Eusebius, and other transcribers. Whether Sanconiathon be a mythe, as some maintain, or whether such a person really lived and wrote between St. Martin's adopted era, 1400 B. C., and Philo Byblius's age, is indifferent; so long as it remains historical, that, under the name "Sanconiathon," we possess some *exuviæ* of Phœnician traditions antedating Christian harmonizings, that cannot have been *written* alphabetically, according to the laws of palæography, earlier than the seventh to tenth century B. C., nor later historically than the second century after the Christian era. We have no hypothesis to sustain beyond establishing, through these fragments, that "Misor" was the ancestor of the Egyptian god Thoth, *Hermes-Trismegistus* (*Her-Mes* = 'begotten of Horus') of the Greeks; and consequently, that this Græco-Phœnician legend is our most valid authority for making a *man* out of the "two *Egypts*" — Upper and Lower — personified in Xth Genesis by commentators as MITZRAIM.

The context of *Ps.* cv. 23, (and wherever else in canonical Hebrew records the singular form MTsUR occurs,) suffices to prove that, by MTsUR, each Jewish writer meant Egypt as a *country*. If the singular number, MTsUR, in Hebrew grammar and history, signifies merely a geographical locality, upon what principle can the dual or plural forms of the same word constitute a *man*?

Among the multitude of appellatives given to Egypt by other foreigners, the present name MUSS'R reappears in the Phœnician Μυαρα — suspected to be an error of copyist for *Musra* — of Stephanus Byzantinus; in the Μεσтραία of George the Syncellus; in the MESSREDJ of the Persian "Boundehesch-Pahlevi"; and so on backwards to the Persepolitan cuneiform inscriptions of Darius, carved at Behistûn early in the fifth century B. C., where it is orthographed M'udrâya. Two centuries earlier, the name MASR, or *Madr* (also Mesrahouan), is chiselled in Assyrian cuneatics on the thresholds of Khorsabad, among the conquests of Asarhaddon, between B. C. 709 and 667; and it may exist perhaps on older sculptures of the ninth century B. C., discovered by Rawlinson.

Albeit, 700 years B. C. are ample for our object; inasmuch as they prove that a singular form of the name *Muss'r* existed in Asia, in days parallel with, and probably anterior to, those passages in the Hebrew Text where MTsUR is its homonyme. Its dual or plural representative in Xth Genesis, MTsRIM, is either a later amplification, or meaning simply *the Muss'rites*, people of *Muss'r*, Egypt, excludes the supernatural idea that MIZRAIM was a *man*.

In this concrete sense of *Egyptians*, we find the correspondent of *Mizraim* in the

Μεστραιοι of Josephus, and of the Syncellus; but the latter uses it in his preface to a document, the *Old Chronicle*, which every scholar repudiates in some mode more or less decisive. Those who now pretend to accept the *Old Chronicle*, or the *Laterculus*, as genuine Egyptian, slur over Letronne's blighting criticisms. The hand of Judaizing Christian imposture stands out undisguisedly in the other portion of the Syncellus's chronography — where he commences his "Laterculus" with Μεστραιμ ο και Μηνης — *Mestraim* (for Mizraim) the same as MENES! That the first Pharaoh of Egypt, Menes, should be metamorphosed into MTsRIM, *the Egyptians*, of Xth Genesis, by a harmonizing monk of Byzantium some 800 years after Christ, and at least 4500 after the death of Menes, is not extraordinary, when one remembers the pious frauds of a school in which the Syncellus was neither the first nor the last ornament; but that writers in our day should reason from such and similar Greek-church literary juggleries, that *Mitsraim* of Xth Genesis was a *man*, instead of an Oriental personification of Egypt, merely proves such writers to possess, as Bunsen has it, "little learning, or less honesty." Our note[596] indicates volume and page wherein complete destruction of τὸ παλαιὸν χρονικόν, 'the Chronicle of the old times, or events,' may be found; and we are content to follow in the wake of Letronne, Biot, Matter, Barucchi, Böckh, Bunsen, Raoul-Rochette, Lepsius, Kenrick, Alfred Maury, &c. — all of whom, more or less earnestly, reject the *Old Chronicle*, uniting with Bunsen's condemnation of it and "similia, quæ hominis sunt Christiani, parum docti, at impudentissimi."

All Grecian antiquity, from Homer to Strabo, has designated Egypt by names in which no form of *Mitsraim* plays a part; nor can it be yet said that any true equivalent for the Semitic *Muss'r* has been discovered amid the numberless appellatives given to their own country by Egyptian hierogrammates. Leaving aside old fanciful analogies that might be retwisted out of Champollion's *Grammaire* and *Dictionnaire*, Dr. Hinck's ingenious TO-MuTeRI, 'Land of the *two* Egypts,' fell beneath the knife of Mr. Davyd W. Nash, who substituted TO-MuRE-KHAFTO, 'the beloved land of the *two* Egypts.' Syncellus's "Mestræans" was supposed by Lenormant to be a compound word — MES-*n*-RE, 'son of the sun': but, 1st, this has not been found as a proper name in hieroglyphics; and, 2dly, the word Μεστραια is but a modern Greek transcriber's corruption (not of an *Egyptian* name, but) of the *Hebrew* and foreign word *Mitsra-im*. Mr. Birch's "*Merter* (Mitzraim), is red under thy sandals," is the nearest approximation to *Muss'r* hitherto suggested; and saves discussion here of the various Hebraical solutions proposed by Rosellini, Portal, or Lanci; some of which would admirably explain why the *Hebrews* gave to Egypt the name of MTsRIM, but none of which prove that the Egyptian natives ever recognized such foreign designation — any nearer, philologically, than "Americus Vespucius" might, by some etymological gladiator, be wrenched out of our "Uncle Sam." We return, therefore, as in so many other instances, to Champollion's fiat of forty years ago: viz., that *Muss'r*, MTsUR, and MTsRIM, in all their forms, were probably alien to the denizens of the Nile, but were names given *to* Egypt and Egyptians by Semitic populations.

But one query remains. In the original idea of the writer of Xth Genesis, was MTsRIM a dual or a plural? The surviving punctuated Text (written or printed in the post-Christian *square-letter*) reads, dualistically, *Mitsraïm;* which would correspond perfectly to the Pharaonic division into "*two* Egypts," Upper and Lower — preserved still in the *Saeêd* and *Bahreêyeh* of the modern Fellaheen. We would submit, notwithstanding, that the *Masorete* diacritical marks float between A. C. 506, and the eleventh century (age of the earliest MSS. extant); and therefore such minute contingencies as a dual or a plural become, archæologically speaking, rather problematical. For ourselves, we think the plural form, *Mitsrîm*, most natural — 1st, because it is the Hebrew literal expression without the later and superfluous points; and, 2d, because the plural MITsRÎM, as the Israelitish name for *Egyptians*, amply satisfied all chorographic and ethnological exigencies whensoever Xth Genesis was projected.

"Misrajim," Bochart declared 200 years ago, "non est nomen *hominis*. Id non patitur forma dualis"; wherefore, denying that there ever was a *man* called "Mizraim," we read simply, for MiTsRIM — *the Egyptians.*[597]

17. פוט — P*h*UT — 'PHUT.'

Hamitic; not the Hebrew 'fat,' 'despicable,' &c.!

That this is Barbary — *i. e.*, the African coast along the Mediterranean west of Egypt — no one doubts. Differences of opinion here resolve themselves into mere conjectures as to space.

The most salient feature of *Phut*, observable in Xth Genesis, is that this personification has no *children* — *i. e.*, colonies, or affiliations; which, coupled with the vague demarcations of *Phut* in other Scriptural passages (*Nahum* iii. 9), shows that to the Hebrews this name meant generally North-western Africa; embracing families of man too remote to be described. The word has since spread very extensively over Africa, if *Foute*, *Fouta*-Toro, *Fouta*-Bondou, *Fouta*-Djallon, &c., names of Fellatah States and tribes, be its derivatives; as *Fâs*, the kingdom of Fez, is, without question; nominally replacing the *Regio Phutensis* of Jerome's time; Ptolemy's city of *Foutis*; and Pliny's river *Phuth* flowing in Mauritania, the country which Josephus considers the equivalent of *Phut*. Indeed, there is no lack of old names, throughout the Moghreb, (part of which containing "*Putea* urbs, *Phut* flumen, *Phthia* portus, *Pythis* extrema," was anciently called *Futeya*), like *Phthamphu*, *Phthemphuti*, *Phautusii*, &c., to establish *Phut's* existence at all recorded ages, close to the *Loubîm*, *Lehabîm*, and similar Libyan designations in Xth Genesis.

Bunsen reads *Phut* as Mauritania; considering that the river *Phut* of Pliny is equivalent to the *Punt* of hieroglyphics; the N or M left out, as in *Moph* for Memphis, or *Shishak* for Sheshonk. Birch holds the hieroglyphical sign (which ascends in antiquity to the earliest monuments) to mean the "*nine bows*. This word has been read *Peti*, and supposed to be the Scriptural *Phut*, the Libyans or Moors; but it must be observed that the hieroglyphical word *Peti* is always applied to a large unstrung bow, in ethnic names." Upon the cuneatic sculptures of Assyria, and among the conquests of Asarhaddon, De Saulcy has read — "Populum *Pout*, hos et gentes fœderatas."

As "P*h*eT-*kah*," or *bow-country*, or as "NiP*h*T — countries," determined by *nine bows*, this name for the last quarter of a century has been identified with *Phut*, (or rather, confounded with the NiP*h*aiaT — true representatives of the *Naphtukhîm* of Gen. x. 13,) in Egyptian sculptures of every epoch; and, without doubt, refers, in hieroglyphics, to Libyan families of *Amazirghs*, *Shillouhs*, &c., that under the present general denomination of *Berbers* stretch westwards from Lower Egypt to the Atlantic.

Deferring some critical minutiæ until we reach the *Naphtukhîm*, our opinion on *Phut* is, that in Xth Genesis it means those countries now called Barbary; while in other biblical texts it covers *Hamitic* affiliations along the Mediterranean face of Africa; to the exclusion of the more inland Negro races, by Hebrew chroniclers unmentioned.[598]

18. כנען — KNA*â*N — 'CANAAN.'

Hamitic; not the Hebrew 'merchant,' 'tribulation,' &c.

Upon no terrestrial personification in Xth Genesis, except CUSH and NIMROD, has more theory been piled upon hypothesis, than in respect to this luckless cognomen and the historical nations that bore it.

Assuming that the *Jehovistic* document of Genesis IXth was penned by the same individuality who compiled the chart of Genesis Xth, orthodox commentators, from the Rabbis and Fathers down to the uninspired annotators of our own generation, sorely vex themselves with Noah's inebriate malediction — "accursed be KANAAN. Let him be *â*BD-*â*BDIM, *slave of slaves*, to his brethren"—(Gen. ix. 25) — whereas, in the Text itself, HAM the father, not KANAAN the son, was the graceless offender. In Hesiod's

Greek version of the same Chaldæan mythe, hapless Οὐρανός, *Cœlus*, had infinitely more serious reasons for swearing at his unnatural son Κρόνος, *Saturnus;* while, as Cahen has duly noted on the Noachian curse, "this is the fourth malediction that one encounters in Genesis: the first being against a snake, the second against the earth, and the third against Cain."

Setting forth thence with a moral *non-sequitur*, commentators next attempt to justify a supposititious extermination of the guiltless grandson's innocent posterity, recorded by "writer 2d" — "but of the cities of these people (the *Canaanites*), which IeHOuaH thy God gives thee for heritage, thou shalt spare nothing alive that breathes" (*Deut.* xx. 16). Yet, despite this and similar omnipotent injunctions to obliterate poor KNAâN, we find "writer 3d" (*Josh.* xv. 63) attesting how "the children of Judah *could not drive* out" the Canaanites from Israel's holiest abode, Jerusalem, even "unto this day!" A fact explained by "writer 4th" (*Jud.* i. 19, 21), "because (the Canaanites) had chariots of iron"; at the same time that "writer 5th" (2 *Sam.* v. 7, 8, 9) bears witness that one band of Canaanites maintained the stronghold of Mt. Zion, *Jebus*, down to the reign of David. Even then, unscrupulously heroic as that monarch was, he was constrained, through political exigencies, chronicled by "writer 6th" (2 *Sam.* xxiv. 18, 24), to *buy* from a Canaanitish land-holder, "Aravna, the Jebusite," the identical "threshing floor" on the site of which Solomon, according to "writer 7th" (2 *Chron.* iii. 1, 3), erected a little paganish temple (smaller than its duplicate at *Hierapolis*) that, although only 90 feet long by 30 front, is estimated to have cost about 4000 *millions* of dollars — United States' currency.

Other sticklers for plenary inspiration who, in direct contravention of the plain words of Genesis IXth (favoring the notion that Ham, and not his son Canaan, was accursed), contend that, in consequence of such malediction, Ham became the progenitor of black (*Negro*) races, may be set aside as entirely ignorant of Scripture. Followers of the learned Dr. Cartwright's "*Canaan identified with the Ethiopian*" may be pleased to refer to the fac-simile portrait [*supra*, p. 127, Fig. 19] for confirmation of a doctrine which has the double misfortune of being physiologically and historically impossible, as well as wholly anti-biblical.

We appeal to the sober author of Xth Genesis for relief from such mental aberrations. His chorography (constructed some time after Joshua the son of Nun, or Nau, had expelled such Canaanitish tribes as survived massacre, or tolerated under the conqueror's yoke, along Israel's roads of march from Mount Sinai to Palestine) attests, *ex post facto*, that already in his time "the families of the KNAâNI (had been) *dispersed*." (*Gen.* x. 18.) Large bodies of these people emigrated to Libya, where their names, traditions, and tongues, exist to this day. Procopius, in the sixth century A. C., mentions an inscription wherein *Phœnicians* recorded their flight into Africa, "from before the face of the brigand Joshua son of Naue:" and in the fourth century, St. Augustine, Bishop of Hippo, relates how, in his diocese, "Our rustics, being asked whence they were, responded, Punically, *Chanani*." Now, it is a fact as certain as any in history, that the Punic-Carthaginians, their parents the Phœnicians, the Canaanites and the Hebrews, spoke one and the same tongue, but with slight idiomatic provincialisms of difference. "The term '*Hebrew* language' does not occur in the Old Testament," says Gesenius, "though it must have been common when part of it was written. Instead of this name, the language is usually termed the *language of Canaan* (*Isa.* xix. 18)." So far, indeed, from *Hebrew*, as philological science nowadays understands the term, deserving honors, owing to its supposititious antiquity, as the "lingua sancta" of Paradise (according to Usher, exactly B. C. 4002–3!), it is positive that Abraham, grandfather of Israel, when he emigrated from "Ur of the Chaldees," spoke, not in *Hebrew*, but, like his Mesopotamian tribe, in an *Aramœan* dialect. Israel's descendants, forgetting their mother-tongue, adopted afterwards, in Palestine, the speech of KNAâN; and, *calling* it "Hebrew," unwittingly sanctified the language of the "slave of slaves," instead of that of the true *Abrahamidœ!* During the Captivity, the

Jews again forgot *Kanaanitish* "Hebrew." Retempered by some seventy years' sojourn in the Euphratic regions of their primitive origin, they brought back with them a later idiom of that *Chaldæan* language which, modified by about 1500 years of time, was a lineal descendant of the pristine speech of Abraham, son of Terah, son of Nahor, son of Serag, son of Reu, son of Peleg; son (that is, *affiliation*) of *Eber* — not a *man*, but the geographical personification symbolized in Xth Genesis (21) by EBR, *êber*; a name which, like its Greek form, *υπερ*, and its Latinized equivalent, *Iberian*, originally meant simply "the *yonder* land;" that is to say, Palestine; a country west of and *beyond* the river Euphrates! "*Hebrews*," as the foreign corruption of EBR, signifies nothing more than *men from* or *of the other side* — the *Yonderers.*

Every effort, therefore, made by orthodox Rabbis, Doctors, or Moolàhs, Jewish, Christian, or Muslim, to enhance the antiquity and holiness of the tongue they call *Hebrew*, only renders more venerable "the language of KNAâN": and thus, by exalting as theologians do, unintentionally, but positively, the "slave of slaves" above the chosen master, they enable the retributive justice of science to make inhumanity and superstition vindicate, in our nineteenth century, the memory of a much-injured people, who called themselves KNAâNI from ante-historical times down to a period far more modern than the Christian era.

The unceasing proclivity of the Israelites to adopt *Canaanitish* customs and worship, to intermarry with *Canaanitish* females, to dwell in peace with or among them — despite denunciations attributed to Moses and the Prophets — no less than the existence of Canaanites everywhere in Palestine after the Christian era: these facts (evident to every possessor of a "Concordance of the Old and New Testaments") merely prove the strong natural affinities of language and of physical organism common to both families. Nay, apart from supernaturalistic caprice, the only satisfactory mode of justifying such vehement declamations of hatred towards KNAâN, found in the writings of Hebrew reformers, is to acknowledge frankly, that human nature, rebelling against these homicidal proscriptions, often rendered them nugatory in practice.

Of the *eleven* affiliations of KNAâN, only five, the *Hethites*, *Yebousites*, *Emorites*, *Guirgasites*, and *Hivites*, were established within the petty territory of Palestine. Add to these the *Canaanites* (possibly descendants of another KNAâN) and the *Pherizites*, who were merely peasants; and we have the *seven* peoples which the Hebrews were enjoined to expel. (*Deut.* vii. 1; *Josh.* iii. 10.) The desire was stronger than the deed, for the Jews never entirely drove the *Canaanites* out, even of Jerusalem.

By classical historians, the KNAâNI were known under the general name of Φοίνικες, *Phœnicians;* and the LXX often substitute the latter name where the Hebrew Text reads *Kanaanites.* Herodotus and later authors assure us, that the Phœnicians came originally from the Persian Gulf; and the *Kanaani*, therefore, would not be indigenous to Palestine; but, nevertheless, they were "*already* in the land" (*Gen.* xii. 5) at the advent of the *Abrahamidæ*, and we regard them as autocthones.

Eusebius quotes Sanconiathon and his translator, Philo Byblius, for the fact that the Phœnicians called their country Χνᾶ, a contraction of KNAâN. On Phœnician coins the city of Laodicea is called *mother of Kanaan.* Older than numismatic record, more ancient than Hebrew annalists (Moses not excepted), more positively authentic than any source to which archæology can appeal, are the Egyptian monuments of Sethei-Meneptha I. and Ramses II.; whereupon KANANA-*land* is frequently mentioned among conquered Asiatic nations, from the seventeenth — sixteenth century B. C. downwards. And it may assuage pruriency in those who fancy the KNAâNI to have been *African* "Æthiopians," (though as "sun-*burned-faces*" they were certainly Asiatic,) to take another look at our portrait of a *Canaanite*, copied from sculptures anterior to the century in which the Mosaic Lawgiver is erroneously believed to have written the book called *Genesis*—a portrait, wherein the features establish that (apart from Canaan's priority of *speech* in the Hebraical "lingua sancta," as, eventually, "beatorum in cœlis") the inestinguishable laws of *type* prove the KNAâNI, as history also testifies, to belong to the same zoological province of creation, though to a lower *gradation* of type, as the Abrahamidæ. Indeed, the root of KNâ meaning 'low,' and that of ABRAM, 'high,' one may perceive the real cause of early antipathy between the *Canaanites* and the *Abrahamidæ* to lie in mutual repugnances between the indigenous "low-lander" and the intrusive "high-lander."

Palestine, in its widest geographical, no less than in its restricted rabbinical sense, is written history's cradle, and natural history's birth-place, for KNAâN.[599]

בני כוש — BNI-KUSh — "Affiliations of KUSH."

19. סבא — SBA — 'SEBA.'

Perplexities are here occasioned by palæographical and phonetic differences between the letters S, S*h*, and S*s*.

Four separate nations or places, as Bochart reminds us, are mentioned in Genesis by names transcribed through *Seba* or *Sheba:* viz. —

A. — *Genesis* x. 7 — סבא — SBA, or *Seba*, affiliation of KUS*h*.
B. — " x. 7 — שבא — S*s*BA, or *Sheba*, affiliation of KUS*h* through RAAMAH.
C. — " x. 28 — שבא — S*s*BA, or *Sheba*, affiliation of S*h*eM through JOKTAN.
D. — " xxv. 3 — שבא — S*s*BA, or *Sheba*, affiliation of S*h*eM through ABRAHAM.

On these discrepancies Fresnel has wisely noted, that post-Mohammedan Arabs have likewise forged genealogies to match some of those in Xth Genesis; at the same time that different Hebrew annalists often contradict themselves, no less than current Arabian traditions. Various are attempts at reconciliation, to be consulted under our references to Volney, Lenormant, Munk, Jomard, and De Wette; but, upon the whole, Forster's appear to be the most successful, viewed geographically. To us, nevertheless, the only apparent difference between the *four* above-cited names is, that one (A.) begins with the letter *sameq*, S; and the other three (B., C., D.) with *sheen*, S*h;* that is, according to the Masorete points added to the modern *square-letter* manuscripts after the sixth century; because, those stripped away, *sheen* remains *Sseen*, or S*s*.

Abraham's grandchild, through Ketoura, the fourth S*h*BA (D.), is excluded from Xth Genesis, and, therefore, appertains not to our researches; except when noticing the confusion he produces in Arabian genealogies. Nor, for similar reasons, do we speculate on which of the four names might apply to the unknown region whence journeyed Solomon's "Queen of *Sheba*"; whom Josephus makes sovereign of Egypt and Ethiopia; and whom the *Abyssinians* have ever claimed as their own; her illegitimate son, by Solomon, being the legendary progenitor of all their kings. The gifts which this "illustrious inquirer after truth" made to King Solomon (1 *Kings* x. 10; 2 *Chron.* ix. 9) — estimated at $2,917,080, of U. S. coinage; besides any quantity of *spices* and *precious stones* — are enlarged upon by Forster, who considers this lady to have been "Queen of Yemen" in Southern Arabia. Indeed, "the offerings of the Queen of *Sheba*" are believed, by Mr. Wathen, to have enabled Rhamsinitus to build "the indestructible masses of the *pyramids*" of Egypt. Hoskins, of course, appoints this ubiquitous dame Queen of African Meroë: but Fresnel, commenting upon inscriptions brought by Dr. Arnaud from the *Hărăm-Bilkis*—a great elliptical temple, considered to be the "Sanctuary of the *Queen of Sheba*"—seems to have determined her Yemenite locality, as well as the name B-*Almakah;* by which, representing a form of Venus, she became subsequently deified by the Sabæans. Oriental tradition has consecrated, elsewhere, the voyages of princesses, about the same period that *Sheba's* queen and King Solomon interchanged affectionate courtesies. So struck, indeed, were the Jesuit missionaries with the resemblance between the journey made, about 1000 B. C., by "a princess named Si-wang-mou, *the Mother of the Western king* (who afterwards went to China, bearing presents to King Mou-wang") and Solomon's "queen of *Sheba*," that these pietists supposed the *Chinese* account to be *a mere travesty* of the Hebrew books of *Kings* or *Chronicles!* The era; many of the presents; the miraculous facilities of transportation over similar immense distances; and the manner in which the "Mother of the Western King and Mou-wang abandoned themselves, even at the end, to all the delights of joy and songs," curiously correspond. Still more singularly;—the Chinese book, in which these parallelisms are recorded, is called *Chi-i* (*i. e.* collection of what is neglected)—a name identical with the Hebrew *Dibrê haiamim*, and the Greek *Paralipomena* (things left out): in which latter volume, under our English designation of "Chronicles," the queen of *Sheba's* visit was registered, like the Chinese story, by far later scribes, until copies became multiplied *ad infinitum*, through the blessing of moveable types.

Deeming, in common with the highest biblical exegetists of our age, Solomon's "queen of *Sheba*" to be less historical than Mou-wang's, we are fain to leave her out of the argument; no less than Josephus's opinion that African *Meroë* was intended by any "*Saba*" of Xth Genesis. Which doubts submitted, let us remember how Pliny assures us that the *Sabæans* stretched from sea to sea; that is, from the Persian to the Arabian Gulf: and, inasmuch as four distinct nations of Arabia are recorded under the appellative *Seba*, *Sheba*, *Sseba*, or *Saba*, it is uncertain whether any one of them can be specially identified at this day. Nevertheless, they are all circumscribed by the "Gezeeret-el-Arab," or *Isle of the Arabs;* and *Seba* (A.), the first of Genesis Xth, as a KUS*hite* affiliation, belongs to the *himyâr* (red), or *dark*-skinned race; — not improbably now represented by the tribes at *Mirbât* and *Zhafâr*, who still speak the old *Ehkèelee* tongue.

No objections militate against Forster's skilfully elaborated conclusion, "that the *Seba* or Sebaim of the Old Testament, and the Sabi or Asabi of (Ptolemy) the Alexandrine, denote one and the same people;" and that "the tract of country between Cape Mussendom and the mountains of Sciorm was originally the seat of Cushite colonies;" because, as Forster's *maps* and reasonings establish, Cape Mussendom was styled, by Ptolemy, "the promontory of the *Asabi*," near which now lies the town of *Cûscan* (*Cushan* of Hebrew writers); and a littoral termed, by Pliny, "the shore of *Ham*," *Littus Hammæum*, now *Maham* [Ma-K*h*aM? *place* of HAM]; adjacent to which is a *Wàdee-Ham*, Valley of Ham; prove that, all around this centre, many local names, commemorative of KUS*hite* settlements, even yet exist.

Not to dogmatize, we conceive that *Omân*, province of Southern Arabia, suffices for the pristine habitat of our *Seba* (A.).[600]

20. חוילה — K*h*UILH — 'HAVILAH.

Two *Havilahs*, both spelt exactly the same way, one KUS*hite* (v. 7), and the other *Joktanide* (v. 29), occurring in Xth Genesis, their separation is difficult: without harassing ourselves about the third — "Land of K*h*UILH," in *Gen.* ii. 11 — which, being ante-diluvian, concerns not human history.

Here again Forster is an excellent guide, because he does little more than copy Bochart. Assigning to the Joktanide *Havilah* the several districts bearing this name in Yemen, he naturally seeks for the KUS*hite* Havilah about the Persian Gulf, fixing upon the Bahrèyn islands as the pivot of inquiry; one of which still retains its original name, *Aval*. "In order to illustrate the ancient from the modern variations of the proper name Havilah, we must begin," he sensibly observes, "by removing the disguise thrown over it, in our English version of the Bible, by its being there spelled according to the Rabbinical pronunciation. The Hebrew word, written *Havilah* by adoption of the points, without points would read *Huile*, or *Hauile;*" and thereby its identity with the *Huaela* of Ptolemy; the *Huala* of Niebuhr; the *Aval*, *Aûal*, *Huale*,

Khau, Khalt, Khaul, Khaulân, of modern Arabic, becomes transparent to general readers.

Thus, enlarging Bochart's ingenious comparisons, the Εὐιλάτ of the LXX; the *Chablasii* of Dionysius (Periegetes); the *Eblitæan* mountains of Ptolemy, still called *Aûal;* the *Chaulothei* of Erastosthenes, and the *Chaldæi* of Pliny; become resolved, by Forster, into the powerful tribe of the *Beni-Khâled:* whose encampments dot the Peninsula from Damascus to the Straits of Bab-el-mandeb; from Mekka, on the Arabian coast, round to the Persian Gulf and Mesopotamia; often on sites where some remembrance of their parental *Havilite* appellatives is traditionally preserved "unto this day."

"Se non è vero, almeno è ben trovato": and, in the present state of knowledge on Central Arabia — wonderfully small, our nineteenth century considered — if Carlyle's "hammer of Thor" might, perhaps, demolish Forster's picturesque edifice, we doubt that Thor himself could erect a substitute more solid.

Albeit, ethnology may well be content when Arabia, and especially the shores and islands of the Persian Gulf, preserve so many reminiscences of *three* "Havilahs;" among which, through closest application of the "doctrine of chances," some local habitation must still exist for the name and lineage of a KUS*hite* KHAUILAH.[601]

21. סבתה — SBT*t*H — 'SABTAH.'

What may have been the origin of the word *Saba,* which, simple or compound, has been preserved in Arabia by Hamitic and Semitic affiliations, from primordial times to the present, there appears to be no means now of ascertaining. Gesenius derives *Sabaism* from *Tsaba,* the heavenly 'host'; which, as concerns the root *Saba,* appears somewhat *ex post facto.* Arab migration carried this name into Abyssinia, if the *Sabæ* of Strabo be now represented by a town called *Essab;* so too Josephus imagines Meroë to have been called *Saba,* previously to its adoption of the name of Cambyses's sister; but Lepsius's Meroïte discoveries prove the whole story to be fabulous. Bochart, cautiously, traced *Sabatha, Sobota,* of Pliny, through *Sophtha,* an island in the Persian Gulf, to the *Massabathæ* on Median frontiers. Pliny, however, says "*Atramitæ* quorum caput *Sobotale* LX templa muris includens"; which fixes this city towards Hadramaut. Of the three Arabian sites where nominal remains of *Sabtah* are now traceable, Volney's adoption of Bochart's index seems most appropriate: that of Ptolemy's city, Σαφθα, *Saphtha, Sabbatha-metropolis,* on the coast of the Persian Gulf, in the province of Bahrèyn; where the *Saab* Arabs roam at present, as Forster's maps confirm.

"The Homeritæ," states the great hydrographer Jomard, "the Hadramitæ, the Chatramotitæ, the Sabæi, the Sapharitæ, the Omanitæ, the Maranitæ, the Miniæi, the Thamudeni, lived where nowadays even are the people of *Hemyar,* the people of *Hadramaut,* the people of *Saba* (or Mariaba), the people of *Dhafâr,* the people of *Oman,* those of *Mahrah,* those of *Mina,* of *Thamoud,* and many other peoples, of which the name, any more than the existence, does not appear to have suffered from time." And it will manifest the pains now bestowed by Orientalists to discover these Arabian localities, to add Fresnel's successes: — "The famous *emporium* of Kana is decidedly identified with Hisn-Ghorâb" — and "the town of *Kharibet,* discovered by M. Arnaud, is the last term of (Ælius Gallus's) Roman expedition (*Caripeta*)."

Though we cannot yet place our finger on the exact spot, there is no reason for seeking *Sabtah* elsewhere than among KUS*hite* affiliations colonized on the Persian Gulf. If not found already, the place and its tribes will soon be recovered by the zeal of Arabian explorers.[602]

22. רעמה — RA*â*MH — 'RAAMAH.'

Bochart's acuteness had settled upon Ρεγμα of the LXX; *Rhegama* of Ptolemy; *Regmapolis* and *Kolpos-Regma* in Steph. Byzantinus. This name is said by Strabo to signify 'straits'; which meaning singularly corresponds to the narrow entrance of the Persian Gulf, on the Arabian side of which Forster's maps fix *Raamah*, and its two colonies *Sheba* and *Dedan;* already grouped together by Ezekiel (xxvii. 20–22).

The inland province of *Mahrah* preserves the phonetic elements of *Raamah;* and there it is that, at *Mirbàt* and *Zhafâr*, Fresnel's discoveries of the *Ehkèelee* tongue, called also *Mahree*, establish the existence of a people, distinct from Semitish Arabs; survivors of the old Himyarite (*red*) stock: the *dark*-skinned Arabians of KUS*hite* lineage, represented by the swarthy *Dowàsir* tribes, as reported by Burckhardt and Wellsted.

These people were called *Rhaminitæ* and *Rhabanitæ* by Roman authors; and *Ramss*, an Arab port just inside the Persian Gulf, perfectly answers to the site of *Raamah* catalogued among KUS*hite* personifications in Xth Genesis.[603]

23. סבתכא — SBT*t*KA — 'Sabtechah.'

"*Sabtaka* is thrown by Josephus into *Abyssinian Ethiopia;* by Bochart, into the Persic Carmania, under pretext of resembling *Samydake:* these two hypotheses seem to us vague and without proofs. Sabtaka has no known trace." So far Volney.

Yet Bochart's suggestion of B for M offers no palæographic difficulties; and if *Samedake* could be identified, S*a*BeTAKe might be *Sabteka*, situate in Kermàn, near the Persian Gulf.

"The *Sabatica Regio* of the ancients, a district apparently in the neighborhood of the Shat-al-Arab, is the only probable vestige I can discover," says Forster, "of the name or settlements of Sabtecha."

For our purposes, this excellent indication is sufficient. Personifying some locality or people of KUS*hite* origin, probably near the mouth of the Euphrates, the chorographic genealogist of Xth Genesis fixes *Sabteka* among Arabians of swarthy hue.[604]

24. שבא — S*s*BA — 'Sheba.' "Affiliation of Raamah."

[Our S*s*BA second (B.), *ubi supra.*]

We have already stated the difficulties of distinguishing which of *four* Arabian SBAs — KUS*hite*, *Yoktanide*, and *Ketourite* or *Jokshanide* — are assignable now to the chart of Xth Genesis, more than twenty-seven centuries subsequently to its projection; but each one, by every process of reasoning upon facts, is circumscribed within *Arabian* denominations. If, on the one hand, time has rendered minute dissections nugatory, on the other it spares us the trouble of seeking elsewhere for historical lights.

Offshoots of Raamah, "*Sheba* and *Dedan*" stand contiguously, not only in Xth Genesis, but in Ezekiel (xxxviii. 13), and belong to the same neighborhoods; whilst Isaiah's KUS*h* and S*e*BA" (xliii. 3), united by a conjunction, serves to fix *Seba* among the dark-skinned Arabs, where the compiler of Xth Genesis had traced this name's genealogical affinities. But, at whatever age (probably *Esdraic; i. e.*, after return from captivity) the fragmentary documents now called "Genesis" were put together, "a sort of spirit of investigation and combination was also at work. We are indebted to this," continues De Wette, "for the genealogical and ethnographical accounts contained in the Pentateuch. They are designed in sober earnest, and are not without some historical foundation, but are rather the result of fancy and conjecture than of genuine historical investigation. To test the accuracy of the table of Genesis Xth, compare the following passages": —

Genesis X.

7. "The sons of KUS*h*, Seba, and Havilah, and Sabtah, and Raamah, and Sabtecha. And the sons of Raamah; *Sheba* and *Dedan.*"

Genesis XXV.

2. "*Abraham* [descendant of S*h*eM] took a wife . . . Ketourah; and she bare him Zimran and *Jokshan*, Medan, and Midian, and Ishbak, and Shuah: and Jokshan begat *Sheba* and *Dedan.*"

Now, both texts concentrate "*Sheba* and *Dedan*" in Arabia. Nevertheless, the unostentatious care evidently bestowed upon his chorography by the practical compiler of Xth Genesis, favors his superior accuracy, and therefore we take his "*Sheba* and *Dedan*" to be the true colonial settlements of KUS*h*.

This is corroborated by Ezekiel (xxvii. 22) — "The merchants of *Sheba* and RAAMAH, they were thy merchants: they occupied in thy fairs with *chief of all spices*:" not merely referring to the rich productions of incense, myrrh, gums, and aromatics, raised in and exported from this part of Arabia then as now, but also to *spiceries* of India and its islands passing in transit through *Sabœan* hands: which, in Joseph's time (*Gen.* xxxvii. 25), were conveyed by inland caravan-portage to Gilead, whence *Ishmaelites* "with their camels bearing spicery and balm and myrrh," carried them to Egypt; and which "maritime merchandisers," under the name of *Tarshish*, had consigned to the Royal Firm of "Solomon, Hyram, & Co." by "coasters" up the Red Sea; and dispatched *via* Petra through this house's factors at Etsion-gaber: (cost of transhipments, freights, camel-hire, insurances, interests, brokerages, commissions, and *grattages*, no less than amount of shares or profits, to us unknown).

Forster skilfully compares the Plinean account of ÆLIUS GALLUS's expedition, "in the words of Gallus himself; the passage being, to all appearance, an extract from the report of that general to his master Augustus:"—"*Sabœos*, ditissimos sylvarum fertilitate odorifera, auri metallis, agrorum riguis, mellis ceroque proventu:" and moreover relates how, "On his arrival before Marsuabæ, the capital of the Rhamanitæ, Ælius Gallus, the Roman geographer informs us, learned from his prisoners that he was within two days' march of *the spice country*:" the very productions for which the Prophet of the Captivity had given celebrity to "*Sheba* and RAAMAH."

Hence, the geographer of Arabia succeeds in identifying the *Saba* of RAAMAH among the "*Sabœi*, with their capital Mar-Suaba or Sabe; whose locality is preserved and determined, in its modern topography, by the town of *Sabbia*, in the district of *Sabiê*;" mapped by him towards the southwestern extremity of the "Isle of the Arabs."

"A highly valuable confirmation of the identity of the modern province of Sabiê, and of its ancient inhabitants, the Rhamanite Sabæans, with the Cushite Raamah and Sheba, arises on our first reference to the 'Description de l'Arabie' [Carsten Niebuhr's]; where we find, in the Djebal, another *Sabbia*, a large town or village, seated in a district retaining, to this day, the patriarchal name of *Beni Khûsi*, or the sons of Cush. Another district, of the same name, *Beni Keis*, is noticed by our author in the Tehama. In the former district occurs a village named Beit el Khûsi [*house* of the KUS*hite*.] A third small district connects the name of Cush with that of his son Raamah; namely, that of Beni Khûsi, in the province or department of Rama. The city of Kusma, south of Rama, M. Niebuhr rightly conjectures to have derived its name and origin from Cush: a conjecture which receives strong light and confirmation from a remote quarter, in the corresponding denomination of Dooat el Kusma, a harbor of the ancient Havilah, near the head of the Persian Gulf; the acknowledged site of the earliest Cushite settlements"—i. e., of the true KUS*him* of all Israelitish chroniclers; affiliated from the personification KUS*h*, by which name the compiler of Xth Genesis figured those *swarthy* races that dwelt *ab initio* exactly where they do now, viz: in *Southern Arabia*.

More conclusive determinations, in primordial ethnology, than in this case of *Sheba* (B.), it would be hard to discover.[605]

25. דדן — DDN — 'DEDAN.'

Leaving aside nice discriminations between the duplex *Shebas* and *Dedans*, the one Hamitic and the other Semitic, we remark that, being a junior colony to *Sheba*, in Rhamanite affiliations, this *Dedan*, through analogy, might be fixed in Arabia, as we have seen in the preceding name, even without the precise words of Isaiah (xxi. 13):—"In the woodlands of *Arabia* shall ye lodge, O ye travelling companies of DDNIM," *Dedanians*: which obviates the necessity for seeking out of the Peninsula.

But the precise location of the geographical son of Raamah, and brother of the preceding Sheba, is fixed at the city and district of *Dadena*, just outside Cape Mussendom, on the Indian Ocean; and taking its natural station among KUSH*ite* tribes of Southern Arabia does not necessitate further research.[606]

With the exception of Nimrod (to be discussed as the next name), who, none will dissent, belonging to Assyrian history, can have no possible relation to *African* theories, here closes the genesiacal catalogue of KUS*hite* affiliations.

The educated reader who has followed us through Hebraical, Greek, Roman, Coptic and hieroglyphical sources, has now beheld every "Ethiopian" postulate on KUS*h* fall, one by one, beneath the knife of historical criticism. As one of the present authors indicated, ten years ago, and as both partially confirmed at a subsequent date by their several researches, the KUS*hites* of Xth Genesis could have been then, as they are now, once for all, glued permanently to *Arabia:* whence to detach them again will be a vain effort, should the reader be pleased to wield in their defence the weapons herein tendered him. That the present tiresome undertaking was needed, the reader can satisfy himself by opening any English *Commentary* on Scripture; and almost every English writer but Forster; who, following Bochart, has consistently vindicated the *Arabian*-claims of *Kush*, to the exclusion of *African* fables: whilst henceforward the Ethnographer may calmly pursue his inquiries without necessarily exclaiming, when he stumbles upon the mistranslation "Æthiopia" in King James' version,

"Hic *niger* est; hunc tu, Romane, caveto."

[To my learned predecessors in KUS*hite* inquiries, who have uttered opinions without first employing archæological processes similar to those herein submitted respectfully to their consideration, I beg leave to quote Letronne: — "One regrets to see erudite and ingenious men, of zeal and perseverance most laudable, thus waste their time in pursuit of such vain chimæras, in allowing themselves to be led astray by assimilations the most whimsical and the most arbitrary. One might say, in truth, that, for them, Winckelmann and Visconti had never appeared on earth, so much do they deviate from the reserved and prudent method of these heroes of archæology; who, not pretending to know in antiquity but that which it is possible to explain through the aid of authentic monuments and of certain testimonies, knew how to stop, the moment they felt the ground fail beneath their tread. It is thereby that they arrived at so many positive results, and not at simple 'jeux d' esprit' or of erudition, that cannot sustain an instant's serious examination. Our new archæologists proceed quite otherwise: they take a monument perfectly obscure [like ÆTHIOPIA]; they *compare* it with a second, with a third, and again with others that are not less so; and, when they have placed side by side all these *obscurities*, they pleasantly figure to them selves that they have created *light*. Upon a first conjecture, they place a second, a third, and a fourth. Then, upon this conjecture, at the fourth generation, they erect an edifice, sometimes of appearance sufficiently goodly, because it is the work of architects who possess talent and imagination. This edifice may even endure, so long as nobody thinks of poking it with the tip of a finger; but the moment that criticism condescends to notice it, she has but to whiff thereon, and down it tumbles like a castle of cards."

To "nos adversaires," as the Abbé Glaire facetiously has it — viz: the biblical dunces in the United States, whose zeal in opposing the long-pondered, long-published views of Morton, Agassiz, Nott, Van Amringe, myself and others, has been more remarkable than literary courtesy, I now turn round for my own part, (after shattering their anti-Scriptural KUS*hite* illusions in regard to *Africa* and *Nigritian* families, for ever), and beg each individuality to accept the following citation; the more pertinent as it emanates from one of themselves: — "But *I* confess that *I* have some considerable dread of the indiscreet friends of religion. *I* tremble," wrote the Rev. Sydney Smith, "at that respectable imbecility which shuffles away the plainest truths, and thinks the strongest of all causes wants the weakest of all aids. *I* shudder at the consequences of fixing the great proofs of religion upon any other basis, than that of the *widest investigation*, and the most *honest* statement of facts. [*Auree parole*, 'golden words,' as Lanci would say]. *I* allow such nervous and timid friends to religion to be the best and most pious of men; but a bad defender of religion is so much the more pernicious person in the whole community, that *I* most humbly hope such friends will evince their zeal for religion, by ceasing to defend it; and remember that not every man is qualified to be the advocate of a cause in which the mediocrity of his understanding may possibly compromise the dearest and must affecting interests of society." And if, in consequence, I discard their *Cushite* suppositions, I can only excuse myself in the words of Strauss: — "Les théologiens trouveront sans doute que l'absence de ces suppositions dans mon livre est peu chrétienne; moi (je) trouve que la présence de ces suppositions dans les leurs est peu scientifique." — G. R. G.]

27. נמרד — NMRD — 'Nimrod.'

Before us stands the sixth and last affiliation of KUS*h* — to whom the writer of Xth Genesis devotes more space than to any other personification secondary to the parental "Shem, Ham, and Japhet"— inasmuch as five of the modern and arbitrary divisions of the text, called *verses*, are especially set apart for Nimrod and his derivations. Hence we may infer that, in the mind of that writer, Nimrod's honor and glory were inherent elements. Now, the associations, the names of *cities* attributed to Nimrod, the language spoken in different dialects throughout the Mesopotamian vicinities of their several locations, and their geographical assemblage in Babylonia and Assyria:—these considerations, we repeat, even were other histories silent, would lead archæology to suspect strong *Chaldæan* biases on the part of the compiler of Xth Genesis; and would increase the probabilities, to be enlarged upon ere we close this discussion, that Xth Genesis is either a transcript of an older Babylonian composition, or else was compiled by some Hebrew imbued, like Daniel for example, with "the learning and tongue of the *Chaldeans*."

Such, *primâ facie*, would be the archæologist's deduction when, disengaging himself from prejudices, no less than from traditions of comparatively recent origin, he had sought to evolve facts from the letter of Xth Genesis itself: especially when to this text he adds the only other passage, (except, of course, the abridged parallel in 1 *Chron.* i. 10), in which Nimrod's name occurs throughout the canonical books, (viz: *Micah* v. 6); wherein "the land of Assyria . . . and the land of Nimrod" are Chaldaic synonymes for the same country.

But, when once the inquirer steps beyond these simple and natural limitations, what pyramids of falsehood and misconception intervene to prevent clear understanding of the words of Xth Genesis? and how baseless the fabrications upon which these pyramids rest!

A "mighty *hunter*," whose imaginary deeds in *venerie* are still proverbial with modern "Nimrods," founds the grandest *cities*. The traditionary builder of a metropolis called *Babel* — BAB-EL, "gate of the *Sun*"; like the Ottoman "Sublime Porte" or the "Celestial Gates" of Chinese autocracy — "presto" becomes constructor of the "*Tower* of Babel;" when, so far as the letter of Genesis Xth and XIth be concerned, neither Nimrod, nor his innocent father KUS*h*, (save as two individuals out of "the whole earth," Gen. xi. 1), were more guilty in such impiety than KUS*h's* grandfather NOAH, who "lived after the flood three hundred and fifty years;" or than anybody else of the seventy-one or two persons — fathers, sons, grand-children, great grand-children, uncles, brothers, cousins, and what not — whose cognomina are enumerated in Xth Genesis.

Cramped within the factitious limits of biblical computation, English writers in particular, following neither Scripture nor true history, but the *Rabbis;* and unable to reconcile supposed Noachic orthodoxy with the sudden rise of so-called "idolatry," have seized, with rapturous eagerness, upon the earliest writer who is conjectured to have known anything more on the subject than we do ourselves; and these authorities behold in Josephus's Greco-Judaic hallucinations a clew to the enigma.

"It is vain we know that Nimrod became mighty, even to a proverb, if the nature and means of his elevation cannot be understood; or that Babylon was the beginning of his kingdom, unless we can find the means of learning for what purposes, and upon what principles, that city was established," reasons, somewhat illogically, the unknown author of *four* very scarce octavo volumes on this speciality,[607] in which we abortively hunted for a *fact:* so that, never having encountered any orthodox commentary on *Nimrod* in which principles of historical criticism were not more or less disregarded, we are reduced to the necessity of attempting to examine for ourselves: notwithstanding that the subjoined "views will doubtless excite astonishment in some, and displeasure in those who," avers Godfrey Higgins, the great Celtic antiquary, "while they deny *infallibility* to the Pope, write, speak, and act, as if they possessed that attribute."

To begin. Let us frankly disavow partialities, in the words which His Eminence, Cardinal Wiseman, aptly borrows from the great Adelung:—"Ich habe keine Lieblingsmeinung, keine Hypothese zum Grunde zu legen. Ich leite nicht alle Sprachen von Einer her. Noah's Arche ist mir eine verschlossene Burg, und Babylon's Schutt bleibt vor mir völlig in seiner Ruhe."

Through the common Oriental mutation of B for M, the word NMRD, of the Hebrew Text, becomes Νέβρωδ in the LXX, and Νεβρώδης in Josephus. Is it a modern or a primeval name? Cuneiform researches, so far as we yet know, have thrown no monumental light on the subject: but hieroglyphical do. Two Pharaonic princes of the XXIId dynasty — between B. C. 936 and 860 — bore this appellative: one, son of OSORKON II., spells his name NIMROT; the other, son of TAKELOTH II., NMURT: and, Mr. Birch observes: — "As the Egyptians had no D, but employed the same homophone of the T to express this sound in foreign names, this name is unequivocally the Assyrian Nimroud, נמרוד, the Νεβρώδης of the Septuagint, a word now known to signify *Lord* in the Assyrian, and unlikely to have been introduced into an Egyptian dynasty, except through intermarriage with an Assyrian house." Subsequent researches have not merely corroborated Mr. Birch's views on the intimate alliances between Egypt and Assyria, during the XXIId dynasty, but Rawlinson and Layard have established that cuneatic writings, and many other arts of Nineveh and Babylon, are long posterior to Egyptian hieroglyphics, and were the natural sequences of Egyptian tuition.

Monumental evidence, then, coetaneous in registration with the events recorded, carries the name NMRD, at a single bound, from its currency in parlance among the present natives of Assyria (as applied to places, such as *Nimroud, Birs Nimroud, Nimroud-dagh,* &c. &c.), back to the *tenth* century B. C., in hieroglyphics: — an age anterior, probably, to that of the Hebrew compiler, or translator, of Xth Genesis; but, while this fact corroborates his accuracy, it serves to sweep away sundry rabbinical and other cobwebs that hang between our generation and the primeval origin of the *word* itself.

What did NMRD, originally, *mean?* No reply can be accepted that does not, in a question involving such vast ramifications, first classify its components adverbially, under distinct heads: —

1st. *Philologically:*—We know not why the translation "Lord" results from arrow-headed investigations, and therefore relinquish discussion, on that ground, to such cuneatic philologues as Rawlinson, Hincks, De Saulcy, and others of the new school.

It may at once be acknowledged that Oriental traditions, of which the *Thalmudic*

Mishna and *Guemaras* of the present Israelites are but one rill out of many streams, concur in representing Nimrod as every thing haughty, tyrannical, and impious; but nothing can be produced to justify these gratuitous assumptions, earlier in date than Josephus; who merely hands us the rabbinical notions of his day (first century after Christ), when he calls Νεβρωδες the leader of those who strove to erect "Babel's tower;" and, as such, that he *rebelled* against Divine Providence. Now, before speculating, in opposition to the express words of Genesis Xth and XIth, what may have been NMRD's performances on that deplorable occasion, it ought to be first shown that the *fragment* termed "Genesis XIth, ver. 1–9," possesses real claims to be considered *historical.* This being as much out of our power as of any body else at the present day, Josephus's modern views upon NMRD's primordial *rebellion* serve merely to illustrate the proneness of the human mind to explain the impossible by inventing the marvellous. So we lay them aside, beyond the only historical fact resulting from Josephus, viz: that, in his age, NMRD was reputed to have been a *rebel.*

Such being the unique source whence flow all later theories upon KUS*h*'s heresies, and his *son's* enormities, we descend the main stream as we find it continued, "even unto this day," by the Rabbis:—"According to the Talmud (tr. *Chagiga*, ch. ii.), the name NMRD, *Nimrod*, is derived from MRD, *marad, to rebel*, because its writers suppose that he induced mankind to rebel against God. This, however, EBN EZRA does not seem willing to admit, but says—'Seek not a cause for every (Scriptural) name, where none is expressly mentioned;' on which his commentator (Ohel Joseph, *in loco*) remarks, 'if the name of Nimrod is derived from the cause stated in the Talmud, it ought to have been, not NMRD, *Nimrod*, but MMRD, *Mamred*.' But, according to Simones (*Onomast.* V. T. p. 472), the name *Nimrod* is composed of NIN, *offspring*, and MRD, *rebellion;* so that NIN-MRD means *filius rebellionis.* A portion of the name NIN survived in *Ninus*, under which appellation he is known to historians as the builder of Nineveh. . . . *He began to be a mighty one in the earth* (*Gen.* x. 8). 'Setting himself up against the Omnipotent, and seducing mankind from their allegiance to the Lord.' (Rashi.) The sacred historian intends here to point out to us the first beginning of those movements and convulsions in society, which led to the formation of states and dominions, especially to that of royalty [!]. And, inasmuch as these movements led to the overthrow of the previous state of things, the name of the man by whom these changes were first introduced, NMRD, *Nimrod*, from MRD, *Marad, to rebel*, is peculiarly expressive." 608

There is—excuse the phrase!—a verdant lucidity about this series of *non-sequiturs* that justifies our tedious extract. In it we perceive the chain of *evidence*, as lawyers would say, through which Christian commentators obtain their first notions upon NMRD—"evidence" upon which each confounder erects his own favorite tower of BBL, *confusion.* "Nous en convenons," concedes the Abbé Glaire; "we agree that the fable of the Titans has some relation to the history of the tower of Babel; but may not one conclude from it that the Greek poets wished to *imitate* the legislator of the Jews, and surpass (enchérir sur) the veracity and simplicity of his recital?"

But, suppose somebody happened to entertain the idea that NMRD may not be derivable from the *Canaanitish* root MRD at all; what, if such case were proved, becomes of Nimrod's *rebellious* propensities?

To ascertain this possibility, a philologist must rise above the level of rabbinical hermeneutics.

We have seen that the word NMRD was a proper name among pharaonico-Assyrian individuals in the tenth century B. C.—an age anterior to most if not to all parts of Hebrew literature extant in our day. This bisyllabic quadriliteral (ceasing to remain any longer mere Hebrew) merges into the vast circumference of *Shemitish* tongues, of which Arabic is the most copious representative.

Now, foremost amid living Semitic lexicographers, stands Michel-Angelo Lanci, and his views are supported by students equally authoritative in their several specialities.

The substance of their researches is: — that the primeval speech whence all Semitish tongues have sprung was, aboriginally, *monosyllabic* in its articulations, and therefore at most *biliteral* in its alphabetical expression; whereas, at the present day, these languages, Hebrew and Arabic essentially, are *dissyllabic* and *triliteral*. "As vowel sounds," holds a supreme authority, Rawlinson, "are now admitted to be of secondary development, and of no real consequence in testing the element of speech, the roots of which are almost universally *biliteral;* the Babylonian and Assyrian [in which languages NMRD's name originated] being found in a more primitive state than any of the Semitic dialects of Asia open to our research [must be older]; inasmuch as the roots are free from the subsidiary element which, in Hebrew, Aramæan, and Arabic, has caused the *triliteral* to be regarded as the true base, and the biliteral as the defective one." Above one hundred examples are given by Lanci; proving how those words which rabbinical scholars suppose to be primordial Hebrew *radicals*, (*i. e.* of three letters), are but a secondary formation along the scale of linguistic chronology; because suffixes, prefixes, or medial elements, have become superposed, or interplaced, upon or within a pristine *monosyllable*. There was, then, a time before the period when the law of *triliterals* became formed; and while on the one hand the Hebrew tongue preserves abundant monosyllabic *reliquiæ* of that remoter age, on the other, the preponderance of *bisyllabic* roots in Jewish literature establishes that such literature arose *after* the law of *triliterals* had already become prevalent. This later age oscillates, it is true, between 700 B. C., and some centuries previously; but cannot, by incontrovertible ratiocination upon historical data, be carried back to *Mosaic* days — fourteenth century B. C. — a linguistic point in which all Oriental *philologers* of the new school coincide.

2*d*. *Archæologically.*—NMRD, therefore, older on Egyptian monuments than any Hebrew writings that have come down to us, was already, in the tenth century B. C., a matured importation from its native Assyria; where, doubtless, this proper name had existed long previously: being distinguished by the, probably-*Chaldæan*, projector of the chart of Xth Genesis, as the earliest traditionary founder of very ancient cities. To explain by a tri-literal verb, MRD, itself susceptible of reduction into an earlier *monosyllable*, the quadriliteral bi-syllabic proper name NMRD, although not absolutely impossible, presents many chances of involving its advocates in anachronisms; and most certainly would never have occurred to modern Orientalists, had it not been for the rabbinical legend current in Josephus's days, which, thousands of years after NMRD's age, and hundreds later than Xth Genesis, endeavored to reconcile Assyrian mythes with a Hierosolymite doctrine of genesaical origins. We have seen above, that the derivation of NMRD from MRD, *to rebel*, is considered speculative even by Talmudists themselves; and, with Gesenius's *Thesaurus*, the writer (G. R. G.) would undertake, upon legitimate principles of Semitic palæography,—such as the commonest mutations of D for N; B for M; L for R; T, T*h*, S, or S*h*, for D, &c. — to draw a dozen, or more, happier, and quite as orthodox, significations for NMRD, Hebraically, than that ungrammatically twisted from MRD, which takes little or no account of the protogramme N.

Hear Lanci's more reasonable etymology. We give it regretfully, because without the ingenious arguments by which the Professor defends it in his *Paralipomeni*, and coupled with all the reservations due to philological intricacies of this archaic nature. The word NMRD is nonsense when wrung out from the verb MRD, *to rebel*. It is a compound of two distinct monosyllables, NM and RD. The former proceeds from the radical, preserved in Arabic, N*e*M, "to spread a good odor:" the latter from R*u*D, "to be responsible." N*i*MR*o*D means, Semitically (whether such was its pristine Assyrian acceptation or not), "*he-whose-royal-actions-correspond-to-the-good-odor* (*of his fame*)."

But, difficulties cease not here! In King James's version, as in all its MS. ancestors back to the LXX (where γίγας κυνηγός, a *hunting-giant*, is its wondrous paraphrase), the next verse (*Gen.* x. 9) states that NMRD was a "mighty *hunter!*" Upon this translation hang chiliads of commentaries. Leaving them in suspension, we again present Lanci's etymologies.

The Hebrew word T*s*ID (translated *hunter*) is not in this case derivable from SÀID, a *huntsman;* but comes from the Arabian verb WSD; instead of Arabicè SUD, Hebraicè T*s*UD, *to hunt.* Now, WaSaD means "to be *firm*," to possess *consistency* and *stability;* which quality, applied to the vast domains assigned in Xth Genesis to Nimrod, makes the words GiBoR-T*s*ID mean "*great-in-landed-tenements*"; and not "vigorous in the chase."

What of Assyrian mythology, on the question of Nimrod, may become exhumed eventually through cuneiform researches, it is useless yet to speculate upon. In the present state of science, Lanci's exegesis, grammatically as to Hebrew, philologically as to *Semitish* tongues, and far more sensibly in connection with the probable meaning of the writer of Xth Genesis, stands of itself, quite as well as, if not better than, the modern rabbinical notion of a "hunter." [Always ready for my own part to surrender any hypothesis the moment its irrationality is proven, I submit (for what I conceive to have been one of the intentions of the compiler of Xth Genesis) the following retranslation of his sentences, accompanied by notes to some extent justificatory. — G. R. G.]

The personage who wrote Xth Genesis is unknown. The language he adopted was *Canaanitish*, afterwards called "Hebrew." The age in which he flourished is obscure: the *alphabet* used by him still more so. His individual biases, beyond a supposable *Chaldaic* tendency, enter, as respects ourselves, into the vast family of human *conjectures.* The *media* through which this document, Xth Genesis, has been handed down, are, in a scientific point of view, suspicious. The vicissitudes (even when restricted to the Hebrew Text) through which the original manuscript has passed, in order to reach our eye in printed copies of King James's version, are not few: because, the oldest Hebrew manuscripts of Xth Genesis now extant do not antedate the tenth century A. C.; the Masorete diacritical marks, upon which orthodox commentaries mainly repose, were not invented before 506 A. C., nor perfected until some 800 years ago; and, finally, the *Ashouri*, *square-letter*, character of present Hebrew MSS. cannot possibly ascend to the second century of our era. It will therefore be conceded that, before the personal ideas of the first editor of Xth Genesis could have reached our individualities, *some* elements of uncertainty intervene; independently of errors of transcribers and of translators, from Hebrew into Alexandrian Greek; from both of these languages into Latin; from the three, in unknown quantities, into English: all conditions of doubt that cannot, nowadays, archæologically (and neither hagiographically nor evangelically) speaking, be altogether dodged. Upon such historical considerations, we opine, the algebraical chances of *mistakes*, in respect to Xth Genesis, are rather more numerous than those of exactitude in interpretation: albeit, Hebraically, the subjoined attempt at an English restoration can withstand criticism quite as well as, according to St. Paul, "Jannes and Jambres withstood Moses."

3d. *Biblically.* — Genesis X.

Verse 8. "And KUS*h* begat NMRD (NEM-RUD = *he-whose-royal-actions-correspond-to-the-good-odor of his fame*); he first began to be mighty upon earth:"

Ver. 9. "He was a *great-landed-proprietor* before (the face of) IeHOuaH; whence the saying — 'like NMRD, *great-landed-proprietor* before (the face of) IeHOuaH:'"

Ver. 10. "And the beginning of his realm was BaBeL; and AReK, and AKaD, and KaLNeH, in the land of S*h*iNA*â*R."

Ver. 11. "From this land he himself (NMRD understood) went forth *(to)* AS*h*UR (*Assyria*), and built NINUeH and ReK*h*oBoT*t*-A*â*IR and KaLaK*h*."

Ver. 12. "And ReSeN between NINUeH and between KaLaK*h*; (he) *she* (Nineveh understood) the great city."

[The text, in verse 11, is ambiguous. It may be read, as in King James's version,

"Out of that land went forth Ashur;" but such rendering leaves out an essential member of the phrase, the word HHUA, '*he* himself,' before the verb "went forth," which can only refer to the antecedent Nimrod. On the other hand, as the literal text has "went forth Ashur," the preposition *to* must be interpolated; but not altogether arbitrarily, because learned Hebraists aver that this preposition is omitted in *Num.* xxxiv. 4, and in *Deut.* iii. 1, and yet its interpolation is obligatory to make sense.

Indifferent to either reading, I will merely mention that three new and distinct translations of Genesis, by eminent Hebraists (Glaire's, Cahen's, and De Sola's), read, "Nimrod went *to* Ashur (Assyria)" — that this last vindicates such explanation by unanswerable arguments, while most of them quote high scholarship in its favor; and, finally, that the Hebraical profundity of "N. M.," who defends this view in *Kitto's Cyclopædia,* is of more Germanic hue, and consequently deeper in Hebrew, if not perhaps in "geological" lore, than that of "J. P. S.," who opposes it. *Non nostrum tantas componere lites:* which future cuneiform discoveries alone can settle.—G. R. G.]

The probable ideas of the constructor of Xth Genesis on NMRD, may now be summed up: —

1st. That *Nimrod* was an affiliation of K*h*aM (Egypt?), swarthy, or *red,* race of mankind, through KUS*hite,* Arabian, lineage.

2d. That, unlike every other proper name, after "Shem, Ham, and Japheth," in Xth Genesis, each of which is a geographico-ethnological personification, NMRD is an *individual;* the only one in the whole chapter. Whether an actual hero, or a mythological personage, cannot be gathered from the text.

3d. That, whether "great in the chase" or not, neither Nimrod's *name* nor his deeds, nor any thing in Scripture, justifies our assumption that the writer of Xth Genesis did not entertain high respect for Nimrod's memory: on the contrary,

4th. This writer distinguishes NMRD from all his geographical compeers, as prominent "before IeHOuaH."

5th. That *Nimrod* was positively the earliest "great-landed-proprietor" known to the writer of Xth Genesis; who ascribes to NMRD the foundation of eight of the proudest cities along the Euphrates and Tigris—*Babel, Erech, Accad, Chalne, Nineveh, Rehoboth-Aïr, Kalah,* and *Resen.*

6th. And, finally, that the practical writer of Xth Genesis is innocent of the sin of causing those incomprehensible delusions about NMRD, which, commencing with Josephus's hypotheses, only 1800 years ago, pervade all biblical literature at the present day.

Two inferences might, however, be drawn from the said writer's peculiarities: — One, that the document, being *Jehovistic,* belongs to a later age than that immediately after Joshua; earlier than which, as shown further on, the mention of *Canaanitish* expulsions renders it archæologically impossible to place the writer: — the other is, that the writer not only was better informed upon *Babylonish* traditions than (to judge by his silence) upon those of other countries, but that he derived pleasure from the elevation of the former above the rest. Would not this imply *Chaldæan* authorship?

Now, whether Nimrod was originally a demigod, a hero, or a "hunting-giant;" whether, under such appellative, lie associations with Ninus, Belus, or Orion; or (were we to "travel out of the record," what we should first examine), whether he was not another form of the *Assyrian* Hercules, to be added to those so skilfully illustrated by Raoul-Rochette—these are speculations foreign to our subject, and we refrain from their present obtrusion.

The compiler of Xth Genesis, whose meaning we strive to comprehend, was satisfied to ascribe to NMRD the foundation of four *Babylonish* and four *Assyrian* cities; and, although the positions of some of these eight are not yet so positively fixed as might be desired, they group together in Mesopotamian vicinities; and thus the last affiliation of KUS*h* becomes placed in Asia—further removed from African "Ethiopia" than the whole, or any, of his geographical brethren.

"Affiliations of the MTsRIM," or *Egyptians*.

27. לודים — LUDIM — 'Ludim.'

We have already seen that *Mitsraïm*, read according to the Masorete punctuation, is a dual referable to the "Two Egypts," Upper and Lower; but, stript of the points which, after all, are but recent and arbitrary embellishments, that MTsRïm is a plural, meaning the *Miss'rites*, or the Egyptians.

The writer of Xth Genesis, therefore, in his system of ethnic geography, deemed these personified off-shoots from *Egypt* to be so many colonies or emigrations from that principal stock; and as such, we perceive that he suffixes to each name the plural termination IM; thereby testifying that he never foresaw modern assumptions in King James's version, that the LUD*s*, the AăNM*s*, the LHB*s*, &c., should have been *men;* one yclept Lud, another Anam, and so forth.

As grand-children of K*h*eM (*Ham*), the hoary ithyphallic divinity of Egypt, these outstreams class themselves under the generic denomination of *Hamitic* families; and their habitats ought naturally to be sought for in regions contiguous to their ascribed focus of primitive radiations: without disregarding either, that the writer of Xth Genesis, by making them *cousins* of Palestinic *Kanaanites*, and of Arabian KUS*hites* (all issues from the same *Hamite* source), never supposed that they were, or could ever become, *Nigritian* races: upon which last "Type of Mankind" he, as well as every other writer in the Old Testament, observes the same judicious silence manifested throughout the Text towards *Tungouses*, *Esquimaux*, *Caribs*, *Patagonians*, *Papuans*, *Oceanians*, *Malays*, *Chinese*, and other human races; the discovery of whose terrestrial existence appertains to centuries posterior to the closure of the Hebrew canon, Xth Genesis inclusive, at some period not earlier than Alexander the Great, B. C. 332; nor posterior to B. C. 130, when the LXX translations were probably complete at Alexandria.

Hence, to judge by existing nomenclatures of tribes and places, LUD appears both on the Asiatic and Libyan flanks of lower Egypt. Thus, on the Syrian frontier, a few miles east of Yaffa, lay the site of *Loud*, Lydda, Diospolis; inhabited afterwards by Benjamites. So also Arabico-*Berber* traditions comprise the L*a*OUT*ah* among Sabian tribes of Yemen, reputed to have immigrated into Barbary. But, whether as exotics or *terrægeniti*, it is on the Libyan side of the Nile, prolonged on the southwestern littoral of the Mediterranean to the Atlantic — districts cut off through the absence of *camels* during primordial ages and by Saharan wastes, from contact with Nigritian families of remote austral latitudes — that the LUDïm have left memorials of ancient occupancy.

Michælis long ago corrected Bochart, and suggested the probabilities that the *Luday*, situate near the river *Laud*, in Tingitana, were the *Ludim:* latterly confirmed by Gräberg de Hemso; who shows that the *Oluti*, *Oloti*, *Louat*, exist among Amazirgh tribes in those Mauritanian neighborhoods to this day; still admitting, too, the national prefix *ait*, "sons of," to their names (like Mac, Fitz, O', Ap, among ourselves), as they did of yore, when the Carthaginian Amon registered in his Periplus the *Ait-o-*LUD, "sons of Lud," or *Aitoloti;* resident in the same Barbaresque vicinities where the *Ludayas* of Spanish writers are now succeeded by the *Beni*-Loud. There is no lack of vestiges of primeval LUDs to be met with in the very regions where analogy would lead us to look for them; and it is surprising that high authorities have altogether overlooked the facts.

[My former "Excursus (in *Otia Ægyptiaca*) on the origin of some of the *Berber* tribes of Nubia and Libya," suggested a ventilation of some disregarded ethnological data, preparatory to that of Xth Genesis, which, after five years' suspension, I am now endeavoring to accomplish. I then submitted authorities on two grand divisions of *Barbaresques* — a noun not derived from *Barbari*, barbarians, but from the aboriginal African name of BRBR — the *Shillouhs*, and the T-*Amazirgh* or Amazirgh-T; both readily traceable through the *Mazices*, *Macii*, &c., of Latin authors, back to the Μαζυες of Herodotus. — G. R. G.]

To render perspicuous the view we take of Barbaresque anthropology, it would be necessary to enlarge here upon generalities before scrutinizing each genesiacal name in detail; but space being wanting, we must curtail our MS. investigations.

Two human families, the *Shillouhs* and the *Mazirghs*, now called *Berbers*, have lain, either aboriginally or from antiquity beyond record, scattered from the Cyrenaica and oases west of Egypt, athwart the northwest face of Africa to the *Moghreb-el-Aksa*, or extremest west, of Marocchine territories on the Atlantic; and formerly even to the *Guanches*, now extinct in the Canary Isles. Estimated by Gräberg de Hemso at four millions of population in Morocco alone, these *Berber* families present differences as well as resemblances comparable to those visible between the *French* and the *Belgians*: they speak dialects of the old "lingua Atalantica," subdivided into *Berber* and *Shilha*; and intermarrying rarely between themselves, have also imbibed little or no alien blood through amalgamation with others.

Anciently they occupied exclusively that Atalantic zone of oases, littoral or inland, which lies between the Sahara deserts and the Mediterranean; now called *Barbary*; "Land of BERBERS," *Berberia*: and the remoteness of their residence along that tract so far surpasses historical negation, that geology alone may decide whether the *Berbers* can have witnessed those epochas when the now-arid Sahara was an inland sea. In any case, we may suppose that, in proportion as its salt-lacustrine barriers to communication with Nigritian plateaux became desiccated, the *Berber* tribes, driven from the coast by Punic, Kanaanitish, Greek, Egyptian, and other early invaders, spread themselves southwards; and, whilst their former invaders have been replaced by successive Roman, Vandal, Saracenic, Ottoman, and French establishments, that they themselves gradually crossed the Sahara; and now, under the name of *Tuaricks*, some offshoots of this main Atalantic stock, modified by the facilities such passage has afforded them of possessing *Negresses* in their hareems, roam along both banks of the Niger and around Lake Tchad.

But the southerly expansion of *Berber* families, except in partial and conjectural instances, is bounded chronologically by one great fact, overlooked though it be by most writers; which is, that, until the *camel* was introduced into Barbary from Arabia, the Saharan wilderness presented obstacles to nomadism almost insurmountable. Now, the *camel* was not imported into Barbary until Ptolemaic times. Mentioned in hieroglyphics only as a foreigner, and never used by the Pharaonic Egyptians, the earliest historical appearance of *camels* in Africa dates in the first century B. C. The vulgar notion of camel-diffusion over Barbary before the Ptolemies, is nowadays archæologically erroneous.[610]

It therefore follows that, whenever Xth Genesis was compiled, the Barbaresque affiliations of the MTsR*i*m could not have penetrated to the latitude of *Negro* races, south of the Sahara, by any other route than up the Nile — *Negroes* never having existed, in a state of nature, north of the limit of *tropical rains*. This long journey was not undertaken by the powerful MTsR*i*m themselves much before the XIIth dynasty, about B. C. 2300: so that the LUD*i*m, for example, like all their uncivilized brethren, driven away from the Nile by the Egyptians; restricted from southerly progress by the Sahara and the absence of *camels*, from northerly by the Mediterranean and the absence of ships (*Berber* habits being the reverse of nautical, and Tyrian privateersmen hovering on those coasts); were, down to Ptolemy Soter, B. C. 320 (as the utmost antiquity), confined in their nomadisms within Barbary between Egypt and the Atlantic littoral of Morocco. The lowest historical age possible for the compilation of Xth Genesis attains to the *Esdraic school*—the earliest (if the document be *Chaldaic*) may antedate Ezra by some centuries: but, logically, the more remote the antiquity claimed for this ethnic geographical chart, the less possible, physically, becomes intercourse between *Berber* tribes (athwart the Sahara and without *camels*) and the true Negro races of Central Africa.

Content with offering this dilemma, we pass onwards, and remark, that the *Berbers* were generically termed *Mauri* by the Romans, and *Moors* by "moyen age" writers; whilst, if we adopt Egypt as the geographical pivot of eccentric radiations, we shall find, that these Mauritanian *Berbers* on the west are to the *Egyptians* what we have shown the Arabian *Kushites* to be on the east, viz., "gentes subfusci coloris"; ÆTHIOPIANS, in its Homeric sense of sun-*burned-faces*. All of them were possibly distinguished by the *red* color on Nilotic monuments; and the term *Hamitic* would be, genesiacally, ethnologically, and geographically, the best designation for these races; were it not for modern *Negro* theories, which ignorance and charlatanism have foisted upon that mystified name we now spell "Ham." "One almost blushes," Agassiz has sarcastically observed, "to state, that the Fathers of the Church, in Northern Africa, have even more recently been quoted as evidence of the high intellectual and moral developments of which the *Negro* race is supposed to be capable, and that the monuments of Egypt have been referred to with the same view. But, we ask, have men who do not know that Egypt and Northern Africa have never been inhabited by Negro tribes, but always by nations of the Caucasian race, any right to express an opinion on this question?"

[Five years ago, Luke Burke's *Ethnological Journal*, and the writer's *Otia Ægyptiaca*, pointed out several analogies between some names of twenty-five *Berber* tribes mentioned by Ebn Khaledoon, and various other ethnic cognomina preserved by the writer of Xth Genesis. The former are certainly reliable, inasmuch as Ebn Khaledoon was a *Berber* himself and the historian of his nation: who contests their common descent from such legendary sources as Abraham, Goliath, Amelek, Afrikis, Himyar, and other fabulous origins; claiming, however, that the *Berbers* "descend from KESLOUJIM (Casluhim), son of MITZRAIM, son of HAM." So, also, through Mohammedan harmonizing, we meet, in the "*Rozit ul Suffa*," with a similar example of pious genealogical frauds—"God bestowed on Ham nine sons: *Hind*, *Sind*, *Zenj*, *Nowba*, *Kanaan*, *Kush*, *Kopt*, *Berber*, and *Habesh!*"

It will be seen, further on, that the CASLUHIM undoubtedly dwelt in Barbary when Xth Genesis was written, as their descendants do "unto this day;" but it need scarcely be insisted upon, with the reader of these pages, that Ebn Khaledoon, an Arabicized Berber, no less than a most learned and conscientious Muslim, naturally felt anxious to connect his own pedigree with that of the genesiacal Patriarchs, to him rendered orthodox and respectable through the *Korân:* and the fact that, overlooking the Hebrew plural terminations, he deemed KESLOUDJIM (the *Shillouhs!*) to be a man, son of MITSRAIM (the *Egyptians!*), another individual, indicates his literary sources; while it serves to illustrate what we have maintained elsewhere, viz.: that the *Berbers* (their own indigenous traditions being unrecorded) appropriated instead the language and religious ideas of their civilizers, the Arabs; who certainly, when the *Korân* was composed, had never taken *Berber* origins into consideration.

Nevertheless, this sentimental bias of Ebn Khaledoon does not touch the archæological fact gained from his pages that, in his time, the LAOUTE are recorded, as one of twenty-five *Berber* tribes then inhabiting Barbary.

"Six hundred lineages of *Berbers*"—the enumeration of Marmol and of Leo Africanus—resolved themselves, about the fifteenth century of our era, into *five* main stems; who, already imbued with longings after Islamite respectabilities, *said* that their progenitors were Sabæans of Yemen: at the same time Leo adds the noteworthy remark, "*subfusci coloris sunt.*" The same quintuple division reappears in the "quinquegentani *Barbari*" of Roman writers of the fourth century; which is important, because it establishes an identical quinary repartition of *Berbers* prior to Mohammedan impressions; and, although it does not contradict, this fact renders it less likely that pagans or semi-Christians should have leaned towards an Arabian origin, before religious motives for such honorary attribution existed in *Berber* minds. To trace whence *Barbari*, or *Berbers*, from about 1400 years ago, through the "Misulani *Sabarbares*, Massylii" of Pliny; the *Sabouboures* of Ptolemy; and possibly, in some instances, the BARBAROI of Strabo, Diodorus, and Herodotus: to resolve the *Zilia, Zilca, Zelis, Salinsi, Zilzactæ, Massyli, Xilohes*, into the Μασσαιλιβυες = AMAZIG-*Libyans*, or the *Massæsylli* into AMAZIG-*Shillouhs;* and then to deduce the *Amazirghs* of the present day from the Μαζυες of Herodotus, B. C. 430: — these are tasks which, following chiefly Castiglione, have been already executed.

History, philology, and analogy unite, therefore, in establishing that the T-*Amazirghs*, or real *Berbers*, distinct in that day from Asiatics or Negroes, existed, about the fifth century B. C., in their own land of *Berberia*, now called Barbary. With the exception of their having embraced Islàm; exchanged the bow, for which they were celebrated long before that age, for the musket; added the *camel* to the horse; and appropriated *Arabic* words to make up for deficiencies in their native vocabulary; the *Berbers* of Mt. Atlas are precisely the same people now that they were twenty-five centuries ago; dwelling in the same spots, speaking the same tongues, and called by the same names, as we shall see presently.

We are now prepared to accept an opinion pronounced by a man of science eminently qualified to judge; which, coupled with Forster's attestation [*supra*, p. 483] of the indelibility of *color* as a criterion of type, when we recall how all *Berbers* "subfusci coloris sunt," ought to possess sufficient weight.

There is but one veritably indigenous race in Barbary, says Bodichon; viz., the GÆTULIAN: — "Ainsi, Atlantes, Atarantes, Lotophages, Occidentaux, Troglodytes, Maurusiens, Maures, Pharusiens, Garamantes, Augéliens, Psylles, Libyens, même Canariens, et toute cette multitude de peuples à qui les anciens donnent l'Afrique septentrionale pour patrie, se confondent en une seule et même race, la GÉTULIENNE." The Arabs, foreigners in Barbary, call the present descendants of this race "*Berbers* and *Kabyles*." Indeed, as tillers of the soil, *i. e.*, as human animals brought into direct contact with the *earth* of Barbary (rank with exhalations so mortiferous, even now, to Europeans), no type of humanity could have outlived, not to say flourished amid, the climatic and geological conditions of Atalantic Africa, but a few furlongs from the sea-beach, except the *Gætulian*. For proofs, read Dr. Boudin's *Lettres sur l'Algérie.*

Cut off from escape on the west by the ocean; on the north by the Mediterranean; on the south by the Sahara (once a sea also), and, until the Christian era, by the absence of *camels;* and on the east by the MTsRIM; these "quinquegentani *Berberi*" have survived the extinction of the *elephant*, together with the depressions of temperature consequent upon the destruction of their primeval forests: and, repugnant through natural constitution to any alien institutions but those of the *Koràn* (construed after their own liberal fashion), they remain now, what they were at their unknown era of creation, *Gætulians*, and nothing else.

Inquire of history.

Phœnicia planted her standards at the Carthaginian ports she occupied: Greece built her strongholds on the littoral of the Cyrenaica: Rome, prostrating all, sent her eagles further into Africa than any Europeans: Persia inscribed her westernmost tablet at Tripoli: Byzantium, after Belisarius's triumph, has been obliterated, even in name: Vandals, massacred in detail, or extinguished by climate more murderous to white races than Numidian arrows, have vanished, physiologically, like other heterogeneous foreigners on the sea-board: Ottoman and Frank invaders still surround their temporary havens with bastions strongest towards the mainland; and French prowess over the *Berber* race is confined to the latter's preparations for the next *razzia*. The *Saracens* alone, themselves "gentes subfusci coloris;" apostles of a genial polygamous religion·

speaking dialects of a tongue long familiar to *Berberic* ears through anterior Punic intercourse: — the *Arabs*, I repeat, cognate with the *Berbers* in nomadic restlessness and social habits, have ridden over the *Gætulians*, through them, and around them: but whilst from the first hour, A. D. 644, that the lances of Islâm penetrated into *Berberia*, the wise policy of its Arabian votaries associated the native *Berbers* in spoils and benefits mutually agreeable; the *Arab* himself, after twelve centuries of Barbaresque sojourn, has become far more *Berberized* as a MOGHRABEE than the *Berbers* have been *Arabicized*. And (asks the reader) what is the "ultima ratio" of all these successive influences upon mankind's Atlantic *type?*

Merely this: — that wherever the *Gætulian* has not (he has in Morocco) revindicated his national supremacy, he rather tolerates Arab encampments in the domains of his birth-right, than hospitably welcomes Arabian presence by practical fusion. "Mohammed" is their moral bond of Barbaresque unity — their common battle-cry. Implacable detestation of *Turks* and *Frenchmen* is the only chord of sympathy between Âbd-el-Kâder (*slave* of the *Puissant*), the heroic and betrayed *Shemite*, and that mulatto-cross between Arabico-Berbers and Negresses, exhibited in a beastly individuality called "the Emperor of *Morocco.*" Hatred to aliens — to anybody but one of themselves, a *Berber* — is still the banner of *Gætulian* instincts.

If, then, Gætulian populations cannot have originated through imaginary importations of *Negroes* from the interior of Africa, nor from imaginary colonizations of *white* races from Europe, whence came they?

History being impartially silent, our alternative lies between *Arabian* immigrations as one possibility, and the autocthonous creation of *Berbers* for Barbary as the other. My own inquiries lend no support to the scientific probabilities of the former contingency. The latter it is not my province to discuss. — G. R. G.]

Viewing, therefore, *Gætulian* families as "une race apart," we proceed to ascertain their relation to the chart of Xth Genesis.

Their present name is *Berbers* in Mauritania, and *Shillouhs* towards the Cyrenaica.

In Ebn Khaledoon's "History of the *Berbers*," we have already noticed that one tribe of this race was called LAOUTE, or LAOUTEH. Cutting off the Arabic plural termination, there remains LAOUT; which, reduced to its simplest expression, vowels being vague, is LUT, or LUD; an appellative, as we have shown, traceable in Barbaresque nomenclatures at all times, back to where history is lost.

In Xth Genesis, the eldest-born of the affiliations of the MTsRîm (or Egyptians), and who, therefore, in the idea of the writer, issued first and went furthest from the supposed parental hive, are the LUDIM. Removing the Hebrew plural suffix IM, there remains LUD. All commentators unite in deeming *Barbary* the geographical sphere of these emigrations.

To have shown that the Laouteh, LUDs, of Ebn Khaledoon, can be no others than the Ludim, LUDs, of Xth Genesis, is likewise to prove that *Gætulian* families are included in that ancient system of geography, and that the LUDIM probably occupied *Mauritania*. A conclusion which our inquiries into the habitats of their fraternal affiliations will fortify. In the meanwhile, we rejoice to learn from Gräberg de Hemso that the *Ludaya* tribe still furnishes the Sultan's body-guard in Morocco, and that their river Tagassa is yet called *Laud* and *Thaluda;* at the same time that it is satisfactory to find such scholarship as Quatremère's sustaining how, "Dans les Loudes de Moïse, je reconnais la grande nation des *Lewata*, la plus puissante des tribus de race Berbère;' and thus ratifying our views upon the LUDîm of Xth Genesis.[611]

28. ענמים — AâNMIM — 'ANAMIM.'

Of course, this is a tribe which (plural termination IM cut off) was called AâNM.

Viewed as *Aânams* the analogies falter, unless we adopt Bochart's speculative idea, that the Semitic word for *sheep*, GNM, be the root of this name. The *Num*-idians,

Nomades, have also furnished comparisons; which we dispute not, because it is in Barbary that commentators locate the people called ANM*ìm*.

Referring the reader to the "causes of verbal obscurity" in Oriental names, ably set forth by Forster and De Saulcy, there are few literal permutations more frequent than those of M and N: and hence it has been long remarked, that ANM is but an anagrammatic form of AMN. Under such view, the AMN-*ìm* become at once *Amonians;* and, from the ancient worshippers of the Egyptian deity AMN-*Kneph*, or NUM, at the "Oasis of Ammon" (now *Seewah*); through the *Nasamonitis, Nasamones;* to the *Amonians*, or the *Garamantes*, whether on the river Cinyphus near Tripoli, or on the *Gir;* the transition is more rapid than the results may appear precise.

Castiglione gives solid reasons why the *Macæ-Ammonii*, or *Macæ-Amnii*, should refer to Amazirgh-Ammonians; which term he supposes became in Greek mouths *Mesammones*, and thence *Nas-ammones*. Hence, the ANM*ìm* would naturally take their places among *Berber* tribes next to the LUDs, their kinsfolk.

The *Nasamones* of Herodotus and of later writers, read by Birch *Nahsu*-Amonians (*Negro*-Amonians?), were a very roving predatory race; who carried their name all over Barbary: but, without insisting upon any one family in whose name AMN is a component, it is for objectors, after perusing what follows, to show that the Barbaresque Anamim of Xth Genesis, cannot be represented by some offshoot of the *Gætulian* stem yet stretching between the Sahara and the Mediterranean.

For ourselves, while descrying the *Anamìm* in the *Berber* tribe of "*Enine*," catalogued by Ebn Khaledoon, we suggest that AâNM may underlie both the words "Nasamones" and "Numidians;" and this for a reason that no Orientalist acquainted with hieroglyphical permutations will disregard. Bunsen, following Ewald, proposed to read the name GUB, *Chub* [which nation Ezekiel (xxx. 5) associates with "KUS*h*, and *Phut* (Barbary) and *Ludìm* (the *Ludayas*, as shown above, No. 27) and all the mingled people,"] as if such name had been written GNUB; and thence to apply it to *Nubia* — a country, we have proved, altogether unmentioned by Hebrew writers. Volney had perceived GUB in the Barbaresque *Cobbii* of Ptolemy, and we adopt his view as by far more natural, according to the context of Ezekiel. Nevertheless, Bunsen's very just remark of the frequent suppression of the N before G or K, in the transfer of Hamitic into Semitic proper names (*ex. gr.*, SHESHONK, *Shishak*), allows us to behold the âNuM of AâNM-IM in the GNUM-*idians* of classical history. If, however, with Bochart, we transcribe the Greek Νασαμονες into Hebrew letters, נשי אמ-ן; NàSI AM-N, or otherwise NàSI-ANuM-*ìm*; we observe that *Nàs* means "people" in Semitish tongues, and thereby such compound name becomes, in English, "*People* of NUM*idia*;" or else, "*People* of (the oasis of) AMoN:" in either case, the *Anamìm* of Xth Genesis.

But Bochart declared that these tribes were "Solinus's *Amantes*, and Pliny's *Hammanientes*, peoples beyond the Greater Syrtis;" and, reminding us that גר, GaR, means "to inhabit," he discloses at once the famed "*Garamantes* near to the fountains of the river Cyniphus." Now, let us add that this river is still called the *Gir*, or *Gar*, by living descendants of these very *Amantes*, who once were the Berber AâMaN-IM alluded to by the ancient Hebrew geographer.[612]

29. להבים — LHBIM — 'LEHABIM.'

The first orthodox English work we chanced to open, in quest of etymological meanings, has, "LEHABIM, *flames; or, which are inflamed; or, the points of a sword!*" and just below, "LIBYA, in Hebrew Lubim, *the heart of the sea;* or, *a nation that has a heart!*"

Let us seek elsewhere. Detaching the plural IM, through which the writer of Xth Genesis indicates that he means a *tribe*, the singular number of whom is LHB, we realize instantaneously how ignorant of Hebrew were the forty-seven translators of King James's version. This may be at once seen by their writing "Mizraim begat Ludim, and Anamim," &c., instead of "*the Luds and the Anams*" and so forth Had they even suspected that IM was already a plural termination, they would not have *doubled* it by printing "Cherubims" for *Cherubs*, or "Seraphims" for *Seraphs!* What should we think of the French scholarship of a person who wrote *tableauxes?*

That these people were *Libyans* no commentator now doubts, although Bochart dissents; and that in LHB, the soft aspirate *he*, H, may be equivalent to such vowels as *a, e, i, o, u*, no palæographer will contest: nor that the LUB*im* of 2 *Chron.* (xii. 3; xvi. 8), of Nahum (iii. 9), and of Daniel (xx. 43), are the same as the LHB*im;* especially in Nahum's text, where a conjunction couples them to P*h*UT; already shown to have been a generic appellative for the whole of Barbary.

Λιβύη of the Homeric Greeks possessed a wider territorial extension than the *Libya* of the Romans; the former signifying Barbary in general; the latter the coast from Egypt to the Greater Syrtis: hence we may infer that the more precise information of Roman geographers rested upon better acquaintance with the localities where the LHBs were domiciled. T-LIBI is the homonyme in Coptic MSS; but perhaps in a sense restricted to tribes on the immediate west of the Nile's alluvium; which also suggests the easternmost limit of Libyan encampments.

Among the *Berber* tribes enumerated by Ebn Khaledoon occur the LeWaTaH; which word in Oriental palæography is the same as LeHaB-*atah;* and its analogies with LeHaB-*im* are salient. Arab tradition invests the present *Beni*-LeWA, of Amazirgh stock, with sufficient correspondences to resolve all these appellatives into the Λευαδαι, Λεβανδαι, of Procopius, about the sixth century B. C.; not forgetting the *Languantan* of Corippus.

Any one investigating such subjects, without preconceptions, will recognise in the LHB*s* of Xth Genesis a nomadic population of *Gætulian* race, and of Barbaresque habitats.[613]

30. נפתחים — NP*h*T*t*K*h*IM — 'NAPHTUHIM.'

Before commencing analyses that arise through new resuscitations of Egyptology, it is desirable to remind the reader of a principle that governs our philological inquiries into 10th Genesis. Extremely simple, it is still, even where known, more or less disregarded by rabbinical writers.

The genesiacal writer's classification of nations is tripartite, under the titular headings "SHEM, HAM, and JAPHETH;" and his lists, therefore, embrace *Semitic, Hamitic*, and *Japethic* families; corresponding [*supra*, pp. 85, 86] to the *yellow*, the *red*, and the *white* colors given by Egyptian ethnographers to such varieties of man as were known to them about the sixteenth century B. C.: but the Hebrew map excludes the *Negro;* which race, the fourth in the quadripartite ethnography of Thebes, is, on the monuments, painted *black*.

Arabian languages are necessarily represented in the proper names of nations belonging to the *Semitic* stock; the Egyptian "sacred tongue" is the most ancient and reliable nucleus for those of the *Hamitic;* while those of the *Japethic*, almost a distinct world, must belong either to the *Indo-Germanic* or to the *Scythic* class of human idioms.

To suppose that the "speech of *Kanaan*" (misnamed *Hebrew*) can answer the purpose of an "open Sessame" to the significations of all proper names in Xth Genesis, which the writer himself has carefully segregated from each other into *three* groups of tongues, spoken by *three* groups of humanity (in his day as in ours, from each other entirely distinct), is one of those aberrations that no educated person of our generation would be likely to boast of; if he reflected that, in considering *Hebrew* as a fitting key to any thing more than to one, the *Semitic*, of these three linguistic portals, he would be as great a dolt as if he sustained that *English* might be contained in a Chinese radical or in a Mandingo root.

No philologist at the present day, when he beholds in Xth Genesis the proper name NP*h*T*t*K*h*IM, would seek for its explanation in a Hebrew vocabulary; because a proper name belonging to the *Hamitic* group of languages ought first to be examined within the sphere of its own positive domiciliations; and it is only when these are wanting, or when comparative philology is the investigator's object, that speculative analogies of such an antique cognomen may be hunted for in the modern Arabic *Qamôos*, or other Shemitish lexicon.

NP*h*T*t*K*h*IM is a plural, of which the singular expression is NP*h*T*t*K*h*.

In Coptic days, according to authentic MSS., the western skirts of Lower Egypt, on the south of Lake Mareotis, *Marea*, *Mariout*, were called NIFAIAT; whence, deducting the plural prefix, NI, we obtain FAIAT as the Coptic vocalization of the hieroglyphical root F-T; or P*h*eT, meaning a *bow;* as we explained under the head P*h*UT. The occupants of these localities, along the desert ridges from *Marea* to *Ptiminhor* (now *Damanhoor*) spoke a *Berber* dialect, and not pure Egyptian; in this, resembling the inhabitants of the nearest oasis, that of Ammon, or Seewah, who, already in the time of Herodotus, 430 B. C., were a mixed "colony of Egyptians and *Ethiopians*," *i. e.*, *sun-burned*-faces; "subfusci coloris," like all *Berber* derivations. We have settled that the preceding affiliations of the MTsR*îm* occupied parts of Barbary, and belonged to branches of the great *Gœtulian* trunk. We shall see that others of the Hamitic brethren did so likewise. What, then, more natural than to find, on the western flank of MT*s*R (Egypt) herself, the NIPHAIAT nomads of that race, speaking their national tongue, the *Berber?*

As usual, Champollion was the first to carry back the NIPHAIAT of Coptic Christian literature to the ancient Pharaonic monuments; confirmed by Rosellini, Peyron, &c., and since universally accepted by Egyptologists as designations of *Libya* and *Libyans*. But, without doubting in the least the Barbaresque application of the word, whether in its Coptic or in its hieroglyphical form, the original name P*h*-T-*kah* sometimes occurs in the singular number, "Bow-country," or plural "Nine-bow-country." Now, the same distinction holds in Xth Genesis, where P*h*UT refers to Barbary as a whole; and NP*h*T*t*K*h*IM, in which the same radical P*h*T is preserved, to *tribes* of the same Hamitic stock. May we not assign "Bow-country" to PHUT, and "Nine-bow-country" to the others? With this reservation, Hengstenberg is right in seizing upon *Niphaiat* as the probable representative of "Naphtuchim." It is easy to prove this identity The Masorete punctuation, through which *Naphtoukhîm* is its present phonetism, commands no reverence; being merely the rabbinical intonation, in the sixth and later centuries after Christ, of a *foreign* proper name antedating them, and the writer of Xth Genesis himself, by unnumbered ages. All that science can now accept are the six letters — NP*h*T*t*K*h*IM.

The hieroglyphical root is P*h*-T; the later Copts added the medial vowels, and it became P*h*aiaT: to make it an Egyptian plural, the NI, or N, was prefixed, and NI P*h*aiaT, thus formed, is simply *the*-P*h*aiaT-s — the proper name, as above shown, of a *Berber* tribe on the western frontier of Lower Egypt. But, Champollion's *Grammaire* tells us how, "in the graphical system, as in the Egyptian spoken tongue, the plural number (of nouns) was expressed by the *désinences* or terminations" — OU, or U: so that, Egyptologically, the name must have been orthographed NI-P*h*aiaTU. Such was the word that presented itself to the researches of the compiler of Xth Genesis, when he classified the MTsR*ite* "affiliations of K*h*aM, after their families, after their *tongues*, in their countries, in their nations" (*Gen.* x. 20). We have only to take the *square-letters* which the later Jews substituted for his own (unknown) calligraphy, and, inserting the omitted vowels, write them below the older Egyptian form — thus,

Ni-P*h*aiaTU,
Ni-P*h*aiaT*t*-uK*h*-IM, } to perceive that this diligent writer (not being conversant, unhappily, with Nilotic syntaxis) has suffixed the Hebrew plural, IM, to a proper name, NIPHAIATU, that was already in its indigenous *plural* form when it reached the chorographic bureau of Jerusalem or Babylon. Hence the following conclusions: —

1st. That *Egyptian* tongues and writings are older than Hebraical transformations of the name *Niphaiatu.*

2d. That the people *Niphaiatu* existed before Xth Genesis was written.

3d. That the Hebrew chorographer must have been unacquainted with the first elements of *Hamitic* tongues; else he could not have appended his own *Semitic* plural, IM, to a foreign name that was already pluralized by its national prefix NI, and suffix U — a blunder to be paralleled in English by the vulgar *Cockneyism* of "post-'ses" for *posts.*

4th. That, as a consequence, the principle laid down at the beginning of this section, of examining *Hamitic, Shemitish,* and *Indo-Germanic* names by their respective languages, is both rational and useful.

But, the less "inspiration" that is required for the construction of an ethnic chart, the more admirable become the human skill and knowledge which, its antiquity considered, compiled such an excellent synopsis of the *nations* existing within the geographical horizon of its day.

The long-chased families of the NiP*h*aiaT*t*U-*kh*-(IM) have been *earthed,* at last, where Bochart indicated his "Naphtuhæi": viz., around Mareotic provinces on the confines of the MTsRIM, or Egyptians. They spoke *Berber* dialects, like the rest of their Barbaresque brethren; and may be safely assumed as ranking among the easternmost representatives of the great *Gœtulian* race.

Nor are their vestiges wanting either in Arabic or in classical geographies. The twelfth tribe catalogued by Ebn Khaledoon is that of the NeP*h*USeH. T and S being palæographically identical, here is the Arabicized form of the same word, precisely; with its plural termination *e*H, in lieu of IM. The same name reappears in the sixth century of our era, and therefore before Arab invasions, in the *Nefusa,* or *Navusi,* of the Latin poet Corippus. And, to back assertions with authority, one of the greatest living Orientalists of France, Quatremère, while commenting on this passage of Xth Genesis, records: "Les *Naftouhis* répondent, je crois, à une des tribus Berbères, celle des *Nafzah,* ou celle des *Nafousah.*" [614]

31. פתרסים — PT*t*RSIM — 'Pathrusim.'

Again stands before us an *Hamitic* word, and again we apply to it our rules of dissection; after lopping away the excrescent Hebrew IM, and thereby restoring this name to its native simplicity — PT*t*RS.

Orthodox lexicography reveals to an inquirer how the Pathros mentioned by Ezekiel (xxix. 14; xxx. 14) means a 'mouthful of dew,' or 'persuasion,' or 'dilatation of ruin'!

The wonted acuteness of Bochart, two centuries ago, perceived that *Pathros,* a district in the Thebaid, would answer very well to the exigenda of PT*t*RS; and the Coptic researches of Champollion and Peyron established that the western side of the Nile, at Thebes, bore the names of *Patoures* (Phaturites), *Tathyrites, Pathurês,* and *Phatrous:* probably orthographed better by Parthey in *Papithourès,* because the name of *Thebes,* "P-API," as the "T*h*o-ReeS," south-land, is preserved in it. But with all deference, and without absolutely denying that the compiler of Xth Genesis may have meant *Pathros* in the Thebaid as the site of his PT*t*RS*im,* we cannot assent to such inference, for the following reason: —

"Dato il caso, e non concesso," that Moses, in the fourteenth century B. C., was the compiler of this chart — and orthodoxy itself claims no date more ancient — the MT*s*R*im* in that age, the XIXth dynasty, had been spread over the Nile's alluvium, for above 2000 years, "from *Migdol* to the Tower of Syene," and far more australly soon after the XIIth dynasty. Consequently, they had left to any people but themselves nothing but the *deserts* on either flank of the alluvials to roam along. *Pathros* was merely a suburban district in the "nome" of Thebes, then at the acme of her glory;

so that to construe the general meaning of Xth Genesis into such a paraphrase as, "out of the MT*s*R*î*m went forth a colony and founded *Pathros*, whence about the seventieth fraction of all humanity known to the Jews was called PT*t*RS*î*m," would be like saying (if for Thebes we read London, and French for Hebrew) that "out of the *Englishmen* went forth a colony and built *Waterloo bridge*, whence arose the grand nation called '*Vaterloos.*'" Besides, Wilkinson has critically noted, that *Pathyris*, or *Tathyris*, was so called after the goddess Athyr; and meant "*the* belonging to ATHYR," as the protectress of the western side of Thebes.

The obstacles to such interpretation increase just in the ratio that the compilation of Genesis Xth is brought down to a more historical epoch. It is evident from the context of the whole paragraph on the "affiliations of the MTsR*î*m," no less than from the ultra-Egyptian areas on which each one of these affiliations is naturally fixed, that such information as the Hebrew writer possessed on the PT*t*RS*î*m had led him to understand this tribe as extraneous to Egypt; and he did not locate their habitats in Egypt itself, because this country was already appropriated by the MTsR*î*m. Quatremère, and before him Golius, had perceived the physical impediments to the location of the PT*t*RS*î*m in upper Egypt:—"Les *Phatrousis* ont été, assez ordinairement, pris pour les habitants de la Thébaïde; mais cette conjecture ne me paraît pas admissible. En effet, Misraïm ayant été le père de l'Égypte inférieure se trouvaient naturellement rangé parmi ces descendants, sans qu'il fût necessaire d'indiquer d'une manière spéciale les habitants de telle ou telle partie de cette contrée. Si je ne me trompe, les *Phatrousis* du récit de Moïse nous representent les *Pharusiens*, qui occupaient une partie de ce qu'on nomme aujourd'hui l'Empire de Maroc."

This identification tallies with our views exactly. In classical geographies the *Pharusii* lie about Mauritania, east of the *Autololes;* and these last are identified with the *Berber* tribes of the AIT-o-LOT, "sons of Lud;" whom we have already proved to have been the genesiacal LUD*î*m. A Persian origin has been ascribed to the *Pharuses* since the time of Sallust; but probably upon no better authority than accidental resemblance of the word *Phars*, coupled with traditions of Achæmenidan invasions of the Cyrenaica; and its claims have been well contested by Lacroix. To behold the PT*t*SR*î*m of Xth Genesis in the *Pharusians* of Barbary is obnoxious to no difficulties, beyond the inconvenient presence of the letter T*t*, "tav" in the Hebrew transcription of the name; and this letter may be the old Hamitic *feminine article;* which clings to *Berber* words as tenaciously as "*atl*" does to proper names in Mexican languages. However, it has been shown above that these people must have resided beyond Egyptian territorial limits; and as one of many *brethren* in genesiacal personifications, the major part of whom are unquestionably *Barbaresques*, the PT*t*RS*î*m must lie to the west of Egypt also; and every reasonable requirement seems fulfilled in the *Pharusii.*

[Albeit, let me revert to a former etymology in "Otia Ægyptiaca;" which, while it does not conflict with a *Pharusian* derivation, exemplifies how a compound *Hamitic* name has become Hebraicized: for, in Berber nomenclature, *Pha*AR*usians*, *Ma-Rusians*, *Ma URi*, and their endless Gætulian homonymes, all inflexions preceding the RA, or AUR, are but demonstrative aggregations to that omnific monosyllable; whose birthplace, according to D'Avezac, might lie among the "Divine AUR*itæ*," and whose tomb is not yet constructed in *MA*R*occo!*

The reduction I formerly proposed of PT*t*RS*î*m was this:—Pi is the universal Hamitic masculine article *the;* T*t* may be T*h*o or To, Coptic and hieroglyphic for *world;* RS, the Coptic RiS and hieroglyphic RiS, meaning the *south;* which connectedly read PiT*to*RiS, *the-world-south*, or "the southern world."

This is a designation appropriate enough to austral populations; and if the PiT*to*RIS-*î*m of Xth Genesis be lineal "affiliations of the MTsR*î*m," their *name* must be resolvable into Egyptian roots. In any case, the Hebrew writer added his plural IM to a word already formed in Northern Africa centuries before his day. — G. R. G.]

Whilst submitting the above dubious solution as preferable to any dependent upon a spurious *Masora*, we nevertheless consider the *Pharusii* of ancient Barbary to be the true PT*t*RS*ì*m of Xth Genesis: confirming such opinion by two prophetic passages; 1st—"They of *Phares* (not Persians, but *Pharusii*) and of *Lud* and of *Phut* were in thine army," says *Ezekiel* (xxvii. 10) to the Tyrian masters of Barbary: 2dly, *Isaiah* (xi. 11) proves that he regarded *Pathros* to be a land entirely distinct from Egypt, when he wrote—"from Assyria, and from Egypt, and from P*h*T*t*uRiS, and from Cush," &c.[615]

32. כסלחים—KSLK*h*IM—'Casluhim.'

The ground here becomes less firm than that whereon we travelled in quest of the preceding tribes; not merely owing to the briars planted in our way by commentators, but also from the ambiguity of the text of Xth Genesis itself.

Let us commence by inquiring into the latter. King James's version, verse 14, has: "And Casluhim, (out of whom came Philistim,) and Caphtorim"; the plain English of which is, that a man called *Philistim* issued from another called *Casluhim*. The commas and parentheses being the conjectural punctuation and interpolation of King James's *translators*, we restore the text to its primitive simplicity, as closely as our alien language permits, thus: "And (the) KSLK*h*IM from whom issued (the) P*h*LST*t*-IM and (the) KP*h*T*t*RIM." Of this the plain English is, that two families, the *Philistìm* and the *Kaphtorìm*, issued from the family of the *Kaslukhìm*.

In psychological speculations, it may not be of the slightest consequence whether either of these families did, or both of them did not. Our English Bible, as Taylor, the erudite translator of *Calmet*, declares, after freely acknowledging its manifold misconstructions, "suffices for all purposes of *piety*." But in matters of archæological, and essentially of anthropographical science, the English Bible is less safe than any standard translation of *Homer*, *Herodotus*, *Cicero*, or *Cæsar;* as our "Introduction to Xth Genesis" abundantly shows.

The question whether the *Casluhìm* were the progenitors of one or both families has amply occupied theological pens, rabbinical as well as Christian; but we may mention that Rosenmüller, Cahen, and Glaire, confirm our reading.

Let us endeavor to ascertain the affinities of the *father*-stock—the KSLK*h*IM. Excepting the Abbé Mignot, followers of the few errors rather than of the many truths of Bochart, had discovered, until latterly, nothing more apposite than that semi-historical Egyptian colony of *Colchians*, planted by one of the Sesostridæ in a section of Mingrelia whence Jason brought the golden fleece. Without doubting the mythico-astronomical basis of the latter event, we summarily dismiss the *Colchians*, as a colony of Egypt, for the very reason given by Herodotus in proof of their extraction: viz., that the former people were "*black* in complexion, and *woolly*-haired," which everybody knows the MTsRIM, or Egyptians, were not.

Now, the "Caucasian" *Egyptians* being impossible procreators for *Negro* Colchians, the former's "children," according to Xth Genesis, cannot have been "woolly-haired *blacks*" either; and, inasmuch as the KSLK*h*IM were "sons of the MTsR*ì*m," they cannot have been the *Negroes* of Colchis. So we are compelled to look elsewhere.

Five of the affiliations of the Mitsrites—the *Ludìm*, *Aãnamìm*, *Lehabìm*, *Nephtukhìm*, and *Pathrusìm*—having already found comfortable homes among Gætulian races in Barbary, it would seem unnatural if the sixth had not left some mementoes of coeval residence in the same regions, between the Sahara and the Mediterranean. Indeed, our *Berber* historiographer, Ebn Khaledoon, has told us [*supra*] that his nation "descends from *Kesloudjìm*," which name is but the Arabicized vocalization of KSLK*h-ì*m. He, therefore, reputed the latter to be a Barbaresque family; and, in consequence, we proceed to test their appellative by an *Hamitic* touchstone.

Its protogramme K is a difficulty, but one of two explanations will remove it. The first is philological: viz., that all Orientalists know how such articulations as K*h*S, KS*h*, KS, glide into one another accordingly as they are enunciated by different tribes. Thus, in the very name before us, that which the native Berbers and Arabs pronounce *Shillouh*, an exotic Spaniard, Marmol, writes *Xilohes*. The writer of Xth Genesis, transcribing a *foreign* name in the unknown Hebrew alphabet he used, from six to *blank* centuries before the present *square-letter* character (in which we now have his text) was invented,—this Hebrew writer, we now repeat, when he placed a *sameq*, S, immediately after the *kaf*, K, probably meant the two letters to represent a *Berber* intonation of KS. In such case, interpolating vowels, we divide the word into KS*h*iLouK*h*-*ìm*, and writing beneath it .... S*h*iLouH—-*s*, we instantly recognize the SHILLOUHS, one of the grand duplex divisions of *Gætulian* families; the other being the *Berbers* [*ubi supra*]. In the Egyptian "sacred tongue" and character, such hieroglyphical signs as the "sieve," or the "garden," equally represent KS and SH; and if, according to orthodox interpretation, an individual yclept *Casluhìm* was really *son* of a man called MTsRaIM, the father's vernacular and writing must have regulated the child's baptismal nomen.

The second explanation is archæological; and although less likely, nay superfluous after the preceding remarks, it is submitted as another proof that the speech of the old MTsRIM, not having been the "lingua sancta" of *Shemite* families, serves to effect that which modern *Hebrew* never can aspire to: viz., a rational solution of the *Hamitic* word KSLK*h*.

"Every name determined by the sign KAH . . . is the proper name of a *province* or *country* more or less extended." This is Champollion's law of hieroglyphical writing; and so familiar to anybody who has read an Egyptological work, that one feels ashamed to pile up authorities.

If an ancient hierogrammateus had written the name of a people called *Shillouh*, he would have spelt it S*h*LUK*h*-KAH; that is, SHILLOUH-*country;* the determinative for *country* being inseparable from a geographical term. It is, then, possible that, on exportation to Jerusalem or Babylon where Xth Genesis was edited, the determinative *kah* may have become transposed from the end to the beginning of the word S*h*LK*h*, in order to suit the *Chaldaic* cuneiform system of writing; in which "determinatives" always *precede* the proper name; just as, in English, we usually say *country* of the SHILLOUHS in lieu of SHILLOUH-*country*. We have only now to suppose that a *Chaldæan* original, written in cuneiform, was transcribed by a Hebrew amanuensis into the *old* alphabet of the Jews; and the copies of this transcription recast, about two or three hundred years A. C., into the modern *square-letter* character — all things possible, and the latter event certain — to perceive that the initial K may be the relic of the sign "kah," now incorporated into a name that (supplying the vowels) we might read K*ah*-S*h*iLUK*h*, *land* of the SHILLOUHS. To which name, inasmuch as the Hebrew writer knew that it referred to a people and not to a man, he added the plural determinative IM, and thus has handed down to us a true signification of *Kasluhìm*, in "country of the SHILLOUHS." Still, we prefer the former explanation, because it is the simplest; and with these new lights continue the inquiry.

The learned Swede, so long Consul-General for his own and the Sardinian government at Tangiers, follows Ebn Khaledoon with his personal corroborative experience, when he deems the *Casluhìm* of Xth Genesis to be no others than the *Shillouhs;* already domiciled in Barbary previously to the intrusion of the first Phœnician colonists: indeed, he favors the opinion that they are autocthones. The conclusions drawn by this eminent scholar from actual Marocchine observation, derive support from another quarter; nor will Orientalists question the vast profundity of Quatremère.

In his judicious critique of Hitzig he observes:—"Quant aux *Kaslouhis*, j'y reconnais les *Schelouh* qui, de nos jours encore, composent une grande division de la nombreuse nation dont les membres sont désignés, d'une manière abusive, par le nom de *Berbères;*

66

on conçoit que ces hommes, qui, dans tous les temps, se montrèrent avides de pillage, avaient, de bonne heure, parcouru l'Afrique pour y excercer leurs brigandages. Que, se trouvant attiré par l'appât des richesses de l'Égypte, ils aient tenté une incursion dans cette contrée, et réussi à s'en rendre maîtres, la chose n'a rien d'improbable. C'est ainsi qu'à des époques plus récentes nous voyons les *Mazices*, qui appartenaient à la même race, infester par leurs brigandages l'Égypte et les contrées voisines."

The *Shillouhs* (sufficiently for the purposes of this essay) have now been started in Morocco and followed to the confines of Egypt. In these wildernesses some of their advanced posts still reside. At the famed oasis of Jupiter Ammon, or *Seewâh*, the same phenomenon is witnessed at the present day for which this oasis was remarkable in the time of Herodotus, viz: the intermixture of Egyptian and *Berber* tribes. And just as its habitants then spoke Coptic and "Ethiopian" dialects, so now their speech is Arabic and *Shilha; i. e.*, the tongue of the *Shillouhs;* into which latter idioms Arabic continues to become the more and more absorbed, in proportion as from oasis to oasis one journeys westwards; until, little beyond words impressed with religious attributes remains of Arabic in the aboriginal tongue of the *Shillouh* votary of Islâm.

The KS*hi*LuK*h*-*ìm* of Xth Genesis resolve themselves, once for all, into the SHILLOUHS: one of the two main branches of the great *Gœtulian* or Libyan family, race, or perhaps "species," of mankind. They inhabited Barbary when the ethnic chart of *Hamitic* stocks was compiled. They do so still, in the nineteenth century A. C.[616]

33. פלשתים — P*h*LST*t*IM — 'PHILISTIM.'

None will dispute that, according to the Text and the versions, these people proceed from out of the KS*h*iLou-K*h*-*ìm*. Ergo, the *Philistîm* were of *Berber* stock, and must have migrated from a Gætulian birthplace into Palestine; a land which, to this day, consecrates in its name the remembrance of one of its earliest occupants, the *Philistines*.

Contrary to the general current of opinion, here we encounter, if the ethnic genealogies of Xth Genesis are historical (as we conceive them to be), a migration from Northern Africa to Asia; that is, from West to East. If we are to be told by "teologastri," that a man yclept *Casluhim*, on his way from Mount Ararat to Mount Atlas, was delivered in Palestine of another called *Philistim*, St. Augustine will reply for us "credo, quia *impossibile*." Can it be shown when the "Philistines" were not in Palestine?

The P*h*LST*t*-IM were in Palestine before the second Pylon of the temple of *Medeenet-Haboo* was erected at Thebes; else Ramses III. could not have recorded, in the thirteenth century B. C., "the POLISITE," among his Asiatic vanquished; by all hierologists recognized as the *Philistines*. They must have been also settled in Palestine before the advent of the *Abrahamidœ*, whose presence the Philistines never quietly tolerated; and these *Philistines* were sufficiently powerful, at the time of the Exode, for Israel's escaping helots to prefer a wearisome desert march by the Sinaic route, lest, peradventure the latter should "see war;" if their valor had tested the right of way through "the land of the P*h*LST*t*-*ìm*, although that was near." And, in their uncompromising abhorrence of later Hebrew domination (which they successfully resisted until Nabuchadnezzar crushed alike the intruder and themselves) the *Philistines* never belied their *Berber* antipathies to an alien yoke. Αλλοφυλοι, *Emigrants*, themselves, they seem never to have comprehended the legality of the charter through which other strangers in the same land claimed its exclusive possession: nor did Jewish holders of this supernatural title-deed ever collect physical force adequate to an eviction.

Leaving aside, as Pundit fabrications, those *Sanscrit* apocryphas through which Wilford traced Palestine to *Pali-stân*, "country of the *Pali*" (Hales's endorsement notwithstanding); and by no means prepossessed in favor of any *Sanscrit* etymology for descendants of Hamitic *Shillouhs* in Palestine or elsewhere, after Quatremère's exposure of their impossibility — leaving aside all these Indomanias, we turn to the Abbé Mignot for some reasonable derivation of PLST*t*.

PLS, or *Felesh,* in Hebrew means *mud;* and the same bisyllable resiles from the Greek πηλος, and the Latin *Palus. Pelusium,* frontier city of Lower Egypt, towards Palestine (surrounded by marshes at the Pelusiac mouth), derived its foreign name from its *muddy* situation; being called SIN, *mud,* in Ezekiel (xxx. 15, 16), and *Teeneh,* mud, by the present Arabs. These coincidences, coupled with the fact that the PLST*t* dwelt between Pelusium and Palestine, led the ingenious Abbé to see, in the *miry* neighborhoods of their abode, the origin of the name *Philistine.* On the other hand, Munk draws the name from FLS, to *emigrate;* being the sense in which the LXX understood PLST*t-ìm,* when they rendered it by αλλοφυλοι. Munk supports this hypothesis by the Ethiopic name of Jewish Abyssinians, the FALASHAS, or *emigrants, if* their name be Semitic.

These appear to be the most rational etymologies of many producible upon the old system, before hieroglyphics were translated; or rather, in Munk's instance, before rumors of Egyptian translations had reached an erudite Conservator of the Royal Library at Paris, even in 1845. Such attempts at solution must be abortive, because, revolving within a vicious and narrow circle of ideas, they all lean upon Hebraical explanations of that which the Hebraicized "language of Kanaan" cannot explain; and for the following reason: —

Upon Egyptian monuments, at a date long anterior to the compilation of Xth Genesis (never supposed by us to be *Mosaic*), the PLST*t-ìm* are recorded. Their name is orthographed "POLIS*i*TE — *men* and *women.*" Allowing vowels to be as vague in hieroglyphics as every one knows they are in *Hebrew,* here, notwithstanding, is a word of three or *four* syllables, represented by at least *four* radical letters, P, L, S, T; as well in the old Egyptian as in the very modern *square-letter* calligraphy. To this primitive name the Jews added IM, in order to make their plural, PLST*t-ìm;* the *Philist-ines:* which word by the Masora is read *Phelesheth* in the singular; the final letter "tau" being inherent: that is, the T was already inseparable from the name thus chronicled at Thebes some three to more centuries before the consolidation of the Hebrew language itself; taking Solomon's era as the earliest and the Captivity as the latest points for pure Hebrew literature. This historical fact thrust before them, rabbinical scholars must pause, and settle with comparative philology the vital question of *biliterals* and monosyllables, ere they can make Egyptologists concede that the triliteral FLS, or PLS, is the root, not of a Semitic, but of an *Hamitic* nomen of this Barbaresque affiliation of the KSiLouK*h-ìm;* because, in the *Hamitic* "language of KNA*â*N" (falsely called *Hebrew*); in cognate *Berber* tongues; and in old Egyptian; the prefix P, P*h,* F, no less than its Berber gradation into *OU, wa, w,* &c., is almost invariably the masculine article *the,* put before the noun it determines. We hold, therefore, that the hieroglyphical POLIS*i*TE is "*the*-OLIS*i*TE," or something similar; and while we pretend not to know either the meaning or the vowelled phonetism of this noun, the presence of the article P hatchets away such fabulous etymons as PLS. *mud,* or FLS. *stranger.* It remains for *Berber* scholars to discover nominal origins of the P-OLIS*i*TE among families of the *Gætulian* race: our part contents itself with suggesting two indications supplied by Quatremère: —

1st. ASHDOD, *Azotus,* was one of the five great cities of Philistia. In the time of Nehemiah (xiii. 23, 24), after return from Captivity, "the *Jews* had married wives of Ashdod," and "their children spake half in *the speech of Ashdod,* and could not speak in the Jews' language."

It is true that the Jews, (who, considering the sanctity of their lineage, have amazingly surpassed all nations in rapidity of linguistic mutation,) in the days of Nehemiah spoke *Chaldee;* but, it would appear from the context that *Hebrew,* i. e. the "speech of Kanaan," was the tongue which their "Pasha" (PK*h*H) sought to reinstil into them by means vehement, not to say singular. "I contended with them, and cursed them, and smote certain of them, and plucked out their hair!" says *Nehemiah* (xiii. 25).

Now, Ashdod's inhabitants were PLST*t-îm* Even as late as Nehemiah, B. C. 520—40, they had preserved their own tongue in Palestine. What more natural, what otherwise possible, than that an "affiliation of the KS*h*iLouK*h*s" should have spoken in some dialect of *Berber?*

2d. — The KS*h*iLouK*h*s, in Xth Genesis, are offshoots of the MTsR*ites*. Hear Quatremère:—"Quant à ce qui concerne l'influence de la langue Égyptienne sur celles des Philistins, nous en trouvons un vestige remarquable. Il existait, sur le rivage de la mer Méditerranée, un lieu situé à peu de distance de la ville de Gaza, dont il formait le port. Ce lieu était nommé *Maïuma.* Comme il avait acquis une grande importance, il fut, sous le règne des empereurs de Constantinople, séparé de l'évêché de Gaza, et devint un siége épiscopal distinct. Ce nom, dont M. Hitzig a cherché l'étymologie dans la langue Sanscrite, appartient indubitablement à la langue de l'Égypte. En retranchant la terminaison grecque, il se composa du mot [Coptic and hieroglyphic] MA *lieu* et de IOM *mer.* Cette denomination, qui désigne *un lieu maritime,* convient parfaitement à un port de mer:" and establishes the *Hamitic* vernacular of the people who named it. Who can these people have been but the *Philistines* who built Gaza?

Another consideration. We have seen that Gætulian races, descendants of K*h*aM, *dark,* are "gentes subfusci coloris;" and also that to half the population of the oasis of Ammon, who were not Egyptians, Herodotus gives the usual Greek name of "*sunburned-faces.*" Emigrants from such stock into Palestine were therefore physiologically *swarthy;* and such were the PTST*t-îm* who founded Joppa, settling along the coast from the Suez Isthmus to Mt. Carmel. Now, as Raoul Rochette has skilfully established, early Greek writers placed the cœlo-piscine adventure of "Perseus and Andromeda" at Joppa; "among the ÆTHI-OP*ians,*" inhabitants of that city of *Philistia.* Had the PLST*t-îm* not been, like all *Berbers,* of the swarthy race, *Joppa* would not have been included in *Æthiopia,* "land of *burnt-faces.*"

Sufficient has been said on the PLST*t-îm* to show that the traditions collected in Xth Genesis accurately ascribe these peoples' origins to Barbary. To reject this deduction is to deny the validity of Xth Genesis, backed as it is by every historical desideratum; without reserving a shadow upon which contrary hypotheses have been erected through imaginary *Sanscrit* analogies that possess, anthropologically speaking, about as much relation to a man of *Philistia,* as to "the man in the moon."

"If, (says Quatremère) as I have attempted to establish, the Philistines were originally of the west of Africa, it is probable that their idiom, primitively, belonged to that speech, improperly termed *Berber,* which is spoken even to-day in northern Africa, from Egypt to the shores of the Atlantic ocean. One may believe that, during their domination (?) in Egypt, the Philistines forgot their own language to adopt that of this country, or made of the two idioms a barbarous mixture. When they were established in Palestine, seeing themselves surrounded by nations that spoke the Semitic dialects, and with whom they had daily relations, either as friends, or as enemies, they must have still more achieved modifications or corruptions of their *lingua propria.*"

Through the "Annals of Thotmes III," a most scientific paper which reaches us while correcting these pages, the antiquity of the *Philistines* can now be carried back to the sixteenth century B. C. Describing the hieroglyphical records of that Pharaoh, Birch reveals how there took place "another campaign against the fortress of Aranatu, that of Kanana, and the land of Tunep; Kadesh was once more attacked, and the campaign extended to Naharaina or Mesopotamia. The Tanai, a *Philistine* tribe who were conquered by Ramses III, the PALUSATA or *Philistines,* and the Gakhil or Galilæans, also contributed to the rent-roll, and the 'silver jug the work of the Kevau' refers to the celebrated metallic works of the Cyprians." Here the reader will recognize various geographical and ethnic names already mentioned in our present disquisition. Mr. Birch's surpassingly-great essay will show him many more.

And this is all we have to say on "P-OLIS*i*TE-*men* and *women;*" — except that orthodox Hebrew dictionaries propose, by way of explanation, "PHILISTINES, *those that dwell in villages!*" 611

34. כפתרים — KP*h*T*t*RIM — 'Caphtorim.'

The first horn of a dilemma (previously stated) displays itself in the absolutely equivocal verse of the ethnic chart itself. Our construction is, that the *Caphtorîm* proceeded (like the Philistines) from out of the KS*h*iLouK*h*s: but if a Lanci were to object that every Mitsrite name, but that of the parenthetical *Philistîm*, is preceded by the demonstrative AT*t*, and were to insist that "W-AT*t*-KP*h*T*t*RIM" means "*and-all*-KP*h*T*t*R*ites*," we should yield at once that, in the Text, the latter are *sons*, not grandsons, of the MT*s*Rîm. In mere hagiography a distinction so minute is of no importance; but in ethnography it makes all the difference whether the KP*h*T*t*Rîm issued primarily from the Egyptians, or whether they are a secondary formation from among the KS*h*iLouK*h*s of Barbary; Gætulians who, like their brethren the *Philistines*, abandoned their birthplace, and went —— whither? Nobody knows!

Bochart pointed out a road to *Cappadocia*, along which English orthodoxy follows him as sheep do their leading-rams — chiefly because, having fixed the Negro *Casluhîm* in Colchis on the Euxine, Protestant divines consider that *his* brother, or *his* son, "*Caphtorim*," naturally took lodgings next door. Our restoration of the KS*h*iLouK*h*s to Barbary shatters that hypothesis, unless Cappadocia, like Colchis, can show to some Halicarnasian a population also "*black* in complexion, and *woolly*-haired." Strabo tells us that the Leuco-Syrians, *white*-skinned-Syrians, resided there. Michaelis thought of *Cyprus*, which Volney rejects; Calmet, first *Crete*, and afterwards *Cyprus*, which second thought is favored in Kitto's cyclopædia by "E. M." *Crete*, however, is adopted by the Germanic scholarship of "J. B. R."; and, based upon similar sources, by that of Munk. One regrets to disturb this happy uniformity; but, let a query or two be propounded — after recalling that, our preceding analyses having vindicated *Barbary* as the region, and *Gætulian* as the race, of *seven* "affiliations of the MTsRîm," the *eighth*, our KP*h*T*t*Rs, whether as offshoots of *Shillouhs* or of *Egyptians*, must have been likewise "gentes subfusci coloris"; speaking a dialect of *Hamitic* tongues; whose birthplace was also Northern Africa.

1st. How, in the remote age of these ante-historical migrations, could *Berber* races have got to Crete? By navigation? Not impossible, certainly; but, it is one thing to suppose a Mr. *Caphtorim* tacking his frail bark, not along shore, but straight out 400 miles (against Etesian gales) to windward, to the Island of Candia; and another to explain the embarkation of a whole tribe of KP*h*T*t*Rs, for aught we know, as numerous as the *Pharusii* or the *Philistines*. Such a voyage, at such unnautical epochas, is rather more difficult to be conceived, in archæology, than some mistake of a copyist in writing that name which, as KPT*t*R (save in the Text, versions, and rabbinical commentors thereon), has never yet been localized.

2d. What vestiges are there in Crete, or in her traditions, of any such Barbaresque visitation? And why, after they had landed at Candia, did the KP*h*T*t*Rs abandon that splendid island *en masse*, and so thoroughly, that not a suspicion of their sojourn is to be found in Cretan, in classical, or in *Hamitic* traditions?

When these two questions have received a reasonable answer, we shall put our

3d, and last interrogatory — How comes it that, after all these improbabilities, the second voyage, from *Crete* to Palestine, is unrecorded?

It is true that three texts are quoted to identify the *Philistines* with Crete: — *Ezek.* xxv. 16, "I will stretch out my hand upon the *Philistines*, and I will cut off the K*h*RT*t*-îm." *Zeph.* ii. 5, "Woe unto the inhabitants of the seacoast, the nation of the K*h*RT*t*-î*m!* the word of IeHOuaH against you; O Kanaan, the land of the *Philistines*." 1 *Sam.* xxx. 14, 16, "We made an invasion south of the K*h*RT*t*-îm, . . the land of the *Philistines*."

Now, if the resemblance of K*h*RT*t*I to *Crete* be the only reason for making those *Shillouh* affiliations, called P-OLISiTE in hieroglyphics, navigate from Barbary to Candia, and thence to Palestine — if this be all, why the same palæographical analogy might bring the K*h*RT*t*-îm from KhaRT*t-oum*, the modern city on the juncture of the

Blue and White Niles! Unluckily for Crete, these texts merely show that K*h*RT*t-ìm* was another name — a nickname perhaps — for a sept of *Philistines* in Palestine. David's life-guards were composed of K*h*RT*t*I and P*h*LT*t*I (2 *Sam.* viii. 18; 1 *Chron.* xviii. 17). They, with the GT*t*I (2 *Sam.* xv. 18), made up a corps of "600 men." Now, the latter being citizens of Gath, the union of all three tribes into a cohort renders their homogeneity, as native Palestinians, more than probable. But, none of these passages touch the *Kaphtorìm;* whose name is distinct from that of the *Kherethìm.*

But, it is said, three other texts confirm the *Cretan* theory: — *Deut.* ii. 23, "The *Âvìm* that dwelled in villages as far as (Gaza?) *Âza*, the KP*h*T*t*Rs who issued from KP*h*T*t*R destroyed them and established themselves in their place." *Jerem.* xlvii. 4, "IeHOuaH will spoil the *Philistines*, the remnant of the *country* of KP*h*T*t*R." *Amos* ix, 7, "The *Philistines* from KP*h*T*t*R."

One must employ double-magnifying spectacles to see anything more here than that *Kaphtor* was some place whence *Philistines* came (far, or near, unrevealed); but, in what does all this concern the "Island of Candia"? Herodotus and Tacitus are quoted. The former merely says, that *Creta* was occupied by barbarous tribes until the time of Minos. This citation does not help *Caphtorim* out of the mire. The latter has "*Judæos, Cretâ insulâ profugos, novissima Libyæ insedisse memorant.*" He speaks of *Jews*, driven out of Candia, taking refuge in *Libya*. What has that incident to do with "*Philistines* from KP*h*T*t*R" in Palestine? Those who fancy that Hitzig or Movers, spite of their immense learning, and dexterity in placing one Indo-Germanic hypothesis alongside of another, have mended matters, will be edified by the perusal of Quatremère's critique of both. From it we translate: "It seems to me probable that the *Kreti* inhabited to the south of the country of the Philistines, upon the shores of the Mediterranean Sea, on the side which looks towards the frontiers of Egypt. And a passage of Herodotus (iii. 5) comes perfectly in support of my opinion. According to the Greek historian, 'from Phœnicia to the environs of Kadytis [Jerusalem], the country is inhabited by Syrians, called *Palestinians.* From Kadytis to the town of Ienusos, the market-places appertain to the Arabs; thence after, to the Lake Serbonis, dwell the Syrians.' This curious passage demonstrates that to the south of the country of the Philistines there was a coast sufficiently considerable occupied by *Arabs.* Now, inasmuch as the passages of the Bible show us these *Kreti* established in the same districts, I think they constituted an *Arab* tribe that the love of gain had fixed upon the shore of the Mediterranean, that they (the *Kreti*) had nothing in common either with the Philistines or with the Cretans."

Orthodox lexicography encourages a searcher with "CAPHTOR — a *sphere*, a *buckle*, a *hand*, a *palm*, *doves*, or *those that seek and inquire.*" We do, "et hinc illæ lachrymæ." The roots Kah-P-T*to*R might signify "*the-Bull-land*"; but neither these, nor any others hitherto offered, having furnished a clew to the genesiacal KaP*h*T*to*R-IM, we humbly place the name upon our "Table" coupled with the word "*unknown.*"

Volney, whose acuteness of perception is beyond all praise, simply says, "les *Kaphtorim* peuvent être les habitans de Gaza." Wherever may have been their abode in Palestine during later times, Xth Genesis makes them so many affiliations of K*ha*M, the *dark* (red) race, through the *Egyptians;* and consequently points to *Barbary* for their origin. Our "Affiliations of the MTsRìm" now arrange themselves as follows:

		Stock and Tongue.	Habitat.	Origin.
1.	*The* LUD,s	Berber	Mauritania	Barbary.
2.	" ÂMaN,s	"	Oases, &c.	"
3.	" LHaB,s	"	Libya	"
4.	" NiP*h*aiaT*t*,s	"	Mareoticum	"
5.	" P*h*T*t*RiS,s	"	Pharusia	"
6.	" KS*h*LouK*h*,s	"	All N.-W. Africa	"
7.	" P*h*iLiST*t*,s	"	Palestine	"?
8.	" KaP*h*T*to*R,s	"?	"	"Unknown."

[All these families of mankind thus re-enter into the grand *Gætulian* group of North-western Africa: of which sundry races, through prehistorical migrations, had partially occupied Palestine in ages anterior to the arrival of the *Abrahamidæ*. The surpassing accuracy of the ancient compiler of Xth Genesis has now been triumphantly vindicated from a new quarter; and that which not a man of the ghostly schools, whence issued his reverence doctor smythe, has ever possessed the knowledge to expound rationally, herein becomes comprehensible through "Gliddon, skeptical views of, — *Index*, p. 401." — G. R. G.] 618

"And KNAâN begat" (*Gen.* x. 15.)

35. צידן — TsIDN — 'SIDON.'

One especial object of our Section A has been achieved in the preceding pages. It was, to rescue the maligned "affiliations of KUS*h*," and the mystified "affiliations of the MTsR*îm*," from the sloughs of despond into which ecclesiastical hands had plunged them. After fixing the former in Southern Arabia among the dark-red *Himyarites*, and the latter in Barbary among the "gentes subfusci coloris" of *Gætulian* origin, we can now look down complacently upon the Egyptian alluvium of the Nile — whether viewed as the true "Land of K*h*eM" (the god), divine procreator of the Egyptian race; or as the "Land of K*h*àM," the *swarthy* people — as the centre-point, whither converge the traditions and the anthropological similitudes of Arabian Asia and of Barbaresque Africa. Our remaining objects will be satisfied by a catalogue of the other cognomina in Xth Genesis, according to the latest views of archæological science; beginning with TsIDoN.

The *city* of Sidon is the simple meaning of our text; not an individual so christened: the vicissitudes of whose *Sidonian* inhabitants, "skilled in many arts," often lauded poetically by Homer, are celebrated prosaically in classic and biblical dictionaries. Its local name was *Sèyda* when the writer (G. R. G.) sojourned there in 1829 and 1830. Orthodox philology replies to our query, as to the signification of the word — "SIDON — *hunting, fishing, venison;*" of which heterodoxy can accept but the second term in this instance; because the Semitic roots of *sàyd*, "to chase," here refer, as Trogus Pompeius tells us, to the icthyologic facilities of the locality; "nam *piscem* Phœnices SIDON vocant." In ethnic classification *Sidon* derives prominence from having once been (*Gen.* x. 19) the easternmost limit of Kanaanitish occupancy; and "after many years," continues Trogus, "the Philistines of Askalon drove out the Sidonians, who sought refuge on the rocky islet upon which they founded *Tyre*."

From Justin, the epitomizer of Trogus's lost volumes, we descend to Bochart, and admire the subdued irony with which he disposes of commentators upon the word TsIDN: — "Quod vir qui in his literis paucos habuit æquales *admirationem* explicat vocem צידון *Sidon*, non sine admiratione legi." The most recent, and incomparably the best qualified archéologue who has journeyed "round the Dead Sea and in the Bible Lands," is De Saulcy. He remarks on "Saydah — This is undoubtedly the Σιδὼν πόλις καὶ λιμὴν (κλειστὸς) of Scylax, the Sidon of Pliny, the Σιδὼν of Strabo, who places it at 400 stadia from Berytus, the *Sidona* of Antonine's Itinerary, the *Sydone* of Peutinger's Table, and, lastly, the *Civitas Sidona* of the Pilgrim from Bordeaux. It would be quite useless to argue this identity, which proves itself."

Conformably to Xth Genesis, KNAâN, parent of *Sidon*, was an affiliation of HAM; but, "according to M. Movers, the Kanaanians, called by the Greeks *Phœnicians*, were a people that appertained to the *Semitic* race; of which some tribes," says he, "at a time which preceded the commencement of our history, marched little by little, some coming from the north, by way of Syria; others, from the south, by way of Arabia; and, according to all appearances, achieved, after several centuries, their establishment, in a permanent manner, in Palestine. Called Kanaanians, from the word *Kanaan*, KNAâN, which means *a low land*, by opposition to the term *Aram*, ARM, which expressed *a high land,* they composed, according to the recital of Moses, a single people, but divided into many nations," &c.

To this theory Quatremère judiciously objects, — that the opinion which attributes a *Semitic* origin to the Kanaanites (aside from its opposition to Xth Genesis, which he considers of *Mosaic* editorship) reposes uniquely upon the resemblance of the tongue spoken by the Kanaanites with the languages in vogue among other peoples to whom general consent now applies the name of *Shemitish.* He holds this basis to be unsafe; because all of the affiliations of Shem did not speak one language; notably the *Elamites,* of Persia; whose tongue differed entirely from that of Aramæans or Arabs: at the same time, surrounded as the KNAâNI ever were by Semitic influences, their language would necessarily imbibe such exotic idioms. Again, it is by Quatremère considered doubtful, either that KNAâN means *a low land,* or ARM a *high* one. Indeed, one might add that the final N in *Kanaan* may be a later addition to an original root, KNâ; said to be the pristine name of the *Phoinikes,* Phœnicians; which is probably preserved through another form, viz.: *Beni-â*NK, "sons of ANAK;" who were not "Giants," as some commentators imagine. Such diversities of scientific opinion are here presented to exhibit some *problemata;* not to solve them.

To us the chart of Xth Genesis has proved a very trustworthy guide so far. It assigns an *Hamitic* origin to KNAâN; and consequently to the foundation of the city of *Sidon.* No facts known to us interfere with this natural view. During the eighth—ninth centuries B. C. the name of Sidon was already sculptured, according to Rawlinson and Layard, upon the monuments of Assyria; but the very conjectural identity, claimed by Osburn, of the S*h*AIRETANA, hieroglyphed on the Egyptian records of Ramses II., with the *Sidonians,* is now overthrown by Hinck's translation of a cuneatic register of Sardanapalus, wherein the "Sharutinian" city becomes situate "between Antioch and Aleppo." We have, moreover [*supra,* p. 239, Fig. 289], identified with Egyptian native soldiery of the royal guard the individual whom Mr. Osburn suspected to be a *Sidonian.* None dispute, however, that *Sidon* must have been a "city" whensoever Xth Genesis was written, so we proceed to the next name.[619]

36. חת — K*h*T*t* — 'HETH.'

The *Hittites* are well known. Of them the patriarchal Abraham (*Gen.* xxiii. 9, 17, 19) purchased not a *double* cavern, called *Machpelah;* but "the field *contracted for.*" Thus, under the magic wand of such scholarship as that of the Vatican Professor of Sacred Philology, multitudes of mistranslated Hebrew words are replaced by their historical meanings. — "I boschi," says Lanci, "diventano *veneri,* le doppie speloncheB spiegansi per *contratti,* i torrenti si cangiano in beneficii, le isole in *popoli* e *stati,* i topi in *virili vergelle,* le rondini in *puledri,* le voragini in *montagne.*"

In hieroglyphics, the K*h*eT, variously euphonized, occur so often, back to the age of Thotmes III., or the sixteenth century B. C., that one need but refer to Mr. Birch's critical papers for authority. The "land of *Kheta*' among Egyptians seems to have meant that part of Palestine where we find the *Hittites* of Scripture; but the name K*h*eT also designated this very wide-spread people; who reappear, through Layard's researches, on the cuneatic inscriptions of Assyria, as the *Khatti* or *Khetta* of Syria. To us, and to the writer of Xth Genesis, K*h*eT*t* is not a man, but a *people* so called.[620]

37. יבוסי — IBUSI — 'JEBUSITE.'

In the book of *Judges* (xix. 10), a flagitious act is recounted, which chronologers assign to about the year 1406 B. C. The date seems too remote, but the earlier it is placed by commentators, the more certain will be the archæological deductions now about to be drawn.

A Levite "rose up and departed, and came over against *Jebus,* which is Jerusalem;" that is to say, the place had been known previously by the name of IBUS; but, in the time of the writer of *Judges*, was called Jerusalem, as a second name for one and the same locality; whence the Benjamites, who gave it this latter appellative, had failed to drive the *Jebusites* out, "even unto this day." (*Jud.* i. 21.) So *Joshua* (xviii. 28), *i. e.* the book so-called, has "and IBUS which is Jerusalem;" and without requiring further information, the following text corroborates what precedes: — (1 *Chron.* xi. 4), "And David and all Israel went to Jerusalem, which is IBUS, where the IBUS*i*m (were then) the inhabitants of the land."

Hence it is certain, that IBUS was a very ancient city, on the site of which the exotic Israelites founded a more recent one they named *Jerusalem* — literally, YeRuS, *heritage*, and S*h*aLaïM, *peace* (in the dual) — written YeRuS*h*aLaïM, and signifying, according to Lanci, "She who inherits two-fold peace."

IBUSI, in Xth Genesis, means therefore "a *man* of, or belonging to, IBUS," a *city;* and not the imaginary *son* of a man of that name. Around this topographical centre clustered the IBUS*i*m before the irruption of Israel's hosts into Kanaan. There the *Jebusites* manfully vindicated their nationality until David stormed their citadel, Mt. Zion; and here some of them remained long after their city was changed into *Jerusalem*, until the invader and the invaded were swallowed up by the Babylonians.

Now, whether a *tribe* called IBUS*i*m built a city and named it after a mythical ancestor, divine or human; or whether the anterior name of a city was adopted by a tribe, is what neither ourselves nor any one else can aver. Xth Genesis speaks of an Ibusian; just as it speaks of an inhabitant of any more celebrated but perhaps not more ancient city than IBUS, already in existence when Joshua entered Palestine.

Mr. Osburn's reading of "Jebusite," among the "thirty-seven prisoners of Beni-Hassan," has not survived criticism [*supra*, p. 173]; but M. De Saulcy recognizes *Gabusa*, or *Jebus*, upon the old cuneiform tablets at Lake Van. We note a "man appertaining to the *city* of *Jebus*" in the IBUSI of Xth Genesis, and pass onwards.[621]

38. אמרי — AMRI — 'AMORITE.'

Around half the circumference of the Lake Asphaltum, and from the Jordan northward to Mt. Hermon, once dwelt a people "of stature high as cedars, and strong as oaks" (*Amos* ii. 9), called the AMORÌM: — cousins to the *Em*i*m*, *Repha*i*m*, *Zuz*i*m*, *Zamzum*i*m*, *Niphil*i*m*, and *Anak*i*m;* falsely rendered "giants" in the versions; all, according to the Vulgate translators, "monstra quædam de genere giganteo" (*Numb.* xiii. 33): some of whom were so tremendously tall, that Caleb's spies reported how "we were in our own eyes as *grasshoppers*, and such were we in *their* eyes." Never theless, astonishing as such human proportions seem, those of a thorough-bred Amorite surpassed them all; according to the orthodox stream of Hebraical traditions supplied by Cahen.

"When Og (the *Amoritish* king of Bashan) saw the Israelite camp, which had six parasangs (twenty-four miles) of extent, he said: I single-handed will undertake the combat with this people, that they do not to me as to Sihon. For this object he detached a mountain six parasangs (twenty-four miles) in breadth, and placed it on his head to heave it upon the Israelites. God caused an insect to come, which, piercing the mountain through the middle, caused *Og's* head to sink therein. He, wishing to disengage himself, could not manage it, because one of his teeth projected in front very considerably. Moses then seized an axe ten cubits (fifteen feet) in length, and jumping into the air to the height of ten cubits (fifteen feet), struck the giant on the ankle-bone of his foot. On falling, the corpse of Og touched the Israelite camp." To similar rabbinical stories Horace replied, "Credat Judæus Apella!" After all, in the Text, another and later writer, during whose day Og's iron bedstead was still exhibited at Rabbath, found, by actual measurement, that this "remnant of giants" had slept within an area of only thirteen and a half feet by six (*Deut.* iii. 11).

Among *Berber* tribes, the name OMARE, *Aâmare,* reappears in Ebn Khaledoon's list; but whether indigenously, or exotically through some ante-historical Kanaanitish or modern Arab affiliation (sons of Omar, or Aâmer?), others may better determine.

It is long since that Rosellini pointed out among the early Asiatic conquests of the XVIIIth dynasty, the "Land of OMAR:" but Birch first suspected this country to be that of the Palestinic *Amorite;* a conclusion enforced by Hincks, and developed by Osburn. There is a question still pending between hierologists and cuneiform decypherers in regard to the "citadel of ATesh" in the land of *Amaru,* which leaves it yet uncertain whether the river *Amoor,* "Jaxartes," or the nation *Amorite* in Palestine, is intended. Nor have the Palestinic travels of De Saulcy ascertained any ruins of a *city* called AMR, whence the AMoRI of Xth Genesis might be derived: although nothing can be more precious to the ethnologist than the "Figure of a *Moabite*" discovered by him on the "hybrid monument, in which the Egyptian and the Assyrian styles are intermingled," at Redjom-el-Aabed. Ignorance of Judaic topography here compels us merely to read an AMoR-*ian;* a man of, or belonging to, the city, country, or tribe, of AMR.[622]

39. גרגשי — GRGSI — 'GIRGASITE.'

This, together with the two preceding and all the following affiliations of KNAâN, has the termination I (*iod*)*;* which in Semitic tongues commonly indicates *the-belonging-to* a place; for instance, *Muss'r* means Cairo; *Muss'r-i,* a Cairine. In Xth Genesis, this adjunct to a geographical proper name has precisely the same grammatical acceptation; and if science cannot always find the *place* alluded to, the fault lies at the door of travellers less qualified than a De Saulcy. GRGS-I signifies nothing more than a man *belonging-to* a locality once called GRGS; although its Palestinic situation still lacks a discoverer. Other books of the Hebrews are silent on this name; which was all that remained of a *Girgasite* even in the time of Josephus, 1800 years ago; unless "the country of the *Gergesenes,*" mentioned by *Matthew* (viii. 28), contained other persons than those "possessed with devils."[623]

40. חוי — K*h*UI — 'HIVITE.'

A man "of, or belonging to," a place called K*h*U; now pronounced, through the modern *Chaldee* substitution of V for U, "K*h*aV." The K*h*UI*tes* rank among the *unexpelled* Kanaanites; because *Joshua* (xi. 19) suffered some of them to deceive him into a peace; and *Solomon* (1 *Kings* ix. 20, 21) exacted "bond-service" from others.

We must never forget, in viewing this name and its fellow-nomina, that time, distance, foreign and obsolete languages now reputed to be "sacred," combined with the singular mixture of scepticism and marvellousness instilled into our minds by juvenile education, lend an enchantment to these Kanaanitish people that would vanish, did we now possess the honor of their acquaintance. They all were petty tribes of a few thousands, at most of fewer myriads of population; comprised within an area so very insignificant, that St. Jerome, who travelled over Palestine (which had previously included the whole of these *nations,* and other people besides), wisely deprecates statistics: — "Pudet dicere latitudinem terræ repromissionis, ne ethnicis occasionem blasphemandi dedisse videamur." That criticism which, precursor of Niebuhr, the author of "Scienza Nuova," applied so successfully to early Roman, might equally well be adapted to early Jewish history—"What we may say about the *poetic geography* of the Greeks suits the *ancient geography* of the Latins. Latium possessed, without doubt, at the commencement, but a petty extent; inasmuch as, while employing two hundred and fifty *years* to conquer twenty different *peoples,* Rome during that time did not stretch out the frontier of her empire further than twenty miles round about." Among "the *cities* of the K*h*U-*im*" (2 *Sam.* xxiv. 7) we cannot yet place a finger upon that particular one whence hailed the "citizen" individualized in Xth Genesis.[624]

41. ערקי — A*â*RKI — 'ARKITE.'

A man of *Arka*, or *Acra;* a city the ruins of which are still seen at *Tel-Arka*, mound of Arka, between Tripoli and Antaradus; but *Akra* must have been already a city when Asar-adan-pal and Temenebar I. recorded its capture in the eighth—ninth century B. C.; else Rawlinson could not have discovered its cuneatic name.

[In former inquiries into the probable origin of some *Berber* names, that certainly present some Kanaanitish coincidences, I indicated the ERKYE of Ebn Khaledoon as homonymous. That some *Kanaanites* sought refuge in Barbary is undoubtedly historical; that some *Berbers* did once occupy Kanaan has been already shown. There is a strange blending of Gætulian and Arabian elements in Palestine anterior to the advent of the *Abrahamidæ*, underlying every record, which the supposition of a creative centre, distinct from that of *Euphratic* tradition, might possibly explain.—G. R. G.][625]

42. סיני — SINI — 'SINITE.'

A man "of, or belonging to the town of SIN," not far from *Acra*, on the slopes of Mount Lebanon. This name reappears among Ebn Khaledoon's *Berber* tribes as the ZIN-ata.[626]

43. ארודי — ARUDI — 'ARVADITE.'

A man of *Rowèyda* (as modern Syrians now designate the little island of *Aradus*), which town, with its continental neighbor *Antaradus*, was a famed Phœnician emporium. Every lexicon explains the familiar locality; but Osburn has the merit of indicating the people and their name hieroglyphed amid the conquests of Sethei I., and Ramses II.; fourteenth—sixteenth centuries B. C.; and Rawlinson that of reading the cuneiform inscriptions in which, during the eighth—ninth centuries B. C., the existence of *Aradus* is chronicled.[627]

44. צמרי — T*s*MRI — 'ZEMARITE.'

A man of the Phœnician town of *Simyra*, not far from Antaradus, on the western spur of Mount Lebanon; afterwards occupied by the Benjamites, who probably expelled its inhabitants—the T*s*MR-*ìm*. A similar name occurs among Ebn Khaledoon's *Berbers;* but, beyond this phonetic and therefore uncertain analogy, we here must emulate the laconic chorography, not merely of Xth Genesis, but of map-makers in general, having nothing to add to the investigations of Bochart.[628]

45. חמתי — K*h*MT*t*I — 'HAMATHITE.'

This is a man "belonging to a *city*" situate on the Orontes at the eastern frontier of Palestine, now called *el-Hámah* by Syrians. Although later Greeks termed it *Epiphaneia* during their dominion, the natives have always preserved its antique nomen. The LXX properly wrote Εμάθ: as did Assyrians, six centuries before them, in cuneatic inscriptions deciphered by Rawlinson; while, at least four hundred years previously, Ramses III. had hieroglyphed the *Hamathites* among his Asiatic vanquished.

We would passingly notice that which, philologically speaking, is incontrovertible in regard to the Hebrew transcription of this name. The letter I, *iod*, has been shown above to be the demonstrative adjunct "of, or belonging to" a locality. T*t*, *tau*, in all ancient *Hamitic* idioms is the feminine article, *the;* prefixed or suffixed even now to abundant Berber nomina—*ex. gr.*, T-Amazirgh or Amazirgh-T. These cut away, the pristine monosyllable of K*h*aMaT*t*I is K*h*M; identical with K*h*eM the name of *Egypt;* and also with K*h*aM the son of Noah, personified symbol of all *Hamitic* families. We have traced the *Philistines* to a Barbaresque source, although history dawns upon them in Palestine. The writer of Xth Genesis, whose authority has been found so unexceptionably safe hitherto, makes a K*h*aM-*ite* citizen on the frontier of Palestine descend from KNAǎN; the figurative son of KUS*h* who was the figurative son of K*h*aM. The *Hamitic* article T is suffixed to the primitive *biliteral* name of a city, whose existence is carried back on Egyptian monuments to Mosaic epochas. There is no historical limit definable for the foundation of the *city;* none, most assuredly, for the antiquity of its *name.* But, archæology may draw, from other data, inferences that appear satisfactory: before considering which, justice to the memory of human greatness suggests a citation: —

"The man who has anticipated by a century the movements of mind towards modern sciences; who has raised up questions which, down to him, were considered to be resolved or to be insoluble; who has carried the investigations of a criticism the most intrepid into documents by all antiquity respected; who never bent himself before established prejudice; who has accomplished the double enterprise of destroying and of reconstructing universal history; who has treated upon all the sciences without being acquainted precisely with any one, and who bequeathed to each of them some fecund teaching; the man who has almost divined all the discoveries of the nineteenth century; who, appertaining to an age [1722] and to a country [Naples] wherein thought was never free, seemed to ignore that the saying of every thing to every body, was to expose himself to be comprehended by nobody; the man whose genius recalls the mighty intellects of Plato and of Aristotle, deserves to be followed step by step in the development of his glorious intelligence and in the vicissitudes of his long and unhappy life." That man was Vico. In "establishing the Principles" of historical criticism, he laid down, for the 107th rule: "the commencements of nations preceded the commencements of *cities.*" A hagiographer smiles at its infantine simplicity — let us raise a laugh at his.

We have seen that, *Sidon, Ibus, Arka, Sin, Aradus, Simyra,* and *Hamath,* were *cities.* We know that the terminal letter I, *iod,* to six of these seven names, produces, in Semitic idioms, exactly the same effect that our addition of an English "*ian*" changes them into a Sidon-*ian,* an Ibus-*ian,* an Ark-*ian,* a Sin-*ian,* an Arad-*ian,* a Simyr-*ian,* and a Hamath-*ian.* Ergo, these people derive their appellatives from *cities;* built, of course, before men could hail from them. What now — let us turn round and ask the smiling querist, as his face augments its longitude while diminishing its risible latitude,—what now becomes of your fables about those MEN called *Sidon, Ibus, Arka, Sin, Aradus, Simyra,* or *Hamath,* whom your schools have dared to find in Xth Genesis, as *sons,* forsooth [!], of another fabulous human being your philologers spell "Canaan"?

But, there is yet another deduction which the reader will draw at once from these premises, viz.: — that, inasmuch as a man could not be a *Hamathian* before the city of Hamath was built, the fact that the writer of Xth Genesis speaks of a K*h*aMaT*t*I, or *Hamathian,* proves that the document called "Xth Genesis" was written *after,* probably long after, this city had existed; and, therefore, that *he* (the writer aforesaid) never dreamed that modern logopœists would metamorphose his *cities* into so many human beings.

The age of the foundation of all these cities receding beyond historical chronology, we have said enough on the *Hamathian* and his compeers: but, while taking leave of the *cities* included in the terrestrial area called KNAǎN, we likewise bid farewell to every commentator who perpetuates rabbinical superstitions about "Canaan" and his gigantic progeny. "These," says the chorographer of Xth Genesis, on closing his *Hamitic* list, — "These are the affiliations of K*h*aM [*i. e.,* the *swarthy*], after their families, after their *tongues,* in their countries, and in their nations." (*Gen.* x. 20.)

Nothing can be plainer, nor more scientifically concise. In our journey from Babylon through Southern Arabia, and round by the shores of the Erythræan (red), Edomite or *Red* Sea, the dark *Himyarites* (red) have accompanied us, over the Suez Isthmus, into Egypt — the true "land of K*h*àM" (dark); its ancient name preserved in *Chemmia* — abode of the *red* people, "par excellence." Thence, towards the west along Barbary we see the prolongations of the same *Hamitic* (dark) families, "gentes subfusci coloris," stretching between the Sahara desert and the Mediterranean, as far as Mauritania: whilst, towards the east, through Palestine, we behold the wrecks of an aboriginal population, linked by traditions and primitive speech to *Egypt* and to Barbary, "tinged with the *red* of Gætulian blood," and *Hamitic* under every aspect.[629]

We next take up the "Affiliations of SHEM."

"And unto S*h*eM (there was) issue." (*Gen.* x. 21—Hebrew Text.)

46. עילם — A*â*ILM — 'ELAM.'

Preceding generations have bent their intelligencies towards the elucidation of *Shemitish* subjects with more zeal, and therefore with more success, than towards that of Japethic or of Hamitic problems.

Owing partly to the fortuitous preservation of this family's chronicles in greater completeness than those of any people except the *Chinese;* still more, to the absence, until this century, of those immortal discoveries epitomized in two names, "CHAMPOLLION and RAWLINSON"; and, beyond any other stimulant of research, to doctrinal biases in favor of a select line that, under the name of Hebrews and Arabs, traces its pedigree backwards to a *biliteral* SM — owing, we repeat, to these historical accidents, we happen to know a little more about some of SM's posterity, their annals, habitats, and associations, than we do concerning other less respectable, because unrecorded, "Types of Mankind."

According to Ainsworth, geologist to the Euphrates Expedition, *Elymais*, country of the *Elymæi* (the capital city of which was also called *Elymais* when classical history first dawns upon its geography), was a Persian province, situate to the south of Media, between the river Tigris and the Persian Appenines, sloping downwards into Susiana and to the Persian Gulf. Tradition, through Polybius and Strabo, ascribes to its *Elymæan* inhabitants a northern origin; and Josephus calls them "the founders of the Persians": with whom they are often confounded in later Hebrew annals; for Persia and Persepolis are both called *Elam* (1 *Maccab.* vi. 12; 2 *id.* ix. 2). They were, however, in the days of Abraham, already occupiers of a *kingdom* called *Elam* (*Gen.* xiv. 1, 9); so that when, more than a thousand years later, the compiler of Xth Genesis registered A*â*ILM on his ethnic chart, he naturally meant the *country* which had been so called from times immemorial before him.

This country (generally, if improperly, included in the sections of territory comprehended by the term *Susiana*), is full of ancient cuneiform remains; both of the Persian and of the older Assyrian period: but, in 1846, one class of the cuneatic inscriptions there discovered, owing to "the number of new characters which they exhibit — characters for which no conjectural equivalent can be found either in the Babylonian or the Assyrian alphabet" — was denominated *Elymæan* by Rawlinson, being monuments distinct from their neighbors.

Under these circumstances, until Rawlinson or his emulous competitors shall breathe upon these "dry bones" of *Elymais*, "and say unto them, O ye dry bones, hear!" it is best not to hazard opinions on the unknown, which the next mail from Europe may perhaps render clear as day. We therefore merely indicate a discrepancy at present evident between modern philological and historical results and the *Semitish* genealogy of A*â*ILM-*ais*, in Xth Genesis. According to the latter, the A*â*ILM-*ites* should have spoken a dialect of the *Aramæan* class of languages: but, according to the former, as interpreted by Lenormant, Quatremère, Movers, and others, the affinities of

AǎILM, cognate if not identical with the Persians, are *Arian*. It seems to us, however, that Löwenstern's solution is satisfactory. He shows how the primitive Elamites were of Semitic extraction, but that, in after times, Scythic conquerors superimposed in Elam their extraneous blood, tongues, and traditions; as the reader can verify in this author's learned papers. In the meanwhile, De Saulcy has read upon cuneatic inscriptions of the age of Asar-haddon, eighth century B. C., that this monarch was "rex populi Assur," and "rex populi *Elam*": and this is confirmed by Layard's Second Expedition, for "Sennacherib speaks of the army which defended the workmen being attacked by the king of *Elam* and the king of Babylon."

Our confidence in the compiler of Xth Genesis stands unshaken. If, as we have proved, his tabulation of the distant *Hamites* is so correct, how much better must a *Chaldæan* chorographer have been acquainted with the legendary origins of a Semitish AǎILM-*ais?* [630]

47. אשור — ASUR — 'Asshur.'

While admitting the equivocal nature of the text of *Genesis* x. 11, we have given reasons [*supra*, p. 509] for reading — "From this land (*Shinar*) he himself (NiMRoD) went forth (to) ASUR (Assyria) and builded Nineveh," &c. Such lesson indicates that we have now before us a *geographical* name.

"It would be strange," critically remarks De Sola, "if Ashur, a son of Shem (*Gen.* x. 22) were mentioned among the descendants of *Cham*, of whom Nimrod was one. It would be equally strange if the deeds of Ashur were spoken of (in verse 11) before his birth and descent had been mentioned." The writer of Xth Genesis, a plain sensible man, compiling the Assyrian department of his chart not impossibly in ASUR itself, was not likely to have committed such a needless anachronism. Let us examine another text.

King James's version, *Genesis* ii. 14—"And the name of the third river *is* Hiddekel: that *is* it which goeth toward the east of Assyria." This text has opportunely received recent ventilation at Paris, in discussions between De Longpérier, an Orientalist as profound in biblical as in all archaic lore, and a learned dogmatist, M. Hoeffer. The ante-diluvian river, miswritten *Hiddekel* in our version, is, in the Text, H-DKL, *the*-DiKLe — a name that, through various historical transmutations, such as DiGLe, DidJLeh, TiGLe, and TiGRE (*Tigrâm*, in Persepolitan inscriptions), is inherited by us in its euphonized Latin form — the TIGRIS.

The Text therefore reads literally — *the* Tigris, "ipse vadens KDMT*t* (*ante*) ASUR;" Parisian debate turned upon the meaning of KDMT*t;* by English interpreters rendered "East;" — a translation which, if true, (as dogmatism had maintained,) would place the city of *Nineveh*, built in the land of ASUR (*Gen.* x. 11), on the west bank of that river; supposing always that the river lay to the *east* of it (Assyria). And thus "Holy Scripture" was triumphantly quoted to prove that, inasmuch as Nineveh was situate *west* of the Tigris, the vast exhumations of Botta, Layard, Place, and Rawlinson, on the *eastern* bank, which people fondly supposed to have been executed in ante-diluvian *Assyria*, not having been made on the site of Nineveh at all, the whole of these discoveries, in regard to *Nineveh*, fell to the ground!

But, Mrs. Rich and St. Jerome naively tell us — "It is one thing to write *history*, and another to write prophecy under the immediate effect of inspiration." If "a prophet is not without honor, but in his own country, and among his own kin, and in his own house" (*Mark* vi. 4); that is, among those mortals who happen to know him best; — the unfortunate scholar alluded to can hope for little elsewhere; since De Longpérier established:—

1st. That Herodotus has nowhere connected the Tigris with Assyria.

2d. That neither the Septuagint, nor the Vulgate, any more than the Hebrew Text, justifies such a reading as "East" in Genesis ii. 14.

3d. That KDMT*t* here meaning simply "en avant vers," the true signification of this passage must be, in English, "the Tigris, flowing *in front towards* (say *opposite*) Assyria."

Our digression introduces another difficulty. Between the land of ASUR in IId Genesis, and ASUR in Genesis Xth, rolls the Flood; which, contrary to the sophistries of the Rev. Dr. J. Pye Smith, we wholly agree with the "Friend of Moses," and the writer of Genesis VIIth, in considering to have been *universal*. If geology, in the XIXth century after Christ, discovers phenomena which prove Diluvian momentaneous universality to be impossible, so much the worse for *geologists*. But to attribute to Hebrew authors living long subsequently to the XIXth century B. C., the intrepid conceptions of modern geology, is to commit a most gross historical anachronism; besides inventing a doctrine utterly irreconcilable with the plain *square-letters* of the Hebrew Text. We would therefore merely inquire of the orthodox geologist whether he considers the land of ASUR, along which ran the river *Tigris* before the universal Flood, to have been specified (by Moses) proleptically or retroleptically? His reply would enlighten us upon one of two propositions. If this Hebrew "scholar and statesman," as the *Friend of Moses* terms him, had before his eyes, as some maintain, certain documents written by ante-diluvian patriarchs, then ASUR, in such manuscripts, must have been the geographical appellative of a *country* existing before the Flood; which country, after the waters had passed away, emerged as ASUR, along with its river *Tigris*, on the same terrestrial area, in order to be catalogued by the writer of Xth Genesis among other countries existing in his later day. Or, if Moses was enlightened upon events anterior to his lifetime through "Divine inspiration," then we possess the authority of the Most High (through Moses) for sustaining that, ASUR, having been the geographical name of a *country* years before the Deluge, and centuries before "ASHUR, son of SHEM," was born, the writer of Xth Genesis was right in mapping the "land of ASUR" as a *country*, according to its ante-fluviatile acceptation in Genesis ii. 14 — a country, too, wherein the masterly geological researches of Ainsworth could discover no traces of any Noachian Flood. That which remains certain is, that ASUR was already a *country*, according to the letter of Scripture itself, whensoever, or by whomsoever, or wheresoever, Xth Genesis was written; and, for our researches, "for us, that is enough." — "That you should wish to *call* MOSES author of the *Pentateuch*, or ESDRAS the restorer of this same work, I do not object," philosophically wrote St. Jerome.

The name of ASUR, in unpunctuated Hebrew, becomes AS*h*UR through rabbinical marks; and passing through different dialects and ages, as AT*t*UR, ATUR, ATUR*ia*, A*th*URA, ASSUR*ia*, &c., it is now written *Assyria* by ourselves. But, while modern Chaldee Jews have preserved in *Athour* the correspondent of *Ashour* as intonated by their forefathers, cuneiform scholars have discovered, in the land of AS*h*UR itself, the indigenous name, petroglyphed *Assour*, upon innumerable records disinterred from the mounds of Khorsabad and Nimroud.

Kings of the "country of ASUR" are now well-known personages to readers of Botta, Layard, Rawlinson, De Longpérier, De Saulcy, Hincks, Birch, Grotefend, Löwenstern, Oppert, Norris, Vaux, Eadie, or Bonomi; and having been found upon sculptures coeval with the epoch of Jehu, king of Israel, ASUR was already the name of Assyria early in the ninth century B. C.: an age, we think, nearly parallel with the compilation of Xth Genesis. These now-familiar topics need no pause; but some of those things which are less so demand notice in tracing ASUR to its primeval source. Rawlinson finds in *Assarac*, (Assarak, Asserah,) "god of *Assyria*" — the deified proto-patriarch of that land — called in the inscriptions "father of the gods," "king of the gods," "great ruler of the gods;" whose mythological characteristics are those of *Kronos* or *Saturn*. "I should suppose him, as head of the Pantheon, to be represented by that particular device of a *winged figure in a circle*, which was subsequently adopted by the Persians to denote ORMUZD, the chief deity of their religious system." And we may now leave hagiography to rejoice over possible connections between the divine *Assarac* and *Ashur* the son of SHEM, among those of other genealogies of Xth Genesis; which document Rawlinson does not consider anything more than "an historical representation of the great and lengthened migrations of the primitive Asiatic race of man." More recently we learn from Layard how — "*Asshur*, the king of the circle of the great gods," heads the list of the thirteen great gods of Assyria, at Nimroud. At Babylon, however, the god *Marduk* is termed "the great lord," "lord of lords," "elder of the gods," &c.; and *Ashur* no longer appears, being the god of upland Assyria, and not of the Babylonian plains.

The cuneiform documents upon which AS*h*UR figures as a native mythological personage approach in antiquity the era of Moses. The hieroglyphical records in which *A-su-ru* occurs as the Egyptian name of *Assyria*, surpass, by two hundred years, the age of the Hebrew lawgiver, because Birch discovers it upon inscriptions of the time of Amunoph III [*supra*, p. 133, fig. 32]. Space now prevents the demonstration that, among its various symbolical meanings, A-SUR signifies also "*the-Bull*-land;" but the writer (G. R. G.) will publish the reasons elsewhere. In the interim, to the author of Xth Genesis, AS*h*UR meant the *country* by us called *Assyria* — nothing more nor less.[632]

48. ארפכשד — ARP*h*KSD — 'Arphaxad.'

"Arphaxad (ARP*ha*KSaD: Sept. 'Αρφαξάδ), the son of Shem, and father of Salah; born one year after the Deluge, and died B. C. 1904, aged 438 years (*Gen.* xi. 12, &c)."

Requiescat in pace!

Such is the terse obituary notice,—unaccompanied by the customary poetical regrets, or general invitation to attend the funeral,—a divinity student encounters when, seeking for instruction about the Savior's genealogy, he opens Kitto's cyclopædia or Taylor's *Calmet* (the best English biblical dictionaries) at the name Arphaxad: and this is *all*. A noble cenotaph! We close those devout, not to say laborious, compendia, and turn to Volney's *Recherches Nouvelles*.

"A fifth people *of Sem is Araf-Kashd*, represented in the canton *Arra-Pachitis* of Ptolemy, which is a mountainous country, at the south of the Lake of Van, whence stream forth the Tigris and the Lycus or great *Zab*. This name signifies *boundary of the Chaldæan*, and seems to indicate that the Chaldæans, before Ninus, had extended themselves even thither. This Araph-Kashd, according to Josephus, was father of the Chaldæans; according to the Hebrew, he produced Shelah, whose trace, as *city*, and *country*, is found in the *Salacha* of Ptolemy. Shelah produced Eber, father of all the peoples *on the other side* of the Euphrates; but if we find him on *this side*, relatively to Judæa, we have the right to say that this antique tradition comes from Chaldæa." Our analyses of Xth Genesis entirely corroborate Volney's deductions of its Chaldaic derivation; and justify Lenormant's orthodox eulogies of him as "un des hommes les plus pénétrants de ce siècle." From the latter we take the following note— "Josephus had made, before Michælis, of *Arphaxad*, the father of the *Casdim* or Chaldæans. M. Bohlen explains *Arrapachitis* by the Sanscrit: *Aryapakschatà*, the country bordering upon *Aria*. This etymology is not unworthy of attention."

There is little to be added to Volney's definition; and that little confirms him. ARP*h*-KaSD — after dividing into two words that which in the Hebrew ancient Text (Synagogue rolls) runs letter after letter, "continuâ serie," along the whole line — yields us, as Michaelis first suggested, ARFA, the Arabic for *boundary*, and KASD, *Chaldæan*. The etymology is in unison with Aramæan origines; and Arphaxad was the *brother* of Aram: while Bochart's identification of it with the province of *Arrapachitis* of Ptolemy's geography also stands; but perhaps not with "nam quod Josephus et alii volunt Chaldæos olim ab eô dictos *Arphaxadæos* merum somnium est."

It is strange how Oriental tradition clings to the vicinities of Ararat as the mountainous birthplace of Chaldaic races. There we find the *Heden* (Eden) of Genesis IId, and "the house of Eden" extant in the time of the prophet Amos (i. 5); while another writer tells us how "Haran Canne, and *Heden*, have made traffic with what came from Seba, and Assyria learned thy traffic" (*Ezek.* xxvii. 23).

There, too, was the *Haïasdan* of the Armenians; and there the *Hadénéche* which Zoroaster ennobled by the title of the "pure Iran" because his birthplace was at Ourmi, on the border of Lake Ourmiah. "There," continues Dubois, "is the antique native-land of *Arpacsad* and of the Hebrews: and their patriarch Abraham, like Zoroaster, was born at Our, on the shores of Lake Ourmiah, in Chaldæa. There touches also Iràn, Arhan, the land of Persian mythes." In which connection let us likewise add, that the river *Akhouréan*, whose sources lie on the same chain, still bears the name of ARPA-TCHAI. But we suggest a melioration.

ARPHAKASD, as a *country* in Xth Genesis, is the parental source, through the province of *Salacha*, of EBER, *the yonderer;* and from the latter, according to the other document (*Gen.* xi. 13–26), sprang ABRAHAM, progenitor of the *Abrahamidæ;* born probably at *Our Kasdìm*, "*Ur* of the Chaldees," whence they issued "to go to the land of Kanâan." It is true that Mr. Loftus considers the enormous ruins of *Werka* to be the real "Ur of the Chaldees," now traditionally called "the birthplace of Abraham;" nor would the establishment of this fact result in any further alteration of our view than by proving (what is very likely) that ARP*ha*-KaSD was a different place from AUR-KaSDIM. The name "Chaldæan" is also ancient enough, having been found in cuneiform on the monuments of Nineveh.

Be all this as it may, there still remains *one* "*Ur* of the Chaldees," AUR-KSDIM in the text, which is unquestionably, as shown by Ritter and by Ainsworth, the present city and district of *Urhoi*, now *Orfa*, or URP*h*A (called, in Greco-Roman times, *Chaldæopolis*, *Antiochia*, *Callirhoe*, and *Edessa*), in *Diàrbekìr*. Allowing very common mutations of vowels, we behold in *Urfa*, or ARP*ha*, ARP*ha*-K*a*SD, "*Orfa* of the *Chaldæan*," the absolute solution of ARPHAXAD, no less than the earliest geographical source of the *Abrahamidæ*.

Thus, at every step, the chorographic exactitude of Xth Genesis is vindicated; and ARP*ha*KaSD, no more a fabulous human being, regains its legitimate heritage among the *countries* of the earth. To the "late *Mr.*" ARPHAXAD, "aged 438 years," we repeat our valedictory, "requiescat in pace!" 633

49. לוד — LUD — 'LUD.'

The high road from Nineveh, in the land of ASUR, Assyria, conducts a traveller towards Asia Minor, through ARFA-KASD, *Chaldæan-Orfa*, into *Lydia;* — a name which, in its Greek spelling of Λυδια, faithfully transcribes the Hebrew LUD-*ia*.

This country derives its name, according to traditions collected by a native of Asia Minor, Herodotus of Halicarnassus, from Lydus, son of Atys; whose crown passed into the keeping of Hercules. This legend indicates the ante-historical ground we tread upon; and probably the intrusion of Hellenic *Hieraclidæ* upon an aboriginal Lydian population, affiliated with the Shemites. The recent explorations of Fellows and the Lycian monuments now rescued from perdition, establish, in the most convincing manner, the transitions of art in all its symbolism, through Asia Minor, from Assyria to Greece; and the mythe of the *Assyrian Hercules* serves as a faithful thread through the mazes of this labyrinth: which mythe, Grote observes, exhibits but the "tendency to universal personification"—being merely "Μυθος, *Saga* — an universal manifestation of the human mind."

But, from the premises, one deduction is solid, viz.: that Herodotus, than whom in Lydian questions there is no higher authority, makes Hercules succeed Lydus — the personified land of *Lydia*. Now, inasmuch as the mythe of *Hercules* antedates all chronology, it follows that Herodotus, who says that *Lydus* preceded the *Hieraclidæ*, looked upon the autocthonous name and traditions of *Lydia* as still more remote from his own day; B. C. 484–430. To us, therefore, the Halicarnassian's testimony, upon the ante-historical affairs of his native Asia Minor, would *ipso facto* outweigh any notices of

Lydia issuing from the "School of Esdras" in Palestine (foreign to Lydian blood, language, and traditions), should the latter contradict him: which, happily, they do not.

The compiler of Xth Genesis, educated, as we now begin to feel assured, amid the "learning of the Chaldees," attributes no affiliations to the geographical locality he designates LUD; any more than, in his classification of the senior *Hamidæ* (*ver.* 6), he ascribes descendants to P*h*UT; which, we have seen, is *Barbary*. This engenders the supposition that he knew little beyond the *names* of either; and that just as to him, composing his ethnic chart in some University of Chaldæa, P*h*UT appeared to be the most western geographical range of *Hamitic* migrations, so LUD probably seemed to lie among the most northerly of *Semitic*. As such, then, he duly registered them in his inestimable chorography.

Some centuries prior to the age of this venerable digest, the *Lydians* are mentioned in Egyptian hieroglyphics. In the Asiatic conquests of Sethei-Meneptha, and of Ramses II., to say nothing of later Pharaohs, associated with *Ionians*, *Riphæans*, and other well-known families of Asia Minor, we find the oft-recurring "Land of *Ludenu*," or "land of the upper *Luden*," and "of the lower *Luden*." This establishes the existence of *Lydia* and of *Lydians* at the XVIIIth dynasty, fourteenth—sixteenth centuries B. C.; in days anterior to and coeval with Moses; *i. e.*, much earlier than the compilation of Xth Genesis. But (to avoid Mosaic conflictions with Egyptian records) it is best perhaps to ascend a few generations beyond modern disputes upon the era of the Hebrew "scholar and statesman;" when by pointing out LUD and *Lydians* in chronicles appertaining to the anterior XVIIth dynasty, we show that Amunoph II., Thotmes III., and Amunoph III., successors of that "new king over Egypt which knew not Joseph" (*Ex.* i. 8), could not readily have heard of Moses's Lydian geography before the great lawgiver was born. Posterior in epoch to the former, and anterior to the latter dignitary, these Pharaohs of the XVIIth dynasty knew nothing about either Joseph or Moses.

Nor is history wanting to support the early spread of Egyptian arms into Asia Minor; for besides a confused aggregation of events of different ages to be met with in every classical lexicon under the head of "Sesostris," we have the authentic account of Tacitus that the Priests of Thebes read to the Emperor Germanicus, from hieroglyphical inscriptions, how "Ramses overcame Libya, Ethiopia, the Medes and the Persians, Bactriana, and Scythia, and held sway over the lands which the Syrians, Armenians, and neighboring Cappadocians, inhabit from Bithynia up to the Lycian Sea."

We cannot quote authority for the discovery of the name LUD in cuneiform writings; unless *Ludenu* be the same as the "Rutennu" of the "Grand Procession of Thotmes III." [*supra*, p. 159], which Birch fixes, in hieroglyphical geography, "north of the Great Sea," and compares with the Assyrian king Sargina's prisoners at Khorsabad.

However, LUD, being identical with *Lydia*, enters, like the rest, as a geographical appellative into the catalogue of Xth Genesis; and the cyclopædic notion that, from a *man* called LUD, "the Lydians in Asia Minor derived their name," ranks among the childish postulates belonging to an age of which science now hopefully discerns "the beginning of the end." [634]

50. ארם — ARM — 'Aram.'

Orthodox lexicography informs us that Aram means "*highness, magnificence; otherwise, one that deceives*, or *their curse*." In this instance the erudition of "N. M." compensates for the meagre article by "J. P. S." in Kitto's cyclopædia.

It has been shown already that Quatremère doubts Mover's derivation of ARM; which the latter considers to mean *a high land*, in juxtaposition to KNAâN, *a low land*. Still, the objection assigned by the former is inconclusive, because RM does actually signify *high*; and with the primeval masculine article *aleph*, A, prefixed, A-RM is *the-high*. Certain it is, also, that the geographical brother of Arpha-Kasd, "*Orfa* of the

Chaldæan," and of *Lydia*, must be sought for along the same Tauric uplands of Asia Minor; where ARM lay among the "mountains of the east" (*Numb.* xxiii, 7). In Punic, also, the same word means *high;* for M. Judas reads on Numidian coins, *Juba* ROUM *melkat* = "Juba, *highness* of the realm."

Diodorus's Αριμα ὄρη or *Arimi Montes*, suggest themselves at once; although authorities disagree upon their location, in Phrygia, Lydia, Mysia, Cilicia, or Syria: but Strabo and Josephus inform us that the Greeks called *Syrians* those people who called themselves *Aramæans:* and when Homer and Hesiod wrote, the Αριμοι extended to Phrygia, which they termed *Arimaïa.* Syria, therefore, in its widest acceptation, seems best to correspond to ARM, because the latter merges into *Mesopotamia;* and in Pliny and Pomponius Mela the name of Syria is applied to provinces even beyond the Euphrates and Tigris.

As the grand centre of *Shemitish* families, Syria still preserves the name of S*h*eM in its Oriental appellative; being known to *Syrians* and the populations around them by no other title than BúR-Es-S*h*àM, *land of Shem.* Arab geography explains this coincidence by reasons worthy of attention. SHAM means the *left hand*, and YEMEEN (*Yemen* in Arabia), the *right;* as, face directed to the East, an Arabian worshipped the rising sun; or looked back to ARM as the traditionary birthplace of his ancestry before, by emigration to Arabia, they had acquired the right to call themselves âRB, *western*-men. Damascus, *Es-Shàm el-kebeer*, "the great Sham," may perhaps be the focus of these ancient radiations: for its identity with ARAM is marked in the passage —"The ARaM*ians* of *Damascus* came to succor Hadadezer king of Sobah, &c. (2 *Sam.* viii. 5. 6) — the versions generally substituting *Syrians* for *Aramæans.*

So extensive was the range of ARM in ancient geography that, to distinguish its divisions, a qualifying name was generally appended to it: thus, *Sedeh*-ARM, the "field of Aram," *Padan*-ARM, the "plain of Aram," and ARM-*Naharaïm*, "Aram of the two rivers," refer to parts of Mesopotamia: ARM-*Damashk* was a Damascene territory; ARM-*Sobah*, probably Cilicia; ARM-*Maakah*, east of the Jordan; and ARM-*beth-Rekhub*, on which authorities vary. ARMI, an *Aramæan*, is a Syrian in one scriptural text (2 *Kings* v. 20). It is a Mesopotamian in another (*Gen.* xxv. 20).

Aramæan was the speech of the patriarchal Abrahamidæ, when abandoning ARP*h*a-KaSD, or its equivalent AUR-KaSD*îm* (Chaldæan *Orfa*, or *Ur* of the Chaldees), they arrived in the land of Kanaan; where, forgetting their ancestral idiom, they adopted and misnamed *Hebrew* "the language of Kanaan," or *Phœnician.*

Thus, from Arabia Deserta to the confines of Lydia, from Syria, over Mesopotamia, to Armenia, do we meet with infinite *reliquiæ* of *Aram:* without being able, after four or five thousand years of migrations, to mark on the quicksands of *Aramæan* geography any more specific locality for ARM, than SYRIA in its most extended sense.

Hieroglyphical researches do not aid us to a more definite ascription of ARM. In the Vatican Museum, the statue of a priest bears the inscription — "His majesty, King Darius, ever living, ordered me to go to Egypt, while his majesty was in ARMA": supposed to be *Assyria.* Nor, in Persepolitan cuneiform records or in those of Assyria, has any more positive identification of ARM been discovered and published than what may exist in *Arm'ina*, *Arama*, &c., considered to be Armenia — a country in whose name ARM is also preserved.

The writer of Xth Genesis may or may not have had more precise views upon ARM; which he set down with its parallels, *Assyria*, *Orfa*, and *Lydia*, on his invaluable chart, and then proceeded to tabulate those tribes of the Semitic stock that looked back upon the *land* of ARM as their birthplace.[635]

"And the affiliations of ARM."

51. עוץ — *â*UT*s* — 'Uz.'

In *Gen.* x. 23, the *four* names after ARM are called BeNI-ARM; *i. e.*, "sons of

Aram"; but, in 1 *Chron.* i. 17, the same four are catalogued as BeNI-S*h*eM; that is, "sons of *Shem.*"

Hence one of two conclusions is submitted to hagiography. Either the writer of Chronicles follows a different genealogical list from that of Xth Genesis — in which case we are at a loss to which document to ascribe "plenary inspiration"—or (as we maintain with every Orientalist) the word BeNI (sons) does not mean, whether in the former or in the latter text, the *bona fide* offspring of a *man* called Aram, or of a *man* called Shem; but simply a general *affiliation;* such as in English we comprehend by Wilkin-*son;* or by *Fitz*-Gerald, *Mc*-Donald, *O'*-Brien, *Ap*-Shenkyn, &c.

ăUT*s*, first of the four, cannot well have been Shem's *son* and *grandson* at one and the same time; unless it be claimed that Shem wedded his own daughter: an escape not provided for in either text; and if it were, what becomes of Aram's paternity? Again, an imaginary human being called S*h*eM could not physically have been progenitor of a *country* called Aram. Common sense, however, based upon the spirit of familiar Oriental personifications, finds no contradiction between the authors of Xth Genesis and of 1 Chronicles; to whom ăUT*s* and his three figurative brethren, as BeNI, "affiliations," were colonies or emigrants from an especial *land* termed ARaM; itself classified generically among countries occupied by *Shemitish* families.

This example, we presume, suffices to show the absurdity of seeing *human* individuals where the writer of Xth Genesis catalogued naught but *countries, cities,* and *tribes*, after the symbolical names "Shem, Ham, and Japheth." — But, our difficulties end not here.

Genesis X.	*Genesis* XXII.
V. 23—And *sons* of ARaM, ăUT*s*, and K*h*UL, and GT*t*R, and MaS*h*.	*V.* 20 — Milcah has also given *sons* to Nahor thy brother.
	" 21 — ăUT*s* his first born, and BUZ his brother, and KMUAL, *Father* of ARaM.
A *third* ăUT*s* occurs among the descendants of Esau (*Gen.* xxxvi. 28).	" 22 — And KaSD—(i. e. *Chaldœa*) &c.

With three distinct personifications (above exhibited), each called ăUT*s*, it is next to impossible for a commentator to avoid equivoques; and the country, or tribe, of one ăUT*s* may be erroneously assigned to either of the two others; even without supposing mistakes in the two later genealogical lists; which discrepancies, however, do not otherwise concern us. Xth Genesis, in every instance, has stood the test of critical geography heretofore; and errors in this case are ours, not its venerable compiler's.

Nevertheless, in the second list (*Gen.* xxii.), ăUT*s* becomes the uncle of ARAM; whereas in Xth Genesis he is the latter's son: while KaSD, *Chesed,* (singular of KaSDIM, *Chaldœans,*) unmentioned by the former author, figures, in the latter's list, among the descendants of Nahor, Abraham's brother.

It is to the *land,* called ăUT*s* in Xth Genesis, that Job's residence is generally assigned, owing to its proximity to Chaldæa; wherefore the latter passage indicates a *country,* rather than a tribe — but in no case a *man.*

These triple chances of error, above noticed, compel archæology to be extremely wary in deciding to which of numerous Arabian resemblances of name we are to attribute the ăUT*s* of Xth Genesis—or really "*land* of ăUT*s*." Bochart ingeniously guessed the *Æsitœ, Ausitis, Ausite,* of Ptolemy, in the Syrian desert towards the Euphrates; where the Idumæan Arabs *Beni-Tamîn* have dwelt; to whom Jeremiah exclaims — "Rejoice thee, daughter of Edom, who livest in the land of ăUTs." Lenormant follows Michælis in selecting Damascus.

In Arab tradition, *Owz* was the parent of the lost Addite tribes*;* and, assuming this wild legend to be historical, by dint of mistranslations Forster has raised a fabric of delusion exceeded only in extravagance by the same enthusiastic divine's *Sinaic inscrip-*

tions! It is in the ill-advised *Appendix* to his excellent Geography, entitled "Hadramûtic Inscriptions," that this erudite Orientalist lost his balance when supposing that, in these very modern *Himyarite* petroglyphs, he found himself "conversing, as it were, with the immediate descendants of Shem and Noah, not through the doubtful medium of ancient history, or the dim light of Oriental tradition, but in their own records of their own annals, 'graven with an iron pen, and lead, in the rock for ever!'" He translates the second line of Wellsted's short inscription as follows: "*Aws* assailed the Beni-Ac, and hunted [them] down, and covered their faces with blackness."

Happy, indeed, though not perhaps to the pious extent of the Rev. Mr. Forster, should we be to recognize *â*UT*s* in these inscriptions; but some trifling obstacles intervene. Suppose, for instance, that the Hadramautic inscription (No. 4), read into *Arabic*, should say nothing of the kind? *Ex. gr.*, that which Forster translates "*Aws* assailed the Beni-Ac," &c., should be, according to Hunt, "the effeminate youths are adorned and perfume their garments and strut proudly"! And suppose, that the language in which these inscriptions of Hisn Ghoràb are written, being the old Ehkèelee or Cushite tongue, does not admit of their being *transcribed* directly into Arabic idioms at all! Fresnel, the Himyarite discoverer "par excellence," gives the same inscription (No. 4), in Arabic *letters*, but has ventured no *translation*. These suppositions Forster, so far as we can learn, has never taken notice of; but goes on translating anything and everything into an Arabic "sui generis," with the same serene composure that Father Kircher, two centuries ago, read off at sight (!) those identical *Sinaic inscriptions* on which Forster has latterly exercised his orthodoxy without mentioning the labors of his Herculean prototype.

*â*UT*s*, under these circumstances, remains on our hands. Probabilities favor the *Æsitæ*, *Ausitis*, of Ptolemy the geographer; and Job's "*land* of *â*UT*s*," on the Arabian frontier of Chaldæa, seems to answer best to the *Aramæan* analogies of Xth Genesis. *â*UT*s*, we infer, was a *tribe*.[636]

52. חול — K*h*UL — 'Hul.'

We enliven the reader with orthodox lexicography as we proceed—"Hul, *pain, infirmity, bringing forth children, sand,* or *expectation!*"

Most authorities abandon K*h*UL in despair: but Grotius indicated that a Cœlo-Syrian city called *Chollæ* by Ptolemy might represent K*h*UL; and Bochart noticed the frequency of this word in the Armenian localities of *Cholua*, *Choluata*, *Cholimma*, and *Cholobetene;* which last might be an Hellenic corruption of K*h*UL-*Beth*, "house of K*h*UL." Recent researches favor the adoption of the "land of *Huleh*," in which is the Lake *Huleh*, at the north of Palestine.[637]

53. גתר — GT*t*R — 'Gether.'

Koranic tradition execrates the memory of "*Thamoud*, son of Gather, son of the *Aram*," among ante-historical tribes distinguished for their idolatry: but nothing can exceed the vagueness of these legends.

Gadara, the metropolis of the Peræa, east of the Jordan, and one of the cities of Decapolis, has been assumed to represent GT*t*R. Here the well-known miracle of the "swine" is said to have been performed. There are many other places whose names, with the slightest modifications, answer equally well: among them, *Katara*, a town and district placed by Ptolemy on the Persian Gulf, sufficiently important to have become the bishopric of *Gadara*.

Gaddir, in Kanaanitish dialects (according to Pliny and Solinus, also in the "Punica lingua") meaning a hedge, limit, boundary, or "a place walled-round," renders the confusion still more perplexing; for in countries traversed by Phœnician caravans, and occupied by their factors, any form of GT*t*R is as likely to have signified *frontier* or *station*, as to be derived from the *tribe* called GT*t*R in Xth Genesis.[638]

54. מש — MS — 'MASH.'

Besides the discrepancy, above removed, between Xth Genesis and the parallel in 1 *Chronicles* (i. 17), in regard to the affiliations of these four names from Shem, or from Aram; here is another, that cannot be explained save through an error of some copyist. Who can really tell whether we should transpose MSK*h* into Xth Genesis, or MS into 1 Chronicles? [*Supra*, p. 473.] Two reasons, however, seem to justify the accuracy of the former text: one that a MSK is already mentioned among the "sons of Japheth" (*ver.* 2); and therefore the repetition of a similar name amid the *Shemites* is improbable: the other that the chart of Xth Genesis is the "editio princeps," of older and more standard authority than the books called *Chronicles.*

The *Macæ*, on the peninsula of the Persian Gulf whereon now stands the derivative city of *Muscat* — the *Masæi* Arabs in Mesopotamia; the *Masani* near the Euphrates; and the *Massonitæ* of Yemen; might entice inquiries: but, we think their habitats somewhat distant from the localities where *Aramæan* tribes appear to group; especially as MSA, *Massa*, descended from Ishmael (*Gen.* xxv. 14), may well assert its right to the latter lineage.

We cannot amend the old view of Bochart and of Grotius, that this Aramæan tribe survives about Mt. *Masius;* along Xenophon's river *Masca;* in the *Masieni* of Stephanus, and perhaps the *Moscheni* of Pliny; all of which point to Upper Mesopotamia as the camping-ground of MaS*h*.[639]

"And ARP*ha*-KaSD engendered SLK*h*, and SLK*h* engendered *â*EBR" (*Gen.* x. 24).

55. שלח —SLK*h* — 'SALAH.'

Orfa in Diàrbekìr has been already demonstrated to be the fountain-source *Arpha-Kasd*, "Chaldæan Urfa," and no other than the true AUR-KaSDIM, "Ur of the Chaldees;" whence flow the earliest traditions of the Abrahamidæ.

*â*EBR, the *yonderer*, third in descent, seems to show either that a displacement had taken place before the *name* itself could well have been assumed; or that the appellative "yonderer" is an *ex post facto* attribution—the consequence of a migration that had previously taken effect.

Between these two names, *Orfa* as a fixed geographical point, and *Êber* "he who has gone *beyond*," stands SLK*h;* transcribed *Salah* in king James's version: perhaps in this instance with more propriety than according to the vulgar Masoretic *Shelah;* which is suggested as the marginal reading.

Sela of Ammianus Marcellinus, or *Sele* of Ptolemy, a city in Susiana, has received the concurrence of many commentators. Others consider SLK*h* unknown. If Volney's suggestion of the *city* and *territory* called *Salacha* by Ptolemy be not the most probable halting-place of the EBERi when they had left Chaldæan *Orfa*, the ignorance of every body consoles us for ours.[640]

56. עבר — ÊBR, or rather êiBR — 'HEBER.'

[The impossibility of transcribing the letter *Gnain* of the Hebrews, *âin* of the Arabs, into any European alphabet, has been noticed by me long ago. As a general principle, I follow the rules of Lane in these substitutions; but unless a European *hears* the sound of *âin* orientally articulated, his imagination can realize its phonetism as little as his adult voice can enunciate it. — G. R. G.]

Etymologically, ÊBR signifies "one of the other side," or "the *yonder-land;*" whilst ÊBRI, a "*yonderer*," or "a man from the other side," has precisely the same radical as the *Greek* Υπερ, Latinized into *Iber* (*Iberes*, Iberian); equivalent to *trans*, *ultra*, &c.

"HEBER (עבר, *one of the other side;* Sept. Ἔβερ and Ἑβερ), son of Salah, who became the father of Peleg at the age of 34 years, and died at the age of 464 (*Gen.* x. 24; xi. 14; 1 *Chron.* i. 25). His name occurs in the genealogy of Christ (*Luke* iii. 35). There is nothing to constitute HEBER *an historical personage;* but there is a degree of interest connected with him from the *notion*, which the Jews themselves entertain, that the name of Hebrews applied to them, was derived from this *alleged* ancestor of Abraham. No *historical* ground appears why this name should be derived from him rather than from any other personage that occurs in the catalogue of Shem's descendants; but there are so much stronger objections to every other *hypothesis*, that this perhaps is still the most probable of any which have yet been started."

If the authors of this volume had written the above scientific *exposé*, it would have been seized upon as another instance of "skeptical views" (save the mark!); but the initials "J. N." appended to the above article in Kitto are those of a profound Germano-Hebraist, the Rev. Dr. JOHN NICHOLSON of Oxford.

Archæologically, the name ÊBR marks a displacement, or dislocation, that must have occurred before such name could have been given or assumed.

Of such dislocation the earliest notice is the march of the *Abrahamidæ* from *Orfa-Chaldee* to Harran (probably *Carræ*), in Mesopotamia, and thence to Kanaan: where the Kanaanites gave to ABRAHAM, probably, the designation of ÊBR, as "he who comes from *yonder-land*,"—*transfluvianus*, or "from the other side" of the Euphrates—whence HEBREW, ÊBRI, became the cognomen of this family. Indeed, it is remarked that the title ÊBRIM, *yonderers*, Hebrews, was given to the Abrahamidæ by foreign nations. They called themselves *Israelites* after Jacob's wrestling match at *Phenuel;* and did not adopt that of "Hebrews" until many centuries later.

We are dealing, therefore, in Xth Genesis—a document compiled at least five, if not ten, hundred years subsequently to the arrival of the earliest Abrahamidæ in Kanaan—with a *people* upon whom the name ÊBR had been imposed, "nolens volens" on their own part. Had the chorographer of Xth Genesis been a man of Abrahamic pedigree, he would probably have designated his own nation by its most honored title, "Israelite;" but, far from that, a *Chaldæan* composing his ethnic map in Chaldæa, naturally gives to ÊBR its radical sense of "yonderer;" either because the Palestinic Abrahamidæ were so termed by surrounding populations, or because they were then, to him, as ÊBeR-*îm*, "people who had gone *beyond*" the Euphrates. That there is no "prefiguration" (*i. e.*, "cart before the horse") in Xth Genesis, has been proven by the names *Sidonian*, *Hamathian*, &c.; folks who could not well have been citizens of those cities, *Sidon*, *Hamath*, &c., until after the houses had been built: and inasmuch as these citizens are catalogued in the same document with ÊBR, the antiquity of the latter's registration is brought down to historical times; long ages after that emigration from Chaldæan *Orfa* into Palestine through which the foreign application of "yonderers," given to Abraham's descendants, had originated.

"Fama crescit eundo;" and Oriental mythos—after Judaism, a little before the Christian era, had penetrated into Arabia; and still more forcibly after Islamism, in the seventh century, had imbued pagan Arabians with extraneous traditions—assimilated ÊBER, now metamorphosed into a *man* and a *patriarch*, to the Arab prophet HOOD: who, in native Arabian tradition, plays a part somewhat like that which MOSES does in Jewish; being their earliest metahistorical Reformer. Who this HOOD probably is, the profound investigations of Fresnel clearly indicate:—

DhU-NUAS, or *Zhu-Nawâz*, is the subject. "CAIRE, 12 *Mars*, 1845.

"The Greeks knew that Bacchus was Arabian, and have sought for the etymology of the name Διόνυσος, *Dionysus*, after their own fashion: they made of it 'the god of Nysa,' Nysa being a city of Arabia, or, as says Herodotus, of Ethiopia, where Bacchus was raised by the Nymphs. . . . . About forty miles to the east of *Zhafâr*, the most ancient of all their (Arabian) metropoles, and the site of the oldest Arabian civilization, is a mountain that Edrisi calls *Loûs*, and that the inhabitants of Mahrah call *Noûs*. . . . . This mountain of *Noûs*, near which is found, not the *Kabr Hoûd*, or tomb of Heber (ÊBR), but the *Kabr Sâleh* (that is to say, the tomb of the FATHER OF HOUD, according to Arab notions) is the point where I place the birth of Bacchus; in other words, the point of departure for those civilizing conquests of which the Arabs have preserved the remembrance. These conquests are not the act of a single man, or if one might so express oneself, 'of a single Bacchus.' *Dhou-Ons* or *Dhou-Noûs* (in the oblique case, *Dhi-Ons* or *Dhi-Noûs*), *Dhou 'l Karneyn* (the man with the two horns), *Afrikis* (the god-father of Africa), *Lokman*, &c., &c., are to me so many personifications of Bacchus; and if you must absolutely have a religious idea pre-existent to Arab kings, a Bacchus outside of Yemenite dynasties, I should venture to tell you to seek for Bacchus in the tomb *Saleh* (SLK*h*) [*Gen.* x. 24] under the *Djabal-Noûs*. Bacchus then will be the father of the patriarch Héber (ÊBR), of the *Abrahamidæ* and of the *Joktanidæ*.

"Will you mount up still higher? Διόνυσος is (Hebraicè) DU-ANOS*h*, *Dhou-Enosh* (the god of the vulgar), or lastly, Enos himself, ENOS, grandson of Adam.

"Agréez, monsieur, &c.,

"F. FRESNEL."

"À M. MOHL, *Journal Asiatique, Paris*."

Our researches do not require our accompanying M. Mohl into antediluvian regions. We are satisfied when shown that EBR in Xth Genesis is the natural appellation of a *tribe;* better known to modern science as source of the *Abrahamidæ*.[641]

"And unto ÊBR were born two sons."

57. פלג — PLG — 'PELEG.'

"And the name of one (was) PLG," explains the author of Xth Genesis, "because in his day the earth was *divided;*" literally, "PLG*ed*," *split*. In modern Arabic even, the identical word FLG means a "split," and "to split;" which again induces a smile at mystifications concerning a "sacred tongue," every third word of which exists in the Arabic *dàrig*, vernacular; every second in the *Nahwee*, or Koranic idiom; every one, in some form or other, by easily recognizable changes of consonant or vowel, in the *Qamoos* — the "Ocean" lexicon of Arabian literature. Any well-educated *Arab*, we fear not to maintain, who could first peruse in some European tongue a few philosophical works on *Hebrew* literature and comparative philology, would master the 5642 words counted (by Leusden) in this exaggerated Kananitish language, after devoting one day to its alphabet, in about a week. This doctrine no Shemitish Orientalist (no Lanci, no De Saulcy, no Quatremère, no Fresnel, no Rawlinson), will deny. "We have remarked in it," comments De Saulcy upon the *Toison d'Or*, a new Phœnician work by the Abbé Bourgade, "a passage the justness of which we ought to applaud; because, in order to write it, one must not have been scared by the scientific anathemas of certain too-exclusive savants. Here is this passage — 'It is therefore rational to make use of Hebrew, and of the other Aramæan idioms to explain the Punic: one may also use Arabic, another ramification of the Semitic family; sometimes even it is indispensable to have recourse to this language, *almost all Hebrew words being found within Arabic*, either without modification, or with very slight modifications, sometimes in the form, at others in the sense, but not *vice-versâ;* the language of the Koràn being incontestably richer than that of the Bible.'"

On the historical monstrosities erected upon this verse of Scripture, it is not for us to dwell. *Pelagos*, the *Pelasgi*, and *Pelargos;* the "Sea," the "fossil people" as Niebuhr beautifully calls them, or the "Stork," do not concern an alien Semitic bisyllable, whose simplest essence is Anglicè a "split." We are loath to reject the Bochartian assimilation of *Phalga*, a town on the Euphrates, near *Charræ*; which town, some say, is *Haran*, built by Abraham's brother, after his own death at *Chaldæan-Orfa:* just in the same way that Moses posthumously describes his own ever-unknown burial-place, his wake of thirty days, &c. (*Deut.* xxxiv. 5-12): but we venture to submit the following doubts: —

1st. If by PLG, or P*h*LG, the editor of Xth Genesis meant what, in every instance but the mythological NMRD, is herein proved to have been a *country*, a *people*, or a *city*, then the parenthetical passage, "because in his day the earth was *split*," may be a gloss by some later hand,—rationally suggested through paronomasia of the triliteral PLG "split," combined with impressions formed upon other documents by such interpolator — the whole having been subsequently recast by the Esdraic school from which we inherit (every possible chance of intervening error and perversion inclusive) this *verse* of Xth Genesis.

2nd. If it were shown that a gloss must be as unlikely as it is dangerous to the claims of plenary inspiration; then, before we can perceive a necessity for supposing that the chorographer of Xth Genesis here alludes to the "Dispersion of mankind," we would inquire whether the words "(was) *split* the earth" do not refer to some local and terrestrial catastrophe—an earthquake, for instance—that, occurring simultaneously, may have become traditionally coupled with a PLG*ian* migration. A similar catastrophe, introduced into Manetho's text in a similar manner, occurred under Bochus, 1st King of the second Egyptian dynasty, when "a huge chasm" was made at Bubastis.

3rd, and lastly—If none of the above possibilities be satisfactory, then, falling back upon the indubitable orthodoxy of the Parisian Professor of Egyptian Archæology, we should perceive in the words "because in his day the earth (was) *split*," merely a partition of territory between the PLG*ian* and the *Joktanide* affiliations of ÊBR the "yonderer."—"Of the two sons of this Patriarch, the first, *Phaleg* (holds Lenormant), indicating that part of the nation that continued to wander in Upper Mesopotamia; *Iectan*, the second, shows us on the contrary the other portion of the same people which first set itself on a march towards the south." The verb "divide" occurs three times in the English version of Xth Genesis (5, 25, 32). It need scarcely be mentioned that, in the Hebrew, the play upon the word PLG "to split" presents itself only in verse 25. The other two passages use a distinct verb, NP*h*RDU, "they *dispersed*."

"Hypotheses non fingo"—and as everything beyond the name of PLG, "split," is an hypothesis, we leave hagiography to "split hairs" on the question; merely insisting here that PLG has no relation whatever to a "Dispersion of mankind."[642]

58. יקטן — IKTN — 'Joktan.'

The compiler of Xth Genesis closed the ancestral line of the *Abrahamidæ*, abruptly, with PeLeG, a "split." Yet to the pedigree of IKTN he devotes particular attention; for, besides cataloguing thirteen of the latter's descendants, he adds, "all these are sons of IKTN": and then fixes their dwelling-places.

Why this difference? Were his partialities Arabian? Did he know all about Arab migrations, and nothing of those of the *Abrahamidæ?* Had the writer been a "Hebrew of the Hebrews," he would scarcely have blocked the "royal line of David" at PLG, "a split"; and thereby left to another hand, in another document (*Gen.* xi 18–26), at a later age, the task of linking Abraham's genealogy to his own ethnic map of nations and places. Here again, a *foreigner* to Judaism and Jews, our conjectural *Chaldæan* chorographer, "laisse percer le bout d'oreille." Such alien would not have greatly concerned himself with the *Abrahamidæ*, a petty tribe that had wandered off to Kanaan; and the writer of Xth Genesis did not: such alien would have taken much interest in the proceedings of the ever restless *Joktanidæ*, always harrying the Mesopotamian frontier; and the writer of Xth Genesis did.

IoKTaN, *Joktan*, *Yoktan*, or correctly *Qahtân*, the *Beni-Kahtân*—most ancient and renowned of all Semitish intruders upon the domains of Cushite-*Himyâr* — need no panegyrist. They have ground their lance-heads upon every pebble "from *Havilah* to *Shur*, that is before Egypt, as thou goest towards Assyria." Their woollen tents are pitched from "*Sephar*, a mount of the east," at the south-western extremity of Arabia, even unto the declivities of Persian Uplands. Their *Nedjdee* horses still chase the wild ass, "gour," over the wildest tracts of Arabia's *hâgar*, "stone," desert: their *dromedaries* are precious at Cairo, Mecca, Aleppo, Bagdàd, and Ispahàn. From them issued Mohammed; whose *Koràn* is the monotheistic code of religious and moral law to above one hundred millions of mankind in Europe, Asia, Africa, and India's islands: their tongue, "the pure *Korèysh*," for twelve centuries has been the envied attainment of poets, historians, and philosophers, of their own exalted race, and of its Arabian contemporaries during consecutive generations.

By "*Beni-Qahtàn*," sons of IKTN, we have hitherto implied the *Joktanides* in general; but the great tribe in Arabia now calling itself *Beni-Kahtàn* claims the direct lineage of this son of ÊBR. They are traced in the *Katanitæ*, *Kithebanitæ*, and *Kottabani*, of Ptolemy; the *Katabeni* of Dionysius; back to the *Cattabanes*, *Kattabanum*, of Eratosthenes in the third century B. C.: while their existence in Arabia is attested by the compiler of Xth Genesis many generations anterior to the age of the Cyrenian geographer.

With the admirable tabulation of the "Settlements of Joktan," and the maps that Forster has appended to his geography, the reader can verify for himself the accuracy of the following schedule of IoKTaN's affiliations.[643]

"And IoKTaN engendered"

59. אלמודד — ALMUDD — 'Almodad.'

The *Allumaeotæ*, *Almodæei*, Ἀλλουμαιῶται, of Ptolemy, a people of central Arabia Felix, represent ALMUDaD by general consent.[644]

60. שלף — SLP — 'Sheleph.'

Ptolemy's *Salapeni*, *Salupeni*, the Greek transposition of "*Beni*-SeLePh," sons of Sheleph, are equally certain: now represented by the tribe of *Metèyr*?[645]

61. חצרמות — K*h*T*s*RMUT*t* — 'Hazarmaveth.'

Who, unacquainted with corrupt Chaldee vocalizations, foisted in the sixth century after Christ upon the old Hebrew Text (under the name *Masoretic points*), would see that the writer of Xth Genesis here wrote *Khàdramaut*? the very name which the Arabs still give to their province of *Hadramàut*, or *Khàzramôt*.

This name, "in the Septuagint version, is written *Sarmoth*, the first syllable being dropped; by St. Jerome (a well-versed Orientalist), in the Vulgate, written *Asarmoth*; the article being incorporated with the name, or the aspirate omitted, conformably with the dialect of the Nabathæans; by Pliny, *Atramitæ*, and *Chatramotitæ*; and by Ptolemy, *Adramitæ*, *Chathramitæ*, and *Chatramotitæ* or *Cathramonitæ*": no less than by Strabo. "So Hadramaut," comments Forster upon Bochart, "is modulated into Hazarmoveth, merely by the use of the diacritic points, . . . an artifice," says this learned and reverend Orientalist, "allowedly, of recent and rabbinical invention."

The tribe and territory of Hadramaut being fully identified in Xth Genesis; the only salient point of interest connected with its later history, is the mission — we follow Mr. Plate — of a "priest of Nagrane, the capital of Christian Hadhramaût," to *China*, in the seventh century of our era; whose successful voyage is attested by the bilinguar stone, in Chinese and Syriac (dated A. D. 782), discovered at *Si-Gan-Fú* in 1625; which inscription is reputed to be genuine.[646]

62. ירח — IRK*h* — 'Jerah.'

This tribe of Arabia, under the Arabic title of *Yàreb-ben-Qahtàn*, "*Yàreb* son of Joktan;" or of *Aboo-l'-Yemèen*, "father of Yemen;" was pointed out by Golius, upon Arab authority, as "Pater populorum Arabiæ Felicis; primus Arabicæ linguæ auctor." Forster, continuing his emendations of Bochart, states that IRK*h* "in the LXX, is written Ἰαρὰχ (*Jarach*); by St. Jerome, *Iare*; by the modern Arabs, *Jerhä* or *Serhä* (pronounced Jer*chä*, Ser*chä*); and also, as shall presently be shown, *Sherah* or *Sheradie*,

Serene or *Zohran:*" — a name thrice registered by Ptolemy, "in his Insula *Jerachæorum*, on the Arabian Gulf, S. of Djedda, and in his Vicus *Jerachæorum*, on the Lar or Zar river, in the vicinity of the Persian Gulf; a town and an island bearing in common this proper name, although separated from each other by a space of 15°, or more than one thousand geographical miles!"

It was Bochart's acuity, as our author honestly remarks, that restored Ptolemy's νῆσος Ἱεράχων, previously rendered *insula accipitrum*, or "the Isle of Hawks," to its patriarchal origin; *insula Jerachæorum, i. e.*, "the island of the Beni Jerah." But this father of European commentators on Xth Genesis did more. He showed that the *Alilæi* of Agatharcides were identical, not merely with the tribe *Beni-Hilal* of the Nubian geographer; but also with Ptolemy's "insula Ierakiorum;" for the reason that *Hilal* means "moon" in Arabic, just as *Ierakh* does in Hebrew.

Most successfully does Forster exhibit the settlements of IeRaK*h* within "a vast triangle, formed by the mouth of the Zar river, on the Persian Gulf; the town of Djar (the *Zaaram reg.* of Ptolemy) on the coast of the Hedjàz, twenty English miles south of Yembo; and the district of Beni Jerah (part of the ancient Katabania), or the southwestern angle of the peninsula, terminating at the Straits of Bab-el-Mandeb;" and the probability that the great tribe, known as the *Minæi* in classical geography, belonged to IeRaK*h-ian* affiliations, is also by him perspicuously elucidated.[647]

63. הדורם — HDURM — 'Hadoram.'

By Fresnel this name is considered to be the same as *Djourhoum;* of whom Arabian tradition reckons an elder branch, the old *Jorhamites*, among extinct, and a younger, the Koranic *Jorhamites*, among existing families. *Jorham* is the "Arabum *Hejazensium* pater" of Pococke; and Bochart associated the name with the *Drimati* of Pliny, and with Cape *Corodamon;* which last, by the facile transposition of D for R, is Cape *Hadoramus*, or of HDURM. Volney accepts *Adrama* for their natural representative; confirmed by Forster in *Hadrama*. and thus, carried onwards through the classical *Chatramis*, *Dacharæmoizæ* of Ptolemy, to the *Dora* and *Dharræ* of Pliny; they are perpetuated in the modern town and tribe of *Dahra:* at the same time that *Ras-el-*Had now preserves one abbreviation of the name, and *Bunder-*Doram another — on the very promontory "Hadoramum" at the mouth of the Persian Gulf.[648]

64. אוזל — AUZL — 'Uzal.'

The native Jews of *Sanaa*, capital of Yemen, have abundantly borne witness that AUZaL was its ancient Arabian appellative, as, to this day, it is among themselves. The "*Javan* from AUZaL" of Ezekiel (xxvii. 19,) must be, therefore, as Volney and Forster unite in indicating, not Grecian *Ionia*, but a town in Yemen, now called *Deifân. Ocelis* of Ptolemy, *Ocila* of Pliny, recognizable in the modern *Cella;* together with *Ausara*, a town of the *Gebanitæ* or Yemenites; are relics of AUZaL long patent through the scholarship of Bochart.[649]

65. דקלה — DKLH — 'Diklah.'

In the *Dulkhelitæ* of Himyar, and the tribe *Dhu-'l-Kalaah* of Yemen, Orientalists perceive this affiliation of *Joktan;* that, perhaps, has carried along with it some remembrance of an ante-historical sojourn on the *Dikle*, or *Tigris:* if, as Bochart suggested, its name have no affinity to *nukhl*, a "palm tree."[650]

66. עובל — *â*UBL — 'Obal.'

Among nine names of existing Arab tribes identified by Fresnel with biblical appellatives (after the rejection of more than forty of the latter as irrecognizable) *Abil* is one. But, it seems more than probable that a branch of these *Joktanidæ* crossed the narrow straits of Bàb-el-Mandeb into Abyssinia, "Arabia Trogloditica;" and gave their patronymic âUBaL, to the *Aualites* Sinus, *Abalites* emporium, *Avalitæ*, and perhaps *Adoulitæ* (D for B), on the African coast of the Red Sea and Indian Ocean, recorded in classical geography. Volney sees them in Edreesee's *Hobal;* or in El-Hamza's *Obil*, that, with nine other tribes, succumbed, about 230 years A. C., in wars with ARDOUAN, Radowàn, king of Persia, better known as the Sassanian ARDISHEER-*Babegàn*.[651]

67. אבימאל — ABIMAL — 'ABIMAEL.'

ABI-MAL, in Arabic, is "*Father* of MAL;" the meaning of which is also "possession of *property*;" in allusion, perhaps, to the wealth accruing to this tribe from their occupancy of the myrrh, incense, balsam, and spice districts of Yemen.

They are the *Mali* of Theophrastus, the *Malichæ* of Ptolemy; surviving in the town *Malai*, or *el-Kheyf;* not far from the tomb of Mohammed at *Medeéneh*.[652]

68. שבא — SBA — 'SHEBA.'

The perplexities accruing to ethnic geography from the presence of *four* SBAs in the book of Genesis, three of them in the Xth chapter, have been set forth in our analysis of the *Hamitic Saba* of Himyar [*ubi supra*, p. 498]: nor is it possible to escape from confounding this *Joktanide's* properties with some of those that appertain to the former's inheritance.

Nothing daunted, Forster says, "the Joktanite Sheba gave its origin, and his own name, to the primeval and renowned kingdom of the Sabæans of Yemen." Perhaps he did. Possibly the *Cushite* SaBA may have done so before him. "Quien sabe?" Nevertheless, "the concurrent testimonies of Eratosthenes, Dionysius Periegetes, Priscian, Festus Avienus, and others of the ancients," collected by Bochart, place the *Sabæans* between the Minæi and the Katabeni, at *Sàba* and *Màreb:* whilst the notice by *Aboo'l-Feda* that "Mareb was inhabited by the *Beni-Kahtan*," or *Joktanidæ*, really favors our author's somewhat peremptory identification of this SBA.[653]

69. אופר — AUPR — 'OPHIR.'

A volume would not suffice to display the aberrations of intelligence printed on this name! Some are exposed in Kitto and in Anthon.

Munk very properly cuts short discussion by reminding those who see *Ophir* at Madagascar, Malacca, or Peru, that the writer of Xth Genesis places AUPR in the midst of the Arabian *Joktanidæ*: which doctrine Volney had previously sustained, and supported by vigorous researches that identified it with the ruined site of *Ophor* on the Persian Gulf.

Bochart and Michælis held the same judicious views; and Forster has left nothing more to be desired; by proving, once for all, that *Ofor*, a town and district of Oman, is the true AUP*h*iR of the Old Testament — that Pliny's "littus Hammæum ubi auri metalli" is the true *Gold Coast* of Solomon's expeditions — and that the whole of them are comprehended within the domains of the *Joktanidæ*.[654]

70. חוילה — K*h*UILH — 'HAVILAH.'

Our prefatory remarks on ASUR, and its ante-diluvian existence, apply with equal force to that "land of *Havilah* where (there is) gold," which, an universal Flood notwithstanding, now reappears exactly where it stood, antefluvially, on the *gold-coast* of Arabia.

We are not free, either, from chances of error in attributing to the present K*h*UILH (the *Joktanide* affiliation of Shem) some possessions that may have belonged to his namesake, K*h*UILH the *Cushite*.

However, the Nubian geographer indicated to Bochart (father of genesiacal geographers) the country of *Chaulan* in Arabia Felix; and Forster, with propriety selects the province of *Khaul*, south-east of Sanaa (*Uzal*); site of Pliny's tribe of *Cagulatæ;* now inhabited by the *Beni*-KHOLÀN. Its topography, moreover, in the immediate proximity of Omanite *gold* regions, satisfies the mineralogical exigenda of the prædiluvian "land of HAVILAH" demanded by the letter of *Gen.* ii. 11, 12; and insisted upon, as a preliminary step towards precision, by Volney.[655]

71. יובב — IUBB — 'JOBAB.'

The *Iobaretai* of Ptolemy, through the ready change of the Greek *b* into the Latin *r*, by a mistake of copyists, revealed themselves to Bochart as the *Jobabitæ* of Xth Genesis. But, "the flexible genius of the Arabic idiom" suffices to explain such difference of pronunciation; and Forster triumphantly points out "the Iobaritæ of Ptolemy, in *Beni*-JUBBAR, the actual name of a tribe or district, in the country of the Beni-Kahtan, south-east of Beishe, or Baisath Joktan, in the direction of Mâreb; and the original, or Scriptural form of this name, in *Beni*-JOBUB or JOBAB, the existing denomination of a tribe and district situated in the ancient Katabania, half-way between Sanaa and Zebid" — *Katabania* being the Greek inversion of *Beni-Qahtàn*, the old JOKTANIDÆ. "All these are sons of JOKTAN;" wrote the venerable compiler of this precious ethnic chart, Xth Genesis, above 2500 years ago.[656]

We have shown that every name (but NIMROD's, which is mythological) in the Xth chapter of Genesis, excepting those of NOAH and "Shem, Ham, and Japheth," is a personification of *countries, nations, tribes*, or *cities*: — that there is not a single "*man*" among the seventy-nine cognomina hitherto examined. [N. B. The number 79 is obtained by adding the 8 *cities*, founded by Nimrod, to the 71 names above enumerated.]

Abundant instances are patent, even in king James's version, where Israel, or Jacob, is put for *all the Jewish community;* and so ASUR, for example, means *Assyria* in such passages as "ASUR shall come as a torrent; ASUR shall arise like a conflagration; JEHOVAH will raise up ASUR against *Moab*, against *Ammon*, against *Judah*, against *Israel*." Now, none will suppose that *Asur, Moab, Ammon*, or *Israel*, are individuals, human beings. It is evident that these are collective names, employed according to the genius of Oriental minds and tongues. And upon whose authority, let us ask, must we modern foreigners offend the spirit of old Oriental writers (apart from common sense itself), in order to find *men* in the seventy-nine ethnico-geographical appellatives of Xth Genesis?

That, in some instances, the name of an ante-historical founder of a nation has been perpetuated by the nation itself, no one denies. Classical history teems with such; *e. g.* HELLAS for the *Hellenes;* DORUS for the Dorians; LYDUS for the Lydians; but they are, in general, about as historical as AFRIKIS of the Arabs; whom the Saracens made the "Father of *Africa*," after they had learned the Latin name of this continent! In most cases, however, the nation or tribe invented a founder; to whom they gave the name of the *country* they happened to occupy: nor does archæology concede to the Hebrews any exemption from this universal law, merely for the sake of conformity to time-honored caprice.

But, if seventy-eight of the seventy-nine names in Xth Genesis are those of *countries, nations, tribes*, or *cities;* such is not the case with four others, catalogued as the parental NuK*h*, NOAH, and his three sons S*h*eM, K*h*aM, and IaP*h*eT*t*.

Our observations on these names limit themselves to guessing, as nearly as we can, what *may* have been meant by the writer of Xth Genesis.

1st. NuK*h* — (Noah), or NUK*h*, in Hebrew lexicons, among its various meanings, signifies *Repose* and also *Cessation*. We place the word "OBSCURITY" beneath it on our Genealogical Tableau. To the chorographer of Xth Genesis this name NK*h* symbolized, probably, a point of time so remote from his own day that he *ceased* to inquire further; and *reposed* from his labors in blissful ignorance, after having comprehended the vanity of human efforts to pierce that primordial gloom. If he did not, we do: and with the less regret, because an expounder (who says he knows all about it) can be met with at every street-corner.

2d. From the *unknown*, then, in the supposed idea of a Chaldæan writer, proceeded three grand divisions of mankind; already distributed, at the age of the compilation of Xth Genesis, each one "after his *tongue*, in their lands, after their nations." It became necessary, for his chorographic and ethnic objects, to classify them. He saw they were apparently divided into *three* cuticular colors; just as the Egyptians before him had perceived the same thing, when they classified *three*, of the four human varieties known to them, by the colors red, yellow, and white.

3d. He gave to them, or adopted through preceding traditions, the three names "S*h*eM K*h*aM and IaP*h*eT*t*"; and called the nations within his horizon of knowledge by these terms, as much for convenience sake, as on account of their several and probable linguistic, physiological, geographical, and traditionary relationship to each other. The meaning which he attached to each of these proper names is utterly unknown; but modern lexicography speculates upon their acceptation as follows: —

A. K*h*aM is the ancient name of Egypt; centre point of the populations which the writer of Xth Genesis classified as BeNI-K*h*aM, "sons of Ham;" and which we call *Hamitic.* In Hebrew, K*h*M means *hot:* but, in Arabic, while HàM has the same acceptation, K*h*àM signifies *dark, swarthy:* perfectly applicable to the peoples that this name embraces in Xth Genesis. The Egyptians designated themselves as the *red* race; wherefore, for Hamitic types, we adopt the red color.

B. S*h*eM, in Hebrew, means *name* "par excellence." It is also supposed to possess the sense of *left* hand, in contrast to *Yemen*, the *right;* but this seems to be an "ex post facto" Arabian commentary. The Egyptians always gave shades of yellow to *Shemitish* races, in accordance with their cuticular color; and we adopt it for our classification.

C. IaP*h*eT*t*. Such rabbinical explanations as "the man of the opening of the tent" belong to the domain of fable.

IAPETUS, son of Cœlus and Terra, was the Titanic progenitor of Greeks in their ante-historical MUTHOI; the "audax genus *Iapeti*" is a symbolical periphrasis for *white* races; and an ancient Greek proverb, του Ιαπετου πρεσβυτερος, "elder than Iapetus," indicates that the sense in which Grecians used it corresponds to our saying "older than Adam." It is not impossible that the writer of Xth Genesis, in his anxiety to discover an ancestor for *white* families, asked some Greek traveller, who replied "Ιαπετος." To ourselves, as anciently to the Egyptians, these families are white.

We conclude in the language of D'Avezac — "Far from admitting that *Genesis* wished to make all the ramifications of the great human family descend from the unique NOAH, we would voluntarily sustain the thesis, that the genesiacal writer only wished to designate the three great branches of *white* races, individualized for us in the three types Greek, Egyptian, and Syriac; whose respective traditions have preserved athwart ages, as an indelible testimony of the veracity of Moses [or, only of that of the unknown *writer* of Xth Genesis], the names of Japheth, of Ham, and of Shem: but, without entering digressionally into a question so vast, let us hasten to say that, to our eyes, the Biblical texts are very disinterested upon any doubts arising from that [doubt] as to the unity or multiplicity of species in the human genus."

Section B.—OBSERVATIONS ON THE ANNEXED GENEALOGICAL TABLEAU OF THE "SONS OF NOAH."

So far as the authors' reading enables them to judge, here, for the first time since Xth Genesis was composed, are tabulated, in a true *genealogical* form, all the ethnic and geographical names contained in that ancient document.

After the foregoing analysis of each name under *Section A.*, the reader requires no prolix remarks to perceive the utility of our Tableau; which, at a glance, exhibits *Father* NuK*h* (Noah), and his three *Sons* — his *Grandsons*, *Great-grandsons*, *Great-great-grandsons*, *Great-great-great-grandsons*, and *Great-great-great-great-grandsons*, according to their natural order. In this manner (the geography of the Hebrew Text being, once for all, defined,) it is to be hoped that science will be relieved from further discussion of main *principles*, whatever may be the light which future Oriental researches cannot fail to shed upon details.

Each *Name* is first displayed in the "square-letter" of the Hebrew Text, without the Masoretic points. Below it, in "Roman" capitals, is placed the conjectural vocalization of our modern, and colloquial, *English* imitation of ancient foreign words. Beneath is put, in "Italics," the spelling of each name as printed in king James's version. This is followed, in "Gothic" letters, with the *geographical* attribution of the several cognomina, conformably to the results attained through our *Section A.* And finally, under every one, in common "Roman" type, is represented the probable *country*, *nation*, *tribe*, *city*, *citizen*, and *personage* historical or mythic, to which the authors' studies ascribe each name.

"*Humanum est errare.*"

[The best parallel I have met with in ancient history of the conversion of symbolical and national names into *personages*, that might be assimilated to the Hebrew map in Genesis Xth, occurs in Tacitus.[657] Speaking of the Germans, he gives one of their antique mythes (which, during his time, was current among them) in explanation of their figurative origins and tripartite distribution into races. "Celebrant carminibus antiquis, quod unum apud illos memoriæ et annalium genus est, TUISCONEM deum, terrâ editum, et filium MANNUM originem gentis conditoresque. MANNO tres filios assignant e quorum nominibus proximi oceano *Ingævones*, medii *Herminones*, cæteris *Istævones* vocantur."

TUISCO is the god *Mars*. MANNUS the Latinized form of our word "Man," in German *Mann*: "*ones*," is the euphonizing suffix to the primitive words *Ingæv*, *Hermin*, *Istæv*.

The learned Zeuss[658] has shown that *Ingæv* is the same as *Yngvi*, "noble;" ancient title of the royal race of Sweden. *Istæv*, also meaning "illustrious," is traced in *Astingi*, royal race of the Visigoths and Vandals: and *Hermin*, in old Gothic *airmun*, meant "the mighty ones."

1. *Hermin*-ones, (in Pliny, *Hermiones*,) comprehended four tribes: the Suevi, Hermudiri, Chatti, and Cherusci. These clans occupied inland Germany.

2. *Ingæv*-ones. These embraced the Cimbri, the Teutones, and the "Chaucorum gentes;" inhabiting west and north-west Germany.
3. *Istæv*-ones — as the *Vindili* of Pliny, included the Burgundiones, Varini, Carini, and Guttones. Their place was north-eastern Germany.

For our purpose of simple illustration, it is not essential to detail the geographical territories assigned to these names; which, mutilated and corrupted by Roman orthography, preserve as little relation to an ancient *German* pronunciation as the Indo-Germanic names of GoMeR, MaGUG, &c., do in our authorized version after passing through Hebrew transcriptions, Septuagint corruptions, and the fabulous vocalizations of Jewish Rabbis of the Masora. What we are driving after becomes evident at once, so soon as we tabulate the *genealogy* of these tribes as we have done that of those in Xth Genesis.

Tuisco
MARS.
|
Mannus
MAN
|

Ingæv.	*Hermin.*	*Istæv.*
"Noble."	"Puissant."	"Illustrious."
North-west Germany.	*Central Germany.*	*North-east Germany.*
Cimbri,	Suevi,	Burgundians,
Teutones,	Hermundiri,	Carini,
Chauci.	Chatti,	Varini,
	Cherusci.	Gothones.

It would be easy to carry this method of illustration, which classifies the mythical, the geographical, and the patronymic personifications of nations in their true historical order, through the traditions of different races all over the world. We content ourselves by indicating to fellow-students the utility of a simple process that has solved many a "vexata quæstio" encountered in our personal researches: especially when studying the Persian genealogies of Firdoosi's *Shah-Nameh;* as we hope to show elsewhere. — G. R. G.]

Section C. — Observations on the accompanying "Map of the World."

1st. The parts in *black* indicate what the writer of Xth Genesis knew not: those *shaded* represent where his knowledge decreases; it being unfair, no less than impossible, to define his information by a sharp line. Other explanations are given on the Map itself.

2d. The great *alteration*, which our results superinduce, is the prolongation of his geographical knowledge (hitherto unsuspected) along the whole of Barbary, between the Mediterranean Sea and the Sahara desert. Former African delusions are curtailed at the First Cataract, *Syene;* southern extremity of the *Egyptians*, MiTsRIM, proper. The compiler of Xth Genesis knew nothing of "Ethiopia" above; nor is any austral land beyond *Egypt* mentioned by a single writer in the Old Testament; because *Chub* (*Ezek.* xxx. 5), GUB, conjectured by Bunsen, after Ewald, to be GNUB, *Nubia*, is an unnecessary effort when we can identify it with the Barbaresque *Cobii* of Ptolemy the geographer [*supra*, p. 515]

MAP OF THE WORLD,

on Mercator's projection, exhibiting the COUNTRIES more or less known to the ancient Writer of 10th GENESIS.

NOTT & GLIDDON'S Types of Mankind, 1853.

3d. The coast of Abyssinia is dotted *red* and *yellow*, because some KUS*hites*, besides the *Joktanide*, *â*UBaL, may have crossed the Red Sea. The latter lent his name to the *Avalites Sinus*, &c., on the African continent.

Section D. — THE XTH CHAPTER OF GENESIS MODERNIZED, IN ITS NOMENCLATURE, TO DISPLAY, POPULARLY AND IN MODERN ENGLISH, THE MEANING OF ITS ANCIENT WRITER.

Verse

1 Now these (are) the T*t*oLDTt-BNI-NuK*h*, (generations of the sons of CESSATION); S*h*eM yellow races, K*h*aM swarthy races, and IaPeT*t* white
2 races: unto them (were) sons after the deluge.* (The) affiliations of IaPeT*t*
white races; — Crimeâ = GoMeR, and Caucasus = MaGUG, and Media
= MeDI, and Ionia = IUN, and Pontus = T*t*uBaL, and Moschia =
3 MeS*h*eK, and Thrace = T*t*IRaS. And (the) affiliations of Crimea =
GoMeR; — Euxine = ASKiNaZ, and Paphlagonia = RIP*h*aT*t*, and Armenia
4 = T*t*oGaRMaH. And (the) affiliations of Ionia = IUN; — Morea = ALISaH,
and Tarsous = TaRSIS, Cypriots = KiT*t*IM, and Rhodians = RoDaNIM.
5 By these were dispersed the settlements of Ha-GOIM the (white barbarian)
hordes in their lands; every one after his tongue, after their families, in their
6 nations. And (the) affiliations of K*h*aM swarthy races; Dark Arabia† =
KUS*h*, and Egyptians = MiT*s*RIM, and Barbary = P*h*UT*h*, and Canaan =
7 KN*â*AN. And (the) affiliations of Dark Arabia = KUS*h*; — Asabia = SeBA,
and Beni-Khàled = K*h*aUILaH, and Saphtha-metròpolis = SaBT*t*aH, and
Rumss = R*â*AMaH, and Sabatïca-regio = SaBT*t*aKA: and (the) affiliations
8 of Rumss = R*â*AMaH; Marsuaba = S*h*eBA, and Dadena = DeDaN. And
Dark Arabia = KUS*h* engendered (the Assyrian Hercules?) = NeM-RuD,
9 he first began to be mighty upon earth. He was a great landed-proprietor
before (the face of) IeHOuaH; whence the saying, *like* NeM-RuD, (a) *great*
10 *landed-proprietor* before (the face of) IeHOuaH.‡ And (the) commencement of his
realm, Babylon = BaBeL, and Erech = AReK, and Accad = AKaD, and
11 Chalne = KaLNeH in the land of Mesopotamia = S*h*iN*â*AR. Out of that
land he (Nimrod) went forth [*to*] Assyria = AS*h*UR, and builded Nineveh =
12 NINUeH, and Rehoboth-*Lion* = ReK*h*oBoT*t*-*â*IR, and Calah = KaLaK*h*, —
and Resen = ReSeN between Nineveh = NINUeH and between Calah = KaLaK*h*
13 (he) *she* (Nineveh?) the great city). And (the) Egyptians = MiT*s*RIM engendered
the Ait-Oloti = LUDIM, and the Ammonians = ÂNaMIM, and the Libyans
14 = LeHaBIM, and the Nefouschs = NiP*h*aiT*t*uK*h*IM, — and the Pharusii =
P*h*aT*t*uRiSIM, and the Shillouhs = KS*h*iLouK*h*IM out of whom issued

* No *translation* is intended by the terms yellow, swarthy, and white races. We use them merely to evolve the ethnological *tripartite* classification of the writer.

† Dark Arabia serves for the dark CUSHITE (red-*Himyâr*) Arabs.

‡ The mention of IeHOuaH makes this copy of the Ethnic Chart *Jehovistic*, and consequently recent, by every rule of exegesis. (PARKER'S *De Wette*, II., pp. 77-145.)

70

15 Philistines = P*h*eLiST*t*IM, and the Caphtors = KaP*h*T*to*RIM. And Canaan
= KN*â*AN engendered Sidon = T*s*IDoN his first born, and Kheth = K*h*eT*t*,
16 and the Jebusian = IBUSI, and the Amorian = AMoRI, and the Girgasian
17 =GiRGaSI, And the Khuian = K*h*UI, and the Accrian = *â*RKI, and the
18 Sinian SINI, — and the Aradian = ARUaDI, and the Simyrian = T*s*iMRI,
19 and the Hamathian = K*h*aMaT*t*I: (Afterwards the families of the Kanaanian
= KN*â*ANI (were) spread *abroad.*) And the boundary of the Kanaanian =
KN*â*ANI (had been) from Sidon = T*s*IDoN, towards *Gerar*, even to *Aâza*,
(round) by *Sodom*, and *Aâmora*, and *Admah*, and *Tseboim*, as far as *Lashâ*.
20 These (the) affiliations of K*h*aM swarthy races, after their families, after
21 their tongues, in their countries, in their nations. And to S*h*eM yellow races
also (there was) issue: he (is) the father of all (the) affiliations of (the)
22 Yonderer = ÊBeR, brother of IaP*h*eT*t* the elder. Affiliations of S*h*eM yellow
races. Elymais = *â*ILaM, and Assyria = AS*h*UR, and Chaldæan Orfa =
23 ARP*h*a-KaSD, and Lydia = LUD, and Aramæa = ARaM; — and (the) affilia-
tions of Aramæa = ARaM; Ausitis = *â*UT*s*, and Hûleh = K*h*UL, and
24 Gatara = GeT*t*uR, and Masonites = MaS. And Chaldæan Orfa = ARP*h*a-
KaSD engendered Salacha? = S*h*eLaK*h*; and Salacha = S*h*eLaK*h* engendered
25 (the) Yonderer = ÊBeR. And unto (the) Yonderer = ÊBeR were born two
affiliations; the name of one (was) (a) Split = PeLeG (because in his days the
earth was split), and (the) name of his brother (was) Joktàn = IoKTaN.
26 And Joktàn = IoKTaN engendered (the) Allumaæotæ = ALMUDàD, and (the)
Salapeni = S*h*eLeP*h*, and Hadramàut = K*h*aT*s*aRaMUT*t*, and (the) Jera-
27 chæi = IeRaK*h*, — and (Cape) Hadoramum = HaDURaM, and San*â*a =
28 AUZAL, and (the) Dhu'-l-Kalààh = DiKLeH, And (the) Abalitæ = *â*UBaL,
29 and Malai (el-Khyèf) = ABIMAL, and Sàba (Màreb) = SaBA, — and Ofor
AUP*h*iR, and (the) Beni-Kholàn = K*h*UILeH, and (the) Beni-Jobàb = IUBaB.
30 All these (are) affiliations of [*Qahtàn*] Joktàn = IoKTaN; — and their dwelling
(was) from Zames Mons = MeS*h*A, towards Mount Zaffàr = SeP*h*aRaH,
31 mountain of the East (or *mountain opposite?*).* These (are) (the) affiliations
of S*h*eM yellow races, after their families, after their tongues, in their lands,
32 after their nations. Such (are the) families of (the) *sons* of CESSATION = NuK*h*,
after their generations, in their nations; and from these were dispersed Ha-GOIM
= the hordes (the peoples) on the earth after the deluge.

(*Here ends the document.*)

The authors cannot but hope, after the evidences herein accumulated, that the impartial reader now agrees with them and with Rosellini, that "la serie dei nomi de' discendenti di Noè è una vera ricenzione geografica delle varie parti della terra;" so far as the world's surface was known to the writer of Xth Genesis.

Viewed by itself, as a document from all others distinct, incorporated by the Esdraic school into the canonical Hebrew writings, Xth Genesis is simply an *ethnic chorograph;* wherein three "Types of Mankind," generically classified as the red, yellow, and white, are mapped out — "after their families, after their tongues, in their countries, in their

* The word here is the same KDM upon which the analysis of De Longpérier was referred to under ASUR [*ubi supra*, p. 534].

nations." In every instance where monumental or written history has enabled us to check the writer's system, his accuracy has been vindicated. In not a few cases exactitudes, so minute as to be relatively marvellous, have been exhibited.

Our genealogical *table* displays the order in which this compiler supposed the different colonies, or affiliations, issued from each of the three parental stems. Our *retranslation* of Xth Genesis, by substituting, as far as possible, modern names for the same nations and countries, has enabled us to comprehend his literal meaning more clearly than when reading Hebraical appellatives now mostly obsolete, no less than veiled by an ancient and foreign mode of spelling them. And lastly, our transfer and redistribution of these seventy-nine cognomina, in a *map*, fix, within a few degrees of latitude and longitude, the boundary of this writer's geographical circumference; and thus indicate the horizon, so to say, of all the knowledge his "gazetteer" contains.

Learned and orthodox works have frequently defined this geography before; and with limitations of area quite as restricted as ours, as regards the sum total of terrestrial superficies. Because, if we have cut off, as not alluded to in Xth Genesis, the whole of *Nubia* above Egypt, and all Africa lying south of the northern limit of the Sahara deserts, our *map*, on the other hand, prolongs the writer's knowledge through Barbary, from Egypt to the Pillars of Hercules. Thus, upon the whole, our restoration is more extensive than that of Volney.

No savant whose opinion is worthy of respectful attention, but excludes all knowledge, on the part of the writer of Xth Genesis, of any portion of *Europe*, except the coasts of the Peloponnesus and of Thracia. All reasonable commentators, by cutting off "Scythia" at a line, drawn from the north-eastern apex of the Black Sea to the Caspian, deny that Xth Genesis includes *Russian Asia;* while none extend the geography of that document beyond a line drawn from the Caspian Sea to the mouth of the Indus, as an extreme; a frontier, to our view, quite unjustifiable, and by far too distant from a *Chaldæan* centre-point.

In consequence, we all agree that Hindostan and its mixed populations; China with her immense Mongol and Tartar hordes; and the Islands of the Indian Ocean; are entirely excluded from Xth Genesis. The lands of Malayana, Oceanica, Australasia, and the Pacific, having been discovered within the last three centuries, were of course unknown to the school of Esdras twenty-three hundred years ago. So was also the "New World;"—the vast American continent and its Islands, prior to the voyages of Columbus, and his successors. The most rigid orthodoxy, therefore, concedes that, upon *Finnish, Samoïde, Tongousian, Tartar, Mongol, Malay, Polynesian, Esquimaux, American*, and many other races, the writer of Xth Genesis is absolutely silent; that, every one of these peoples lay very far beyond the utmost area demonstrable through his chorography.

Nothing "heretical," then, accrues from our simple demonstration of the truth of that which the educated of all Christendom now-a-days insist upon.

But, the orthodox will even allow a little more. Beginning at the Cape of Good Hope, they will admit, that the compiler of Xth Genesis does not embrace that region, nor its inhabitants, the *Bosjesmans, Hottentots, Kaffres*, and *Foolahs*, in this ethnic geography. They will voluntarily renounce also, in the name of this genesiacal writer, acquaintance with any part of Africa more austral than a line drawn athwart its continent from *Senegal* on the western to Cape Gardafui on the eastern or *Abyssinian* coast. Thus much, we opine, no one "nisi imperitus" can hesitate to grant.

Upon reflection, in view of the impassabilities of the immense Sahara desert (first, geologically, when it was an inland *sea;* and secondly, zoologically, until the *camel* was introduced and propagated in Barbary, after the first century, B. C.), all scholars, we presume, will coincide with our limitation; and, by way of compensation for the additional knowledge which our analyses have secured for the author of Xth Genesis, along *Berberia*, Barbary, they will not insist upon his acquaintance with anything south of the northern edge of the *Sahara:*—the *oases* of Séewah, El-Khârgheh, &c., remaining, between orthodox readings and ours, "sub judice."

So far, to judge by published commentaries, there are no insurmountable obstacles to harmony between the most catholic interpreter of Xth Genesis and ourselves. "Nos adversaires" will now fairly confess that the battle-ground, upon which their and our opinions have to be fought, lies on a miserable strip of the *Nile's* deposits; along the countries we term, in common, the *Nubias.*

Yet, even here, reasonable persons—those who have of their own accord, and for the sake of truth, already abandoned the *Tchoudes, Finns, Samoïdes, Tongousians, Tartars, Mongols, Malays, Polynesians, Esquimaux, American-aborigines, Hottentots, Bosjesmans, Kaffres, Foolahs, Senegalians, Abyssinians,* the Sahara desert, &c., &c., as not included in Xth Genesis—such reasonable persons, we think, cannot make out, legally, a "casus belli" between our results and their individual preconceptions, upon matters so pitiful in geography as the *Nubias.*

They have read our analysis of KUS*h*. They have seen every affiliation of KUS*h* settled in *Arabia.* Now, if every affiliation of KUS*h* in Xth Genesis be *Arabian*, why must we seek for these KUS*h-ites* elsewhere? Indeed, if we both agree in classification, neither party has any other genesiacal *names* to dispute about.

KUS*h* and its affiliations being irrevocably determined in *Arabia,* and proved to have been generally of the Himyar-*red* stock, it would be as absurd to look for them in Nubia as on the Caucasian mountains. We know that until the XIIth and perhaps the XIth dynasty, the boundary of the MT*s*R*îm*, *Egyptians*, was the 1st Cataract of Syene: and inasmuch as the *Nubias* were then little known to Egyptians, they were undoubtedly far less known to Asiatics.

Consequently, there was a time when *Nubia* herself was a "terra incognita." We have only to continue this Asiatic ignorance of Africa for a few centuries, and every one will allow that there is no improbability involved in the assertion that the *Nubias* were unrevealed to the compiler of Xth Genesis at Jerusalem, or at Babylon. His *map* proves that they were so; and, thus far, discussion is at an end.

With the *Nubias* vanishes the last possibility that *Negro* races were known to the writer of Xth Genesis. He never mentions them; nor indeed does any other writer in the canonical Scriptures, from *Genesis* to *Malachi.*

Negroes are, therefore, excluded from mention in the Old Testament; together with *Finns, Uralians, Mongols, Tartars, Malays, Polynesians, Esquimaux, American*-Indians, &c., &c. The *map* of Xth Genesis, under the heads "Shem, Ham, and Japheth," merely covers those families of mankind classified by the Egyptians, in the days of SETHEI-MENEPTHA, 15th–16th centuries B. C., into the yellow, the red, and the white human types.

Such is our conclusion. Science and reason confirm it. Xth Genesis proves it. Nevertheless, few persons beyond the circle of education exempt from ecclesiastical prejudice, will, for some time to come, accept this result! *Why?*

[Our *manuscripts* comprise critical answers to this query viewed in all its bearings upon the *Ante-Diluvian* Patriarchs, and upon the two pedigrees of ST. JOSEPH recorded in *Matthew* and *Luke.* Inasmuch, however, as their production here would necessitate a second volume to this work, we postpone their publication; remembering St. Paul's sage admonishments to Timothy and to Titus—"not to give heed to fables and endless genealogies"—"but avoid foolish questions and genealogies." (1 *Tim.* i. 4; *Titus* iii. 9: Sharpe's *New Testament* "translated from Griesbach's Text;" London, 1844, pp. 380, 392–3).—G. R. G.]

CHAPTER XV.

BIBLICAL ETHNOGRAPHY.

Section E.—TERMS, UNIVERSAL AND SPECIFIC.

THERE is nothing in the language of the Bible which illustrates more strongly the danger of a too rigid enforcement of literal construction than the very loose manner in which *universal terms* are employed. Those who have studied the phraseology of Scripture need not be told that these terms are used to signify only a *very large amount* in number or quantity. *All*, *every one*, *the whole*, and such like expressions, are often used to denote a *great many*, or a *large portion*, &c. Examples may be found on almost every page of the Old Testament, but we will first select a few from the many scattered through the New. And we beg the reader to bear in mind the fact already established, viz., that neither the writers of the Old or New Testament knew anything of the geography of the earth much beyond the limits of the Roman empire, nor had they any idea of the spheroidal shape of the globe. Be it noted also that, in order to avoid the mistakes of the English AUTHORIZED VERSION, our quotations are borrowed from Sharpe's *New Testament* as closest to the original Greek.

In the account given by Matthew (iv. 8, 9) of the temptation of Christ, we have these words:

"Again the Devil taketh him on to a very high mountain, and showeth him *all the kingdoms of the world*, and their glory; and saith unto him; 'All these will I give thee, if thou wilt fall down and worship me.'"

Before accepting such words as "all the kingdoms of the world" in a *literal* sense, it may be well to peruse the commentary of Strauss, in his *Life of Jesus*:—

"But that which is the veritable stumbling-block, is the personal apparition of the Devil with his temptations. If even there could be a personal Devil, 't is said, he cannot appear visibly; and, if even he could, he would not have behaved himself as our Gospels recount it. . . . The three temptations are operated in three different places, and even far apart. It is asked, how Jesus passed with the Devil from one to the other? . . . The expressions, *the Devil takes him*, . . . *places him*, in Matthew—the expressions, *fetching*, *he conducted*, *he placed*, in Luke, indicate incontestably a displacement operated by the Devil himself; furthermore, Luke (iv. 5) saying that the Devil showed Jesus 'all the kingdoms of the world *in a moment of time*;' this trait indicates something magical. . . . Where is the mountain from the summit of which one can discover all the kingdoms of the earth? Some interpreters reply that by the *world*, COSMOS, one must understand Palestine only, and by *the kingdoms*,

BASILEIAIS, the isolated provinces and the tetrarchies of that country: a reply which is not less ridiculous than the explanation of those who say that the Devil showed to Jesus the world on a geographical map."[659]

In reference to these diabolical powers we may also be permitted to rejoice with our readers over the following fact, recently announced by the Rev. John Oxlee (Rector of Molesworth, Hunts, England) in his "Letters to the Lord Archbishop of Canterbury:"—

"In the *Chronicon Syriacum* of Bar Hebræus, we have it duly recorded, that, in the year of the Hegira 455, or of our Lord 1063, certain Curdean hunters, in the desert, brought a report into Bagdad; how that, as they were hunting in the desert, they saw black tents, with the voice of lamentation, weeping, and yelling; that, on their approaching them, they heard a voice saying: '*To-day died* BEELZEBUB, the Prince of the Devils; and every place where there is not lamentation for three days, we will erase from its very foundation.' . . . Hence it is apparent, even on the indubitable testimony of the devils themselves, that Beelzebub, the Prince of the Devils, died a natural death, nearly eight hundred years ago; and was lamented and bewailed, with all due honors, by the municipal authorities of Bagdad, Mosul, and other cities in the land of Senaar. There, then, let his mortal remains peaceably rest, never more to be disturbed, in the future, by human curiosity."[660]

We have a repetition of the previous passage in Luke, which should probably be taken in a figurative or allegorical sense; for although the evangelists had little idea of the extent or the shape of the earth, yet it cannot be maintained that Jesus or the devil were so ignorant as to suppose that a view of the world could be greatly extended by ascending a mountain. If we could take this language in a literal sense, it would at once settle the question as to the amount of geographical and ethnological knowledge of the evangelists. Here are some more instances of "universal terms" used loosely in a vague or general sense:—

(*Mat.* xii. 42)—"The queen of the South . . . . came from *the ends of the earth* to hear the wisdom of Solomon."

(*Luke* ii. 1)—"And it came to pass in those days that a decree went forth from Cæsar Augustus that *all the world should be registered.*"

(*John* xxi. 25)—"And there are also many other things which Jesus did, which if they should be written one by one, I do not think that *the world itself* would contain the written books.

(*Acts* ii. 5)—"And there were dwelling in Jerusalem Jews, devout men, *from every nation under heaven.*"

(*Acts* xiii. 47—quoting *Isaiah* xlix. 6)—"I have set thee to be a light of the Gentiles, that thou shouldest be for salvation *to the ends of the earth.*"

(*Rom.* x. 18—quoting *Ps.* xix, 4)—"Yes, verily, their sound went into *all the earth*, and their words *unto the ends of the world.*"

These examples will be quite sufficient to show the manner in which "*universal terms*" were used, and the necessity for measuring their extent by a proper standard. We now present a remarkable text, and the only one in the New Testament which alludes directly to the dogma of *unity* of races.

(*Acts* xvii. 26) — "And [God] hath made of one blood *all nations* of men to dwell on *all the face of the earth*, and hath determined the appointed seasons, and the bounds of their habitation." It will be noted that this saying of Paul is not autographed in his *Epistles*; but, as Hennell critically annotates, "rests mainly on the testimony of the author of *Acts*, who himself intimates that he is the same as the author of the third Gospel." 661

Now, can any reason be assigned why a wider signification should be given to "universal terms" here than in the previous examples? Have we not seen, too, in the quotation just preceding this, the loose manner in which the same writer (St. Paul) uses such terms? Should not this paragraph, also, deserve the less credit, inasmuch as it has no parallel? It should be remembered that when St. Paul stood upon Mars's Hill and preached to the men of Athens, his knowledge of nations and of races did not extend beyond that of his hearers; and the expression, "*hath made of one blood all nations of men*," was certainly meant to apply only to those nations about which he was informed; that is, merely the *Roman Empire*.

Leaving the New Testament we take up the Old, and such passages as these meet our eye: —

(1 *Kings*, xviii. 10) — As "IeHOuaH thy God liveth [most sacred form of Jewish oath], there is no nation or kingdom, whither my Lord hath not sent to seek thee; and when they said, '*He is* not there,' he took an oath [a certificate] of the kingdom, that they found thee not." If this text were to be taken literally, Obadiah's most solemn affidavit is here given that Ahab's emissaries had visited *China*, *Norway*, *Peru*, *Congo*,—in short, circumnavigated the whole globe, besides traversing it in every direction, during the tenth century B. C., in quest of Elijah!

(1 *Kings*, x. 24) — "And all the earth sought the face of Solomon, to hear his wisdom." Is this to be accepted *verbatim et litteratim?* Must no allowance for poetic license be made, when David says, — "And the channels of the sea appeared, the *foundations of the world were discovered*" (2 *Sam.* xxii. 16).

Receding to previous chapters (that is, not written during earlier ages, but merely bound up in books placed anteriorly to *Kings* and *Samuel* in the present order of arrangement), we come to—"AND NOW KuL-HAReT*s* (the WHOLE earth) was of one LIP and of DeBeRIM AK*ha*DIM." — The last two words, plurals in Hebrew, cannot be literally rendered into English, as *ones words*; but the sense is "one language."

The whole context refers to an idea purely *Chaldæan*, and to a preternatural event exclusively *Babylonish*; viz., the city and the tower of BaBeL, which IeHOuaH "descended to see" after they were *built*. The two things, *tower* and *city*, are inseparable; and we perceive that the people "ceased to build the *city*," after they were "dispersed thence over the face of the WHOLE EARTH."

(*Gen.* xi. 1) — "On that account it was called BaBeL, because IeHOuaH there BeLeL (confounded) the LIP (speech) of the WHOLE EARTH." The root BLL means *to mingle*, to *talk-gibberish*; and, conformably to the favorite genius of Semitic description, the writer avails himself of a play upon words—*i. e.*, really "perpetrates a *pun*"—because the monosyllabic etymon of BaBeL, itself meaning "confusion," is the same as that of BeLeL.—We might say in English, "BABEL-*babble*," and thus realize part of the alliteration of BaBeL-BeLeL, while losing half its *double entendre*; because, BaBeL does not mean in English what it does in Semitish idioms, viz., "gibberish" as well as *confusion*. Another mode of conveying an idea of this play upon words would be, to translate BaBeL-BeLeL by "higgledy-piggledy." Poor, dreary, and mis-timed though such jocularity may seem to us, and inconsonant with the sanctity of the volume in which it is now found, nevertheless, no Orientalist will dispute the assertion, that similar *rebuses*, or *riddles*, are the delight of Eastern narrators;[662] while, by the Talmudic Rabbis, this pun was supposed to cover awful mysteries. Few persons are aware that, as the Text says nothing about the *destruction* of either city or tower, theologians derive their notions in this respect, not from the Bible, but from the spurious and modern tales of Hestiæus, of Polyhistor, of Eupolemus, and of the "Sibylline Oracles." The classical texts may be found in Cory's *Ancient Fragments.*

The reader, who has comprehended the principles of criticism, established further on in the *Archæological Introduction to Xth Genesis,* can now seize the historical value of this document (*Gen.* xi. 1–9) in a moment.

1st. It has no connection with what precedes or succeeds it; but breaks in, parenthetically, between what is now printed as the 32d *verse* of Chap. X. and the 10th of Chap. XI.: its apparent relation to either originating solely through modern, arbitrary, and therefore unauthorized, divisions into *chapters* and *verses.*

2d. Age and authorship unknown, its antiquity cannot ascend beyond the seventh—eighth century B. C., because its divine ascriptions are *Jehovistic;* nor could it well have been embodied into the book called "Genesis," earlier than about B. C. 420, by the Esdraic School; because, the mention of "the land of *Shinar*"—of "*brick* they had for stone (or rather L-ABNi, *for building*) and *bitumen* they had for mortar"[663] of the "*city;*—therefore the name of it was BaBeL (Babylon)"—carries us at once to plains between the *Shinar hills* and the Euphrates-river; to the *bricks* of Chaldæan mounds; to the bituminous springs of *Hit* (*His* of Herodotus, and hieroglyphic IS);[664] and to the Babylon of Nebuchadnezzar; than whom, although the name of a place called BBL is as old as Thotmes III. of the XVIIIth Theban dynasty, 1500—1600 B. C., nothing *cuneiform* yet found at Babylon is anterior.[665]

3d. What connections BàB-eL[666] "*Gate* of the SUN" (like the Chinese "celestial gates;" or their Mongol derivative, the Ottoman "Sublime Porte"), may have with this name's origin: whether BELUS the king; BAAL the god; or "*Bel* and the dragon;" are to be taken into consideration:—these curious inquiries, if familiar to our studies, are foreign to our present purposes and objects. But, "in sober sadness," let us ask—Can such words as KuL-Ha-AReT*s* (the *whole* earth) be accepted, by ethnological science in the nineteenth century, when contained in such an unhistorical document? At any rate, "Types of Mankind" must respectfully leave them aside.

"Isis! dea infelix, Nili remanebis ad amnem
Sola, carens et voce!"

The ignorant of all races and ages, especially inland-populations such as the Jews were, when a foreign tongue strikes their auricular nerves, do not suppose that the speaker is uttering sense, but believe that he is merely exercising his vocal muscles instinctively, in the same manner that *geese* "talk." The writer of *Matthew* is not free from this illusion; because, where our authorized mistranslation has "Use not vain repetitions, as the heathen *do;*" the original *Greek* reads—"And when ye pray, *babble* not as the heathen do" (*Mat.* vi. 7:—Sharpe, *N. T.*, p. 10). In the idea of the Hebrews, vouched for, according to De Sola, even by such mighty commentators as Rashi and Mendelssohn,[667] the "One language" at Babel was merely the "lingua sancta;" that is to say, all mankind there talked *Hebrew* at first; but (after the dispersion thence, when their speech was "confounded"), only SHEM's *sons* miraculously preserved the Hebrew tongue immaculate; "the rest of mankind" BABEL-*babbled* in gibberish!

The above hints are furnished to others. We feel as charitably disposed as Josephus did when writing.—"Now, as to myself, I have so described these matters as I have found them and read them; but if any one is inclined to another opinion about them, let him enjoy his different sentiments without any blame from me."[668]

Section F. — Structure of Genesis I., II., and III.

Far more important, at an ethnological point of view, are the first *three* chapters of the book called "Genesis;" and to them we can here devote but a paragraph or two.

Our *Archæological Introduction*, in Part III., has pointed out their Esdraic age, and the *Persic* origin of some of the mythes they contain. All modern divisions into chapters and verses, of course, are to be abstracted; being mere European addenda. *Jewish* divisions of the book of Genesis are entirely different. They are *twelve* in number; of which the first SeDR — *Chapter* I. to *Chapter* VI., *verse* 9 — is called the "Bereshith," *beginning.*[671]

To understand this "structural analysis of the book of Genesis," according to exegetical principles now universally recognized by Hebraists, we refer the reader to a masterly critique by Luke Burke,[672] and to the solid evidences supplied by De Wette.[673] The more salient characteristics distinguishing the two documents are, the words ELoHIM, in king James's version replaced by "God;" and IeHOuaH, for which our appellative "Lord" is substituted; neither of these two Hebrew divine names being *translated;* as the writer will demonstrate in some future treatise. The relative order of these documents becomes intelligible to the reader by being placed in juxtaposition. Our purpose now being merely the exhibition of some *structural* peculiarities not generally known, it is unnecessary to retranslate the whole three chapters, and impossible to justify herein our verbal interpretations. With Cahen's *Bible*, the reader can easily fill up gaps for himself in the former case: adequate explanations in the latter would require the publication of a volume of results which, obtained through ten years' incessant travel and study, G. R. G.'s *manuscripts* embrace. To the anthropologist, however, it will be satisfactory to behold the true place of the word A-DaM in these texts — אדם, says Cahen, "l'espèce humaine, singulier collectif." And, as concerns other questions, we must be content for the present to submit an observation written by the great Hellenist, R. Payne Knight, to his colleagues Sir Joseph Bankes and Sir W. Hamilton: —

"It must be observed that, when the ancients speak of *Creation* and destruction, they mean only formation and dissolution; it being universally allowed, through all systems of religion or sects of philosophy, *that nothing could come from nothing, and that no power whatever could annihilate that which really existed.* The bold and magnificent idea of a creation from nothing was reserved for the more vigorous faith, and more enlightened minds of the moderns; who need seek no authority to confirm their belief; for, as that which is self evident admits of no proof, so that which is in itself impossible admits of no refutation."[674]

DOCUMENT No. I. — GENESIS I.; II. 3.

"Elohim."

Harmonical Ode of Creative cosmogony — antique and scientific.

"In the beginning, ELoHIM created the (universality of) skies, and the (universality of) earth. And the earth was T*to*HU — and — BoHU (literally — masculine and feminine principles dislocated, or confounded; paraphrastically — "*without form and* a *confused mass*"), and darkness was upon the face of the abyss, and the (breath) spirit of ELoHIM hovered (like a descending bird) over the face of the waters—

[*V.* 3, 4.]

(SUN'S daeg. — SOL. — E *mi.* — *Chorus 1st.*)

"And it was ĕReB (*western* twilight) and it was BeKR (early dawn) — *Day* ONE!

[*V.* 6, 7.]

(MOON'S daeg. — LUNA. — A *la.* — *Chorus 2d.*)

"And it was ĕReB (*western* twilight) and it was BeKR (early dawn) — *Day* SECOND!

[*V.* 9—12.]

(TIWES'S daeg. — MARS. — D *re.* — *Chorus 3d.*)

"And it was ĕReB (*western* twilight) and it was BeKR (early dawn) — *Day* THIRD!

[*V.* 14—18.]

(WODN'S daeg. — MERCURY. — G *sol.* — *Chorus 4th.*)

"And it was ĕReB (*western* twilight) and it was BeKR (early dawn) — *Day* FOURTH!

[*V.* 20—22.]

(THOR'S daeg. — JUPITER. — C *ut.* — *Chorus 5th.*)

"And it was ĕReB (*western* twilight) and it was BeKR (early dawn) — *Day* FIFTH!

"And ELoHIM said, 'Let us make (the universality of) the A-DaM (THE-RED-man) after our image, like our likeness, and let him rule over the fish of the seas and over the bird of the skies and over the cattle and over all the [whole] earth and over all the crawler crawling upon the earth.' And ELoHIM created (the universality of) the A-DaM (THE-RED-man) after his image, after the image of ELoHIM created (he) them. And ELoHIM blessed them and ELoHIM said to them 'Be fruitful and multiply, and fill the (universality of) earth and subject it, and rule over fish of the seas and over bird of the skies and over all the living that crawls upon the earth.'

[*V.* 29—30.]

(FRIGA'S daeg. — VENUS. — F *fa.* — *Chorus 6th.*)

"And it was ĕReB (*western* twilight) and it was BeKR (early dawn) — *Day* the SIXTH!

[*Ch.* ii. *v.* 1, 2.]

(SATUR'S daeg. — SATURN. — B *si.* — *Benediction*)

"And ELoHIM blessed the (universality of) *day*-the-SEVENTH and sanctified it, because he S*h*aBaT*t* (rested, and *seventhed*) from all his work which ELoHIM created to act" — (*i. e.*, by its own organism henceforward).

"SABBATH," *Saturday*; commencing at sunset on Friday, and ending at sunset on Saturday.

FINIS.

(Margin headings: SUN'S daeg. etc. — Pagan Saxon names of the *week-days*. SOL. etc. — Ancient Oriental Planetary System. E *mi.* etc. — MUSICAL NOTE.)

DOCUMENT No. II. — GENESIS II. 4; III. 24.

"Jehovah."

Popular Creation of the World — later, and Persic.

"Such (the) generations (literally, *bringing-forths*) of the skies and the earth according to their creation, on (the) day IeHOuaH-ELOHIM made earth and skies.

[*V.* 5, 6.]

"And IeHOuaH-ELOHIM formed the (universality of) A-DaM (THE-RED-man) of dust from the A-DaMaH (THE-RED-earth) and breathed in (his) nostrils breath of life, and the A-DaM (THE-RED-man) became (a) living creature. And IeHOuaH-ELOHIM planted (a) garden in ĕDeN (or, *in*-DELIGHT) to (the) East, and there placed the (universality of) A-DaM (THE-RED-man) whom he had formed.

[*V.* 9—14.]

"And IeHOuaH-ELOHIM took the (universality of) A-DaM and placed him in (the) garden of ĕDeN (or, DELIGHT) to cultivate it and to guard it.

[*V.* 16—20.]

"And IeHOuaH-ELOHIM made the A-DaM (THE-RED-man) to fall (into a) great drowsiness, and he slept; and he took one of his ribs and filled-in flesh in place thereof. And IeHOuaH-ELOHIM constructed the rib which he had taken from the A-DaM (THE-RED-man) into AiS*h*iH (woman — or ISE, ISIS) and brought her to the A-DaM (THE-RED-man).

[*V.* 20. *Ch.* iii. *v.* 19.]

"And the A-DaM (THE-RED-man) called (the) name of AiS*h*aT*t*U (his wife, or ISeT, ISIS) K*h*iUaH (*life*), because she was (the) mother of all K*h*aIa (*living*).

[*V.* 21—23.]

"So he drove-out the (universality of) A-DaM (THE-RED-man); and he placed at (the) East to (the) garden of ĕDeN (*delight*) the (universality of) KeRuBIM (FIERY-DISKS), of which he made the CENTRAL-FLAME revolve to guard the road to (the) tree of the K*h*aIaIM (*lives*).

FINIS.

Our present object limiting itself to the *Creation of Man*, as set forth in the above two documents — each, the reader now perceives, distinct altogether the one from the other — we withhold (contrary to our habit) authorities for our arrangement of the "document *Elohim*." The Hebraist will concede that we have adhered with rigid fidelity to the Text; and that suffices until we resume biblical mysteries on a future occasion, when *authority* enough shall be forthcoming. Yet, to the curious investigator, we feel tempted to offer the "Air" of the *Music of the Spheres:*

If he be a musician, he can play it on a piano; if he is a geometrician, he will find its corresponding notes on the sides of an *equilateral triangle* added to the angles of a *square;* if he loves metaphysics, Plato will explain the import of *unity, matter, logos, perfection, imperfect, justice, repose;* while Pythagoras will class for him *monad, duad, triad, quaternary, quinary, senary*, and *septenary.* We hope to strike the OCTAVE note some day ourselves; but, in the meanwhile, should the reader be profound in astronomical history, and if he can determine the exact time when the ancients possessed *neither more nor less* than "five planets, besides the Sun and Moon," there are two archæological problems his acumen will have solved — 1st, the arithmetico-harmonical antiquity of the *number* 7; and 2d, the precise era beyond which it will thenceforward be impossible to carry back the composition of that ancient *Ode* we term "*Genesis* i—ii. 3."

Being of an epoch much more recent; arranged upon a geographical basis purely *Chaldæan;* and containing allusions to a *garden* of DELIGHT (like the famed "hanging-gardens" of Babylon, and the *paradisiacal* parks of Persia); the "Jehovistic document" throws little or no light upon ancient ethnography. A-DaM, as we shall see, never was intended by the Jehovistic writer, to be the *proper-name* "Adam," as the versions pretend. The woman AiS*h*aH (when the masoretic *points* or other arbitrary and modern diacritical marks are removed) becomes ASH, or (vowels being vague) ISE: identified with the Coptic ISE, as well as with the hieroglyphical appellative of that primordial ISI, whom the Greeks (through the addition of their euphonizing *Sigma*) made into the goddess ISIS: "for," says CLEMENS *Alexandrinus*, "in that which belongs to the *occult* the enigmas of the Egyptians are similar to those of the Hebrews."[675] One of the titles of this myrionymed goddess was "the universal *mother*;" and naturally so, "because she was the *mother* of all living" (*Gen.* iii. 20).

"I am," says ISIS, "*Nature;* parent of all things, the sovereign of the elements, the primary progeny of Time, the most exalted of the deities, the first of the heavenly gods and goddesses, the queen of the shades, the uniform countenance; who dispose with my rod the numerous lights of heaven, the salubrious breezes of the sea, and the mournful silence of the dead; whose single deity the whole world venerates in *many forms*, with various rites and many *names.* The Egyptians, skilled in ancient lore, worship me with proper ceremonies, and call me by my true name, Queen ISIS."[676]

In consequence, the "document JEHOVAH" does not especially concern our present subject; and it is incomparable with the grander conception of the more ancient and unknown writer of Genesis Ist. With extreme felicity of diction and conciseness of plan, the latter has defined the most philosophical views of antiquity upon *cosmogony;* in fact so well, that it has required the palæontological discoveries of the XIXth century — at least 2500 years after his death — to overthrow his *septenary* arrangement of "Creation;" which, after all, would still be correct enough in general principles, were it not for one individual oversight, and one unlucky blunder; not exposed, however, until long after his era, by post-Copernican astronomy. The oversight is where he wrote (*Gen.* i. 6—8): "Let there be RaQIê;" i. e., a *firmament;* which proves that his notions of "sky" (solid like the concavity of a copper basin with *stars* set as brilliants in the metal),[677] were the same as those of adjacent people of his time: indeed, of all men before the publication of NEWTON'S *Principia* and of LAPLACE'S

Mécanique Céleste. The blunder is where he conceives that AUR, "light," and IOM, "day" (*Gen.* i. 14—18), could have been physically possible *three whole days* before the "two great luminaries," *Sun* and *Moon*, were created. These venial errors deducted, his majestic song beautifully illustrates the simple process of ratiocination through which—often without the slightest historical proof of intercourse—different "Types of Mankind," at distinct epochas, and in countries widely apart, had arrived, naturally, at cosmogonic conclusions similar to the doctrines of that Hebraical school of which his harmonic and melodious *numbers* remain a magnificent memento.

That process seems to have been the following. The ancients knew, as we do, that man *is* upon the earth; and they were persuaded, as we are, that his appearance was preceded by unfathomable depths of time. Unable (as we are still) to measure periods antecedent to man by any *chronological* standard, the ancients rationally reached the tabulation of some events anterior to man, through *induction* — a method not original with Lord Bacon, because known to St. Paul; "for his unseen things from the creation of the world, his eternal power and godhead, are clearly seen, *being understood by the things that are made*" (*Rom.* i. 20). Man, they felt, could not have lived upon earth without *animal* food; ergo, "cattle" preceded him; together with birds, reptiles, fishes, &c. Nothing living, they knew, could have existed without light and heat; ergo, the *solar system* antedated animal life, no less than the *vegetation* indispensable for animal support. But terrestrial plants cannot grow without *earth;* ergo, dry land had to be separated from pre-existent "waters." Their geological speculations inclining rather to the *Neptunian* than to the *Plutonian* theory — for Werner ever preceded Hutton — the ancients found it difficult to "divide the waters from the waters" without interposing a metallic substance that "divided the waters which were *under* the firmament from the waters that were *above* the firmament;" so they inferred, logically, that a *firmament* must have been actually created for this object. [*E. g.*, "The *windows* of the skies" (*Gen.* vii. 11); "the waters *above* the skies" (*Ps.* cxlviii. 4).] Before the "waters" (and here is the peculiar error of the genesiacal bard), some of the ancients claimed the pre-existence of *light* (a view adopted by the writer of Genesis Ist); whilst others asserted that "chaos" prevailed. Both schools united, however, in the conviction that DARKNESS — *Erebus*[678] — anteceded all other *created things.* What, said these ancients, can have existed before the "darkness?" ENS ENTIUM, the CREATOR, was the humbled reply. ELoHIM is the Hebrew vocal expression of that climax; to define whose attributes, save through the phenomena of creation, is an attempt we leave to others more presumptuous than ourselves.

"GOD," nobly exclaims De Brotonne, "has no need to strike our ears materially to make himself heard, our eyes to make himself seen. The first act of triumph of the spirit over matter is the discredit of emblems that have disguised the infinite God; and the first step towards truth is to recognize him *without image*, after having, for so long a period, modelled him after our own."[679]

What definition of the Godhead more sublime than that in the Hindoo *Vedas?* —

> "He who surpasses speech, and through the power of whom speech is expressed, "know, O thou! that He is BRAHMA, and not these perishable things that man adores.
>
> "He who cannot be comprehended by intelligence, and he alone, say the sages, "through the power of whom the nature of intelligence can be understood, know, "O thou! that He is BRAHMA, and not these perishable things that man adores.
>
> "He who cannot be seen by the organ of vision, and through the power of whom the "organ of seeing sees, know, O thou! that He is BRAHMA, and not these perishable "things that man adores.
>
> "He who cannot be heard by the organ of audition, and through the power of "whom the organ of hearing hears, know, O thou! that He is BRAHMA, and not "these perishable things that man adores.
>
> "He who cannot be perceived by the organ of scent, and through the power of "whom the organ of smelling smells, know, O thou! that He is BRAHMA, and not "these perishable things that man adores."[680]

Phœnician, Chaldæan, and many other nations' cosmogonies present both striking resemblances and divergences. Some of them are compared with *Genesis*, very ably, by Palfrey;[681] from whom we borrow these words of the *Alexandrian* cosmogony of DIODORUS SICULUS — "This is not unlike what Euripides says, who was a disciple of Anaxagoras. For this is his language in the Melanippe:

> 'There was one aspect to sky and earth;
> Then the secret powers doing their office
> Produced all things unto the regions of light,
> Beasts, birds, trees, the sea-flock,
> *Finally*, men themselves.'"

But that which ancient philosophers attained through the laws of inductive reasoning, if to themselves clear and satisfactory, could not be conveyed in a form so indefinite to the intelligence of the illiterate, nor to children. Such undeveloped minds require *dogmatical* tuition. The teachers, so to say, had inductively ascended along an imaginary ladder, from *man* as its basis; until, having established some facts in nature antecedent to his terrestrial advent, they reached its top, when they recognized that there must be a FIRST CAUSE anterior to the "beginning:" but, so soon as these scientific results were to be conveyed to pupils, the dogmatical method became necessary: wherefore the preceptors reversed the order; and, commencing at the top of the supposititious ladder, they taught — "In the beginning ELoHIM *created*." Each rung, as they came down, marked, like degrees on a scale, the order in which previous induction had established the relative places of events; and thus every intellectual nation possessed *a* "Genesis." That of the Hebrew Elohistic writer possesses the superior merit of being a scientific hymn,[682] arranged in true accordance with the *septenary* scale of numerical harmonies.

Viewed as a literary work of ancient humanity's loftiest conception of Creative Power, it is sublime beyond all cosmogonies known in the world's history. Viewed as a narrative inspired by the Most High, its conceits would be pitiful and its revelations false; because telescopic astronomy has ruined its celestial structure, physics have negatived its cosmic organism, and geology has stultified the fabulous terrestrial mechanism upon which its assumptions are based. How, then, are its crude and juvenile hypotheses about *Human Creation* to be received?

Before answering this interrogatory, it may be instructive to peruse some Fathers of the Church:

1st. ORIGEN. — "To what man of sense, I beg of you, could one make believe, that the first, the second, and the third *day* of creation, in which notwithstanding an evening and a morning are named, could have existed without *sun*, without *moon*, and without *stars?* —that, during the first day, there was not even a *sky!* Who shall be found so idiotic as to admit that God delivered himself up like a man to agriculture, by planting trees in the garden of Eden situate towards the East; that one of those trees was that of life, and that another could give the science of good and evil? No one, I think, can hesitate to regard these things as *figures*, beneath which mysteries are hidden."[683]

The same patristic scholar adds elsewhere—"Were it necessary to attach ourselves to the letter, and to understand that which is written in the Law after the manner of the Jews or the populace, I should blush (*erubesco dicere*) to say aloud that it is God who has given us such laws: I should find even more grandeur and reason in human legislations; for example, in those of the Athenians, of Romans, or of Lacedæmonians."[684]

2d. CLEMENS *Alexandrinus* — "For your *Genesis* in particular was never the work of Moses."[685] — "Horum ergo scripta (Orphei et Hesiodi) in duas partes intelligentiæ dividuntur; id est, secundum litteram sunt ignobilis vulgi turba confluxit, ea vero quæ secundum allegoriam constant omnis philosophorum et eruditorum loquacitas admirata est."[686] St. Clement applies exactly the same principles to *Genesis* (xxvi.), where he exclaims — "O divine jesting! It is the same that Heraclitus attributes to Jupiter.

Abimelech is Jesus Christ, our king, who, from the heavens above, considers our sports, our actions of grace, our transports of joy."[687]

3d. St. AUGUSTINE — "There is no way of preserving the true sense of the first *three* chapters of Genesis, without attributing to God things unworthy of him, and for which one *must have recourse* to allegory."[688]

4th. St. JEROME — who, in his commentary upon Jeremiah, enforces the allegorical method — "Sive MOSEN dicere volueris auctorem Pentateuchi, sive Esdram ejusdem instauratorem operis, non recuso."[689]

Let the most philosophic of many truly-learned Rabbis close the list:—

MAIMONIDES — "There are some persons to whom it is repugnant to perceive a motive in a given law of the (divine) laws; they love better to find no rational sense in the commandments and prohibitions. That which leads them to this, is a certain feebleness they feel in their souls, but upon which they are *unable to reason*, and of which they know not how to give any account. This is what they think. If the laws should profit us in this (temporal) existence, and that they had been given to us for such or such a motive, it might very well be that they are the product of the reflection and of the intelligence of a *man of genius:* if, on the contrary, a thing possesses no comprehensible sense and that it produces no advantage whatever, it emanates, without doubt, from the Deity, because human thought could not lead to such a thing. One would say that, according to these weak minds, man is greater than his Creator; because *man* (according to them) speaks and acts while aiming at a certain object; whereas God, far from acting similarly, would order us, on the contrary, to do that which to ourselves is not of the least utility, and would forbid us from actions that cannot cause us the slightest damage." (Arabicè, *'Dellàlat el Khàyereen;* Hebraicè, *More Neboukhìm;* "Guide to the Strayers," ch. xxxi.: MUNK'S Translation, Paris, 1833.)

They all — i. e., the Fathers of the first centuries — attributed a *double* sense to the words of Scripture, the one obvious and literal, the other hidden and mystical, which lay concealed as it were under the outward letter. The former they treated with the utmost neglect;[690] following St. Paul's authority — "For the *letter* killeth, but the spirit giveth life." — (2 *Corinth.* iii. 6.)

Section G. — COSMAS-INDICOPLEUSTES.

But, in the proportion that Hellenic learning faded in Alexandrian schools, so patristic talent and scholarship also deteriorated. That "Genesis" which, by the earlier Fathers, had been ascribed to EZRA rather than to MOSES, and the language of which, to more refined Grecian intellects, appeared too contemptible for Divinity unless construed in an *allegorical* sense, at length began to be accepted *verbatim et litteratim* by Christian writers: the strenuousness of orthodoxy, in any creed, increasing always in the ratio that mental culture declines. At last, arose a Monk who, unjustly forgotten by the Church though he be now, did more to petrify theological stolidity in Europe, for 800 years, with respect to the first three chapters of *Genesis*, than any human being but himself — COSMAS-*Indicopleustes*.

"He is," says the learned Mr. Sharpe, "of the dogmatical school which forbids all inquiry as heretical. He fights the battle which has been so often fought before and since, and is even still fought so resolutely, the battle of religious ignorance against scientific knowledge. He sets the words of the Bible against the results of science; he denies that the world is a sphere, and quotes the Old Testament against the pagan philosophers, to show that it is a plane, covered by the firmament as a roof, above which he places the kingdom of heaven. . . . The arguments employed by Cosmas were unfortunately but too often used by the Christian world in general, who were even willing to see learning itself fall with the overthrow of paganism. All knowledge was divided into sacred and profane, and whatever was not drawn from the Scriptures was slighted and neglected; and this perhaps was one of the chief causes of the darkness which overspread the world during the middle ages." [691]

To comprehend the force of these observations it may be well to preface our description of the *Topographia Christiana* by a few excerpts from Matter.[692]

The only Christian Father whose writings evince the humblest acquaintance with Egyptian studies, CLEMENS *Alexandrinus*, expressly says, that the "Egyptians taught the Greeks the movement of the planets round the sun;" and, since 1848, Egyptology can proudly add the extraordinary discoveries of Lepsius in hieroglyphical Astronomy, which are likely to be carried to results little expected, through Biot.[693]

About B. C. 603, Thales had observed an eclipse of the sun. He taught the *spheroidity* if not the sphericity of the earth; he knew the obliquity of the ecliptic; knew that the moon was illumined by the sun; and explained solar eclipses by the intervention of the lunar disc between the earth and the sun. In the succeeding century, Pythagoras sustained the *sphericity* of the earth, and its movement, with the planets, round the sun; and his disciples Leucippus and Democritus added some acquaintance with the rotary motion of the earth upon its axis. Eudoxus advocated similar doctrines. Now, Thales, Pythagoras, and Eudoxus, had studied under genuine hierogrammatists in Egypt.

The grand Stagyrite (who had not drunk of Nilotic waters) maintained the contrary; viz., that the sun revolved around the earth. In vain did Aristarchus strive to bring science back to truer principles. His voice was unheard for sixteen centuries. Hipparchus determined the precession of the equinoxes, &c., during the 2d century B. C.; but, his more important works being lost, "tulit alter honores;" because Ptolemy, a far better geographer than astronomer, has not revealed what of his great predecessor's views militated against his own celestial dogmas. In the early part of the 2d century, after C., Ptolemy had wofully retrograded from ancient Greco-Egyptian science; for he held to the absolute immobility of the earth, and made the sun revolve around our globe. Denouncing the contrary system as too ridiculous to merit attention, he gives his own reason for opposing it, viz., "that one always sees the *same* half of the sky"! "The earth," says Claudius Ptolemy, "is not only central, but also stationary. If it had an individual motion (upon its axis) such movement would be proportioned to its mass. It would, therefore, leave behind it the animals and other bodies, which would be carried into the air,—it would fly away from them, and escape from the sky! No object not fixed to the earth, no bird, could advance to the eastward with the same rapidity as the globe"! Unsuspected before Newton, the laws of gravitation and attraction could not ease Ptolemy's perplexities.

We have seen that the older and wiser Fathers of the Church (who must have been more or less read in the higher Grecian classics), unable to reconcile the *letter* of "Genesis" with what they well knew to be positive philosophy, had recourse, like Philo, to *allegorical* explanations: which means, simply, that they disbelieved genesiacal stories as revealed in the *Septuagint*, and therefore nullified them by inventing mystic hypotheses. They sustained, however, in their writings, no especial theory upon astronomy or geography: but, that with which Clemens, and Origen, and Anatolius, and Synesius, and Theophilus, and even Cyril, had refrained from meddling, was grasped, with Promethean audacity, by an itinerant trader of the sixth century after C.; whose temerarious zeal, when he had adopted monastic vows, was exceeded merely by his delicious stupidity; as we now proceed to prove. Cosmas, setting a Greek copy of "Genesis" before him, composed, upon that poor version's literal language, his *Topographia Christiana*.[694] Of *Hebrew* he had not an idea.

Fig. 357.

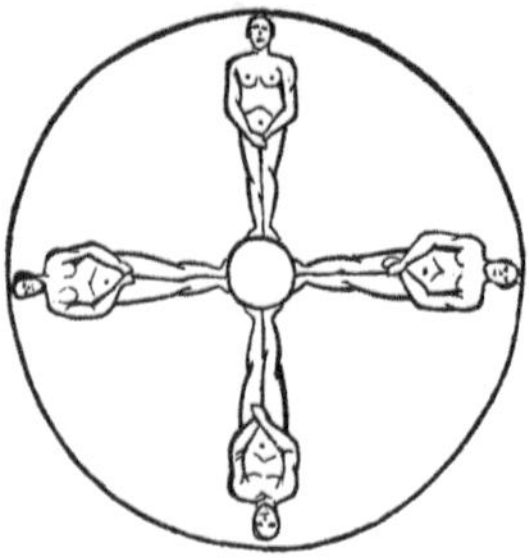

He, Cosmas aforesaid, commences with a practical demonstration of the absurdity of "Antipodes," — by drawing a figure like this —

He then acutely observes:—"Cum figura hominis recta sit, qui fit ut quatuor illi eodem tempore stantes recti non sint; sed quocumque vertas eos, quatuor illi simul nunquam videantur; quomodo ergo fieri potest ut vanas illas mendacesque hypotheses admittamus? Quomodo ergo fieri potest ut eodem tempore pluvia in quatuor illos decidat? Quod ergo nec natura nec *mens nostra* admittere potest, id cur frustra supponitis?"—"Thus," continues Montfaucon, "Cosmas here and throughout Topographia Christiana, *ut et multi alii ex SS. PP. qui nec gravitatis centrum, nec astronomicas observationes, callebant.*" 695

St. Augustine it was who had "*seen* folks with an *eye* in the pit of their stomachs;" so his testimony is unsafe; but Lactantius had beheld fewer marvels, and we quote him:—"Ineptum credere esse homines quorum vestigia sint superiora quam capita, aut ibi quæ apud nos jacent inversa pendere, fruges et arbores deorsum versus crescere. . . . Hujus erroris originem *philosophis* fuisse quod existimarint rotundum esse mundum."

For the sake of contrast with later patristric orthodoxy, let justice be meted out to some old rabbinical capacities. The most ancient authors of the *Guemara* were acquainted with the spherical form of the earth; for they say, in the Jerusalem Talmud, that Alexander the Great, going over the earth to conquer it, ascertained that it was *round;* and it is on that account that statuary represents him with a globe in his hand.696 Albeit, there are Judaical authorities of higher antiquity in the *Zohar* — a book which probably antedates, but in any case approximates to, the Christian era 697 — whose knowledge of the more ancient systems of cosmogony led them to write as follows:—"In the book of Chamnouna *the Old* one learns, through extended explanations, that the earth turns upon itself in the form of a circle; that some (people) are above, and others below; that the aspect of all creatures changes according to the appearance of each place, while preserving nevertheless the same position; that such a country of the earth there is that is lighted, whilst such others are in darkness; the former have day when to others it is night; and there are some countries where it is constantly day, or, at least, where night lasts but a few instants." 698

But such profanity was unintelligible to Cosmas. No ray of light, from scientific sources, could penetrate into a blockhead.

To him, the habitable earth is a plane surface, having the form of a parallelogram, of which the sides are double in length to the top and bottom. Inside this oblong square are *four* basins, the Mediterranean, the Caspian, the Red Sea, and the Persian Gulf. Outside the parallelogram the circumambient ocean surrounds the inner oblong-square, and separates it from the outer continents (primitively inhabited by Adam's family), from *paradise,* and from the "garden of Eden," which are situate upon a mountain at the East. Here dwelt our first parents, until the *ark* of Noah, during the deluge, ferried them over to the inner continent where we ourselves reside unto this day. Cosmas ignored whatever he could not find *in the Bible;* and, wiser than our modern theologers, this modest pattern for prurient orthodoxy never discovered *China,* Northern *Europe,* Central *Africa, America, Polynesia,* or *Australia,* in the canonical Scriptures. Let his *map,* and his own perspicuous language, explain true Mosaic cosmology. He begins with the exact Greek letter of *Genesis* i. 1: but his editor kindly furnishes the *Vulgate*:—"Scriptum est In principio fecit Deus cœlum et terram. Primum itaque *cœlum* fornicatum." 699

[N. B. My own tracing (made at the British Museum, in 1848, for personal remembrance) being too rough, we are indebted to the accomplished Mrs. Luke Burke for the *fac-simile* transcript, of which the above is a copy; reduced slightly more than one half. Typographical exigenda compel us also to transfer Cosmas's explanations from the *map*

COSMAS'S *Map.* — FIG. 358. — "I. TABULA."

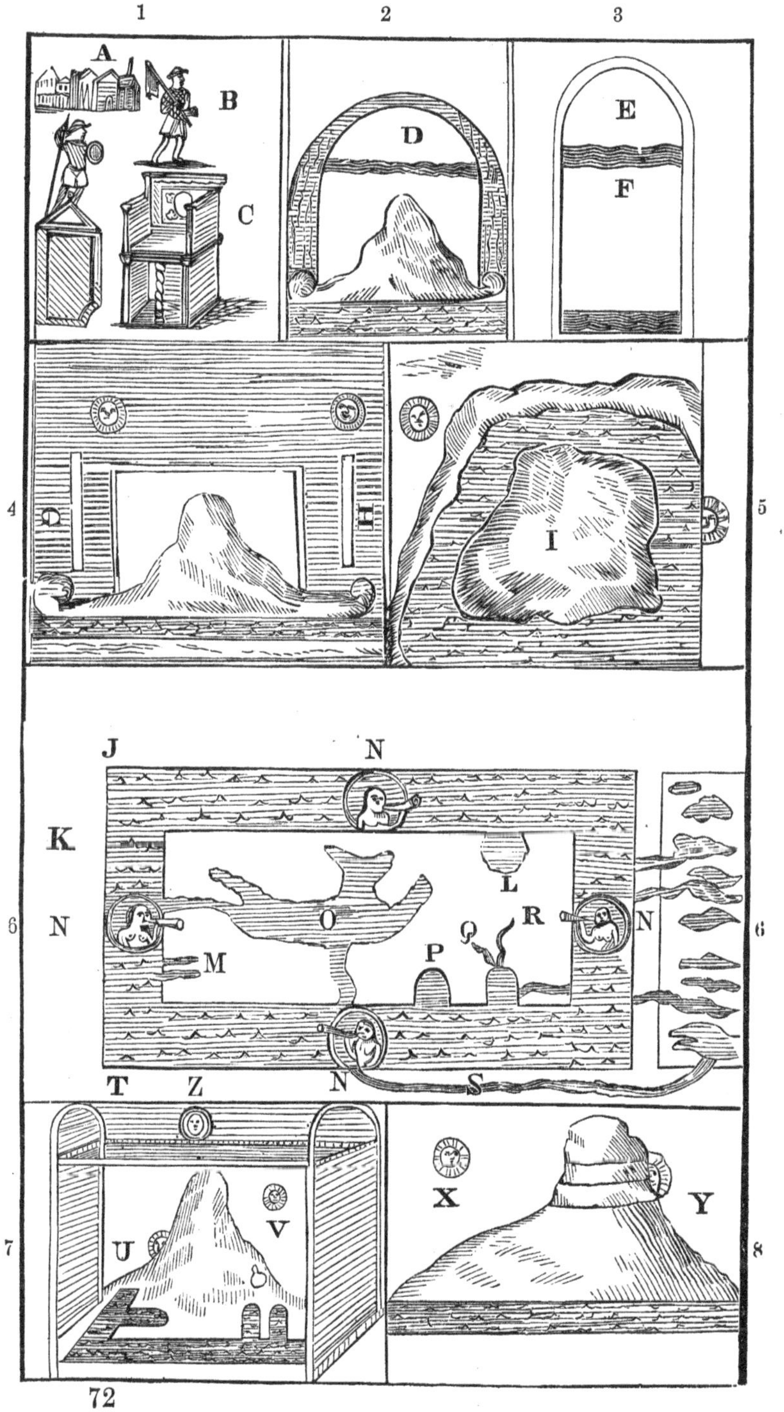

itself into our text; but the letters A, B, C, &c., indicate the place of each. As the work of Cosmas is exceedingly rare, we hope theological students will appreciate the pains taken to furnish them with so clear an illustration of what they still call "Mosaic" cosmogony. —G. R. G.]

COSMAS'S GREEK EXPLANATIONS.

A — Adulis city (*Abyssinia*).
B — the road from Adulis to the East — Ethiopians travelling.
C — Ptolemy's chair.
D — Firmament.
E, F — Waters which are above the Firmament.
G, H — Columns (to support the Firmament).
I — inhabited earth.
J — land beyond the Ocean, where men dwelt before the Deluge.
K — land beyond the Ocean.
L — Caspian Sea.
M — River Phison.
N — 4 Points of the compass.
O — Mediterranean Sea.
P — Arabian Gulf.
Q — Tigris.
R — Euphrates.
S — River Gihon.
T — land beyond the Ocean.
U — the Sun Occident.
V — the Sun Orient.
X — the Sun Occident.
Y — the Sun Orient.
Z — is Cosmas's picture of the Almighty looking down, and seeing that "it was good."

In the IVth book of "Topographia Christiana," the pious Cosmas describes his hydrographic and ecclesiastical principles; but, rich as they are, his argumentation is too prolix for our purposes, which are served by translating Montfaucon's synopsis of his author's elucidation of *Plate* I.

"*Fig.* 1. In the first figure, the city ADOULI or *Adulis* [in Abyssinia] (for it is so called in both ways by Cosmas) is shown. *Axumis*, which is two miles distant from the Red Sea, is situated to the East; for which reason an *Ethiopian* is represented, in his Ethiopian costume, taking the Axumis road to Adulis. Then Ptolemy's chair is delineated in the form it is said to have had by Cosmas. That [part of the chair] however, sculptured all over in characters, had only the last portion of the inscription added. But the inscription on the stone tablet placed opposite was finished—a fragment of which from the lower part together with its characters or letters had been destroyed. Above the stone tablet king Ptolemy EVERGETES himself is represented in his military attire as he appears in the picture. These things you will find more fully explained in page 140 and the following.

"*Fig.* 2. In the second figure the shape of heaven and earth is delineated according to the opinion of Cosmas and the old Fathers, who thought the earth, as it were, a *flat surface*, extending beneath and inclosed by *walls* on all sides; and that these walls were raised to an immense height, and finally arranged themselves into the form of a vault; while the *firmament* pervaded the higher part of the vault so that it (beatorum sedes) might be the seat of the Blest. [The same idea ('firmament,' Hebraicè SK*h*KIM K*h*ZKIM — literally, *solid skies*) occurs in *Job* xxxvii. 18. Thus Cahen renders — 'As-tu étendu avec lui les *cieux, solides* comme un miroir métallique?' And Noyes —

— 'Canst thou like him spread out the *sky*
Which is *firm* like a molten mirror?' 700

But, under the firmament, they thought the sun, moon, and stars, were put in motion; and that a *conical* mountain of wondrous height rose up in the northern parts of the earth; and while the sun, performing his *circuit round the earth*, stood behind this mountain, there was night to those inhabiting the earth; but, on the other hand, it was *day* when the sun shone upon us on the reverse [i. e., on *our* side] of the mountain: and, in a similar way Cosmas reasons with respect to the moon and stars; see page 186 and the following.

"*Fig.* 3. Exhibits a prospective view of the universe; that is to say, of the heavens and the earth in the part where they are more closely drawn together; for Cosmas thought the earth was *square* and *oblong*, and the same is assumed with respect to the heavens. See page 186 and following.

"*Fig.* 4. Represents a conical mountain, and the earth, together with the sun and moon, *under* the firmament. But on the sides [*Job* ix. 6—âMUDIH—'*Pillars* (of the earth)'; *Job* xxvi. 11—'*pillars* of the skies'] are represented the *pillars* of heaven, with an inscription [in *Greek!*] upon the plan here presented—*οἱ ςύλοι τοῦ οὐρανου*—*the columns of the sky;* which columns, according to the opinion of Cosmas, I think to be those walls which arise on the sides from the earth up to the heavens (*Psalms* cxlviii. 4—'Ye WATERS that be *above* the *skies*').

"*Fig.* 5. The outline of the earth and its *ἐκνογραφίαν* are traced out. You may observe that Cosmas conjectured that the immensely-high conical mountain presented an obstacle where our earth could not, at the northern part, be so well inclosed by a right line; because its foundations on that side are *round*, as if they proceeded from a great promontory in the ocean.

"*Fig.* 6. Displays the rugged plain of the earth, such as Cosmas explains in many places; for he thought, as we have said before, that the earth was oblong, and its length twice as long as its breadth, and that an *ocean* surrounded the entire earth, as is here represented. But, beyond the ocean, there was yet *another* land adhering closely, on all sides, to the walls of heaven. Upon the eastern side of this *transmarine* land he judges that MAN WAS CREATED; and that there the *paradise of gladness* was located, such as here, on the eastern edge, is described: where it received our first parents, driven out of paradise to that extreme point of land on the sea-shore. Hence, upon the coming of the *deluge*, Noah with his sons was borne by the ark to this earth we now inhabit. The four *rivers*, he supposes, to be gushing up the spouts in paradise; with subterranean channels through the ocean, to our earth, and in certain places that they gush out anew. He considers that the Hyrcanian Sea [Caspian] is joined to the ocean; which we have elsewhere shown was the opinion of certain ancients.

"*Fig.* 7. He briefly dispatches the whole machinery of the world, which, as the ancients thought, was composed of the *sky* and the *earth*. Its form he represents, with the conical mountain above alluded to. But Cosmas-Ægypticus deemed that the earth which we inhabit was always inclining from the north to the south. Albeit Cosmas contradicts himself. How can such a mass as that of heaven and earth stand, supported by nothing, since it is always pressed downward? He answers—the earth, inasmuch as it is ponderous matter by nature, seeks the bottom; but the igneous parts tend upward; therefore, when sky and earth are thus joined and cannot be torn asunder, the one pressing from above and the other from below, neither yielding to the other, the whole machine remains immovable and *suspended*. ['This is a grand argument,' says Mr. Burke, commenting in a private letter, 'and beats the Newtonian theory out and out! Only fancy; two forces shut up in a box, one pulling up, and the other pulling down, and the BOX, *in consequence*, remaining 'immota et suspensa!' This is, beyond exception, the brightest mechanical idea I have ever come across'].

"*Fig.* 8. He represents the conical mountain on that side which is turned adversely to the earth; where, when the sun arrives, night is produced to the earth's inhabitants. In the same place the *revolutions of the sun* are indicated by lines [upon the conical mountain]; whereby the various *seasons* of the year are caused. When, therefore, the sun arrives at the lower line, the nights then are longer, and it makes *winter*, *τροπη*, or revolution: the sun performing the major portion of his course *behind* the mountain. When, however, the sun comes to the middle line of the mountain, then the *equinox* is produced; the sun in performing his course having reached the equinoctial line When, finally, the sun touches the uppermost line, then the *summer* revolution takes place, and he attains to the *tropic*. This is in conformity with the opinion of Cosmas. who describes the revolutions of the sun in these words—*μεγάλη νὺξ*, *great night; μέση νὺξ, middle night; μικρὰ νὺξ little night;* as you behold in the picture."

Through the above parody upon nature, Cosmas explained all celestial phenomena — the course of the moon, its phases and eclipses, as well as the sun's rotation round the earth's flat plain. The *Topographia Christiana* became the text-book of ecclesiastical orthodoxy, for above 800 years, down to Galileo; and Cosmas's caricature on the one hand, coupled with ignorance of the Hebrew text of Joshua (x. 12–14) on the other, induced the murder of Giordano Bruno.

Nevertheless, according to the literal language of the first IX chapters of "Genesis," Cosmas was not far from the truth. Were the ancient writers of those chapters to arise from the grave, and were they respectfully requested to indicate which commentary best represented their meaning — that of the *Topographia Christiana;* or those recent attempts "to make Moses sound in the faith of the geological section of the British Association for the Advancement of Science"[701] — they would unanimously claim the former as their own.

Happy middle-ages; when Europe made up in credulity what it lacked in intelligence! "They had neither looked into heaven, nor earth; neither into the sea, nor the land, as has been done since. They had philosophy without scale, astronomy without demonstration. They made war without powder, shot, cannon, or mortars; nay, the mob made bonfires without squibs or crackers. They went to sea without compass, and sailed lacking chronometers. They viewed the stars without telescopes, and measured altitudes without barometers. Learning had no printing-press, writing no paper, paper no ink; magnetism no telegraph, iron no rails, steam no boilers. The lover was forced to send his mistress a deal-board for a love-letter, and a billet-doux might be of the size of a trencher. They were clothed without manufactures, and the richest robes were the skins of formidable monsters. They carried on trade without books, and correspondence without postage: their merchants kept no ledgers; their shopkeepers no cash-books. They had surgery without anatomy, physicians without materia-medica; who gave emetics without ipecacuanha, and cured agues without quinine. They dispensed with lucifer-matches, coffee, sugar, tea, and tobacco"[702] — and, never having heard of the first three chapters of "Genesis," they believed in *Topographia Christiana!*

The *book* is scarcely known, now-a-days, to theologers; but its commentary (orally transmitted from father to son) survives all around us. We have conceived it our duty not to let the one continue without the other; and therefore have rescued from further oblivion the Mosaic chart of Cosmas.

Section H. — ANTIQUITY OF THE NAME "ADaM."

After what has been already set forth, there seems scarcely reason to answer an interrogatory, above propounded, relative to "human creation" as narrated in *Genesis.* Archæological criticism might finally rest upon *one* Hebrew word; viz. ADaM.

The philological law of *triliterals*, in Semitic tongues, has been touched upon during previous examinations of Xth Genesis. "Non omnia possumus" — and the authors must reiterate that, in order to keep within *one* volume, they have been forced to expurgate redundancies, often, they fear, at the sacrifice of perspicuity. In lieu of *extracts* from the pages of Lanci, Meyer, Gesenius, Neumann, Ewald, Wilhelm von Humboldt, Prichard, Bunsen,—in addition to those previously drawn from Rawlinson, De Saulcy, &c. — all corroborating our correctness, we must substitute *references* to their authoritative works.

The reader will observe, notwithstanding, that the bisyllable ADM cannot be a primitive but must be a *secondary* formation, according to the progressive scale of linguistic development. To reach the primary root, or monosyllable, within this triliteral word contained, an *affix*, a *suffix*, or a *medial*-letter, must be first removed. Among Hebraists of the highest modern school, on the European continent, the fact that "Adam" is a *dissyllabic* name alone suffices to prove that its possessor appeared on earth thousands of years subsequently to the primordial ages of humanity; because *in principio* man articulated but *monosyllables.* Or else (what is the same thing in result, no less than more positive) the Israelite who (in some form of *coin*-letter) wrote the word ADM, of *Genesis*, lived at a philological epoch when the pristine *monosyllables* had already (organically through development) merged into words of *two* syllables; and therefore, that writer committed an egregious anachronism when he retro-leptically ascribed a *triliteral* proper-name, or rather noun, to his first human progenitor.

The word ADM, or with an additional vowel, ADaM, is consequently to be divided into two separate words, A and DaM; or A-DaM. Now, A, *aleph*, is the primeval, Semitic, masculine article $A =$ "the":[703] an article that, in Scripture, is prefixed to above forty masculine substantives; although, until recently, the fact was unperceived by Hebrew grammarians, or Jewish lexicographers.

In the next place, the word ADaM does not proceed, as the Rabbis suppose, from ADaMaH (*Gen.* ii. 7)—a *bisyllable* from a *trisyllable!*—but the latter is an extension of the former root, DaM (Arabicè, *Dem*), meaning *blood;* the color of which, being *red*, originated the secondary signification of DaM, as "red;" and "to be *red.*"

Consequently, A, the letter "*aleph*," being the masculine article *the;* and the noun DaM meaning *blood*, or "red," we have only to unite these two words into A-DaM, to read *the-blood*, or THE-RED, in "Genesis;" which duplex substantive, applied to man, naturally signifies "*the-red*-man;" and, when applied to the *ground*, ADaMaH ("out of the dust" of which this *the-red*-man, ADaM, was moulded), it means *the-red-earth: i. e.*, that rubescent soil out of which the Jehovistic writer of Genesis IId imagined Hebrew man to have been fashioned by Creative artisanship. The BeNi-ADaM also, in Psalms (xlix. 2. Comp. *Ps.* lxii. 9: and contrast with BeNoT*t*-HaADaM, *Gen.* vi. 2), are reputed to be *patricians* of the pure Abrahamic stock; whereas the plebeians (including all those who are, like Anglo-Saxons, mere GOIM, *Gentiles*) belong altogether to a different and lower level . . . in the eye of IeHOuaH.

We adopt entirely the Italian rendering of the great interpreter of Sacred Philology at the Vatican; and think, with Lanci, that *il-rossicante*, "the-Blusher," is the happiest translation of the old Semitic particle and noun A-DaM.

How does this interpretation bear upon ethnography?

Reader! simply thus. As no "Type of Mankind" but the *white race* can be said (physiologically) to *blush;* it follows, that, according to the conception of the writers of Genesis (who were *Jews* and of the "white race"), not only did the first human pair converse between themselves, no less than with God and with the serpent, in pure *Hebrew*, but they were essentially A-DaM*ites* (*red*-man and woman) "blushers:"—and therefore, these Hebrew writers, never supposed that A-DaM and ISE (vulgaricè, Adam and Eve) could have been of any stock than of the *white* type—in short, Hebrews, *Abrahamidæ*, like themselves—these writers aforesaid.

Thus, through a few cuts of an archæological scalpel, vanishes the last illusion that any but *white* "Types of Mankind" are to be found in the first *three chapters* of the book called "Genesis."

The "Chinese" having been carefully removed further on from connection with the Mesopotamian SINIM of Isaiah (xlix. 12), nothing remains but to refer the reader to the *map* [*supra, p.* 552] we have given of Xth *Genesis* for the whole of Ethnography comprehended by the writers of the Old Testament: Strabo, who followed Eratosthenes about B. C. 15, furnishing every possible information upon what of geography was attainable, in the first century after C., by the writers of the New.

The present authors have asserted these results before.

"That part of the *map* colored deep-red includes all the world known to the inspired writers of the Old Testament; and this, with the part colored pale-red, includes all known to St. Paul and the Evangelists. — As we have no evidence that their inspiration extended to matters of science, and we know that they were ignorant of Astronomy, Geology, Natural History, Geography, &c. — what evidence is there that they knew anything of the INHABITANTS of countries unknown to them, viz.: Americans, Chinese, Hindoos, Australians, Polynesians, and other contemporary races?" — (J. C. N.: *Bibl. and Phys. Hist. of Man;* New York, 1849; "Map" and pp. 54–67.)

"These unhistorical *origines* of nations are now adverted to, as a prelude to the discussion of the Xth chapter of Genesis (see *Ethnological Journal*, No. VI., note, page 254), whereby it will be demonstrated that, under the *personifications* of "Shem, Ham, and Japheth," their fifteen *sons*, and seventy-one *grand-children*, the Hebrew geographers, whose ken of the earth's superficies was even more limited than that of Eratosthenes, about B. C. 240, have never alluded to, nor intended, Mongolian, Malayan, Polynesian, American, or Nigritian races." — (G. R. G.: *Otia Ægyptiaca;* London, 1849: p. 124, "note.")

Five years have since elapsed. Most of the conclusions advanced by the authors have been challenged. Whether those conclusions were based, or not, upon thorough investigation of each department of the subject, the reader of the present volume is now best qualified to decide.

PART III.

Supplement.

BY GEO. R. GLIDDON.

ESSAY I.

ARCHÆOLOGICAL INTRODUCTION TO THE Xth CHAPTER OF GENESIS.

"Scriptura primum intelligi debet grammaticè antequam possit explicari theologicè."
(LUTHER.)

"THE XTH CHAPTER OF GENESIS — *Archæological Introduction to its Study*" — is the heading given, in our "Prospectus," to Part III. of this work.

To the generality of readers, educated under convictions that every process calculated to probe the historical evidences of the Hebrew Scriptures has heretofore been rigorously applied to them, an *Introduction* termed "archæological" may seem, to say the least, superfluous at the present day — while to not a few persons, the proposed method of examination may, at first sight, even wear the aspect of presumptuousness. Nevertheless, having announced the intention, it behooves us to justify it.

In common with other Protestants, since our earliest childhood, we have been assured that the Bible is the *word of God* — and that the inspiration of the writers of both Old and New Testaments rests upon testimony the most irrefragable. We have also been admonished in the language of the Apostle (1) to "*search* the Scriptures;" coupled with the corroborative exhortation, (2) "seek, and ye will find; knock, and it will be opened unto you."

Thus, on the one hand, asseverations the most positive fortify the inquirer who conscienticusly examines whether the divine revelation of the Bible and the inspiration of its penmen are "built upon a rock;" at the same time that, on the other, the Gospels themselves invite him to search, seek, and scrutinize.

Supported by such authority, no legitimate objection can be sustained, by Protestants, against the employment of what we conceive to be the only method through which the historical validity of a given proposition can be thoroughly tested; nor will logical orthodoxy contest Vater's axiom—"*Faith in Christ can set no limits to critical inquiries; otherwise he would hinder the knowledge of Truth.*"

(1) *The good Tidings according to* JOHN v. 39.

(2) *The good Tidings according to* MATTHEW, vii. 7; copied in *The good Tidings according to* LUKE, xi. 9. We follow SHARPE: *The New Testament, translated from Griesbach's Text;* wherein "will" is substituted for the "shall" of king James's version.

Homo, according to Bacon, *naturæ minister et interpres, tantum facit et intelligit quantum de naturæ ordine re vel mente observaverit; nec amplius scit, aut potest.* A finite being, circumscribed within the intellectual horizon of the mundane age in which each individual lives, man can reason merely upon phenomena. *Quicquid enim*, wrote the immortal Newton, *ex phenomenis non deducitur hypothesis vocanda est; et hypotheses vel metaphysicæ, vel physicæ, vel qualitatum occultarum seu mechanicæ, in philosophia locum non habent.*

What is *Philosophy?* Etymologically, the "love of wisdom," and paraphrastically, the "love of knowledge;" multiform are the significations through which this sublime Greek word has travelled. From the ablest English historian (3) of its phases, we extract such paragraphs as will convey to the reader our individual perceptions of its import at this day.

"We shall find some obscurities cleared up, if we can master an accurate and comprehensive definition of Philosophy. The definition I have finally settled upon is this:—

"*Philosophy is the explanation of the Phenomena of the Universe.* By the term explanation, the subject is restricted to the domain of the intellect, and is thereby demarcated from religion, though not from theology.

"Philosophy is inherent in man's nature. It is not a caprice, it is not a plaything, it is a necessity; for our life is a mystery, surrounded by mysteries: we are encompassed by wonder. The myriad aspects of Nature *without*, the strange fluctuations of feeling *within*, all demand from us an explanation. Standing upon this ball of earth, so infinite to *us*, so trivial in the infinitude of the universe, we look forth into nature with reverent awe, with irrepressible curiosity. We must have explanations. And thus it is that Philosophy, in some rude shape, is a visible effort in every condition of man — in the rudest phase of half-developed capacity, as in the highest conditions of culture: it is found among the sugar-canes of the West Indies, and in the tangled pathless forest of America. Take man where you will—hunting the buffalo on the prairies, or immovable in meditation on the hot banks of the Ganges, priest or peasant, soldier or student, man never escapes from the pressure of the burden of that mystery which forces him to seek, and readily to accept, some explanation of it. The savage, startled by the muttering of distant thunder, asks, 'What is that?' and is restless till he knows, or fancies he knows. If told it is the voice of a restless demon, that is enough; the explanation is given. If he then be told that, to propitiate the demon, the sacrifice of some human being is necessary, his slave, his enemy, his friend, perhaps even his child, falls a victim to the credulous terror. The *childhood of man* enables us to retrace [archæologically] the infancy of nations. No one can live with children without being struck by their restless questioning, and unquenchable desire to have everything explained; no less than by the facility with which every authoritative assertion is accepted as an explanation. The History of Philosophy is the study of man's successive attempts to explain the phenomena around and within him.

"The first explanations were naturally enough drawn from analogies, afforded by consciousness. Men saw around them activity, change, force; they felt within them a mysterious power, which made them active, changing, potent: they explained what they saw, by what they felt. Hence the fetichism of barbarians, the mythologies of more advanced races. Oreads and nymphs, demons and beneficent powers, moved among the ceaseless activities of Nature. Man knows that in his anger he storms, shouts, destroys. What, then, is thunder but the anger of some invisible being? Moreover, man knows that a *present* will assuage his anger against an enemy, and it is but natural that he should believe the offended thunderer will also be appeased by some offering. As soon as another conception of the nature of thunder has been elaborated by observation and the study of its phenomena, the supposed Deity vanishes, and, with it, all the false conceptions it originated, till, at last, Science takes a rod, and draws the terrible lightning from the heavens, rendering it so harmless that it will not tear away a spider's web!

"But long centuries of patient observation and impatient guessing, controlled by logic, were necessary, before such changes could take place. The development of Philosophy, like the development of organic life, has been through the slow additions of thousands upon thousands of years; for humanity is a growth, as our globe is, and the laws of its growth are still to be discovered. . . . One of the great fundamental laws has been discovered by Auguste Comte — viz: *the law of mental Evolution* . . . which he has not only discovered,

(3) G. H. Lewes: *Biographical History of Philosophy;* London, 1846. The substance of our remarks may be found in vol. iv. pp. 245–262, under the heading of Auguste Comte, "the Bacon of the nineteenth century," and author of *Cours de Philosophie Positive.* The original source of this abstract may be found in Comte, vol. i. edit. Paris, 1830, "Exposition," pp. 3–5, 63, &c.; but we take Mr. Lewes's later definitions from *The Leader;* London, 1852; April 17, 24, and May 1. A profound thinker has recently done full honor to Mr. Lewes's work. (Vide McCulloh: *Credibility of the Scriptures;* Baltimore, 1852, vol. ii. pp. 454–458.)

but applied historically. . . . This law may be thus stated: "Every branch of knowledge passes successively through three stages: 1st, the *supernatural*, or fictitious; 2d, the *metaphysical*, or abstract; 3d, the *positive*, or scientific. The first is the necessary point of departure taken by human intelligence; the second is merely a stage of transition from the supernatural to the positive; and the third is the fixed and definite condition in which knowledge is alone capable of progressive development.

"In the attempt made by man to explain the varied phenomena of the universe, history reveals to us," therefore, "three distinct and characteristic stages, the *theological*, the *metaphysical*, and the *positive*. In the first, man explains phenomena by some fanciful conception suggested in the analogies of his own consciousness; in the second, he explains phenomena by some *à priori* conception of inherent or superadded entities, suggested in the constancy observable in phenomena, which constancy leads him to suspect that they are not produced by any *intervention* on the part of an external being, but are owing to the nature of the things themselves; in the third, he explains phenomena by adhering solely to these constancies of succession and co-existence ascertained inductively, and recognized as the *laws* of Nature.

Consequently, "in the *theological* stage, Nature is regarded as the theatre whereon the arbitrary wills and momentary caprices of Superior Powers play their varying and variable parts. . . . In the *metaphysical* stage the notion of capricious divinities is replaced by that of *abstract entities*, whose modes of action are, however, invariable. . . . In the *positive* stage, the invariableness of phenomena under similar conditions is recognized as the sum total of human investigation; and, beyond the laws which regulate phenomena, it is considered idle to penetrate."

"Although every branch of knowledge must pass through these three stages, in obedience to the law of evolution, nevertheless the process is not strictly chronological. Some sciences are more rapid in their evolutions than others; some individuals pass through these evolutions more quickly than others; so also of nations. The present intellectual anarchy results from that difference; some sciences being in the *positive*, some in the *supernatural* [or *theological*], some in the *metaphysical* stage: and this is further to be subdivided into individual differences; for in a science which, on the whole, may be fairly admitted as being positive, there will be found some cultivators still in the metaphysical stage. Astronomy is now in so positive a condition, that we need nothing but the laws of dynamics and gravitation to explain all celestial phenomena; and this explanation we know to be correct, as far as anything can be known, because we can predict the return of a comet with the nicest accuracy, or can enable the mariner to discover his latitude, and find his way amidst the 'waste of waters.' This is a *positive* science. But so far is meteorology from such a condition, that prayers for dry or rainy weather are still offered up in churches; whereas if once the *laws* of these phenomena were traced, there would be no more prayers for rain than for the sun to rise at midnight."

We have only to reverse the order, and apply its triple classification to individuals, and in the natural arrangement of the strata, tracing backwards from the *positive* to the *metaphysical*, from the latter down to the *supernatural*, we shall perceive that this last, at once the oldest stage and unhappily the most common, represents the least mature, the least educated, the most antiquated, state of human intelligence. In consequence, the mere *supernaturalist* believes anything and everything, however impossible.

"The *Metaphysician* believes he can penetrate into the *causes* and *essences* of the phenomena around him; while the *Positivist*, recognizing his own incompetency, limits his efforts to the ascertainment of those laws which regulate the succession of these phenomena."

In the quintuple classification of those sciences into which *Positive Philosophy* has hitherto been successfully introduced, M. Comte (1832–40) admits only Astronomy, Physics, Chemistry, Physiology, and Sociology. It strikes us that, at the present day, this division is more exclusive than the progression of knowledge any longer warrants. *Archæology*, for instance, we claim to have arrived at its *positive* grade; and although its laws are by no means popularly appreciated, to have become as certain in its results as any other human science. A brief exposition of its attributes may prepare the reader for a just recognition of its utility.

Αρχαιος, *antiquus*, "ancient," and Λογος, a "discourse," are Hellenic words—meaning, when *united*, in general acceptation, "discourse or treatise on the opinions, customs, and manners of the ancients." This is the definition of *Archæology* proposed by the sage Millin, (4),

(4) *Introduction à l'étude de l'Archéologie;* Paris, 1796; pp. 2, 20, 22.

adopted by Lenormant, (5) and recognized by all true scholars from Niebuhr to Letronne; especially among those intellectual giants who since Champollion's era have solved the chief enigmas of hieroglyphical and cuneatic records. *Archæography*, as distinct from archæology, according to Fabricius, (6) is a term which should be limited to the study of ancient monuments especially, whereas archæology embraces every process of investigation into all historical subjects. Dionysius Halicarnassensis, in the first century before C., and Josephus in the first century after, treated upon *Archæology*, but entirely neglected Archæography, or the study of monuments; whence their several incoherencies: the former, however, had some clear perceptions of the truth when he named Archæology "the science of primitive origins."

Albeit, the word has deviated somewhat from its pristine sense; for among the Greeks an *archæologist* signified a man who brought together the most ancient recollections of a given country; whereas, at the present day, the name is applied exclusively to him who, possessing intimate acquaintance with the *monuments* of a given ancient people, strives through the study of their characteristics to evolve facts, and thence to deduce logical conclusions upon the ideas, tastes, propensities, habits, and history of departed nations; many of the greatest and most essential of whom having left but fragmentary pages of their *stone-books*, out of which we their successors must reconstruct for ourselves such portions of their chronicles as are lost; no less than confirm, modify, or refute such others as have reached us through original, transcribed, or translated annals.

Archæology, so to say, has now become the "backbone" of ancient history; its relation to human traditions being similar to that of Osteology to Comparative Anatomy; or to what fossil remains are in geological science. An *Antiquary* is rather a collector of ancient relics of art, than one who understands them; but an *Archæologist* is of necessity an Antiquary who brings every science to bear upon the vestiges of ancient man, and thus invests them with true historical value. In short, an Archæologist is the *monumental* historian — the more or less critical dealer in and discoverer of historical facts, according as by mental discipline, diversified attainments, and the study of *things*, he acquires thorough knowledge of each particle preserved to his research among the *débris* of antique humanity.

Were the simplest rules of this science popularly taught, we should not have to prolong the lamentations of Millin at errors prevalent for want of a little archæological knowledge. He narrates how Baronius took a statue of Isis for the Virgin Mary — how the apotheosis of the Emperor Germanicus was mistaken for St. John the Baptist's translation to heaven — and how a cameo called "the agate of Tiberius," which represents the triumphs of this prince and the apotheosis of Augustus, came to be long regarded as the triumphal march of Joseph! Neptune and Minerva giving the horse and olive to man would not have been metamorphosed into Adam and Eve eating the forbidden apple; nor would a trumpery pottery toy have been considered by His Eminence Cardinal Wiseman (7) as a Roman memento of Noah's Ark after the universal flood, although among its animals were "thirty-five human figures!" Without archæology, says Millin, one is liable with the historian Rollin to speak of the *Laocoon* as a lost monument — to dress up Greek heroes in Roman garments — to adorn Hercules with a *perruque à la Louis XIV!* Æsop, at the court of Crœsus, would hardly have addressed himself to a colonel in French uniform; nor Strabo, in "Démocrite Amoureux," have pointed his quizzing-glass at steeples, and amused his leisure by making almanacs; neither would Horace call Servius Tullius "Sire;" nor Racine have invoked a goddess as "Madame" in his classic plays. (8)

More than half a century has elapsed since Millin wrote. Hundreds of archæologists have made their works accessible to the literary public. Yet so slow is the diffusion of

(5) *Archéologie*, par M. Ch. Lenormant, de l'Institut: *Revue Archéol.*; Paris, 1844; 1re partie, pp. 1–17.

(6) *Bibliotheca Antiquaria;* p. 181.

(7) *Connection between Science and Revealed Religion;* 1849; vol. ii. pp. 139–143.

(8) See many recent instances of antiquarian shams exposed by Letronne — "L'amulette de Jules César, le cachet de Sépullius Macer, le médaillon de Zénobie, le coffret d'Antinoüs, le sabre de Vespasien, et d'autres antiquités *modernes*" — *Mémoires et Documents; Rev. Archéol.;* Paris, 1849; pp. 192–223.

critical knowledge, that in our own land and hour, there are still some not uncultivated minds who imagine the *Aborigines* of this American continent to have descended from the "Lost Tribes of Israel"(9)—who see the *Runic* scribblings of Norsemen upon the Indian-scratched Rock of Dighton (10)—who, regardless of Squier's exposure,(11) yet suppose the local pebble manufactured for that *museum* since 1838, to attest *Phœnician* intercourse with the mound-builders of Grave Creek Flat (12)—and who, disdaining to refer to the long-published determination of its pseudo-antiquity, (13) still believe that the *gold seal-ring* of RA-NEFER-HET, a functionary attached to a building called, about the sixth century B. C., *after* King SHOOPHU, should have once adorned the finger of CHEOPS, builder of the Great Pyramid in the thirty-fourth century B. C. (14); thereby becoming 5300 instead of only some 2500 years old!

The instances around us of the misconceptions, which the slightest acquaintance with the rudiments of archæology would consign forever to oblivion, are inexhaustible. Would that some of them were less pernicious to moral rectitude! They offend our vision under the prostituted names of "*Portraits* of CHRIST" (15) — they excite one's derision in the ludicrous anachronisms of modern art current as "*Pictorial Bibles*" (16) — they bear witness to theological ignorance when *Chinese* are asserted to be referred to in the SINIM of Isaiah (17)—and they amount to idiocy when ecclesiastics continue disputing whether MOSES wrote a *resh*, R, or a *daleth*, D, in a given word of the Hebrew Pentateuch, notwithstanding that every archæologist knows that the *square-letter* characters of the present Hebrew Text (18) were not invented by the Rabbis before the second century *after* Christ; or 1600 years posterior to the vague age when IeHOuaH buried the Lawgiver "in a valley in the land of Moab opposite to Beth-peor; but no man has known his sepulchre *unto this day*."(19) But—"point de fanatisme même contre le fanatisme: la philosophie a eu le sien dans le siècle dernier; il semble que la gloire du nôtre devrait être de n'en connaître aucun." (20)

The above illustrations suffice to indicate some of the utilitarian objects of the science termed "Archæology;" which furnishes the only logical methods of attaining historical certainties. Its indispensableness to correct appreciations of biblical no less than of all other history, nevertheless, remains to be proved by its application. We shall endeavor to be precise in our experiments; but, must not forget that "precision is one thing, certainty another. An absurd or false proposition may be made very precise; and, on the other hand, although the sciences vary in degree of precision, they all present results equally certain." We propose to test the principles of archæological criteria by applying them to biblical studies, and to test the authenticity of *one chapter* of the Hebrew records through the former's application: and inasmuch as Truth must necessarily harmonize with itself, if archæology be a true science the Scriptures will prove it to be so incontestably; and if the Bible be absolute truth, archæology will demonstrate the fact. We need not perplex ourselves with apprehensions. It would imply but small faith in the *Bible* were we to suppose that arch-

(9) DELAFIELD: *American Antiquities.*

(10) *Transactions of the Royal Society of Antiquaries of Copenhagen*, 1840–'43. *Antiquitates Americanæ*, 1837; sect. xv.

(11) London *Ethnological Journal:* "Monumental Evidence of the Discovery of America by the Northmen critically examined" — Dec. 1848; pp. 313–324.

(12) SCHOOLCRAFT: New York *Ethnological Society's* Trans. 1845; vol. i. pp. 386–397.

(13) See "A Card": New York *Courier and Enquirer*, 12 Feb. 1853.

(14) ABBOTT: *Catalogue of a Collection of Egyptian Antiquities*, now exhibiting at the Stuyvesant Institute; New York, 1853; plate No. 1051, p. 64.

(15) Founded exclusively upon no more historical bases than the spurious "Letter of LENTULUS" — or derived from "Veronica's Sudarium"; ALBERT DURER, 1510,—vide COLE: *Passion of our Lord;* London, 1844.

(16) HARPERS', for instance; New York, 1842–'45.

(17) Rev. Dr. SMYTHE: *Unity of the Human Races;* 1840 — "And while even China (*Is.* li. [*sic*] 12, Sinim, a remote country in the S. E. extremity of the earth, as the context intimates) and the islands of the sea are specified" — p. 43. and note.

(18) GLIDDON: *Otia Ægyptiaca;* p. 112; and *infra*, further on.

(19) *Deuteronomy* xxxiv. 6 — CAHEN's translation.

(20) AMPÈRE; *Recherches*, &c.; Rev. des Deux Mondes; Sept. 1846, p. 738.

æological scrutiny could affect the divine origin insisted upon for the book itself by those who make it the unique standard of all scientific as well as of all moral knowledge.

Instead, however, of the ordinary mode in which biblical history is presented to us in books bearing the authoritative title of professed "Christian Evidences," the requirements of archæology demand that we should reverse the order of examination. In lieu, for instance, of asserting *à priori* that the *Creation of the world* took place exactly "on October 20th, B. C. 4005, the year of the *creation*" (21) — or sustaining, *ex cathedra*, with universal orthodoxy, that MOSES *wrote the Pentateuch* — it is incumbent upon us, while we deny nothing, to take as little for granted. If such be the era revealed by the Text, our process will lead us to that date, with at least the same precision through which Lightfoot (by what method is unknown), ascertained that *Anno Mundi* I, "VIth day of creation . . . his (Adam's) wife the weaker vessell: she not yet knowing that there were any devils at all . . . sinned, and drew her husband into the same transgression with her; this was about *high noone*, the time of *eating*. And in this lost condition into which Adam and Eve had now brought themselves, did they lie comfortlesse till towards the cōol of the day, or *three o'clock afternoon.*" (22) If the *Pentateuch* was originally penned in the Mosaic autograph, the proof will resile to our view, through archæological deductions, with the force of an Euclidean demonstration.

The analytical instruments of archæology are purely *Baconian;* viz: proceeding from the known to the unknown; through a patient retrogressive march from to-day to yesterday, from yesterday to the day before; and so on, step by step, backwards along the stream of time. Each fact, when verified, thus falls naturally into its proper place in the world's history; each event, as ascertained, will be found tabulated in its respective stratum. It is only when our footsteps falter, owing to surrounding darkness or to treacherous soil, that we may begin to suspect historical inaccuracies; but, at present, we have no right to anticipate any such doubts, considering the averments of œcumenic Protestantism, of the orthodox sects, that the *Bible is the revealed word of God.*

Our inquiries are directed to a single point. We desire to ascertain the origin, epoch, writer, characteristics, and historical value of but one document: viz.—*The* Xth *Chapter of Genesis;* familiar to every reader. It is presented, however, to our inspection as one of *fifty* chapters of a book called "Genesis"—this book being the first of *thirty-nine* (23) books that constitute the compendium entitled the "Old Testament;" and the latter is bound up in the same volume with another collection to which the name of "New Testament" is given: the whole forming together that literary work to which the designation of "*The* BIBLE" is reverentially applied in the English tongue — a name derived from *byblos*, the Greek name for *papyrus*, being the most ancient material out of which its derivative *paper* was made. *Byblus*, the Egyptian plant, gave to the Greeks their name for paper, and paper their name for "the book" in *το βιβλειον*. On adopting Christianity, the Greeks designated their earliest translation of the Hebrew Scriptures, TO BIBLEION, as *the book* — "par excellence;" which words we moderns have adopted into our national tongue in the form of "Bible."

With every desire on our part to obtain solution of our queries by the most direct road and in the shortest method, we do not perceive the possibility of detaching a solitary *chapter* of the Bible from the volume itself, until by archæological dissection we are enabled to demonstrate that such separation is feasible. In consequence, it behooves us to examine, with as much brevity as is consistent with perspicuity, the entire *Bible;* and, if we hold "all the books of the Bible (24) to be equally true," the Xth *chapter* of the first book will be found unquestionably to be true likewise.

Soliciting that the reader should divest his mind, as far as in him lies, of preconceived biases; we invite him to accompany us patiently through an investigation, in which the

(21) Rev. Dr. NOLAN: *The Egyptian Chronology Analyzed;* London, 1848, p. 392.

(22) *Harmony, Chronicle, and Order of the Old Testament,* &c.; London, 1647, p. 5.

(23) Mystic origin of the XXXIX "Articles" of the Anglican Church.

(24) POOLE: London *Literary Gazette,* 1849, p. 432 — unaccountably suppressed in *Horæ Ægyptiacæ,* 1851.

subject banishes all ornament, but that cannot fail to elicit some portions of the truth.

The incipient steps of our analysis do not call for much expenditure of erudition. In popular Encyclopædias most of the preliminary information may be verified by the curious reader; for Calmet, Kitto, and Horne, contain catalogues of the various editions of the *Bible*, done into English, that have been put forth, during the last four centuries, from A. D. 1526 down to the present year.

At the sight of such catalogues of different *translations* said to proceed from one and the same original, few can refrain from asking, in all humbleness, why, if any one of them were absolutely correct, should there have been a necessity for the others? In the course of studies carried over many years, we have been at pains to compare sundry of the most prominent English translations (among them ancient as well as modern editions), not only with themselves, but often with the Latin, Greek, or Hebrew originals, of which each purports to supply a faithful rendering. They all differ! some more than others; but in each one may be found passages the sense of which varies essentially from that published by the others. Hence arose in our minds the following among other doubts.

Some of these *Translators* can have known little or nothing of *Hebrew* — or they must have translated from different originals — or, they did not consult the *Hebrew* Text at all, but rendered from the Latin or the Greek versions — or (what recurs with far more frequency), each *translator*, wherever the original was ambiguous, rendered a given passage in accordance with his own individual biases, or with the object of fortifying the peculiar tenets of his Church, Kirk, Conventicle, Chapel, or Meeting-house. Now, these discordant *Bibles* being thrust upon us, each *one* as the only and true "Word of God," it is humanly inconceivable that God should have uttered that Word in so many different ways, and thereby have rendered nugatory the comprehension of one passage, by permitting a translation, in significance totally distinct, of the self-same passage in other modern editions. For instance, that the reader may at once seize our meaning: there are few texts more frequently quoted, especially under circumstances where consolation is administered; there are none perhaps that have originated such Demosthenian efforts at pulpit-oratory, or have produced in some minds more of those extatic emotions "that the world cannot give," than the verse wherein Job ejaculates — "For I know *that* my Redeemer liveth." (xix. 25). The "*Multitude* of those who are called Christians," as Origen termed them in A. D. 253 (25); the "Simpletons, not to say the imprudent and the idiotic," of Tertullian, A. D. 245; (26) the "Ignorant" of St. Athanasius, A. D. 373 (27); and the "Simple believers" of the milder St. Jerome, A. D. 385 (28); have always imagined, in accordance with the lower scholarship of orthodoxy, that Job here foreshadows the Messianic advent of Christ. (29)

The context does not appear, philologically or grammatically, to justify such conclusion; inasmuch as the preceding verses (1 to 22) exhibit Job — forsaken by his kindred, forgotten by his bosom friends, alien in the eyes of his guests and of his own servants — overwhelmed with anguish at the acrid loquacity of Bildad the *Shuhite*, protesting vehemently against these accusations, and wishing that his last burning words should be preserved to posterity in one of *three* ways. To support our view, and to furnish at the same time evidences of different *translations*, we lay before the reader three renderings of verses 23 to 26. He can, by opening other translators, readily verify the adage that "doctors differ," although the *Hebrew* Text is identically the same throughout.

---

(25) *Commentary upon John:* and *Contra Cels.*, lib. viii..

(26) *Ad Praxeam*, sec. iii.

(27) *De Incarn. Verb. — contra Paul. Samosatæ.*

(28) *Comm. in Es.* xxxii.

(29) Noyes: *Op. cit.*, p. 147 — "That there is no allusion to Christ in the term [*redeemer*], nor to the resurrection to a life of happiness, in the passage, has been the opinion of the most judicious and learned critics for the last three hundred years; such as Calvin, Mercier, Grotius, Le Clerc, Patrick, Warburton, Durell, Heath, Kennicott, Doederlein, Dathe, Eichhorn, Jahn, De Wette, and many others."

I. King James's *Version*. The *italicized* words are the Translators'.

23 " Oh that my words were now written! oh that they were printed [*sic!*] in a book!
24 That they were graven with an iron pen and lead in the rock for ever!
25 For I know *that* my redeemer liveth, and *that* he shall stand at the latter *day* upon the earth.
26 And *though* after my skin *worms* destroy this *body*, yet in my flesh shall I see God."

The marginal reading, authority unknown, substitutes—" Or, *After I shall awake, though this* body *be destroyed, yet out of my flesh shall I see God.*" In the authorized version, by the interpolation of "worms," Job is made a believer in the resurrection of the body: in the margin, he believes that he shall behold God " out of the flesh;" that is, in the spirit! What did he believe?

II. Noyes, *New Translation of the Book of Job;* Boston, 1838; p. 37.

23 " O that my words were now written!
O that they were inscribed in a register!
24 That with an iron pen, and with lead,
They were engraven upon the rock for ever!
25 Yet I know my Vindicator liveth,
And will stand up at length on the earth;
26 And though with my skin this body be wasted away,
Yet in my flesh shall I see God."

Noyes (*Notes*, pp. 144–6) says—" Or we may render, *Yet without flesh I shall see God*"—and enumerates cogent " objections to the supposition that Job here expresses his confident expectation of a resurrection."

III. Cahen, "Job;" *La Bible*, Traduction Nouvelle, avec l'Hébreu en regard; Paris, 1851; pp. 86–7. We render the French literally into English.

23 " Would to God that my words were written! Would to God that they were traced in a book
24 With a burin of iron and with lead! that they were engraved for ever in the rock!
25 But I, I know that my 'redemptor' is living, and will remain the last upon the earth:
26 And after that my skin shall have been destroyed, this delivered from the flesh, I shall see God."

In the foot-note, Cahen explains that the Hebrew word גאלי, GALI, which he renders " mon rédempteur," proceeds from the verb GAL, " to deliver;" meaning likewise "revendiquer;" which corresponds to the *Vindicator* of Noyes. The idea of Job's hope of a *resurrection*, itself a mythological anachronism, is popularly derived from the LXX and the Greek Fathers, with ideas developed in the Latin Church after St. Jerome.

Thus the reader has now before him three specimens, amid the wilderness of *Translations*, wherein are involved theological dogmas of "resurrection of the body," "redemption of the soul," and the antiquity of "Messianic prefigurations"—questions of no slight religious importance; and yet, withal, unless he be profound in *Hebrew*, his opinion upon the merits of either rendering is alike worthless to himself and to others; nor can he conscientiously distinguish which is veritably the "word of God" among these triple contradictions. The ridiculous anachronism perpetrated in king James's *version* (*v.* 23) that makes Job wish that his words were "printed" (probably 2500 years before the art was invented!) (30) has long ago been pointed out; and is alone sufficient to destroy the alleged inspiration of that "authorized" verse. For ourselves we mourn that want of space compels the suppression of some archæological remarks on the "book of Job" (*ây*IUB—meaning " L'uomo iracondo che rientra con rossore in se stesso"). We derive them from studies at Paris, under our honored preceptor Michel-angelo Lanci, to whom we here renew the warmest tribute of respect and admiration.

To Anglo-Saxon Protestantism the biblical profundities of the "Professor of Sacred and Interpreter of Oriental Tongues at the Vatican"(31) since the year 1820, are entirely un-

(30) Nott: *Biblical and Physical History of Man;* 1849; pp. 136, 137.
(31) Gaetano Deminicis: *Biografia del Cavaliere D. Michel-angelo Lanci*, Fermo, 1840; p. 10.

known. Written in the purest Italian exclusively for the lettered—restricted to one edition of 125 copies for each work, at a cost of 125 *francs* ($25) per copy—and, for manifold reasons, artistically fashioned upon a plan not easily comprehended without an *oral key*—Lanci's enormous labors upon Semitic palæography, to the "profanum vulgus" of theology, must long remain sealed books. In 1848–9, no copy of the *Paralipomeni,* (32) nor of the *Seconda Opera Cufica,* (33) both published during 1845–7, at Paris (the latter at the expense of Nicholas, Czar of Muscovy), existed within the Library of the British Museum: notwithstanding that Lanci's volumes were for sale at two leading booksellers' in London; and that their absence at the Museum-Library had been formally notified to its unnational "Powers that be." (34) The *Vie Simboliche della Bibbia* (known to us in its author's manuscript) will not be published for a period incalculable, because dependent upon human longevity. Our mutual friend, Mr. R. K. Haight of New York, is, in the United States, the sole possessor of Lanci's works that we know of. (35)

History records that it was in consequence of the discrepancies, notorious among such *translations* into English as existed at the beginning of the seventeenth century, that, in the reign of king James, a new version of the Scriptures was published: which duly received the royal, ecclesiastical, parliamentary, and national sanction, and is now consecrated amongst us Anglo-Saxons as the unique and immaculate "Word of God"—the standard of faith among Protestant communities of our race throughout the world. It is, and ought to be, in the hands of every one; so that no obstacles to the verification of such quotations, as we shall have occasion to make, exist at the present day among readers of English. As the document we are in quest of, Xth Genesis, is contained within this volume, we are compelled by the rules of archæology first to examine the book itself; in order to obtain some preliminary insight into its history, its literary merits as a *Translation*, and the repute in which the latter point is held by those most qualified to judge.

To avoid mistakes arising from confusion of editions, we quote the title-page of the copy before us.—"THE HOLY BIBLE, containing the *Old* and *New Testaments:* translated out of the original Tongues; and with the former Translations diligently compared and revised, by His Majesty's Special Command. Appointed to be read in Churches. London:

---

(32) *Paralipomeni all' Illustrazione della Sagra Scrittura;* Paris, qto. 2 vols.; 1845.

(33) *Seconda Opera Cufica — Trattato delle simboliche rappresentanze Arabiche e della varia generazione de' Musulmani caratteri sopra differenti materie operati;* Parigi, 1846–'47; qto. 2 vols.

(34) GLIDDON: *Otia Ægyptiaca;* London, 1849; p. 17, note; see also p. 110.

(35) Through the Chevalier's epistolary kindness, I am enabled to correct a former mistake, into which other authority had led me; and I gladly seize occasion to quote from one of numerous Italian autographs in my possession:—

"ROMA, 18 *Ottobre*, 1851.

"*Car*mo *Amico!*

"You say, in *Otia Ægyptiaca* (p. 31), that 'pyramid' is derived from *pi* and *haram;* the former being a Coptic article, the latter an Arabic word, combined even nowadays among the Arabs in [their name, EL-HaRaM, for] *pyramid.* This is not according to grammatical exactness; because *haram* is not altogether radical. The demonstrative [letter H] *he* is prefixed to it, which serves in lieu of the Coptic *pi.* *Ram* [Arabicè], RM, is the root (*altitude*). *Haram,* HRM, says, therefore, *the-altitude;* and it is a synonyme of the Coptic *pi-ram,* in which the *he*, H, that you have yoked to it, plays no part. The word *ram,* besides being a Semitic, is also a Coptic word, with the sense of *height*... But very huge seems to me the error of Ewald, in Bunsen, who presumes to explain a text of *Job* (iii. 14) by changing a *b* into *m*, and making a *Hara*Mòt of his own out of the biblical *Hara*Bòt.... I transcribe for you the complete article of mine, which on some occasion may be of aid to you:

"*Article taken from the 'Vie Simboliche del Vecchio e Nuovo Testamento' regarding a passage in Job....* [We have not two pages to spare, and therefore are compelled to omit the acute philological reasonings of our valued preceptor.—G. R. G.] The said two verses, most entangled in the versions of others, through my inquiries now read—'Now should I have quiet with the kings and mighty-ones of the earth who already repose in their subterranean habitations; or with the princes who had gold and (who) caused their sepulchres to be filled with silver.' [Comp. CAHEN, xv. p. 12.] ... I will not leave this argument without first giving you an illustration of that arduous verse 6 of Psalm ix.; in which, it appears to me, interpreters have strayed away from truth. Here recurs that *charabôt* which I explained. Now, if philologers are wise enough to accept my discovery, they will see that this sentence of the Psalm, in the place above-named, speaks with vibratory locution—'They closed to the enemy the subterranean abode in perpetuity: thou destroyedst the cities, and with these the memorial of those perished.'" [Compare *King James's Version!*] ...

"Affmo vostro, MICHEL-ANGELO LANCI."

Printed by George E. Eyre and Andrew Spottiswoode, Printers to the Queen's Most Excellent Majesty, and sold at their Warehouse, 189, Fleet Street, 1844. [Nonpareil Reference, 12mo.]" The Dedication "To the most high and mighty Prince, James," states that His "Highness had once out of deep judgment apprehended how convenient it was, that out of the Original Sacred Tongues, together with comparing of the labours, both in our own, and other foreign Languages, of many worthy men who went before us, there should be one *more exact Translation* of the Holy Scriptures into the *English Tongue.*"

It thus becomes patent that our copy is not printed in one of "the Original Sacred Tongues," but merely professes to be a "more exact *Translation*" into English than, at the date of its publication, 242 years ago, had previously appeared. Even conceding that the Holy Scriptures in the "Original Sacred Tongues" may have been revealed word for word by the Almighty, and granting that their *editio princeps* was a manuscript in the autographs of divinely-inspired Scribes, no reasonable person will deny the possibility that this English *translation* may embrace some errors—none among the educated will be so unreasonable as to insist upon the *infallibility* of its English translators, however erudite, however conscientious; nor perchance will claim *inspiration* for these worthies. Childishly credulous as we are by nature, and uncritical though the generality of us remain through education, no sane Anglo-Saxons, since the middle ages, allow "divine *inspiration*" to men of *their own race.* We accord the possibility of "inspiration" solely to members of a single family that lived a long time ago, and a great way off; whose descendants (although nowadays ranking among the best citizens of our cis-Atlantic Republic) are still abused by our kinsfolk across the water; and who, although contributors to our own and the latter's welfare and glory, are yet debarred, as unworthy, from a voice in the British Parliament: and all this, forsooth, in the same breath of acknowledgment that we derive our most sacred Code of Religion, Morals, and Laws, from their *inspired* ancestors! and whilst, based upon our modern notions of their ancient creed, we nasally vociferate that they and ourselves are "of one *blood* as brothers"!

Our copy, such as it is, may be accepted without hesitation as a lineal descendant of the primary *authorized* version in the English language, wrested from the Lords Spiritual and Temporal through the intelligence of our ancestors, quickened by the Reformation; who bled for the same rights that we their posterity can now assert, in the free United States of America and in Great Britain (without even the merit of boldness), viz. the right to examine the Scriptures, and everything else, for ourselves, and to express our opinions thereon in the broad light of heaven.

Archæologically speaking, in order to insure minute exactness, it would be imperative to collate, year by year, and edition by edition, the whole succession of copies of our "authorized version"; and, by retracing from the exemplar on our table backwards to that first printed in black-letter during the reign of king James, to ascertain whether any and what changes, beyond variations in typography, may have been introduced. But such dreadful labor is, to the writer, impossible for want of the series; ungenial to his tastes as well as unnecessary for his objects. He contents himself with the assertion that there are many differences between such copies of divers editions that have fallen in his way, although considered by others of little or no moment; being chiefly *marginal*, as in the superadded and spurious chronology; or *capitular*, as in the apocryphal headings to chapters, &c.; neither of which can have any more to do with the original "word of God," than the printer's name, the binding, or the paper.

As *positivists* in Philosophy while archæologists in method, we clear the table of these comparatively-trivial disputations; and bounding retrogressively over the interval that divides our generation from that of His Majesty King James, the reader is requested to take with us the historical era of the promulgation of the "authorized version" as a common point of departure; viz.: A. D. 1611.

The most ancient printed copy of king James's *version*, that has been accessible to us, lies in the British Museum. It contains a memorandum by the Rev. Dr. Horne to the effect that the *title-pages* are of the primary edition of the year 1611, but that the rest appertains to that of 1613. The whole folio is printed in black-letter. Its frontispieces are literary gems; and so faithfully portraying the symbolism of Europe's "moyen age" in their astrologico-theological emblems, that every antiquary must deplore that castigating zeal which has effaced such quaint expressions of ancestral piety, to substitute for them, in some of our current copies, typographical whims that cannot pretend even to the venerable halo of bygone days. The title-page to the Old Testament is embellished by vignettes, among which figure the *Lion*, *Man*, *Bull*, and *Eagle*; (36) ancient signs for the solstices and equinoxes. Moses is truthfully represented, as in Michel-angelo's statue, with his characteristic horns; according to the *Vulgate* of Exod. (xxxiv. 29, 30, 35), "*cornuta* esset facies sua," which preserves one sense of the Hebrew KRN, *horn*. The zodiaco-heraldic arms of the "12 Tribes" of Israel are also preserved; (37) together with a variety of other symbols, archæologically precious. That of the New Testament is still more curious, inasmuch as it exhibits the esoteric transmission (perceived even as late as at that time by learned reformers in England) of certain antique symbolisms of Hebrew Scriptures into those of the Orientalized Greeks or Hellenized Jews. The "4" solstitial and equinoctial signs of the "4 *seasons*" remain, but are now attached to the figures of the "4" Evangelists; while the zodiaco-heraldic arms of the "12 *Sons of Jacob*" (*Gen.* xlix. 1, 28), whence the "12 *Tribes of Israel*," lie parallel with and officiate as "pendants" to the "12 Apostles," each with his symbolical relation to the "12 months" of the year, &c.—the whole, indeed, saving its uncouth artistic execution, so vividly solar and astral in conception, as to betray that primeval Ægypto-*Chaldaic* source whence students of hieroglyphical and cuneiform monuments,—exhumed and translated more than two centuries subsequently to the publication of our English "editio princeps"—now know that the types of this imagery are derived. The reader, who seeks throughout our modern editions in vain for the once-consecrated embellishments of ages past, may now perceive that we are not altogether ill-advised when hinting that great liberties have been taken with the authorized English Bible between A. D. 1611, era of its first promulgation, and those copies ostensibly represented in the current year (1853) to be its lineal and unmutilated offspring. Theologically, however, these variants through omission or commission are not of the same importance as they seem to be archæologically, nor need we dwell upon them now.

The accuracy of this English version, and its fidelity to the original *Hebrew* and *Greek* MSS., must rest upon the opinion we can form of its Translators; legalized by the royal seal and confirmed by an act of Parliament. With the value of the two last authorities, regal or parliamentary, in questions of purely-philological criticism and of strictly-literary knowledge, we American Republicans may be excused in declaring that we have nothing to do. Until it is proved to our comprehension that the acquaintance of those worthy M. P.'s with the "original sacred tongues" was profound, and that they devoted one or more Sessions to the verification of the minute exactness of the volume they endorsed, their fiat upon the literary merit of the book itself carries with it no more weight in science than, to bring the case home, could the Presidential signature to an act of Congress authorizing the printing in *Arabic*, at national expense, of the Mohammedan *Korân*, in the year 1853, be accepted as a criterion or even voucher of such huge folio's historical or philological correctness.

To us the only admissible evidence of the exactitude of king James's version, as a faithful exponent of the "word of God" (originally written, and closed some 1500 years before that monarch's reign, in *Hebrew* and in *Greek*), must be twofold—*historical*, and *exegetical*; the former, by establishing the learning, oriental knowledge, critical skill, and integrity of the men; the latter, by demonstrating that rigid examination will fail to detect errors in the performance itself. Of this duplex evidence we now go in quest; remarking at the outset,

(36) Conf. SALVERTE: *Sciences Occultes;* i. pp. 46, 47. Comp. *Ezekiel* i. 10, with *Apocalypse* iv. 7. RIGHELLINI: *Franc-maçonnerie;* Paris, 1842; i. p. 324, pl. 4, fig. 1.

(37) Conf. KIRCHER: *Œdipus Ægyptiacus;* Rome, 1653; vol. ii. part 1. p. 21. DRUMMOND: *Œdipus Judaicus;* London, 1811; plate 15—"Dissertation on XLIXth Chapter of Genesis":—and LANCI: *Paralipomeni, passim.*

74

that, inasmuch as (precise date unknown) the gift of "divine inspiration" is said by Protestants to have ceased about 1750 years ago with the last *Apostle*, nobody claims for these English Translators any supernatural assistance during the progress of their pious labors; and, therefore, in matters appertaining to the merely-human department of linguistic scholarship (whilst we doubt not their excellence as men, their attainments, nor their good faith), we must concede the chance that their production, owing to man's proneness to err, may be found to fall short, in a literary point of view, of the standard by which a similar performance would be judged were a *new Translation* of the Old Testament "authorized," after the same fashion, at the middle of this XIXth century.

I. The Historical Testimony.

In the year 1603, owing to the enormous defects recognized in all popular *translations* then current, the revision that had been ordered in the days of Elizabeth was carried into effect by James. *Fifty-four* of the most learned graduates of the Universities of Oxford and Cambridge were appointed to the task, *seven* of whom died before the work was completed: (38) among the last, Lively, (39) the best if not the only *Hebraist* on the translation, whose labors were of short duration; and, "much weight of the work lying upon his skill in the Oriental tongues," his loss was irreparable; because the surviving *forty-seven* translators rejected the assistance of the only remaining Hebraist in England, viz., "Hugh Broughton, fellow of Christ College, Cambridge, who had certainly attained a great knowledge in the *Hebrew* and Greek tongues." Indeed, says the very learned Bellamy, (40) from whom we derive the fact, "it was well known that there was not a critical Hebrew scholar among them; the Hebrew language, so indispensably necessary for the accomplishment of this important work, having been most shamefully neglected in our Universities; and, as at this day [1818], candidates for orders were admitted without a knowledge of this primary, this most essential branch of biblical learning. It was, as it is at present, totally neglected in our schools, and a few lessons taken from a Jew in term-time, whose business is to *Judaize*[!], and not to Christianize, serve to give the character of the Hebrew scholar," in England.

In consequence, then, of the inability of the *forty-seven* translators to read one (and the *oldest*, the *aboriginal* "divine word") of those "sacred tongues" of which their servile dedication makes parade, "it appears they confined themselves to the Septuagint (Greek) and the Vulgate (Latin); so that this was only working in the harness of the first translators; no translation (excepting perhaps Luther's, 1530—1545), from the original Hebrew only, having been made for 1400 years," says Bellamy.

"If we turn," continues elsewhere this outspeaking writer (whose erudition *nemo nisi imperitus* will contest), "to the translations made in the early ages of the Christian Church, we approach no nearer the truth; for as the common translations in the European languages were made from the modern Septuagint and the Vulgate, where errors are found in these early versions they must necessarily be found in all the translations made from them."

Whether the *Vulgate* and the *Septuagint* versions are faultless will be considered anon. Our present affair is with king James's *translation*, and certainly appearances are not flattering.

We learn from Fuller, (41) how at once, on its first apparition, objections were raised against its accuracy in England; but as these emanated chiefly from Romanist scholarship, in those days of reformation at a discount, their validity is slurred over by Protestant ecclesiastics. Gradually, as Hebraical scholarship struggled into existence — that such

(38) Fuller: *Church History;* 1655; pp. 44–46.

(39) *Ibid*, p. 47 — and Horne: *Introd. to the Crit. Stud. of H. Scrip.;* 1838; ii. pp. 70, 80; note 5.

(40) *The Holy Bible, newly translated from the Original Hebrew; with notes critical and explanatory;* London, 1818, 4to — published by the subscriptions of Royalty, Nobility, and Clergy; but never completed, and now out of print. Our quotations are from the "general preface."

(41) *Church History;* pp. 58, 59 — also Horne: *Introd.;* ii. pp. 76–78.

giants as Walton,(42) 1657, had redeemed the Oriental wisdom of Oxford — the voice of the great Dr. Kennicott (43) was uplifted a century later, 1753–9, protesting vehemently against the perpetuation of fallacies which the *forty-seven* translators' ignorance of Hebrew had spread over the land through king James's *version.* He commences — "The reader will be pleased to observe, that, as the study of the Hebrew language has only been *reviving* during the last hundred years," (44) &c. — that is, only since the time of Walton, his predecessor: — which passage implies that fifty years previously to the latter's epoch, 1657, (*i. e.*, at the time of the forty-seven translators, 1603–11), the study of Hebrew was all but defunct, or rather it had scarcely yet begun to exist; that is, in *England.*

This point was considered so familiar to every general reader, that no hesitation was felt when stating it, 1849, with reference to the same question, (45) in the following words: "Now the Hebrew language in 1611 had been a dead language for more than two thousand years, and though these men (the forty-seven translators aforesaid) were renowned for their piety and learning, yet very few, if any of them, were competent to so important a task. In fact, the Hebrew language may be said only to have been recovered within the last century by modern Orientalists: and from the ignorance of these very translators of the original language, the Old Testament was taken mostly from the Greek and Latin versions, viz: the *Septuagint* and *Vulgate.* Being, then, a translation of bad translations, which had passed through numerous copyings, how could it come down to us without errors?"

Nevertheless, want of ordinary information on Scriptural literature prompted a reviewer, (with intrepidity characteristic of that undeveloped stage of the reasoning faculties which, in accordance with Comte's positive philosophy, has been already classed as "the theological,") to indite these remarks: — "Dr. Nott, again, speaks disrespectfully of the English version of the Scriptures. He makes the astounding assertion that 'the Hebrew language may be said only to have been recovered within the last century, by modern Orientalists.' Most surprising is it that any one should believe that the Jews should have wholly lost a knowledge of their ancient and sacred tongue; and that a knowledge of it should only have been recovered by modern Orientalists, displays an amazing want of reading and scholar-like accuracy, and a credulity exceedingly rare, *except in an unbeliever.*" (46)

"Mutato nomine, de te fabula narratur!" Under the head of KNâAN [*supra*, p. 496], the "Association" may find a series of facts on the permutations, which the so-called "Lingua Sancta" of the Israelites has undergone, still more "astounding," where we took occasion to repeat and enlarge upon the positions of Dr. Nott's "Reply." In the meanwhile, the "ipse dixit" above quoted of Kennicott, that a century and a half posterior to the *forty-seven* translators of king James's version, the study of Hebrew was only "reviving," may, by some, be considered as authoritative as that put forth, in 1850, in proof of the united scholarship of an "Association."

"This only is certain, that, in Nehemiah's time, the people still spoke Hebrew; that, in the time of Antiochus Epiphanes and the Maccabees, the Hebrew was still written, though the Aramæan was the prevalent language; and, on the contrary, about this time, and shortly after Alexander the Great, even the learned Jews found it hard to understand difficult passages of the old writings, *because the language had ceased to be a living speech.* The reign of the Seleucidæ, and the new influence of an Aramæan people, seem gradually to have destroyed the last traces of it;" (47) and this about two thousand years ago!

(42) *Biblia Sacra Polyglotta* — complutentia Textus Originalis — Hebraicos cum Pentat. Samarit., Chaldaicos, Græcos, Versionumque Antiquarum — Samarit., Græc. Sept., Chaldaicæ, Seriacæ, Lat. Vulg., Arabicæ, Æthiopicæ, Persicæ.

(43) Author of *Vetus Testamentum Hebraicum;* cum variis Lectionibus; Oxon. 1780; and of *Dissertatio Generalis in Vetus Test. Heb.;* 1780.

(44) I. *Dissertation — State of the printed Hebrew Text of the O. Test. considered;* Oxford, 1753; p. 307.

(45) NOTT: *Op. cit.*; p. 134.

(46) The Rev. Dr HOWE, in *The Southern Presbyterian Review,* "conducted by an Association of Ministers;" Columbia, S. C.; vol. iii. No. 3.; Jan. 1850 — refuted by Dr. NOTT: "Chronology, Ancient and Scriptural," in *Southern Quarterly Review;* Nov. 1850.

(47) GESENIUS, apud *Parker's De Wette:* i., *Appendix,* p. 457 — compare also p. 221.

Such is the position of *Hebrew* in the world's philological history as a *spoken* tongue: yet, "a knowledge of that language which is contained in the scanty relics of the Old Testament has been preserved, though but imperfectly, by means of *tradition*. Some time after the destruction of Jerusalem in the Palestine and Babylonian schools, and after the eleventh century in those of Spain, this tradition was aided by the study of the Arabic language and its grammar. Jerome learned the Hebrew from Jewish scholars. Their pupils were the restorers of Hebrew learning among the Christians of the sixteenth century;" (48) that is, on the *continent;* for, with the exception of Lively, who died, and Hugh Broughton, whose aid was refused, history does not record any man deserving the name of a *Hebraist* in England, even during 1603–11. Finally, "the name *lingua sancta* was first given to the ancient Hebrew in the Chaldee version [made long after the Christian era, when *Hebrew* had orally expired,] of the Old Testament, because it was the language of the sacred books, in distinction from the Chaldee, the popular language, which was called *lingua profana.*" (49)

These citations here seem indispensable, lest dogmatism, peeping from out of its theological chrysalis, should feel itself again called upon to "astound" a reader by charging us with errors of its own commission: otherwise an apology would be due for this excursus. We return to Dr. Kennicott.

After setting forth the causes of mistaken renderings in king James's version, he declares—"A *New Translation*, therefore, prudently undertaken and religiously executed, is a blessing, which we make no doubt but the Legislature [!] within a few years will grant us." (50) Six years later, finding his humble prayer unheeded, he comes out clamorously against "our authorized version": claiming that some of the earlier English translations were more faithful and literal, (51) and backing his appeal with the subjoined among other examples:

Luke xxiii. 32. CHRIST made a *malefactor!* "And there were also two other malefactors led with him to be put to death;" instead of "two *others*, malefactors." The *Greek* reads simply, "And two others, evil-doers." (52)

Judges xv. 4. Three hundred *foxes* tied tail to tail, instead of wheaten *sheaves* placed end to end! "And Samson went and caught three hundred foxes, and took firebrands, and turned tail to tail, and put a firebrand in the midst between two tails." The *Hebrew* is, "And Samson went and gathered three hundred sheaves of wheat, and taking torches and turning (the sheaves) end to end, set a torch in the midst between two ends." (53)

1 *Kings* xvii. 6. ELIJAH not fed by *ravens*, but by *Arabs!* "And the ravens brought him bread and flesh," &c. In the *Hebrew*, "And the ÔRBIM (âRaB-*ìm*) brought him bread and flesh." Kennicott thinks *Orbìm*, inhabitants of Oreb, or Orbo—"villæ in finibus Arabum," says St. Jerome: but, *Arabs* seem to us more natural and correct. In no contingency "crows"! (54)

It is superfluous now to continue our excerpta from Kennicott, or narrate how it comes to pass that, owing to nice appreciations of the Text that none of them could construe, the *forty-seven* (in Psalms cix.) have made pious king David (disputed author of that

---

(48) DE WETTE: *Parker's transl.;* Boston, 1843; i. p. 128—cited by NOTT, in the "Reply." Comp. also, PALFREY: *Academical Lectures on the Jewish Scriptures;* Boston, 1838; i. pp. 8–20—"It is out of the question for any man to suppose, that he can be acquainted with Hebrew as familiarly and thoroughly, as he may be with Latin and Greek."

(49) CONANT'S *Gesenius:* Hebrew Grammar; New York, 1846; p. 23.

(50) *Op. cit.;* p. 567. Cf., also, MUNK: *Palestine;* Paris, 1845; pp. 433–436.

(51) II. *Dissertation;* Oxford, 1759; pp. 579, 580, *seq.*

(52) SHARPE: *N. Test.;* p. 165.

(53) JOHN DOVE: *Vindication of the Hebrew Scriptures*; London, 1771—in his furious assault upon the "Authorized Version," and lamentations at English ignorance of Hebrew, also derides the "foxes"; p. 71, *seq.* GLAIRE: *Livres Saints Vengés;* Paris, 1845; ii. pp. 57, 58, contests the "fagots"—but vide CAHEN: vi. pp. 68, 69, note 4.

(54) GLAIRE: *Op. cit.;* ii. p. 85, reads "Arabes"; but CAHEN, viii. p. 77, "corbeaux"—acutely adding, "Universa historia fabularum plena est."

rhapsody) (55) utter such fearful imprecations against his foes; when, in the "original sacred tongue," he actually complains that his enemies are heaping these outrageous maledictions upon himself!

Well might the Reverend Doctor quote Michælis — "I am amazed when I hear some men vindicate our common readings with as much zeal as if the editors had been inspired by the Holy Ghost!" Still better does he terminate his earnest work with supplications for a *new* Hebrew Text, and for a *new* English "authorized" *translation.*

Reader, these things were published at Oxford and disseminated over Great Britain about ninety-four years ago — not in expensive folios veiled through the dead languages, but in two English *octavos* — not by a "skeptic" whose indignation at any kind of imposture impels him to spurn it, but by that Church of England Divine, collator of six hundred and ninety-two ancient Hebrew biblical manuscripts, (56) whose folios, together with the *Biblia Polyglotta* of his illustrious precursor, Walton, are the only *English* labors on the Scriptures that receive homage from continental erudition, as performances on a par with the colossal researches of Germans, Frenchmen, and Italians, even unto this day!

Kennicott passed away. Other scholars followed in his footsteps. From a few of the latter we extract what they have left in print respecting king James's version, with a prefatory citation from Bellamy, to whom we owe the collection.(57)

"It is allowed by the learned in this day and every Christian nation, that the authorized translations of the sacred Scriptures, in many places, are not consistent with the original Hebrew. A few extracts are here given, from some of our most learned and distinguished writers, who were decidedly of opinion, that a New Translation of the Scriptures was absolutely necessary; not only on account of the great improvement in our language, but because the *Translators* have erred respecting things most essential. The following are some of the eminent men who have left their testimony concerning the necessity of a new translation:—

'Were a version of the Bible executed in a manner suitable to the magnitude of the undertaking, such a measure would have a direct tendency to establish the faith of thousands. . . . Let the Hebrew and Christian prophets appear in their proper garb: let us make them *holy garments* for glory and for beauty; . . . *the attempts of individuals should be promoted by the natural patrons of sacred learning.*' — (Bishop NEWCOMBE.)

'Innumerable instances might be given of faulty translations of the divine original. . . . An accurate translation, proved and supported by sacred criticism, would quash and silence most of the objections of pert and profane cavillers.' — (BLACKWELL'S *Sac. Class. Pref.*, 1731.)

'Our English version is undoubtedly capable of very great improvements.' — (WATERLAND'S *Script. Vindicated,* Part 3, p. 64.)

'Nothing would more effectually conduce to this end, than the exhibiting the Holy Scriptures themselves in a more advantageous and just light, by an accurate revisal of our vulgar translation.' — (Dr. LOWTH'S *Visitat. Sermon,* at Durham, 1758.)

'The common version has many considerable faults, and very much needs another review.' — (*Biblioth. Lit.*, 1723, p. 72.)

'The Old Testament has suffered much more than the New, in our Translation.' — (DODDRIDGE'S *Pref. to Family Expositor.*)

'Many of the inconsistencies, improprieties, and obscurities, are occasioned by the translators' misunderstanding the true import of the Hebrew words and phrases, showing the benefit and expediency of a more correct and intelligent translation of the Bible.' — (PILKINGTON'S *Remarks,* 1759, p. 77.)

'The version now in use in many places does not exhibit the sense of the Text; and mistakes it, besides, in an infinite number of instances.' — (DURELL'S *Crit. on Job,* 1772, *Pref.*)

'That necessary work, a New Translation of the Holy Scriptures.' — (LOWTH'S *Prelim Dissert. to Isaiah,* p. 69.)

(55) Cf. DE WETTE: ii. pp. 520–529 — and CAHEN: xiii. p. 247, "Sommaire," and p. 249, note 20.

(56) *Diss. Gen. in Vet. T. Heb.*; 1790; Tables, pp. 110–112.

(57) *Op. cit.*: "General Preface"; 1818.

'Whoever examines our version in present use, will find that it is ambiguous and incorrect, even in matters of the highest importance.' — (Prof. SYMOND'S *Observations on the Expediency of revising the present Version*, 1789.)

'At this time, a New Translation is much wanted, and universally called for.'—(GREEN'S *Preface to Poetical Parts of the New Test.*)

'Great improvements might now be made, because the Hebrew and Greek languages have been much better cultivated, and far better understood, since the year 1600.' — (Dr. KENNICOTT'S *Remarks*, &c., 1787, p. 6.)

'The common version has mistaken the true sense of the Hebrew in not a few places. Is it nothing to deprive the people of that edification which they might have received, had a fair and just exposition been substituted for a false one? Do we not know the advantages commonly taken by the enemies of Revelation, of triumphing in objections plausibly raised against the Divine Word, upon the basis of an unsound text or wrong translation?' —(BLANEY'S *Prelim. Disc. to Jeremiah*, 1789.)

'They [the *forty-seven*] are not acquainted with the Hebrew, without which no man should pretend to be a critic upon the writings of the Old Testament. It has some peculiar properties and idioms which no other language has, with which every critic should be acquainted. . . . The Hebrew is fixed in nature, and cannot change. . . . He should be acquainted with the genius of the Hebrew tongue, and with its manner of expressing spiritual things, under their appointed images in nature.' — (ROMAINE'S *Works*, vol. v. p. xvi.)

'It is necessary that translations should be made from one time to another, accommodated to the present use of speaking or writing. This deference is paid to the heathen classics, and why should the Scriptures meet with less regard?' — (PURVER.)

'The common English translation, though the best I have seen, is capable of being brought, in several places, nearer to the original.' — (WESLEY.)

For other arguments, continues our author, see Bishop Newcombe's "Chief reasons in support of a corrected English translation of the Scriptures for national use:" adding on his own account: —

"Notwithstanding all that has been done, the translators have left it [our version] defective in *mood*, *tense*, *person*, *gender*, *infinitive*, *imperative*, *participles*, *conjunctions*, &c.; and in many instances, almost in every page, we find verses consisting in a great part of *italics;* in some, a third part; in others, nearly half; as may be seen in the Bibles where the words *for which there is not any authority in the original* are always so marked."

Descending into works of less exclusive circulation, what do we encounter?

"It is not to be denied that a translation of Holy Scripture, if undertaken in the present day, would have many advantages superior to those which attended king James's translation. The state of knowledge is much improved. . . . Our language has undergone some changes in the course of two centuries, by which it has varied from being precisely the same as when our translators wrote. Many words which were then polite and elegant, are now vulgar, to say the least. . . . Nor can we refrain from complaining also of the negligent manner in which the press has been conducted in all our public editions: what should be printed in poetry is set as prose; what should be marked as a quotation, or a speech, reads like a common narrative. . . . And this perplexity is occasionally increased by improper divisions of chapters and verses, which but too often separate immediate connection. . . . Undoubtedly, the present version is sufficient to all purposes of *piety*."—(TAYLOR'S *Calmet's Dictionary of the Holy Bible* — voce "Bible.")

"It is needless to pronounce a formal encomium on our authorized version. The time, learning, and labor expended on it were well bestowed. It far surpasses every other *English* version of the entire Bible in the characteristic qualities of simplicity, energy, purity of style, as also in *uniform fidelity* [*!*] to the original. A revision of it, however, is wanted, or rather a *new translation* from the Hebrew and Greek, based *upon it* [*!*]"—("S. D.," in KITTO, ii. p. 919.)

"No less than 30,000 various readings (58) of the Old and New Testament have been

(58) Say rather, with the Rev. Prof. MOSES STUART — "Investigation has dissipated this pleasant dream. In the Hebrew MSS., that have been examined, some 80,000 *various readings* actually occur, as to the Hebrew consonants. How many as to the vowel-points and accents, no man knows. And the like to this is true of the New Testament"—(*Crit. Hist. and Defence of the O. Test. Canon;* Andover, 1835; p. 192.) "Nemo est, qui in uno aliquo codice, sive MSto sive impresso, textum incorruptum exhiberi arbitratur. Riderent docti; si quis codicem aliquem cum istis Apostolorum autographis, in omnibus, consentire dixerit"—(KENNICOTT: *Dissert. Gen.;* par. 13, p. 6.)

discovered; . . . and putting alterations made knowingly, for the purpose of corrupting the text, out of the question, we must admit, that from the circumstances connected with transcribing, some *errata* may have found their way into it; and that the sacred Scriptures have in this case suffered the same fate as other productions of antiquity. . . . In the last 220 years, critical learning has so much improved, and so many *new manuscripts* have come to light, as to call for a revision of the present authorized version." — (SEARS, *Hist. of the Bible*, 1844, pp. 651, 665.)

"The *second* thing which I would strongly recommend, is constantly to study and peruse the Original Scriptures; the Old Testament in the Hebrew, and the New Testament in the Greek. . . . There is no such thing as any *written* Word of God independent on the word of man. The Lord Jehovah may have uttered the whole Law from Mount Sinai; and, yet, Moses may not have accurately recorded it. . . . In like manner, the Gospel may have been fully preached by Christ; and, yet, the Evangelists may not have fully recorded it. . . . One painful conviction is, that the plain import of the Word of God has been most *fantastically*, *ignorantly*, and *wilfully* perverted, as well in the translation as in the interpolations. . . . Many gross *perversions*, not to say mistranslations, of the Sacred Text have been occasioned by dogmatical prejudices and sectarian zeal."—(REV. JOHN OXLEE, *Letters to the Archbishop of Canterbury*, London, Hatchard, 1845; pp. 117, 137–8.)

Fuerunt autem, relates Kennicott, *qui de hac re aliter senserunt:* among the non-extinct is the Rev. Dr. Horne, who makes the fiercest battle in defence of "our authorized version;" quoting many writers on the opposite side to ours, whose combined "association," like the one prelauded, fails in authority for want of *Hebraical* knowledge in its parts; but, when the best is done for it, he naively remarks on our translation — "It is readily admitted that it is not immaculate; and that a revision, or *correction*, of it is an object of desire to the friends of religion" — and then the reverend gentleman breaks forth in rhapsodical glorifications and thanksgivings, *that it is not worse!* (59)

Nor are the erudite among Christians alone the denouncers of king James's version. Anglicized Israelites hold it in estimation equally low, to judge by the following Editorial:

"*What we should like to see at the World's Fair.*—It would give us a great deal of pleasure to see at the World's Fair a correct English version of the Bible, resting upon the solid fundament of the results of modern criticism; reaching the elevation of modern science, and being accomplished by men of a thorough scholastic education, and free from every foreign influence, who take the letter for what it is without paying any regard to authorities, and without coming to the task with a certain quantity of prejudices. Such a work would reconcile science and religion; it would reclaim many an erring wanderer to the straight path of truth; it would evaporate many a prejudice and a superstition; it would greatly modify many sectarian views, and would closely unite the men of opposite nations. It appears, however, that the men for this task are not yet among the mortals; for the theologians come to the Bible with an established system, which must lead them away from the true import of letters, where they find again their own system whenever it can be done conveniently; and where their sentiments frequently overbalance their critical judgment." —(*The Asmonean*, New York, July 22, 1853.)

Thus we might go on, citing work after work wherein, if king James's version is not denounced for its *perversions* of the "original sacred tongues," its erroneous readings are more or less apologetically but thoroughly confirmed by many instances in which the erudition and fairness of the authors compel them to *substitute* their own translations for those of our "authorized" copy. Notable examples may be seen in the recent work of our much-honored fellow-citizen, Dr. McCulloh. (60).

Albeit, as said before, if our version were decently accurate, why should so many laborious men run the risks of incurring some theological obloquy, coupled with pecuniary loss, in efforts to *correct* the false renderings of that superannuated edition by publishing emendatory *retranslations* in *English?* Among the many we have consulted may be cited:

"The HOLY BIBLE, according to the established Versions, with the exception of the substitution of the *original Hebrew names*, in place of the words LORD or GOD, and of a few corrections thereby rendered necessary. (London, 1830; Westley and Davis.)"

This book, however, seems to have closed at 2 *Kings*. The uninitiated may be informed

(59) *Op. cit.;* ii. pp. 77–83.

(60) *Credibility of the Scriptures;* Baltimore, 1852. See particularly vol. ii. Appendix, "On the Human Soul" pp. 465–489.

that the word "Lord" of our version renders merely the *Dominus* of the Vulgate, and the Κυριος of the Septuagint, and does not directly *translate* the original Hebrew word IeHOuaH; the latter being suppressed, by "His Majesty's special command," in the "authorized" copies, only 6846 *times!* The number of times it occurs in the *Hebrew Text* are 6855: (61) on which hereafter. Another is: —

"The Holy Bible, containing the *authorized* version of the Old and New Testaments, with *twenty thousand* [*!*] emendations. (London, 1841; Longman, Brown & Co.)"

Its *title* attracted our notice, as savoring of a *Tauric* genus known as Hibernian; aptly illustrated in that "same old knife which belonged to 'my grandfather,' after having received thirteen new handles and seventeen new blades." The *preface* justified our first impressions, when we read—"This is our authorized English version, which is characterized by unequalled fidelity, perspicuity, simplicity, dignity, and power. . . . No one has *yet detected a single error* [in it!!!] in reference to those great and vital truths in which all Christians agree." After which, where the utility of 20,000 *emendations?* Suffice it, that, maugre this huge amount, not perceiving any of the catalogue of "emendations" hereinafter submitted to the reader, we refrained from its purchase, after a morning's examination.

A third, which we have long possessed through the kindness of its publishers, merits attention, and is ushered by a most excellent preface: —

"The Holy Bible, being the English version of the Old and New Testaments, made by order of King James I., carefully *revised and amended*, by several Biblical Scholars. (Sixth edition, Philadelphia: J. B. Lippincott & Co., 1847.)"

After a brief sketch of preceding translations into English, from 1290 to 1611, the *preface* states — "From these facts, and from comparing the translation of king James with those which preceded it, nothing is more obvious, than that the common version is but a revision of those executed by Tindal, Coverdale, and others, and that, however excellent it may be, the paramount praise, under God, is due to William Tindal and Miles Coverdale." In the above sentiments we heartily concur; having enjoyed opportunities, in the course of our studies, of comparing some points in both of the latters' self-sacrificing editions with the so-called "revision" of the *forty-seven.* Αυτοδιδακτοι, however, like Abderitan Democritus, in some branches of Oriental philology; and possessing, furthermore, an apparatus tolerably complete of *continental* criticism in biblical matters; we prefer direct references to the *Hebrew Text,* now rendered accessible in a very handy form, and illumined by Cahen's most useful parallel French translation.(62)

From the nature of these premises it will be seen that, save under the scientific point of view and for the general cause of human enlightenment, the writer, as an individual, is not urgent in exacting another "authorized" version of Texts to which he has acquired (what any man who really is serious in such matters can acquire as he has) access for himself. At the present day that in Protestant countries, such as Great Britain and the United States, it has become a common practice to *worship* king James's translation, and "study divinity;" that our English version, with all the unnecessary deviations from its Hebrew prototype, is reverenced by the masses as a "fetiche," or viewed with a relic of that semi-idolatrous awe refused by Protestants to crucifixes, pictures, or images, our observations may perhaps seem indecorous to those who choose to cramp their intellects and continue to ignore the splendid results of continental exegesis. We should regret the fact, the more so because offence is unintentional; but, "the epoch of constraint has passed away [in these United States] for ever: a freeman will be free in all things; material and political emancipation suffice no longer for him. He knows that there is a sublimer liberty, that of thought and belief. It is with sorrow that he beholds those sweet illusions fleeting away

(61) Walton: *Bibl. Polyg.;* Prolog. C. 8, § 8, p. 275. Horne: *Op. cit.;* i. p. 38. But, above all, Lanci: *Paralipomena;* 1845; *passim.*

(62) La Bible: *Traduction Nouvelle;* 22 octavo volumes; Paris, 1831–'51.

that whilom had been the charm of his childhood; but reason exacts it, and he sacrifices his illusions upon the altar of *truth*." (63)

Of that wherein the aspirations of a Newcombe, a Lowth, and a Kennicott (to say nothing about others of the best of England's biblical critics), have been baulked, it would be at this day egregious folly to entertain further hopes, viz: that the British Lords, Spiritual and Temporal, will, in our generation at least, permit such a radically-correct *re-translation* of the Hebrew Scriptures as would supersede the vulgar version "appointed to be read in churches." The Universities, especially the *Oxonian*, — part of whose support depends, like some institutions on this side of the water,—upon a "Book Concern," would oppose such violation of vested privileges. By the evangelical dissenting sects, sundry of whose various hierarchies derive subsistence from those very linguistic quibbles that a *new* standard version would obliterate, such a proposition would be repelled with devout horror. *Exeter Hall* shudders, even at the thought: "Bible Societies" whine that the reign of Anti-Christ is come indeed. As positivists we lament not that our brief span of life will have been measured, long before a *new* English version may be "authorized;" because, through the slow but unerring laws of human advancement in knowledge, by the time that *theologists* shall have accomplished their metaphysical transition and have awakened to the stern realities of the case, the development of science will have rendered any *new translation* altogether supererogatory among the educated who are creating *new religions* for themselves.

In the utterance of these long-pondered thoughts, though written years ago, we have been somewhat anticipated by our learned friend McCulloh; (64) with a quotation from whose admirable chapter on the "Value of Translations" we conclude this *historical* division of the two-fold evidence.

"No emendation however of our common translation would affect the revelations made in the Scripture, upon any subject which Jehovah has directly addressed to the understanding or consciences of mankind, whether as regards their faith or practice. That a new translation would considerably affect our theological creeds, or our ecclesiastical institutions, there is no doubt; but this again is a most desirable object if such things are not accordant to the undoubted word of God. No Christian in his senses can wish to remain under any error respecting the import of Jehovah's revelations; and hence nothing can be more absurd than to oppose a correction of our common translation, on the ground that it would overturn some of the inventions that theologians have heretofore constructed upon the comparatively defective Hebrew or Greek Texts upon which that translation has been made.

"The popular objections of unlearned persons to the amendment of our present translation, however, are often, unfortunately for Christianity, sustained by learned men and accomplished scholars, whose interests or whose prejudices are too deeply involved in the present condition of things to be willing to admit of any innovation. Their creeds, institutions, and ecclesiastical establishments, for the most part, were constructed contemporaneously by divines or statesmen of similar theological or ecclesiastical views with those who made our authorized version. To change the terms or texts of Scripture that have been heretofore used as the basis for ecclesiastical institutions, or theological assumptions concerning divine truths, are shocks too violent, either for the pride or self-interests of men, to acquiesce in willingly. . . . . Dr. Vicesimus Knox, (65) of the Church of England, says, 'For my own part, if I may venture to give an opinion *contrary to that of the profound collators of Hebrew Manuscripts*, I cannot help thinking a new translation of the Bible an attempt *extremely dangerous and quite unnecessary*. Instead of serving the cause of religion, which *is the ostensible motive* for the wish, *I am convinced that nothing would tend more immediately to shake the basis of the* ESTABLISHMENT' (*i. e.*, of the Church of England). 'Time,' says the reverend gentleman, 'gives a venerable *air* to all things. Sacred things acquire *peculiar* sanctity by long duration.'"

And finally, the unlettered dogmatist who, possessing no knowledge of the real merits of the topics before us, would thrust into court "his" *opinion*, may as well be told by the reader, that: —

"At the rational point of view, a sentiment such as is termed Christian conscience, a

(63) MUNK: *Examen*, in CAHEN'S *Exodus*; p. iv.

(64) *Op. cit.*; i. pp. 281, 283.

(65) *Annual Obituary*; vi. p. 352; — *Op. cit.*; p. 283, note.

75

sentiment that reposes upon suppositions, has no voice in scientific discussions; and, every time that it would meddle with them, it ought to be called to order through the simple dictum: *Taceat mulier in ecclesia.*" (66)

II.—The exegetical Evidence.

"Eh! datevi pace, o teologoni di vecchia scuola, che la verità vuol risplendere anche a traverso di quel denso velo che la ignoranza di alcuni di voi si presume di opporle. Intanto per apprendimento vostro fatevi or meco a leggere qualche altro versetto in cui . . . sarà pure una di quell' esse novità che a' preoccupati leggitori fanno strabuzzare occhi e naso aggrinzare." (67)

The foregoing section has prepared the reader for the "experimentum crucis" to which we now propose submitting various passages of king James's version, by way of testing the vaunted accuracy of its *forty-seven* translators. Three of these instances have been already indicated; (68) one of which, wherein Job longed that his speech should be "*printed* in a *book*," was noticed above.

For convenience sake, having now a few more of these literary curiosities to present, we will tabulate them under alphabetical signs, and prefix to this initial gem the letter

A.—*Job* xix. 23.

One almost blushes to make this imbecility more palpable to general intelligence by recalling to mind that *block*-printing was unknown to Europe prior to A. D. 1423, and printing in *types* before 1457—although the former invention existed, according to Stanislas Julien, (69) in *China* at A. D. 593, and the latter about 1041. Yet, by this "translation," the patriarch must have foreshadowed the art six to ten centuries previously to the advent of Christ! Like every writer comprised in the Old Testament Canon, Job knew as much of *China* as they all did of America; that is, to be frank, just nothing at all. How *forty-seven* able-bodied men could have overlooked this blunder while "correcting proof," surpasses comprehension; unless we ourselves perpetrate another anachronism, as well as a pitiful conundrum, and suppose that "Job-printing" may have suggested some inappreciable affinity between the Anglo-corrupted name of that venerable Arab and the glorious art. What more simple than to have *printed* what the "original sacred tongues" read, "*inscribed* in a *register?*"

B.—*Job* xxxi. 35. [N. B. The first citations always present the textualities of king James's version.]

> "Oh that one would hear me! behold, my desire *is, that* the Almighty would answer me, and *that* mine adversary had written a book."

Can human intelligence understand what possible connection Job's supplication, that God should reply to him, can have with his individual craving that his own unnamed enemy should have indited a *book?* If this text be "divinely inspired" in king James's version, then "the Lord have mercy upon his creature" *archæology!* Because, were these words authentic, logic could prove:—

1. That, at least 2500 years ago, polemical works in the form of "books" were not unknown even in Arabia.
2 That, inasmuch as Job could have no benevolent motive in such wish, vexed as he felt at the aggravations heaped upon his distressing afflictions by his proverbial *comforters*, and knowing, as he must necessarily have done, the power which a Reviewer has over an *author*, he longed, with vindictive refinement, as the most terrible retribution to be inflicted upon an adversary, that his particular enemy should actually write a *book*, in order that Job might *review* him; probably, as Horace Smith conjectured, "in the *Jerusalem Quarterly*."

---

(66) Paul. 1 *Corinthians* xiv. 34;—Strauss: *Vie de Jesus*; Littré's transl., Paris, 1840; ii. p. 378.

(67) Lanci: *Op. cit.;* i. p. 150.

(68) Nott; *Op. cit.;* pp. 136, 137.

(69) Communication to *L'Académie;* June 7—London *Athenæum;* 19 June, 1847.

Cahen renders —

> "Alas! that I have not one who hears! Behold my *writing* — let the Almighty answer me — and the book edited by my adverse party." (70)

This version (for reasons to be elaborated elsewhere) is unsatisfactory, like all we have seen, but Lanci's; because among other oversights it does not afford due weight to the word TaU; vaguely rendered "sign" or "mark" in *Ezekiel* ix. 4. TaU is the name of the last letter in the post-christian *square-letter* alphabet of the Jews; which 142 years B. C., on the earlier Maccabee coinage was cruciform; sometimes like the *Latin*, at others like the *Greek* cross. (71) At the time when Ezekiel wrote in Chaldea, during the sixth century B. C., this *cruciform* letter was the one he must have used, no less than the shape of that "mark" which should be stamped upon the foreheads of the righteous. Its etymological and figurative meaning was "benediction" or "absolution;" just what its descendant, the "baptismal sign" (drawn with water on the foreheads of infants) signifies at this day. Ezekiel's TaU had no direct relation, beyond a distant resemblance in shape and perhaps an occult one in hierophantic mysteries, to the "Crux Ansata," or the sign for "Ankh," *eternal life*, of the more ancient Egyptian hieroglyphics; but its original is now-a-days producible from the *cuneiform* monuments of Assyria; though our demonstration of the fact must be reserved to other opportunities.

It is one thing to prove that the *forty-seven* were wrong in their appreciation of the "word of God:" quite another to emulate the presumptuous part of theologians and dictate dogmatically the English sense of ancient texts in themselves obscure. Our task limits itself to the former office in this essay; but, not to shrink from the utterance of what little we know, the following *free rendering* indicates a probable solution of this tortured passage, and combines Lanci's with other views:— says JOB, "Who will give me one that will listen to me? [i. e., as my judge]. Behold! (here is) my TaU [i. e., he holds up masonically the *cruciform* emblem, as his "absolution"]. The Omnipotent will answer for me [i. e., guarantee me, be my surety, become responsible for me —"that I seek not to evade," *understood*]. And now let my opponent write down his charge [i. e., let my accuser, my calumniator, put his accusations into writing—"that everybody may see them," *understood*].

And, while on the subject of TaU, we may continue our expurgations with other examples.

C. — *Psalms* lxxviii. 41.

> "Yea, they turned back and tempted God, and limited the Holy One of Israel."

Bad as the Jews were, in this case they did precisely the contrary! "The Psalmist," says Lanci, (72) "celebrates in this canticle the marvels which the Lord had done in behalf of rebellious Israel; nevertheless, as the latter finished by conversion, God pardons him and spreads over the culprit the most ample bounties. Conversion, therefore, is the import of this verse, and then it is said—"they (became) converted, they supplicated the Puissant, and implored TaU [i. e., "absolution," or "benediction"] of the Holy of Israel."

D. — 1 *Samuel* xxi. 10—15.

> "And David arose; and fled that day for fear of Saul, and went to Achish the King of Gath. — And the servants of Achish said unto him, *Is* not this David the king of the land? did they not sing one to another of him in dances, saying, Saul hath slain his thousands, and David his ten thousands? — And David laid up these words in his heart, and was sore afraid of Achish the King of Gath. — And he changed his behavior before them, and feigned himself mad in their hands, and scrabbled on the doors of the gate, and let his spittle fall down upon his beard. — Then said Achish unto his servants, Lo, ye see the man is mad: wherefore *then* have ye brought him to me? — Have I need of madmen, that ye have brought this *fellow* to play the madman in my presence? shall this *fellow* come into my house?"

Reminding the reader that DAVID, besides being the warrior-king, was Israel's *bard*, we let Lanci speak for himself:—"The LXX (Greek) made a periphrasis at the first verse, and

---

(70) *Op. cit.;* vol. xv. p. 143.
(71) LETRONNE: *Examen Archéologique;* 1846; plate i., and pp. 11-18.
(72) *Sagra Scrittura Illustrata;* Roma, 1827; ch. ix. CAHEN, xiii. p. 175, note.

added to the (Hebrew) Text by twice mentioning the gates of the city, first to make David play upon his harp, and afterwards to cause him to fall against the said gates. There is perhaps no passage in Scripture that has been more completely denaturalized through the obscurity of a single word. It is evident that David had altogether a part more dignified, more reasonable, to adopt than to counterfeit a lunatic; and moreover that Achish did not display great esteem for his court by saying that madmen were not wanting in it. But the famous TaU, misunderstood, has thrown all interpreters into error. So we will give to it its veritable sense of *to bless;* to this we add that SHÂAR [in Hebrew, as in vulgar Arabic now] does not signify 'door' in this passage, but *poetry*, as its Arabic root teaches: DALETH has the value of 'door' in the same sense that Chaldees and Arabs call 'doors' [*bàb, bibàn*] or 'houses' [*bèyt, beyoòt*] the *strophes;* that is, those *commencements of chapters* and of *strophes* that we [Italians] call 'stanze' [and that in English is adopted for poetry in our word *stanzas;* a word that in Italian, like the above nouns in Oriental speech, has the double meaning of 'stanza' and 'chamber']. If it be insisted that David was raving; it will be, then, with *poetic* furor — the prophetic transport that animated him: but the Arabic root SHAGIÂ, which signifies *to exhibit valor, bravery, courage*, accords much better with the context. These few rays of light ought to be sufficient to dissipate the thick tenebrosities which Translators have piled upon this divine narrative. We may thenceforward give to these verses a reasonable translation and worthy of the majesty of Scripture: — 'David arose, and fleeing on that day from the presence of Saul, came to Achish the king of Gath.—Then the servants of Achish said to him, 'And is not this David king of the earth? is it not in his honor that it was sung in chorus [not, at ancient *Fandangos!*]: Saul has killed a thousand, and David ten thousand!' — David weighing these words in his heart, feared greatly in presence of Achish king of Gath.—It was for this that in his presence, he [David] celebrated their power in a varied *hymn* and in inspired *verses;* and, at each *commencement of a strophe* he made TaU [i. e., he made 'benedictions' — he *blessed them*]; and already the *sweat* was dripping upon the chin's honor [i. e., upon his *beard*, in Oriental phraseology] when Achish interrupted him, and said to his servant: 'hearken to this man who affects inspiration [literally, '*comes* the inspired']; are *poets* [*bards, improvisatori*] wanting to me, that you must bring this one to celebrate my power? and shall (such as) he come into my house?' Nevertheless, David escaped, and took the road that conducted to the cavern of Adulla." (73)

Who seem most "cracked," David, or the bibliolaters of king James's version?

E. — *Leviticus* xi. 20.

"All fowls that creep, going upon *all* four, *shall be* an abomination to you."

To us, likewise! "Raræ aves," invaluable however to museums of Natural History. Not merely, were this prohibition authentic, did *four-legged-fowls* exist in the days of Moses, but the inhibition to eat them would now be worthless to a *Caraïte* Jew, because the breed is extinct. Cahen renders — "Every winged-insect [or literally, *flying-creeping* thing] that walks upon four [*claws, feet,* understood] is an abomination unto you."

Dwelling not upon *verse* 21, although marvelling how "legs" could be placed anatomically elsewhere than "above their feet," we refreshen ourselves with

F. — 2 *Kings*, vi. 25.

"And there was a great famine in Samaria: and, behold, they besieged it, until an ass's head was *sold* for fourscore *pieces* of silver, and the fourth part of a cab of dove's dung for five *pieces* of silver."

"Sternhold and Hopkins had great qualms
When they translated David's psalms";

but the sufferings of these poor men were infinitesimally small compared to those the *forty-seven* would have experienced had they partaken of that delicate repast, for about two-thirds of a pint of which the starving Samaritans paid such monstrous prices! *Pigeon's dung*, or "doves'-dung," owing to the quantity of ammonia it contains, is still used throughout

(73) *Op cit.;* Ch. ix. § 3. CAHEN: vii. p. 86, preserves the old mistakes.

the East, in the absence of modern chemistry, to give temper to Damascene sword-blades, &c. It sharpens weapons, not appetites! Can one conceive a human stomach, however depraved by want, alimented upon "guano?" Bochart, (74) two centuries ago, showed that "pois chiches," in Italian *ceci*, in English "chick-pea," — the commonest Oriental vetch, or pea, — is the rational interpretation of the word; and thus the only enigma preserved is, how *forty-seven* Englishmen could have committed a mistake so extraordinary. The obsolete word "cab" aptly illustrates how imperative it has become, through unavoidable changes of language within 250 years, to issue a re-translation in our current vernacular, lest the illiterate should think that "*cab*-riolets," 26 centuries ago, plied in the streets of Samaria! Superstition is gradually elevating the vulgar Cockney speech of the age of King James into our "lingua sancta;" and the translation authorized in his reign will some day become unintelligible and useless in the "Far West," except to those who possess glossaries wherewith to read it. Theologers would act wisely to consider these things, while we pass on to

G. — *Leviticus* xxi. 18 and 17.

> "He that hath a flat nose" — [is forbidden] — "approach to offer the bread of his God."

A *flat nose*, in the Abrahamic type of mankind, among their "Cohenîm" or priesthood, was, in the days of the Hebrew Lawgiver, as it is now among Israel's far-scattered descendants, too great a deviation of physical lineaments from the indelible standard of the race (portrayed as we exhibit them in our present work from the monuments of that epoch, and as we daily see them in our streets) not to excite suspicion that such cases testified to admixtures of foreign (75) and consequently of "impure blood"; and therefore to debar a priest with a "flat nose" from the Tabernacle was rational at their point of view. *Negro* families [as already demonstrated, *supra*] are unmentioned throughout the Hebrew Text; and negrophilism may accordingly rejoice that the rendering selected by the *forty-seven* cannot now be applied to the former "de jure," where it is notoriously (in the *Free* States of this Federation, especially) "de facto."

Happily — no thanks to our translators — "Snubs" of universal humanity may legally officiate at sanctuaries; the word K*h*RM (76) meaning only a "*mutilated* nose:" and the inhibition referring to noses injured by deformity, accident, disease, or law, (77) our apprehensions were futile, like their translation.

An ethnological item has been touched upon involuntarily, and now we may as well give ventilation to another much-abused text.

H. — *Song of Solomon*, i. 5, 6.

> "I *am* black, but comely, . . . Look not upon me because I *am* black, because the sun hath looked upon me: my mother's children were angry with me; they made me keeper of the vineyards; *but* mine own vineyard have I not kept."

The apocryphal "prologue" at the head of this chapter tells us that here the *Church* "confesseth her deformity"! It were well if, before printing this acknowledgment (which it is not for us to dispute), the "Establishment" had corrected the deformity of their *translation:* which has led our anglicized Nigritians to claim this supposititious bride of Solomon as a Venus of their own species! With equal reason, some commentators, even of modern

(74) SALVERTE; *Sciences Occultes;* i. p. 44. CAHEN (whose *notes* are infinitely more valuable than his textual translations), viii. p. 127, *note*, adds — "Selon plusieurs commentateurs, il s'agit ici d'une nourriture misérable, de quelque herbe à vil prix," &c.

(75) On returning from the Captivity, "the children of Habaiah, the children of Koz, the children of Barzillai, which took *one* [SIC, in our version!] of the daughters of Barzillai the Gileadite to wife, and was [*! idem*] called after their name," were, "as polluted, put from the priesthood" — (NEHEMIAH vii. 63, 64.)

(76) CAHEN: vol. iii. pp. 99, 100.

(77) "I cut off both his *nose* and ears," proclaims DARIUS, of Phraortes, and of Sitratachmes, at Behistun. (RAWLINSON: *Persian Cuneif. Inscrip.;* 1846; part i. p. 34.) Philanthropy need not shudder at atrocities of the fifth century B. C., for in Turkey such punishment is as common now as it was 3300 years ago, if MOSES wrote this passage.

times, (78) infer that she was "an Egyptian princess;" while others identify the lady with "Pharaoh's daughter;" for "King Solomon loved many strange women. . . . Moabites, Ammonites, Edomites, Zidonians, *and* Hittites," and what not! (79) It need hardly be mentioned that, the dynasty out of which the sage king selected additions to his *hareem* being yet unfound in hieroglyphics, the monuments of Egypt throw no light upon this otherwise very probable amalgamation. (80)

The "*Canticle of Canticles of which of Solomon*, that is to say, *one of the Canticles of Solomon*," as Lanci literally interprets its epigraph, (81) has suffered much at the hands of the *forty-seven*. They, and others, lost sight of the simple fact (to be exemplified in its place), that, in the ancient Hebrew Text, divisions into *chapters*, *verses*, *words*, or by *punctuations*, are absolutely unknown; while, paralleled to this day in Arabic calligraphy, no *notes* of admiration, interrogation, &c., mark inflections of the sense. The context alone can indicate a query; so that a "crooked little thing which asks a question," added to fidelity of construction and acquaintance with Levant usages of the present hour, rescues our pretty Shulamite *brunette* from all Ethiopian hallucinations [*supra*, p. 483].

"I am *brown* (Italicè "fosca," dark, *tanned*) but pretty," says the girl coquettishly; then [deprecatingly to her swain], "Do not mind that I am *browned*, because the sun has tanned me; [which she explains by adding] the male-children of my mother [*i. e.* my *step-brothers;* who, in the East, control their maiden sisters after the father's death] having become free to dispose of me, placed me watcher of vines: ["don't you see?" *understood*] my own vine, have I not watched it?" (82)

One improvement heralds another: it is so in machinery: it is equally true in biblical hermeneutics, the moment a man's mind soars above the supernatural grade of ratiocination. From the simple proposition that they who expound the Scriptures should understand them, we hold that no one is competent to impugn these deductions who is unacquainted, not merely with the original Hebrew and Greek languages, but with the noble achievements of Continental exegesis. Hear a living Church of England dignitary: —

> "Those who advocate the free use of philology in the interpretation of the Scriptures, find their fiercest and most uncompromising opponents in the ranks of those who are slaves to the Puritanical *Bibliolatry*, so common in this country. According to this school, every word in the canonical books of the Old and New Testament (in king James's version) proceeds from a divine and miraculous inspiration. . . . By those who believe in the plenary and verbal inspiration of the (English) Scriptures, science in general and philological science in particular, are viewed with distrust, if not with abhorrence; and the more so, if this *bibliolatry* is combined with a certain amount of *ecclesiolatry*." (83)

It is a pity, certainly; for if some expounders possessed the intelligence they would deplore their want of education: but we continue.

I. — *Habakkuk* ii. 11.

> "For the stone shall cry out of the wall, and the beam out of the timber shall answer it."

That a stone should cry out from a wall is an idea consonant with Oriental hyperbole; but that a *beam* should answer out of timber seems to be an unpoetical and far-fetched conception, as it presupposes the proximity of a "timber-yard" to the wall aforesaid. It furthermore is not in unison with the context; wherein the prophet, who "surpasses all which Hebrew poesy can offer in this department," (84) declaims against Chaldæan flagitiousness. The propriety of his metaphor resiles to view through Lanci's rendering and notes of interrogation.

---

(78) *The Friend of Moses;* New York, 1852; p. 468, note.

(79) 1 *Kings* iii. 1; xi. 1.

(80) Rosellini: on Osorchor of Manetho's XXIst dynasty.

(81) *La Sagra Scrittura;* ch. v. § 4. Cahen: xiv. 3, 4, has not seized the poet's meaning.

(82) Lanci: *Paralipomeni;* ii. p. 45.

(83) Phileleutherus Anglicanus: *A Vindication of Protestant Principles;* London, 1847; pp. 43, 44; — Gliddon: *Otia Ægyptiaca*; 1849; p. 93.

(84) De Wette: ii. p. 466.

"Peradventure, shall the *statue of stone* [an *Assyrian* bas-relief?] from the wall cry out? The *cricket* [scarabæus, or beetle] from out of the wood will it respond?" (85)

There is a verse of another prophet that Lanci restores, in which our *forty-seven* have metamorphosed *famines* into "young men," and *sorrows* into "maids."

J.— ZECHARIAH ix. 17.

"Corn shall make the young men cheerful, and new wine the maids."

The "Sons of Temperance" may not be pleased with the moral, but the Daughters will not fail to appreciate an emendation that relieves their antique sisters from the charge of unfeminine indulgences.

The old *Vulgate* had translated — "For, what is the goodness of God, what is his glory, if not the corn of the elect, and the wine which fecundates the virgins?" Vatablus and Pagnini make "confusion worse confounded" by reading — "The corn which makes the young men sing, and the new wine of the girls." But, based upon *radicals* preserved in Arabic, our teacher proposes: —

"What is more sweet and more agreeable than corn in scarcities, and wine that fortifies in afflictions?" (86)

"Per saltum," inasmuch as in the chaos of our memoranda of *false-translations* orderly classification is inconvenient, while to our objects quite unnecessary, we open —

K. — *Genesis* xxiii. 9, 17, 19.

"The cave of Machpela" —

purchased by Abraham for Sarah's inhumation — to remark, that the word *Machpela* which, according to our authorized verity, seems a "proper name," is grammatically, in Semitic tongues, "a thing *contracted-for*;" so that, it is as vain for tourists in Palestine to search for *Machpela*, as for biblical chorographers to define its latitude and longitude. (87)

L. — 1 *Samuel* xix. 13.

"And Michal took an image, and laid *it* in the bed, and put a pillow of goat's *hair* for his bolster, and covered *it* with a cloth."

Manifold were the sins of David, but idolatry was certainly not one of the number; although scandalous suspicions have been rife in regard to this *image*. Commentators have likewise expounded how the *image* being laid in the bed, and covered up with the bed-clothes, the messengers supposed that the invalid whom they were sent to slay (*v.* 11) was asleep therein: but we are told: —

M. — 1 *Samuel* xix. 16.

"And when the messengers were come in, behold, *there was* an image in the bed, with a pillow of goat's *hair* for his bolster:" —

whence it is evident that the *forty-seven* deemed the "image" to be of the masculine gender. Their notions of an Oriental *bed* too must have been peculiar, in England, two hundred and fifty years ago, when a "pillow" was made to serve for a "bolster;" and such a hirsute contrivance! However, having commenced rolling down hill, they reach the bottom through a series of cascades that would excite Homeric smiles were not "God's word" the sufferer: as may be seen by the subjoined restitution; after comprehending that Michal, the astute daughter of king Saul, was a princess in whose "trousseau" were doubtless many of the crown regalia: —

"Michal took her casket full of jewels, and placed it upon the bed; whence were reflected magnificent splendors; and she hid them with a curtain [? *coverlid*]." . . . "The messengers having arrived, O surprise! the jewels [being] upon the bed, from their summits was thrown out a magnificence of splendors." (88)

---

(85) *Op. cit.;* i. p. 283; — CAHEN, xii. p. 115, also reads differently from our version; but see his *note* 11.

(86) *Sag. Scrit.;* ch. ii. § 1; — CAHEN, xii. p. 156, follows the Rabbis.

(87) *Paralip.;* i. p. 144.

(88) *Sag. Scrit.;* ch. vii. 4. The *note*, 13, of CAHEN, vii. p. 76, shows how the text puzzled him. LANCI, *op. cit.*, proves that in no place are TteRaP*h*IM "idols."

Humiliated at this sight, the assassins remembered that Michal was a royal daughter whose husband, escaped from their clutches, was just the man to reward them with a hempen neckcloth on his accession to the throne; so, apologizing for their intrusion, the emissaries withdrew.

Goats appear to have been favorites with our translators. Not content with transmuting jewels into "goat's *hair*" and filling the royal "bolster" with this rare, elastic, and odoriferous article, they must needs metamorphose one of the sublimest Hebrew names of Deity into a "*scape-goat*"!

N.—*Leviticus* xvi. 8, 10, 26.

> "And Aaron shall cast lots upon the two goats; one lot for the Lord, and the other for the scapegoat.... But the goat, on which the lot fell to be the scapegoat, shall be presented alive before the Lord, to make an atonement with him, *and* to let him go for a scapegoat into the wilderness.... And he that let go the goat for the scapegoat, shall wash his clothes," &c.

ÂZAZL—*âzazel*—is the Hebrew word. "This terrible and venerable name of God (says Lanci) through the pens of biblical glossers has been a *devil*, a *mountain*, a *wilderness*, and a *he-goat!*" (89)

It will give an idea of the lucidity of Rabbinical criticism, to quote the following:—

"Aben Esra, according to his habitual manner when he is in trouble, enunciates in the style of an oracle: 'If thou art capable of comprehending the mystery of *Azâzêl*, thou wilt learn also the mystery of his name; for it has similar associates in Scripture; I will tell thee by allusion one portion of the mystery; when thou shalt have thirty-three years, thou wilt comprehend us.' He finishes abruptly without saying anything more allegorically or otherwise." (90)

The ante-Christian Hebrew text was undivided into *words*. Our preceptor re-divides ÂZAZ*e*L into two distinct nouns; ÂZAZ and EL. The latter, every sciolist knows, means the *strong*, the *puissant* par excellence, the *Omnipotent*. ÂZAZ, identical with the Arabic *âzàz*, has its radical monosyllable in *â*Z, "to conquer" and "to be victorious;" wherefore, ÂZAZ-*E*L signifies the "God of *victory*"—here used in the sense of the "*Author* of *death*," in juxta-position to I*e*HO*ua*H, the "*Author* of *life*:" to the latter of which *Authors* the Jews were enjoined to offer a *dead* goat; while, by contrast, to the former they were to offer a *live* one. Thus, death to the *Life-giver*—life to the *Death-dealer*. The symbolical antithesis is grand and beautiful.

For the sake of perspicuity we submit a *free* translation to the reader:—"And Aaron shall place lots upon the two he-goats; one lot to I*e*HO*ua*H, and one lot to ÂZAZ-*E*L.... And the he-goat upon which the lot has fallen to ÂZAZ-*E*L shall be placed *alive* before I*e*HO*ua*H, to become exempted by him, to be sent forth to ÂZAZ-*E*L in the desert.... And he who shall have led forth the he-goat to ÂZAZ-*E*L shall cleanse his clothes," &c. In verse 9, the other he-goat offered to I*e*HO*ua*H was to be *killed*.

Having thus entirely misapprehended the sense of the above passages, it was quite natural that our gifted translators, one *Divine Name* having vanished through their skill, should have been blinded to many others. Here is one of them:—

O.—*Job* xxi. 15.

> "What *is* the Almighty, that we should serve him? and what profit should we have, if we pray unto him?"

We have illustrated, under the preceding letter N, the splendor of antithesis which Hebrew literature conceived in the selection of *Divine Names;* and herein leniency may be accorded to the English interpreters, because neither they nor early or later scholiasts, could have anticipated a discovery due to the profoundest Semitic savant of our genera-

---

(89) *Sagra Scrittura;* ch. iii. § 1;—*Paralipomeni;* ii. p. 354.

(90) CAHEN: iii. p. 68. It may be well to warn cavillers that this subject has been studied. We do not agree in HENGSTENBERG's idea (*Egypt and the Books of Moses;* pp. 169–184), that *âzazel* is "Satan." For parallelisms on the sacrifice of he-goats to the God-Preserver and the God-Destroyer, conf. RIGHELLINI (*Examen;* ii. p. 246); MOVERS (*Die Phœnizier;* i. p. 367); and MAURY (*Génies Psychopompes;* Aug. 1845; pp. 295, 296—and *Personnage de la Mort;* Aug. 1847; pp. 325, 326) in the *Revue Archéologique.*

tion, the affable Professor (for thirty-nine years) of Sacred Philology at the Roman Vatican.(91)

The original of the substantive rendered "profit" is NUâIL — a noun which, occurring but once amid the 5642 (92) words preserved, in the Hebrew and Chaldee Bibles, to our day (fragments, so to say, of the ancient tongue) — is unique; and consequently its signification is recoverable solely through its extant radical in Arabian dialects. Its true root is *wâal*, "to be eminent"; and its sense, "the most *sublime*." The prototype of "Almighty" is textually S*h*aDaI; literally, "the most *valorous*." Let the reader now compare king James's version with the subjoined:—

"Who is the *most* VALOROUS (S*h*aDaI), that to him we must be servants? who the *most* SUBLIME (NUâIL), that we should go [out of our way] to meet him?"

Variety is pleasing, so we skip over to

P. — *Micah*, v. 2.

> "But thou Beth-lehem Ephrata, *though* thou be little among the thousands of Judah, *yet* out of thee shall he come forth unto me *that is* to be ruler in Israel."

The emendation suggested relates principally to the word rendered "thousands," of which the singular, in the unpunctuated Hebrew, is ALUP*h*.

ALeP*h*, א, first letter of the Hebrew alphabet, in its Phœnician original is the tachygraph of a *Bull's head;* and its name is derived from that of the animal, because the *bull* is "leader" of the herd.(93) Hence ALeP*h* became a title as the "leader," general, dux, or *chief;* of which examples are numerous in the discrepant so-called "Dukes" of Edom, &c.; corruption of the Latin "dux, duces"; which, with more propriety in English, should be rendered *chiefs*. Copying the Latin and Greek versions, without archæological knowledge of the Hebrew tongue, our translators have read *Elf-îm* "thousands," when *Chiefs* is its real meaning; thus:—

"And thou Bethlehem of Euphrata, [even] if thou art little among the CHIEFS of Juda, I will cause to issue from thee the dominator of Israel." (94)

Without regard to the fantastical and spurious headings to this Chapter in our version, we may add, that the reading of CHIEFS is as old as the second century B. C, when the LXX Greek version was made by the Hellenistic Jews of Alexandria; because about 68–69 A. D. the author of the "*Good Tidings according to Matthew*," in citing the above passage from Micah, read "Princes"; (95) and he does not appear to have been acquainted (96) with the Hebrew Text. Paulus and De Rossi even contend that the speech of Christ, Χριστος, was Greek. (97) But, we wander from our theme.

Q. — *Isaiah* xviii. 1, 2.

> "Woe to the land shadowing with wings, which *is* beyond the rivers of Ethiopia; — That sendeth ambassadors by the sea, even in vessels of bulrushes upon the waters, *saying*, Go, ye swift messengers, to a nation scattered and peeled, to a people terrible from their beginning hitherto; a nation meted out and trodden down, whose land the rivers have spoiled."

We cite this passage not with a view of destroying the interpretation of the *forty-seven*, in this instance excusable enough, but by way of elucidating how meritorious it would be to reconstruct their time-worn edifice, guided by the lights which Oriental, and particularly *Egyptian*, researches of our living generation cast upon subjects until this century utterly dark.

All interpreters here have been at fault. The LXX render 'Οναὶ γῆς πλοίων πτέρυγες — i. e. *Væ terræ navium alis.* The Vulgate — *Væ terræ cymbalo alarum.* Cahen substitutes — "Ah!

(91) LANCI: *Op. cit.;* p. 354, &c.
(92) LEUSDEN, apud GESENIUS, in *Parker's De Wette;* i. p. 459; — MUNK: *Palestine;* p. 436.
(93) GESENIUS: *Script. Ling. Phœniciæ;* 1838; p. 19.
(94) *Sagra Scrit.;* ch. i. § 2; — "Trop petit pour être parmi les *chefs* de Iehouda," CAHEN: XII. pp. 96, 97 — see *note* 1.
(95) *Matt.* ii. 6; SHARPE'S *New Test.;* p. 3.
(96) HENNELL: *Origin of Christianity;* 1845; pp. 123, 124: and *Christian Theism;* pp. 82, 83.
(97) GESENIUS; *Heb. Sprache und Schrift;* 1815; p. 46.

pays sous l'ombrage des voiles'; (98) and the late Major Mordecai Noah actually read — "Hail! Land of the (American) *Eagle*"!

Rosellini (99) was the first to indicate that here the prophet apostrophizes *Egypt* under the metaphor of her national symbol — ☞ — the "winged globe"; as Birch defines it, "emblem of KHEPER, the *Creator Sun*".(100) We subjoin the learned Pisan's emendation, with a few additions: —

FIG. 359.

"Ho! Land of the *Winged Globe* [Egypt]! which art beyond the rivers of KUS*h* [i. e. the "torrens Ægypti," on the Isthmus of Suez; *supra*, p. 484]: that sendest into the sea, as messengers, the canals of thy waters; and that navigatest with boats of *papyrus* on the face of the waves. Go, ye light messengers, to the *elongated* people [i. e. stretched out along the narrow alluvials of the Nile,] and *shaved* nation [the Egyptians were essentially a *shaven* population — vide *Genesis* xli. 14,]; to a people terrible from the time that was, and also previously; to the *geometrical* people [Geometry originated in Egypt], who treading [with their feet cultivate their fields]; whose lands the rivers will devastate [referring to some unfulfilled prophecy]."

R. — *Ecclesiastes* xi. 1—2.

> "Cast thy bread upon the waters, for thou shalt find it after many days. . . . Give a portion to seven, and also to eight; for thou knowest not what evil shall be upon the earth."

Unless there was some cabalistic *key* to the latter portion of these sentences, through which the Translators understood what *they* wrote, the super-refined meaning they attached to the numerals 7 and 8 surpasses our feeble comprehension: even Solomon, reputed author and great magician, could not unravel their knot. Let us substitute: —

"Cast thy bread where fruits are borne, because time will restore it with usury. . . . Give the measure (*porzione*) even to saturity and abundance, because thou knowest not what evil may come upon the earth." Here, comments Lanci, (101) the sage exhorts man to do good, and to charitable acts towards the poor who, satiated with abundant food, will cause to rain upon him, through the fervor of their prayers, ample benedictions during bad seasons. But, what can be expected from men who translate "*Tor*, *Sus*, and *Agùr*" — *ve* T*t*UR *ve* SUS *ve* âGUR,

S. — *Jeremiah* viii. 7, — by

> "the turtle and the crane and the swallow,"

— when the prophet meant "the *bull* and the *horse* and the *colt*"? (102)

T. — *Zechariah* v. 1, 2, 3.

> "Then I turned, and lifted up mine eyes, and looked, and behold a flying roll. . . . And he said to me, What seest thou? And I answered, I see a flying roll; the length whereof *is* twenty cubits, and the breadth thereof ten cubits. . . . Then said he unto me, This *is* the curse that goeth forth over the face of the whole earth; for every one that stealeth shall be cut off, *as* on this side according to it; and every one that sweareth shall be cut off *as* on that side according to it."

If the prophet had been so unfortunate as to receive the words of this angelic vision in *English*, he would have required a second revelation to understand its Translators' impenetrable meaning.

A "flying roll"! Think of a parchment synagogue *roll* (MeGiLaH, *Meghillà*), of such proportions, actually *flying* through the air! Consider the amount of inspiration it must

---

(98) IX. pp. 66, 67.

(99) *Monumenti Civili;* ii. pp. 394–403.

(100) GLIDDON: *Otia Ægypt.;* pp. 95, 96: — "It is *the Morning Sun:* it is often called *the beam of light which rises*, or 'comes out,' *of the horizon*" — BIRCH: *Egyptian Inscription at the Bibliothéque Nationale;* R. Soc. Lit.; 1852; iv. p. 3.

(101) *Sag. Scrit.;* ch. iv. § 64. CAHEN: xvi. p. 129, *notes* 1, 2.

(102) *Paralip.;* ii. p. 391. The "seasons" should be "rutting-times — although CAHEN, x. pp. 30, 31, prefers the old reading.

have required to comprehend *which* side was mortiferous to thieves, which to swearers; for in Aristotelian logic, "if the one is the other, the other must be the one:" and remember that in the phrase "according to it" lies lost, forgotten, and entombed, *one-half* of the *ineffable Tetragrammaton* IHOH (JEHOVAH)! that most terrible, the most occult monosyllable of the palindromic name vocalized as ADONAI, the "Lord"! Here is the sense, *verbatim et litteratim*: —

"And turning myself, I raised my eyes, and saw: and behold a *whirling disk* [of fire—having a mystic relation to the Egyptian 'winged-globe,' emblem of KHEPER, the *Creator-Sun*]. (103) Then the angel said to me: 'What seest thou?' I answered, 'I see a *whirling disk* of twenty cubits in length and of ten in height' [its *wings* enlarging the lateral diameter]. And he said to me: 'This is the malediction [of God] which spreads itself upon the surface of the whole earth; verily, every thief by this [the *whirling disk*] *as (if) by* OH [deuterosyllable of IH-OH] shall be destroyed; and every perjurer by this [the *whirling disk*] *as (if) by* OH shall be destroyed.'" (104)

"The which, philologers will recognize as common sense and justness, if as much was not perceived by those wretched theologists (*teologastri*) who, in philological knowledge not surpassing the Hebrew alphabet, go hunting about through lexicons in order thence to spit forth a doctoral decision in people's faces"; says Lanci. (105)

But, as the time for the exposition of these recondite biblical arcana has not yet arrived, our meaning is best conveyed to the *Illuminati* (106) by amending

U. — *Psalms* xxxvii. 7,

> "Rest in the Lord, and wait patiently for him; fret not thyself because of him who prospereth in his way, because of the man who bringeth wicked devices to pass" —

as follows: — "Keep silence in (the secret of) IHOH, and take delight in it: dispute not with him who seeks to penetrate into the acquiring of it, nor with any vain man who attempts it." (107)

V. — *Psalms* cx. 1—7.

> "The Lord said unto my Lord, Sit thou at my right hand, until I make thine enemies thy footstool. — The Lord shall send the rod of thy strength out of Zion; rule thou in the midst of thine enemies. — Thy people *shall be* willing in the day of thy power, in the beauties of holiness from the womb of the morning; thou hast the dew of thy youth. — The Lord hath sworn, and will not repent, Thou *art* a priest for ever after the order of Melchisedek.—The Lord at thy right hand shall strike through kings in the day of his wrath.— He shall judge among the heathen, he shall fill *the places* with the dead bodies; he shall wound the heads over many countries.— He shall drink of the brook in the way: therefore shall he lift up the head."

This superb ode has by some been suspected to have been derived from hymns of pagan origin, sung during the season that Ezekiel (viii. 14) saw the "woman weeping for T*ta*M-UZ," about the winter solstice, or 21st December, where the Church almanacs place the anniversary of the *unbelieving* St. Thomas. They refer to the fact that St. Jerome's Vulgate renders T*ta*M-UZ by *Adonis*, favorite god of the Phœnicians in Palestine and Syria, to justify their reading of "Says JEHOVAH to ADONIS" (108)! Others, again, take MELCHISEDEK to be the *Melek-Sadyc*, the "just king," whose name SYDYC, with the title of "just" is preserved, by Sanconiathon, as the father of the *Cabiri*, &c. (109) St. Paul, however, cites this Psalm frequently in his *Epistle to the Hebrews*; and whoever put the headings to the former in our authorized version has asserted that its language can apply to no other than the Messiah. With all deference, the subjoined paraphrase of Lanci's close Italian

(103) See preceding page, under Q.

(104) LANCI: *Sag. Scrit.*; ch. iii. § 7; — *Paralipomeni*; i. p. 97, *seq.*; ii. p. 354; and *Lettre à M. Prisse*; 1847, p. 33. These views are later than CAHEN'S, xii. p. 144.

(105) *Paralip.*; i. p. 3.

(106) MACKAY: *Free-Mason's Lexicon*; 2d edit.; Charleston, S. C.; 1852; voce *Jehovah*, and *Name*: — also, ROCKWELL: *Discourse* before the G. L. of Georgia; Oct. 30, 1851; p. 27.

(107) *Paralip.*; i. p. 149; — CAHEN: xiii. p. 84, *note* 7.

(108) Compare PARKHURST: *Hebrew Lexicon*; voce "Adonai"; with ANTHON: *Class. Dict.*; 1841; pp. 26, 27; — also R. P. KNIGHT, to be cited hereafter.

(109) CORY: *Anc. Frag.*; pp. 8, 9, 13, 16; "Sanconiatho."

translation of the "Dixit Dominus," while it removes the senilities of the *forty-seven*, shows that the composer of that ode dedicated it to some contemporary *priest* called MELCHISEDEK, living at the time of its composition.

"Said IeHOuaH to my Lord: 'Sit thou on my right until I make of thy foemen a stool for thy feet'.—IeHOuaH from Zion will send the wand of thy glory: go, rule in the midst of thy foes.—Thy people will behold spontaneously, when thou shalt understand thy powerful qualifications for the splendor of the priesthood; from the womb, the germ of thy birth was mysterious.—IeHOuaH swore, nor does he retract his oaths: '*Thou, O Melchisedek, shalt be, upon my word, Priest* (a Cohen) *forever!*'—My Lord at thy right hand slew kings in the day of his furor—At the ruling amid the Gentiles, the confines having been passed by force, the chief of vastest land swooned—He will pour himself out more than a torrent through (its) course; wherefore will he raise his head." (110)

As every departure from the literal Italian entails another remove from the original Hebrew, grace is here purposely sacrificed to fidelity; but, from the general tenor of the context, owing to the distinctions observed by the writer between the use of the terms "Jehovah" and "my Lord," one might infer, that this poetical effusion commemorates some conquest over foreigners, with which the composer and his sacerdotal friend MELCHISEDEK were familiar; scenes in which the latter personage (named after the long-anterior "King of Salem") (111) had been an actor. We must console ourselves (under the expected charge that all this is mere conjecture) by reflecting how, if Lanci's shaft may have missed the bull's eye, the arrows of forty-seven able-bodied men flew wide of the target; and that another nail has been driven into the latters' version, which we shall have the satisfaction of "clinching" under the succeeding letters.

According to Cruden's laborious work, (112) the words "grove" and "groves" are "authorized" to re-appear in the English Bible about thirty-six times. Theologians of the lower grade naturally suppose that, in the "original sacred tongue," one single noun, repeated throughout the Text, as its substitute is in our version, must be the latter's representative. Vain illusion!

W. — *Genesis* xxi. 33.

> "And *Abraham* planted a grove in Beer-sheba, and called there on the name of the Lord, the everlasting God."

He did nothing of the kind! He, ABRAHAM, "set up (אשל, ASeL) a *tablet* (or *stele*) in Beersheba, and (קרא, KaRA, *read; also, wrote*) *engraved* it with the name of IeHOuaH to perpetual duration." (113) Here, take note, the original for "grove" is ASeL.

X. — 2 *Kings* xxiii. 6.

> "And he brought out the grove from the house of the Lord, without Jerusalem, unto the brook Kidron, and burned it at the brook Kidron, and stamped *it* small to powder," &c.

A word occurs frequently in the Text, written in two ways, âST*t*URT*t*, and âS*h*T*t*RUT*t*; which is punctuated, by the Massora, *Astôret*, and *Ashtarôt*. At other times, according to the peculiar provincialism (*patois*) of each biblical writer, the same word appears in the form of ASeRA, or plural AS*h*eR-IM. These are all proper names of one person; and that person is no other than the *goddess* ASTARTE of the Palestinians; HATHOR of the Egyptians; âTYR of the Himyaritic Arabs; the VENUS of Græco-Roman mythology, and of our vernacular. Now, here the word for "grove" is AS*h*eRaH: and our Translators' deed in rendering ASeL by "grove" in one place, and AS*h*eRaH by "grove" in another,

(110) *Paralip.;* ii. p. 110. How extensively obscure is the sense of this Psalm may be seen from CAHEN'S *notes*. xiii. pp. 251–256, 355, 356.

(111) *Genesis* xiv. 18. "Salem," commentators tell us, was the name of *Jerusalem*—YeRuS*h*aLaïM, from *Yerus*, "heritage," and *Shalaïm*, "peace," in the dual; literally, "She who inherits twofold peace" (*Paralip.; in loc.*). They also tell us that MOSES wrote *Genesis*, about the 14th—15th century B. C. Perhaps their archæological ingenuity will explain how it came to pass that the old town of *Jebus* was called "Salem" before it was taken by the Jews of JOSHUA (*Josh.* xviii. 28; *Judges* i. 21; xix. 10, 11; &c.), long after MOSES'S death? Until they do, that MOSES wrote XIVth *Genesis* is simply impossible; as likewise the contemporaneousness of ABRAHAM with a "King of Salem." Such anachronisms betray the modern age of this chapter; and render the *older* MELCHIZEDEK very like the Phœnicians' "SADYC the *Just*," whose place in history is mythological."

(112) *Concordance*—from 10th Lond. edit.; Philadelphia, 1841; p. 254.

(113) *Paralip.;* i. p. 97, *seq.*

is cecity, if not worse. We pass over, therefore, the extraordinary circumstance how JOSIAH could find a "grove" in a *house*, unless that grove was very small, or the house very large, which SOLOMON's *temple*, only ninety feet by thirty, was assuredly not—and how he could carry about and break up with facility an entire "grove" seems inexplicable. Not so when we read — "And he dragged the (wooden *statue* of) VENUS (AS*h*eRaH) (114) out of the house of IeHOuaH:"—a proceeding which begins to reveal to us, what some "teologastri" have ventured recently to doubt, (115) viz., the infamous *atrocities* of ancient Jewish templar worship; that we propose to lay bare in another place. "Ex abundantiâ," we give a correct but *modest* restoration of *verse* 7 of the same chapter, which intelligent readers can compare with the blundering performance of the *forty-seven*: — "And he (JOSIAH) broke down the little chapels of the *shameless* priests that were in the house of IeHOuaH, where the women spread perfumes before the *niches* of VENUS"—for, says *verse* 5 — the Jews "had burned incense to BAAL, to SHEMS, to the *Moon*, and to the *Signs of the Zodiac*, and to all the *Asterisms* of Heaven!"

It was the *discovery* (about 620 B. C.), to say the least, of the "Book of the Law" of MOSES, (116) lost and forgotten for some 700 years, which instigated the reforming JOSIAH to these vigorous measures: but pious iconoclasts had been shocked at similar abominations before; as the following text clearly exhibits; while it also relieves poor JOASH, the worthy father of the valiant GIDEON, from the accusation of idolatry that *forty-seven* men stimulate "simple believers" to hurl at his innocent head.

Y. — *Judges* vi. 25, 26.

> "And it came to pass the same night, that the Lord said unto him, Take thy father's young bullock, even the second bullock of seven years old, and throw down the altar of Baal that thy father hath, and cut down the grove that *is* by it:—And build an altar unto the Lord thy God upon the top of the rock, in the ordered place, and take the second bullock and offer a burnt sacrifice with the wood of the grove which thou shalt cut down."

Decency forbids that we should explain the sculptural obscenities that GIDEON's eyes beheld. Orientalists, whose studies may have led them into antique *pornography*, will comprehend us and the exactitude of the venerable Lanci's translation, (117) of which we submit a close but softened paraphrase: —

"And it was in that night that IeHOuaH said to him [Gideon]: 'Take the young bullock of thy father, and another bullock of seven years, and thou shalt fell, with the *altar* [supporter] of BAAL [the obscene God] that [bullock] which is thy father's; afterwards thou shalt break down the VENUS [ASHERA, the foul goddess] which was above it. Then thou shalt build up, in regular proportion [i. e., according to Mosaic rules], an altar to IeHOuaH, thy ELOH, on the summit of that [yonder] rock; and, taking the second bullock, thou shalt burn it in holocaust with the wood of the VENUS by thee broken up.'"

We may now inquire of the reader, in all good faith, whether, in every instance laid hitherto before his acumen, our emendations have not made plain sense of that which was utter nonsense; and whether the *Bible*, properly translated, is not a much loftier book, far grander, as regards mere literary excellence, than the version, "authorized" exactly 250 years ago, has ever made it appear?

If such be his candid opinion, he will feel a high gratification at the revisal, through the application of pure grammar and philology, of that imaginary text, on the authority of which the *Copernican* system was traduced by ecclesiastical ignorance; while the telescopic discoveries of the immortal Galileo, A. D. 1615, condemned, as "absurd, false in philosophy, and formally heretical, being contrary to the express word of God," nearly brought him to those fagots whereupon, only fifteen years before, Giordano Bruno's living

(114) CAHEN preserves "Aschera" in his translation (viii. p. 190, &c.); accurately remarking that, if the Rabbis bestowed more attention on "*Antiquités bibliques*" — "there would not be then less respect for the sacred writings, but they would no longer be regarded as the Pillars of Hercules of all civilization" (p. 205).

(115) *Inter alios*, the Rev. Dr. SMYTHE of Charleston, S. C.: *Unity*; p. 112, note.

(116) 2 *Kings* xxii. 8; and 2 *Chron.* xxxiv. 14.

(117) *Paralip.*; ii. 28–31. CAHEN: vi. p. 31, "Aschera."

body was calcined "ut quam clementissime et citra sanguinis effusionem, puniretur." (118) Had Lanci never turned his vast Semitic acquirements to any other Scriptural text but Joshua Xth, 12, 13, astronomical posterity should weave for him a wreath of laurels. But, to appreciate his labors, one must bestow a final smile of pity upon the *forty-seven.*

Z. — *Joshua* x. 12, 13, 14.

> "Then spake Joshua to the Lord in the day when the Lord delivered up the Amorites before the children of Israel, and he said in the sight of Israel, Sun, stand thou still upon Gibeon, and thou, Moon, in the valley of Ajalon. . . . And the sun stood still, and the moon stayed, until the people had avenged themselves upon their enemies. *Is* not this written in the book of Jasher? So the sun stood still in the midst of heaven, and hasted not to go down about a whole day. . . . And there was no day like that before it or after it, that the Lord hearkened to the voice of a man: for the Lord fought for Israel."

So far "authorized version!" and, in lieu of examining whether the ancient Text has been truthfully rendered, those among whom knowledge has not yet advanced beyond the theological grade are lavishly vituperative of scholars who, knowing the English translation of this passage to be an absurdity, despise the commentaries upon it as a sham.

To place the reader at our point of view, let us first ask the question—what is this "book of Jasher?" One of the twenty *lost* books of the Hebrews cited in the Old Testament, is the facile reply. "*The book of Jasher*, that is, *the Righteous.* (*Josh.* x. 13; 2 *Sam.* i. 18.) This book must have been of no very ancient date, for it contained the Lamentations of David on the death of Saul and Jonathan. A spurious work with this title has come down to us, containing the history recorded in the first seven books of the Old Testament." (119) According to Cahen (vii, pp. 121–124; 2 *Samuel* i. 17–27), the verse runs —

> "17. David composed this lament upon Saul and upon Jonathan his son. — 18. And ordered to be taught to the children of Judah [the elegiac Lament called] *the Bow;* behold, it is written in the book of Jasher."

Then *follows* the lament itself, from *verse* 19 to 27: in which David, in poetic strain, says (*v.* 22, 23) —

> "The *bow* of Jonathan never retreated;
> The sword of Saul never returned empty:
> (Oh) Saul and Jonathan!"

Consequently, David, about B. C. 1056, had composed this beautiful ode; and a later writer says, "behold, it is written in the *Book of Jasher;*" that is, David's ode is. Ergo, this *Book of Jasher* was a collection of poems compiled *after* B. C. 1056. Now, the writer of "Joshua Xth" quotes, from this same *Book of Jasher*, the passage which in king James's version runs — "So the sun stood still in the midst of heaven and hasted not to go down about a whole day;" continuing his citation down to "the Lord fought for Israel." Hence it is positive that "Joshua-*ben*-NUN," could not have been the author of the "Book of Joshua;" because, having departed this life about B. C. 1426, he could have known nothing of a subsequent collection of poems that contained the lamentations of David upon events that happened some 370 years after Joshua himself was dead and buried. Moses is the only man who is privileged by orthodoxy to describe his own demise: (120) a second instance cannot be tolerated. Now, this author of "the Book of Joshua" is utterly unknown, and its date is very *modern*, perhaps as low as the sixth century B. C.; (121) as are likewise the "Books of *Samuel.*"

The next point, to which attention is invited, regards the sentence—"*Is* not this written in the Book of Jasher?" *What* was written in the said book? Commentators, ignorant of Oriental usages, concur in the notion that those passages which *precede* the book cited, were contained in the said book. Such opinion is fallacious, because, as Orientalists know, it is the universal custom of Semitic writers to quote the authorities they introduce *before*

(118) HUMBOLDT. *Cosmos;* transl. *Otté;* New York, 1851; iii. p. 17.

(119) DE WETTE: i. p. 411.

(120) *Deuteron.* xxxiv. 5–12. N. B. The *dates* are from the margin of our English Bible.

(121) DE WETTE: ii. pp. 186–191; and p. 228, for *Samuel.*

the extracts or citations they make from the latter's works; so that, what *follows* the words "Book of Jasher" must be the quotation from that book.

The literary criticism of age, manner, and authorship, being briefly defined, we glance next at the topography; observing, that any proposed verifications of the latitude and longitude of *Gibeon* and *Ajalon* by tourists in modern Palestine are mere "traveller's tales:" for *Gabâ*-ON, "occultation of the sun," and *Aial*-ON, (122) "dawning of the sun," refer respectively, the former to the *West*, the latter to the *East*, as points of the compass. Now, suppose two towns, one on either side of a valley, opposite to each other; the one, *Gabâ*-ON, on the western summit; the other, *Aial*-ON, on the eastern; while a battle was raging between Israelites and Ammonites in the valley between and beneath. Suppose, again, by anticipation of the text (and you have as much right to suppositions, in this case, as the *forty-seven* collectively), that the twenty-four hours during which this fight went on occurred at an *equinox;* and that it so happened, by a singular juncture of the solar and lunar motions, that, at six o'clock P. M. precisely, the sun *set* in the West at the same apparent moment that a full moon *rose* in the East; you would have light for twenty-four hours in the valley; or twelve hours of sunlight through the day, and twelve hours of moonlight through the night. Such combinations are so natural, although rare, that if any tourist were to furnish an astronomer with the exact latitude and longitude of such a *valley* in Palestine, the latter could calculate the precise *day* when such celestial combinations occurred, and thus fix the *era* alluded to in the "Book of Joshua." Finally, in the Hebrew, these two lines are rhythmical, besides containing a play upon the words GBâUN and AILUN, by poetic license: —

"To the eyes of Israel, O Sun! in the *hills* [B-GBâUN] even hide thyself:
But thou, O Moon! be most resplendent in the [B-âMKAILUN] *valley*."

We conclude with the lesson of that sage from whom both text and commentary are derived. (123)

"In precisely that day that IeHOuaH [the document is *Jehovistic*] delivered up the Amorean in face of the children of Israel, Joshua spake to IeHOuaH and said: TO THE EYES OF ISRAEL, O SUN! IN THE HILLS EVEN HIDE THYSELF: BUT THOU, O MOON! BE MOST RESPLENDENT IN THE VALLEY. And the sun set, and the moon endured until the multitude glutted (their) vengeance upon their enemies:—And is it not written in the book [entitled] *the Just?* [here follows the quotation] 'The sun which, running *along the meridional partition of the heavens* [*i. e.* along the equinoctial line], goes down [sets], was not as precise [true, exact], as by day, intent upon new-birth?' For certainly there was not before, nor after, a day equal to that in which, IeHOuaH having listened to the voice of man, IeHOuaH (himself) fought for Israel."

It may be prudent to observe that a passage in *Isaiah*, and another in *Ecclesiastes*, properly translated, lend no support to the supernaturalist commentary. That of *Habakkuk* (iii. 11) has no relation to the event; as, with "one longing, lingering look" at king James's *translation*, we prove by the subjoined rendering: — "Sun and moon set at their season; by the light of thy arrows they shall march, by the splendor of the lightning of thy lance." (Referring probably to a *night* attack.)

Thus vanishes "JOSHUA's miracle!" The late Rev. Moses Stuart, than whom as a Hebraist, and upright champion of theology, none superior have yet appeared in these United States, supplies this definition of a "miracle" — "I have it before me, in a letter from one of the first philologists and antiquarians that Germany has produced. It is this: 'The laws of nature are merely developments of the Godhead. God cannot contradict, or be inconsistent with himself. But inasmuch as a miracle is a contradiction of the laws of nature, or at least an inconsistency with them, *therefore a miracle is impossible.*'" (124)

Reader! We have submitted *seriatim* to your judgment a positive example of the errors of our truly-vulgar version for every letter of the English alphabet. We have kept no

(122) Like *Beth*-ON — "House of the Sun"; or ON, the *Sun*, Hebrew name for *Heliopolis*.

(123) LANCI: *Paralipomeni;* ii. pp. 381–390. It is of no use to consult CAHEN on these passages, except for the *text* (points deducted); vi. pp. 38, 39.

(124) *Crit Hist. and Defence*, &c.; Andover; 1845; p. 19.

account of digressional instances of other blunders, made by the *forty-seven* translators 250 years ago; although these are numerous, they are thrown in to make weight. The whole are taken, almost promiscuously, from our biblical portfolio, referred to years gone by.(125) You may now begin to think that we may be serious, when we affirm that our theological armory contains *hundreds* more, to prove that king James's translators were not "inspired;" and that, whatever may be the fact as regards the "original tongues," the English version cannot be accepted by science as a *criterion* in matters concerning anthropology.

The ladder of time has been ascended to the year 1600, when our "authorized version" was not; but when many English translations, some in MSS., others in print, required but an act of Parliament to make them orthodox. With the former, chiefly *Saxon* versions, from Alfred the Great down to John Wycliff, our inquiries do not meddle; none of them having been seen by us: nor, indeed, do we take intense interest in the latter, save to remember how William Tyndal, "homo doctus, pius, et bonus," for *printing* the earliest English translation of the New Testament, in 1526, and of parts of the Old, was rewarded by strangulation and cineration in the year 1536. Copies of his work, together with that of Myles Coverdale, 1535, have been before us for examination; and it is a singular fact that, in the majority of cases, where king James's translators departed from the version of Tyndal, or more particularly from that of Coverdale, they commenced floundering in the mire; and that where they have appropriated the readings of either, it has been done without acknowledgment. Fuller, the Church historian of those times, says of Tyndal that "his skille in Hebrew was not considerable: yea, generally, learning in languages was then in y[e] infancie thereof"—and we have shown (*ubi supra*) that *Hebrew* scholarship was all but unknown in England until the generation of Walton; that is, half a century later than the emission of king James's standard version.

The period of English history embraced within the sixteenth century is distinguished on the one hand by the successive intellectual upheavals of the educated classes, each surge towering higher and higher; and on the other by the mind-compressing enactments of the "Lords Spiritual and Temporal" in the repeated erection of barriers that gradually sunk lower and lower. Tyndal's body was burnt; that of Grafton, (126) guilty of printing "Matthew's Bible," was incarcerated; the Inquisition at Paris merely confiscated 2500 copies of the edition afterwards known as "Cranmer's;" in 1546, an act of Parliament only forbade the possession and reading of either "Tyndal's" or "Coverdale's." The reaction now began to feel its weakness, the progressives their strength: and so long as the sacerdotal caste could keep before the popular mind a parliamentary idea that Tyndal's version was "crafty, false, and untrue," its sages, satisfied that resistance had begun to endanger the "Establishment," as it is still called, were preparing to give way. Unhappy Tyndal, as the first Englishman to trample upon theological impediments through *publication*, has ever remained the "bête noire" of High Church orthodoxy; nor, owing to the obfuscations of history by ecclesiastical writers, has his memory yet received from posterity the justice that it merits.

About 1542, an act permitting certain persons to possess the "Word of God," as we term it now, "*not being of Tyndal's translation*," was graciously issued. It provides—

"That no manner of person or persons after the first day of October, the next ensuing, should take upon him or them to read openly to others in any church or open assembly, within any of the king's dominions, the Bible or any part of the Scripture in English, unless he was so appointed thereunto by the king, or any ordinarie, on pain of suffering a month's imprisonment. Provided, that the Chancellor of England, captaines of the warres, the king's justices, the recorders of any city, borough, or town, the speaker of parliament, &c., which heretofore had been accustomed to declare or teach any good, virtuous, or godly exhortations in anie assemblies, may use any part of the Bible or holie Scriptures as they have been wont; and that every nobleman and gentleman, being a householder, may read,

---

(125) Nott: *Bibl. and Phys. Hist.;* 1849; p. 135.

(126) See Hunt, *History of Journalism*, 1850, for the legal barbarities then perpetrated upon Printers generally—*mutilations, hangings, drawings and quarterings, gibbets,* and *fagots!*

or cause to be read by any of his familie servants in his house, orchards, or garden, and to his own familie, anie text of the Bible or New Testament, and also every merchant-man, being a householder, and any other persons other than *women*, prentises, &c., might read to themselves privately the Bible. But no woman [except *noble-women* and *gentle-women*, who might read to themselves alone, and not to others, any texts of the Bible], nor artificers, prentises, journeymen, serving-men of the degrees of yomen or under, husband-men, or laborers, were to read the Bible or New Testament in Englishe to himself, or any other, privately or openly, upon paine of one month's imprisonment."

Three hundred years have effaced even the remembrance of such legislative prohibitions. The "general reader" of our day never dreams that "my Bible" was once forbidden to his plebeian use. He claps his hands at Missionary Meetings when it is triumphantly announced that myriads of *translations* of the Scriptures are yearly diffused among the Muslims, the Pagans, and other "heathen," printed in more languages than are spoken, in more alphabets than there are readers. Has it never struck him to inquire, when the clamor of gratulation has subsided, whether these myrionymed versions are correct? If they are, what is commonly the case, mere servile paraphrases of king James's *English* translation, as we have proven the latter's woeful corruptions (*ubi supra*), must not the mistranslations of that text be perpetuated and increased by transfer into another tongue? and if so, is not that one of the providential reasons why the spiritual effect of these versions among the "heathen" falls below that material one produced by drops of rain on the Atlantic? Or, if the Missionary translators of the Scriptures into *Feejee*, *Kamtchadale*, or *Patagonian*, possess (what is so rare, as to be a pleasant proverb) sufficient Hebraical erudition to translate into the above, or any other tongue, direct from the *Text*, do not these excellent men "ipso facto" confirm all we have asserted in regard to our "authorized" version, by leaving its interpretations aside?

There are (although few Anglo-saxons know it) human dialects, orally extant, wherein there is no name for "God," no appellative for "Heaven," because such ideas never entered the brain of those low "Types of Mankind" for which a *Missionary* version has been manufactured. The highly-cultivated Chinese remained impenetrable to the disputes, sustained by the learned Jesuits and the evangelical Dominicans with the quintessence of "odium theologicum," on the following heads: —

"1st., if, by the words *Thian*, and *Chang-ti*, the Chinese understand but the material sky, or if they understand the Lord of Heaven? — 2d., if the ceremonies made by the Chinese in honor of their ancestors or of their national philosopher *Khoung-tseu*, are religious observances or civil and political practices?" (127)

Unable to settle the first problem by reference to Chinese lexicons, those Catholic Missionaries submitted it to the decision of the Emperor Khang-hi; and the solution of the second dilemma was referred to the Pope!

Regarding this "Foreign Missionary" discussion from the same point of view, as here in the United States we should look upon a dispute between Chinese Bonzes as to what *we* mean by "Providence," or in what light *we* celebrate the "Anniversary of Washington"; and feeling the same sort of astonishment that would fill ourselves were we told, that by one Chinaman the first doubt had been submitted to His Excellency the President, and that the settlement of the latter had been left by the other Chinaman to His Holiness the *Dalai-Lama* of Thibet: — the wise and jocular Emperor wrote in autograph beneath the Pope's *Constitution*; —

"This species of decree concerns none but vile Europeans: how can it decide anything upon the grand doctrine of the *Chinese*, of whom these people in Europe do not understand even the language?"

And then enforced his jest by banishing both Jesuits and Dominicans, about 1721, to Macao. Protestant successors in the Celestial Empire are still perplexed with the same linguistic obstacle; for about 1844, it was proposed to invent a *new* name for Deity, (that is, neither

(127) PAUTHIER: *Chine*; pp. 446—448.

Chinese nor English,) and compromise the matter by writing YAH; (128) while the papers have since held out hopes that the scruples of converted neophytes in China are about to be overcome by adopting "*Shin.*"

On the African coast the *Sooahelee* dialect, so restricted in its barbarous jargon that all its vocables implying civilization are borrowed from the *Arabic*, (129) a Missionary, who *translates* the "First three Chapters of Genesis" into the native tongue, can find no more euphonious rendering of our word "God" than MOOIGNIAZIMOONGO. (130) And, in America, no idea of "Original Sin" can be conveyed to an *Ottomi*-Indian, without the agglutination of monosyllables into TLACATZINTILIZTLATLACOLLI; nor will the last *Delaware's* heart experience "Repentance" until his mind has perceived the meaning of SCHIWELENDAMOWITCHEWAGAN. (131) But, we apologize for the digression.

During the second half of the sixteenth century, the frail hedge planted around the popular accessibility of the Scriptures vanished beneath the spades of the accumulating delvers for knowledge. At the Convocation of Hampton Court, in 1603, those measures were adopted that have placed the *Bible* before the people. Far, far, be it from us to undervalue the "Great Fact" — still farther to contest its vast educational utility. Would that *all* the "Sacred Books" of the East were equally accessible and equally read! The canonical literature of the Hebrews would be elevated infinitely beyond its present scientific estimation by such free comparisons; but not so its *English* "authorized" translation, and that is the only point for which these paragraphs contend.

In the years 1603–11, then, our Forty-seven Translators had before their eyes *many* English translations of the Old Testament. They possessed, furthermore, the *Latin* Vulgates, first printed in 1462, and revised in the *Sextine* edition of 1590, and the *Clementine* in 1592: together with numerous editions of the *Greek* Septuagint, both printed and manuscript. Their critical apparatus was copious enough wherewith to study the Original *Hebrew* Text, which lay before them in a variety of editions, more or less accurate, printed between the years 1488 and 1661; besides Jewish *Manuscripts.* If to their unquestioned knowledge of Latin and Greek, had been added a little Hebrew of the genuine school, which might very easily have been imported from the Continent, their version would have been better; but the confession of ignorance to themselves was as irksome, as to their race and country anti-national. They completed their labors without the contemporary aids within call; and "His Majesty's Special Command" has consecrated them for two hundred and forty-two years. "Undoubtedly, the present version is sufficient to all purposes of *piety*"; (132) our part is to show that it has long ceased to be adequate to the requirements of science.

It seems, therefore, considering the facilities they enjoyed, and still more the many they disdained, that errors so tremendous as those which modern criticism exposes should have been backed by orthodoxy with praises less extravagant; because, their *Hebraical* qualifications for the task being *nil*, the multiplicity of foreign versions, without that discriminating criterion, could but augment the multiplicities of their mistakes. (133)

The earlier English versions, if here and there superior to readings adopted by the Forty-Seven, were radically defective, owing to the same natural causes that precluded the possibility of making a direct translation from the Hebrew in 1611; viz.; small acquaintance with the vocabulary and grammar of the language itself. Fuller, for instance, infers that poor Tyndal rendered the Old Testament from the Latin, "as his friends allowed that he had no skille in Hebrew"; and the same authority explains that the reason why king James

(128) Dr. Bowring: in London *Literary Gazette.*

(129) Gliddon: *Otia;* p. 126.

(130) Rev. Dr. Krapf: *Jour. Amer. Oriental Soc.;* iii.; Boston 1847; pp. 261–274.

(131) Gallatin: *Trans. Amer. Ethnological Soc.;* New York, 1845; i. pp. 28–35.

(132) Taylor: in both the English and American editions of *Calmet's Dictionary*; voce "Bible."

(133) After this was written, a friend asked us to read "*The Translators Revived; a Biographical Memoir of the Authors of the English Version*"; by A. W. McClure; 12mo; New York, 1853. It merits nothing here beyond this mention, but a *review* in any newspaper is much at its author's service.

appointed Fifty-Four Translators was because "many and great faults" were already notorious amid the earlier translations.

The *Samaritan* text was unavailable to them for two reasons; one, that no copy had reached Europe until 1623, or twelve years later than the publication of king James's version; (134) the other, that those whose Hebraical accomplishments were so slender could have elicited nothing from any cognate Oriental idiom. It is superfluous, therefore, to speculate upon what philological feats our Forty-Seven might have performed through Samaritan contexts.

As the oldest of all "printed" books, A. D. 1462, the Latin *Vulgate* must have riveted the attention of men whose reverence for the invention induced them to carry the antiquity of moveable types back to the age of Job (xix. 23; *ubi supra*). With the numerous Latin versions, (135) made prior to St. Jerome, from the Greek, our translators did not trouble themselves; nor need we, because this first of Hebraists among the Fathers declares— "For the most part, among the Latins, there are as many different Bibles as copies of the Bible; for every man has added or subtracted, according to his own caprice, as he saw fit."

To remedy this evil, Jerome completed a retranslation of the Old Testament, directly from the Hebrew, between the years 385 and 405. (136) His contemporaries loudly protested against such profanity, lest it should sacrilegiously disturb that bibliolatry with which Christian communities then regarded the *Septuagint;* but, about 605, Pope Gregory invested it with respectability, by adopting its lections along with the old *Italic* version. The consequence was that the monastic scribes, having equal authority for either, began to correct the first by the second indiscriminately; and succeeded in fusing them both so inextricably into one, that the emendations of Alcuin in the ninth, of Lanfranc in the eleventh, and of Nicolaus in the twelfth centuries, failed to establish any uniformity among *manuscripts* which, in the words of Roger Bacon, "every reader alters to suit his own whim." Such was the state of the *Latin* version current until the sixteenth century, when Stephens undertook to castigate its errors in his printed editions: Clarius, in the meantime, submitting a schedule of 80,000 mistakes for the edification of the Council of Trent. However, on the unlettered side, fanciful substitutions; on that of scholarship, ruthless expurgations; impelled Sixtus V. to volunteer the office of "proof-reader:" and, in 1589, a copy of the Vulgate issued from the Vatican, wherein "eaque res quo magis incorrupte perficeretur, *nostra* nos *ipsi manu* correximus:" *i. e.*, the Vicar of God corrected the press himself. Alas! Such condescension only made the innumerable faults of that edition "notorious as ludicrous. Bellarmine luckily hit upon a plan to correct the errors, and save the infallibility of the Pontiff." New recensions were executed, "quod vix incredibile videbatur," in *nineteen days;* and the year 1592, during the apostolic vicarage of Clement VIII., brought out a standard Papal copy, wherein the odium of all errors patent in the former Pope's edition was charged upon the "printer's devil."

This Romanist *finality* abounds with misinterpretations if collated with the Hebrew Text; and when placed before the Forty-Seven, some ten years after its appearance, could only have served to lead them more astray; even if the fear of Papistry did not prevent adoption of such of its readings as attracted rather their fancy than their septi-quadrigentesimal criticisms. Consequently, the *Divine Afflatus* did not penetrate into king James's version through the *Vulgate;* which fact renders nugatory, as regards the Latin language, any inference derivable from their Preface in favor of the peculiar sanctity of this among the "Original Sacred Tongues" whence "one more exact translation" was by them made. Perhaps some streams of the apostolic imponderable reached our translators by transmission through the *Greek?*

At least three, and probably more, *printed* editions of the Greek *Septuagint* (137) were procurable by our Translators in the year 1603; independently of such *manuscripts* as they may have consulted; from the number of which last must be deducted the Codex-*Alexan-*

(134) KENNICOTT; *Dissert. Gen.*; p. 475.
(135) DE WETTE: i. pp. 183–191.
(136) *Ibid.*; i. p. 257, *seq.*
(137) DE WETTE: i. pp. 81–82.

drinus, (138) now in the British Museum; because it did not arrive in England until the year 1628. (139) The *printed* editions issued during the sixteenth century were naturally copies resulting from the collation of such *manuscripts* as to their respective editors were more or less accessible; and if the originals were defective the transcriptions must be still more so. We can utter no opinions on the critical value of the printed editions, before ascertaining what scholarship may have decided upon the archæological merits of the *manuscripts* themselves; nor is it in our power to enumerate what copies of the latter may or may not have been consulted by our translators; chiefly because our own note-books do not afford the *dates* at which many celebrated Greek MSS. were known throughout Europe. (140) We presume they used copies of the Codex-*Vaticanus* (printed in 1587, by Cardinal Caraffa), of which the antiquity is estimated by Kennicott at A. D. 387, while others suppose "a few years later;" (141) among them Montfaucon and Blanchini, who refer it to the fifth century. None of other Greek Codices extant can possibly antedate, in any case, the fourth century; for even the oldest, the Codex-*Cottonianus,* once conjectured to have been Origen's property, is now proved to have been calligraphed towards the end of the fourth or the commencement of the fifth century. Its fragments lie in the British Museum. (142) This falls within the lifetime of St. Jerome, A. D. 331–422; (143) who laments that, in his day, "the common (Greek) edition is different in different places, all the world over;" and reiterates, "It is corrupted everywhere to meet the views of the place and time, or the caprice of the transcribers." (144)

> "Thus it seems that, in the time of Jerome, *three* different editions of the LXX were in use under the sanction of the several churches, and with their authority, viz.: Origen's Hexapla in Palestine, the text of Hesychius in Egypt, and that of Lucian in Constantinople and its vicinity. No wonder the existing *manuscripts* have come down to us with so many corruptions." (145)

Such asseverations, when once recognized to be true in fact, suffice to damage the accredited uniformity of the *Greek* versions; but a little further inquiry will evince that it was impossible, through the very nature of human things, that any Hellenic translation from the Hebrew could be "inspired."

If, then, only four centuries after the Christian era, the Greek translation (finished about the year 130 B. C., at Alexandria) no longer existed in its "editio princeps," but its later recensions alone had flowed down to St. Jerome's time in three turgid streams, each one essentially *corrupt,* it follows that all MSS. now extant, no less than all *printed* editions made from such MSS., must be still more blemished, owing to later mistakes, than even the best exemplar known to St. Jerome. It is in this vitiated state that the *Septuagint* reached our translators in the year 1603: —

> "No one of these recensions is found pure; for they have flowed together, and become mixed also with the other Greek versions. . . . The criticism of the Seventy has hitherto advanced no farther — and perhaps it never can — than to a collection of the various readings. The editions hitherto published do not afford the true and exact text of the manuscripts." (146)

But, not merely does the *Greek* version falter in its historical traditions. Its deviations from the *Hebrew* original render objections to its plenary authenticity unanswerable.

> "As a whole, this version is chargeable with want of literalness, and also with an arbi-

---

(138) WOIDE thinks its age to lie towards the end of the fourth; but if KENNICOTT selects A. D. 395, he reports other opinions as low as the ninth century (Ist *Dissert.*, pp. 306, 307).

(139) TAYLOR'S *Calmet;* voce "Bible."

(140) PORTER (*Principles of Textual Criticism*, Dublin, 1848) might supply deficiencies; but memory is treacherous, and we have not now his most excellent work: *vide Otia*, pp. 111–113.

(141) KENNICOTT: IId *Dissertation;* p. 407.

(142) HORNE: *Introd.;* i. pp. 105–107.

(143) ANTHON: *Class. Dict.;* voce "Hieronymus"; p. 625.

(144) DE WETTE; i. p. 181.

(145) *Ibid.;* p. 180.

(146) DE WETTE; i. pp. 181–183.

trary method, whereby something foreign to the text is brought in. In general, it betrays the want of an accurate acquaintance with the Hebrew language, though it furnishes many good explanations. (147)

"The character of this version is different, according to the different books. It is easy to distinguish five or six different translators. . . . Indeed, the real value of the Septuagint, as a version, stands in no sort of relation to its reputation. All the translators engaged in it appear to have been wanting in a proper knowledge of the two languages, and in a due attention to grammar, etymology and orthography. Hence they often confound proper names, and appellations, kindred verbs, similar words and letters, &c., and this in cases where we are not at liberty to conjecture various readings. The whole version is rather free than literal," &c. . . . The *Text* of the Septuagint has suffered greatly. Through the multitude of copies, which the very general usage rendered necessary, and by means of ignorant critics, the text of this version, in the third century, had fallen into the most lamentable state." (148)

"Although we cannot say from whom it (the LXX) emanated, it is certain that it is the work of one or several Jews of Egypt, of Greek education (if always our version called *the Seventy* be exactly the same as the one that was made at that epoch); because one may discover in it traces of that philosophy which afterwards developed itself among the Alexandrian Jews, and of which PHILO is for us the principal representative. It does not appertain to us to characterize here the translation under its philological aspect; we must content ourselves with establishing that, in many places, it differs sensibly from our Hebrew text, and that very often its variants agree better with the text of the Samaritans. Nevertheless, the latter does not sufficiently conform to the version of the Seventy, that one could imagine a common source for both compilations." (149)

It results from Talmudic exegesis that its authors, beyond vague impressions of errors contained in the Greek version, not only did not know, save through hearsay, the *Septuagint* themselves (although they suppose its Translators to have been seventy-two), but that it was impossible for the Palestinic Jewish Rabbis to read it, owing to their ignorance of the Greek tongue.(150) Not a word in the *Mishna* and the two *Guemeras* refers to Aristobulus, or Philo, or to the Apochryphal books; neither to the *Essenes*, nor to the *Therapeutæ*. The Jews of Palestine were separate people from those of Alexandria; and it was a concern exclusively interesting to the latter to defend the many false renderings of the Septuagint, of which remarkable examples are exhibited in the learned treatise of Franck, whence we condense some facts into a foot-note.(151) But hear Sharpe: —

"It will be enough to quote two passages from this (LXX) translation, to show how the Alexandrian Jews, by a refinement of criticism, often found more meaning in their Scriptures than ever entered the minds of the writers. Thus when the Psalmist, speaking of the power of Jehovah, says with a truly Eastern figure (*Psalms* civ. 4, *Text*), '*He maketh the winds his messengers, and the lightning his servants*,' (152) these translators change the

(147) *Ibid.*; p. 147.

(148) TAYLOR'S *Calmet*; voce "Versions."

(149) MUNK: *Palestine*; p. 487. Cf. also, AMPÈRE: *Recherches en Égypte*, &c., 2de part.; Rev. des D. Mondes, 1846.

(150) FRANCK: *La Kabbale*: Paris, 1843; pp. 273, 329.

(151) "Already the Thalmud had a vague knowledge (*Thalm. Babyl. Tract. Meguillah*; fol. 9, ch. i.) of the numerous infidelities of this antique translation [viz., of the LXX]. . . . Thus, when the sacred Text says positively (*Exod.* xxiv. 9, 10) that Moses, his brother, and the seventy elders, saw the God of Israel upon a throne of sapphire; according to the (Greek) translation, it is not God who was seen, but *the place which he inhabits*. When another prophet, Isaiah, sees the Lord seated on his throne and filling the temple with the folds of his robe (*Isaiah*, vi. 1), this too-material image is replaced by the *glory of God*. . . . When it concerns Adam and Eve, (the Greek interpreter) would carefully avoid saying, with the Text, that God created them male and female (*Gen.* i. 27); but this double character, these two halves of humanity, are united in *one and the same being* — Ἄρσεν καὶ θῆλυ ἐποίησεν αὐτόν. . . . . 'Who has created all things?' asks the Hebrew prophet (*Isaiah* lx. 26); 'Who has rendered them *invisible*?' says the Alexandrian interpreter" (FRANCK: *La Kabbale*; Paris, 1843; pp. 329–331). Our author furnishes several other examples of downright perversions committed by those Alexandrines called "the LXX": of which our space denies insertion. After our own conclusions were formed, it was most gratifying to find them all confirmed by RUBENSOHN ("Origin and Structure of the Septuagint" — *Christian Examiner*; Boston, March, 1853; pp. 165–187), who truthfully observes — "Such a version — if it should be thus designated — is not only conformable to the spirit of those times, but there are many indications that the Greek version was originally intended only as an auxiliary book for the use of the Alexandrian Jews."

(152) So also CAHEN, xiii. p. 229, and *note* 4 — "des flammes brûlantes, ses ministres." St. PAUL too, although said to have been "a Hebrew of the Hebrews," follows the *Septuagint* in quoting this passage (*Epist. to the Hebrews*; i. 7) even to *Jews!* (SHARPE'S *New Test.*; p. 395) — a passage non-existent in the *Hebrew* Text.

sentence into a philosophical description of the spiritual nature of angelic beings, and say (in the Greek), '*He maketh his angels into spirits, and his servants into a flame of fire.*' Again, when the Hebrew text, in opposition to the polytheism with which the Jews were surrounded, says (Text, *Deut.* vi. 4), '*The Lord is our God, the Lord alone*' [literally, 'Hear, O Israel! IeHOuaH, our God, IeHOuaH (is) *one!*']; the translators turn it to contradict the Egyptian doctrine of a plurality of persons in the unity of the Godhead, (153) by which the priests said that their numerous divinities only made one God; and in the Alexandrian Greek this text says, '*The Lord our God is one Lord.*'" (154)

Should the reader now turn to the above passages in our "authorized" version, he will perceive that the *forty-seven* have rendered into English the exact words of the *Greek;* and thus he will behold a little of the damning evidence produceable that these worthies could not construe a simple line of the Hebrew Text; but have palmed off upon us, as genuine "inspiration," language that, being Alexandrian forgeries, cannot be Divine; confessions of creed that, not being in the original Hebrew, cannot be "inspired."

Here, as concerns king James's translation in its relations to the *Greek* versions, we might bring our inquiries to a close: the seal of condemnation has been so legibly stamped upon it. But, inasmuch as some data respecting the origin of these Grecian documents may be useful to our researches into the Hebrew Text, it is desirable to reach that epoch when the *Septuagint* had not yet been manufactured.

Ascending from St. Jerome in the IVth century to the great Origen in the IId, we find him complaining of the corruptions manifest in the Greek MSS. of his day — "But now there is obviously a great diversity of the copies, which has arisen either from the negligence of some transcribers, or the boldness of others—or from others still, who added or took away, as they saw fit, in making their corrections." (155)

"From the time of the birth of Christ to that of Origen," continues Eichhorn, "the Text of the Alexandrian version was lamentably disfigured by arbitrary alterations, interpolations, omissions, and mistakes. Justin Martyr had a very corrupt Text, at least in the minor Prophets." (156) He was decapitated in A. D. 164, having been converted about the year 132; thus sealing his convictions with his blood.

The works of Origen's predecessors in the first century, Flavius Josephus, born A. D. 37, and of Philo Judæus, who flourished about A. D. 40, exhibit through their citations, (both being Hellenized Jews writing in Greek rather for Grecian and Roman readers than for their own countrymen,) that some alterations had already been made in the copies of the Septuagint respectively used by them: at the same time that the writers of the New Testament, by quoting the *Greek* version, in lieu of the Hebrew, have invested the former with a traditionary sanctity, fabulous when claimed for extracts from the Old Testament not cited directly from the Hebrew Text. (157). Its discussion would lead us astray from the inquiry as to when and by whom the Original *Greek* translations were made; and the fact is noted merely to establish the existence of the latter, in what state of literal preservation no man can tell, at the Christian era.

"All we can determine with certainty is, — that the whole, or the greater part of the Old Testament, was extant in the Greek language in the time of Jesus the son of Sirach. [Sirach presupposes that 'the Law and the Prophets, and the rest of the books,' were already extant in his time; that is, in the 38th year, which is probably the 38th year of Evergetes II., about 130 B. C.]" (158)

This year before Christ 130 is recognized, nowadays, by all biblical scholars, to be the *minimum* epoch at which Greek versions of certain books of the Old Testament canon were already in circulation at Alexandria. Tradition, itself, claims no date for the existence of

---

(153) Compare BURNAP: *Expository Lectures;* Boston, 1845; p. 9; — and CHENEVIÈRE: *Système Théologique de la Trinité;* Geneva, 1831; *passim.*

(154) SHARPE: *Hist. of Egypt;* 1846; p. 196.

(155) DE WETTE: i. p. 165.

(156) DE WETTE: i. p. 166.

(157) STRAUSS: *Vie de Jesus;* and HENNELL: *Origin,* &c.; enlarge upon these themes.

(158) DE WETTE: p. 146; — also, STUART; *Crit. Hist. and Defence;* pp. 241, 423.

same circumstances earlier, as the *maximum*, than the reign of Ptolemy Philadelphus; and about 260 years B. C. suffice for a chronological stand-point that reconciles scientific probabilities. The medium suits well with the dispersion of some Hebrew exemplars after the saccage of the temple by Antiochus, B. C. 164; and is parallel with the literary restorations of the *Maccabees.*

To read (as we ourselves formerly did with confidence) the works of some leading English Divines in quest of information about the *Septuagint*, and the chronology erected upon its numerations, one would actually suppose, from the positive manner in which statements are put forward, that they had studied the subject! Hales, (159) for instance, assures us that *Seventy*, or Seventy-two, elders of the Jewish congregation, after the reception by the king of a copy of *Law* from Jerusalem *written in letters of gold*, sat down at Alexandria, and did the Hebrew into Greek in 72 days, "d'una sola tirata"; with many episodes equally romantic. Half a century has elapsed since any Continental critic of biblical literature who ventured to give further currency to such wretched stories would have been jeered into silence and overwhelmed with literary obloquy. The reader is referred to De Wette for facts and authorities, (160) and to Bunsen (161) for endorsement of the following sketch; after remarking that wherever the number "70," or its cabalistic equivalent "72," occurs in Jewish connections, it carries with it more cogent evidences of historical untruth than even the *forties*, or "Erbaïnàt," so common in Hebraical literature. (162)

The origin of the *Greek* version, stripped of verbiage and exaggerated traditions, was the natural consequence of the great influx of Jews — a people ever partial to the fleshpots of Egypt — into Alexandria, immediately upon the foundation of that city by Alexander the Great, about B. C. 332. Enjoying privileges under the early Ptolemies, the number of Jewish colonists constantly augmented: at the same time that incipient intercourse with their Greek fellow-citizens superinduced first the disuse and next the oblivion of that *Syro-Chaldee* idiom the Israelites had brought back with them, from Babylonish bondage, in lieu of the Old Hebrew orally forgotten; and led their Alexandrine descendants to adopt the *Greek* tongue, together with much of Grecian usages and Philosophy. They became *Hellenizing*-Jews (163) at Alexandria, without ceasing to be Hebrews in lineage or religion; just as their present descendants are *Germanizing*, *Italianizing*, or *Americanizing* Israelites, according to the country of their birthplace or adoption.

The conquests of the Macedonian are to us the most salient causes of the transmutations that took place throughout the Levant owing to the wide-spread of Grecian influences; but Pythagoras, Plato, and Herodotus, are earlier prominent expressions of Greek infiltration into Babylonia and Egypt during the fifth and sixth centuries B. C., which was far more exten-

(159) *Analysis of Chronology.*

(160) *Op. cit.*; i. pp. 136–144.

(161) *Egypt's Place in Universal Hist.*; 1848; i. pp. 184, 185.

(162) LEPSIUS: *Chronologie der Ægypter*; 1849; i. p. 365. We find the subjoined to the purpose among "Talmudical statements: — In *Megilla*, ix. *a*, we read the following account: 'Ptolemy the king called seventy-two old and wise men to Alexandria, and confined each in a separate room, without telling them the reason of their being called. He afterwards visited each of them, and directed them to write down in Greek the words of Moses. God inspired them with a sameness of ideas, so that their translations literally agreed.' In *Sophrim*, § 1, we read another passage: 'Five sages were called to Alexandria by the king Ptolemy, to translate the law into the Greek language; this day was as oppressive to Israel as the one when the golden calf was made, for they were unable to do justice to the subject. Then the king assembled seventy-two sages, and set them in seventy-two cells,' &c. . . . . In *Taanith* occurs the following passage, which also DE ROSSI quotes (*Imrai Binah*, § 7): 'There are certain days on which we fast on account of the law: such a day is the eighth day of Thebeth, because on that day the law was translated into the Greek under the second Ptolemy, king of Egypt, and darkness covered the earth for three days.'" — ("*Greek Versions of the Bible* — the passages extracted from LANDAU'S *Vorwort zum Aruch*" — *The Asmonean*; New York, 5 Aug. 1853.) Little historical criticism is required to perceive that the writers of these Talmudic legends, several centuries after Josephus, had merely given another shape to the same baseless tradition of the false Aristeas: and we may class JUSTIN MARTYR'S evidence (*Admonitione ad Græcos*) that "he *saw* the 72 cells into which the translators were locked up"; and EPIPHANIUS'S (*De mensuris et ponderibus*) that these cells were 36, each for two translators; — with St. AUGUSTINE'S, where he says "*Vidimus* — we have seen" men with an eye in the pit of their stomachs.

(163) According to Philo, the Jews exceeded a million at Alexandria alone (RAPAPORT'S *Erech Milin*; quoted in *The Asmonean*; New York, July 20, 1853).

sive commercially than until recently accredited; while Greek *condottieri* had been employed in Egypt from the seventh century by Psametticus: nor was Xenophon the first General, nor Ctesias the first Doctor, who volunteered their services to the Achæmenidæ of Persia. Into Jerusalem itself, Greek ideas had penetrated very soon after the erection of the Second Temple in the fifth century. These result from the history, and are stamped upon the *proper names* of the Jews of Palestine, particularly after Alexander's era. Nor were such Hellenic infiltrations without a certain influence upon the canonical literature of Judaism; for the "political satire" (164) entitled the "*Book* of DANIEL" betrays, through its *Greek* words, as much as by its exegetical adaptations, an author of the age of Antiochus Epiphanes, not earlier than the plunder of Jerusalem by that king about 164 years B. C. Continental scholarship long ago placed this fact beyond dispute; (165) and the Hebraical erudition of the late Rev. Moses Stuart (166) induced him to fortify it with his customary skilfulness.

So much nonsense still passes currently, in regard to the various dialects spoken by the Jews after their return from the Captivity, that we must here digress for a moment. Independently of books read and others cited, we have sought for information on these subjects from some of the most cultivated Hebrew citizens of the United States, and have invariably met with the kindest readiness to enlighten us. We possess not (merely because we omitted to ask for it) the sanction, of the many very learned Israelites consulted, to publish their honored names; but not on that account are the hints with which all have favored us the less appreciated by ourselves nor the less useful to readers. No interdict being laid by one of the writer's valued friends, Mr. J. C. Levy of Savannah, upon the many indices to knowledge for which his goodness has rendered us his debtor, we condense the substance of two recent communications; coupled with regrets that certain inexorable limits of typographical space should compress what ought to be in "Brevier" into "Nonpareil." (167)

---

(164) New York *Daily Tribune;* Feb. 10, 1853. The attribution to "Discoveries" at Babylon is fabulous. For that of the *Decalogue*, conf. GLIDDON, *Otia*, 1849; p. 19: — extended in New York *Sun*, "Historical Sketches of Egypt," Nos. 6, 7; Jan. 19 and 25, 1850.

(165) MUNK: *Palestine;* p. 420; — DE WETTE: ii. pp. 483–512; — CAHEN: *Notes on Daniel.*

(166) *Hints on the Interpretation of Prophecy;* Andover, 1842; pp. 71–108.

(167) EXTRACT 1.—"The information I promised barely is, that the Babylonian Captivity lasted from 538—486 B. C., when Zerubabel, with 50,000 men, went to Palestine with the permission of Cyrus. A second colony followed in the year 458, led by Ezra, under the reign of Artaxerxes Longimanus. He was, again, followed by Nehemiah, 444. During the Captivity, by good treatment, they adopted Babylonian customs and manners, and amalgamated with their conquerors (*Ezra* v.; *Nehemiah* xiii. 1–3), and forgot their native Hebrew. Besides this, the Samaritans speaking an Aramaic (Chaldaic) dialect, as well as the Syrians who ruled for a long time in Palestine, exercised great influence over the Jews; so that the Hebrew soon disappeared as the vernacular (*Nehemiah* xiii. 24) to yield to the Chaldaic, and the mother-tongue probably was the language of their real mothers. This may be best proved by the fact, that all civil acts, official documents, and legal formulas, were written in that language, and that the Talmud itself is written, to a great extent, in this tongue. Furthermore, numerous proverbs originating at this time, and popular books of that age, are all in the same language. The chief prayers of the Jewish Service, composed by Ezra, are in the Chaldaic language. Already at the consecration of the Temple on the 1st of the 9th month and in the 24 days of its duration, it was found necessary to accompany the reading of the Law with translations and *explanations* (*Nehemiah* viii. 8, 12); the latter being the beginnings and foundation of the Talmud, or traditional oral law, which was first prohibited to be written down, in order to preserve life and motion for the letter of holy writ. That this prohibition was afterwards transgressed much to the injury of the development of Judaism, and caused all schisms among the Jews, is well known. Had these explanations which are mostly contradictory of each other, not been collected and made a code of, all strife might have been avoided.

"*Written Chaldaic translations* were in existence in the time of the Maccabees — the first known is that of ONKELOS, disciple of R. Gamaliel (53 after X), and fellow-student of the Apostle Paul. This translation is paraphrastical, especially in the prophetic and poetical parts of the Bible. More explanatory is that of JONATHAN-BEN-NGOOZIEL. A third translation is the *Targum Jerushalme* (Jerusalem translation), fragmentary, and exhibiting a commentary in accordance with the reigning ideas of the age. Macedonian and Egyptian rule in Palestine produced among the Jews Grecian manners, customs, and *ideas*, also language; so that translations of the Bible were soon necessary. The oldest mentioned is that of AKILAS, often referred to in ancient writings, to explain Chaldaic parts of the Bible; there you have the Greek translation of the LXX. Philo, Josephus, and other Jewish authors wrote in Greek, proving their ignorance of Hebrew by the blunders in translation and explanation of the Text. Greek technical terms are even to be found abundantly in the Talmud."

EXTRACT 2. — "I am not satisfied with the meagre reference given you regarding the ignorance of the Jews

Returning to the LXX. — Some precursory events had prepared Jewish Alexandrian immigrants for the adoption "nolens volens" of the Greek tongue and alphabet, consequent upon the oblivion of the Aramæan dialect which their progenitors had re-imported into Palestine. The children were growing up in ignorance of a "Law" their Alexandrian parents could no longer read in *Hebrew*. To have paraphrased that "Law" into *Syro-Chaldee*, like their brethren in Palestine and Babylonia, would at Alexandria have been useless; because the parents had forgotten Syro-Chaldee, and the children already talked *Greek*, by the reign of Ptolemy Philadelphus, B. C. 284–45. What more in unison with the instinctive characteristics of that "Type of Mankind" which, beyond all others (from the days of Abraham), changes its *language* with most facility, while it repels admixture of alien blood and tenaciously adheres to its own religion, than that one of its branches, the Alexandrian Hebrews, should cause the sacred writings of their forefathers to be translated into Greek? This was precisely that which they did, although the exact year of the commencement of such translations can no longer be fixed; but the style and idioms of the several books, to which, after collection into one canon, the name of *Septuagint* was subsequently given, indicate different times and divers hands. (168)

While confined to Judaism in Alexandria, this Greek translation was reputed orthodox by the Hellenizing Rabbis as much as the Hebrew Scriptures themselves; and more authoritative, because they could read no other. It was read in the Synagogues of that city, and wherever Jewish congregations were planted under similar Grecian circumstances; but a Greek version was of no use, and therefore of little value, to the Jews of Palestine, Syria, and Persia; who understood not the Greek tongue, but spoke *Chaldaic* "patois." The Greeks themselves, regarding all languages but their own as barbarous, Hebrew inclusive, never troubled their heads about the *Septuagint* until after apostolic missions had propagated the *New Testament*, composed in Greek by Hellenized Jews also; when the recurrence of quotations from the Old Testament, in the evangelical books, instigated its readers to reference to that Code; and as these Christianized readers were ignorant of Oriental idioms, of course the *Septuagint* version was the only one accessible to them: while, to give it an air of antiquity and of royal respectability of origin, both Græcized Jews and Judaizing Christians coincided in attributing its authorship to "70" translators, appointed (like our *forty-seven* English translators by king James) under the hand and seal of Philadelphus; whose encouragement of literature was testified by munificent donations (cost to himself, nothing) to the Alexandrian Library. A pseudo-Aristeas "reported" a fable so flattering to Alexandrine pride, to Jewish respectabilities, and to Christian orthodoxy; while the real tradition seems to have reached us in an account that the authors of the *Septuagint* were but "*five*:" (169). and so, veneration for the *Septuagint* increased from day to day in the ratio that time rolled onward, and that the remembrance of its natural origin faded from the "memory of the oldest inhabitant" of Alexandria; nor would the harmless legend have been disturbed, had not proselyting furor on the part of new converts to Christianity led them to provoke rabbinical susceptibility by appeals to the *Greek* version of the Old Testament in support of novel doctrines promulgated in the *New*: the two texts

everywhere of Hebrew after the Captivity. . . . I offer you what your opponents cannot object to — that is, the XIIIth Chapter of NEHEMIAH (the chronology of the book you know better then I do). Jewish or Christian chronology make it about 450 before X. This chapter will show you, that the Dragoman [Arabicè *Turgemân*, "Interpreter"] was necessary in reading the Book of the Law. Gibbon (vi. vol. chap. 50, p. 262) quotes, in a note, Walton (*Prolegomena ad Bibl. polyglot.*, pp. 34, 93, 97; also, Simon, *Hist. Critique du V. et du N. Testament*), to illustrate that the Bible was translated into Arabic at a much earlier period than the time he is treating of (about 550 after X); and he proves the fact 'from the perpetual practice of the Synagogue of expounding the Hebrew Lesson by a paraphrase of the vulgar tongue of the country.' . . . I think these very respectable authorities, if you need them." Mr. Levy's views are amply supported by GESENIUS (*Geschichte der Heb. Sprache*, &c.; p. 198).

(168) DE WETTE: i. p. 145; — TAYLOR'S *Calmet*; voce "Versions."

(169) *Ibid.*; p. 150 — *note* from the *Talmud*, Tract Sopherim, ch. i. — "The work of the *five* elders, who wrote the Law in Greek, in the time of King Ptolemy": unless they meant the *Pentateuch*, attributing one book to each elder? Conferre, also, the high Jewish authority of RAPAPORT, in "*Erech Milin*" — New York *Asmonean*; July 29, 1853.

having been made singularly harmonious; owing to scrupulous care on the part of the apostles to cite each passage according to its *Greek* coloring in the Septuagint; for a long time held in common to be canonical as well by Jews as by Greeks.

Bewildered for a time by these dexterous sophisms, and mystified through literary ambuscades which it required a Grecian intellect to comprehend, the worthy old Rabbis (taken in reverse) had no resource but to proscribe the *Septuagint*, and ostracize its readers. "The law in Greek! Darkness! *Three days fast!*" (170) Because, says the *Talmud*, "on that day, in the time of King Ptolemy, the Law was written in Greek, and darkness came upon the earth for three days." (171) Little by little, however, their perceptive faculties expanded to the true posture of affairs; and by proving incontinently that many things, which looked one way in the *Greek*, looked quite another in the Hebrew, the Rabbis soon defeated their assailants; routing them so repeatedly, that gradually the latter thought it safer to let such doughty controversialists alone: a method of repulsion continued with never-failing success by Israel's wide-spread posterity even now; who, when summoned by anxious "Missionaries for the Conversion of the Jews" to adopt a Trinitarian faith which Semitic monotheism (172) despises, have merely to show such well-meaning persons that king James's version does really copy the *Septuagint* rather than the Hebrew, to see these itinerant simplicities pocket their English Bibles and slink off. Some day, perhaps, when the rules of archæology through popular diffusion have augmented, all over Anglo-Saxondom, that mental element termed "common sense," sundry excellent persons, in the language of Letronne, "sentiront, je pense, l'inutilité, la vanité de leurs efforts." (173)

The above conclusions on the *Septuagint*, long known to scholars, if not previously expressed in print with the same "brutale franchise" habitual to writers who believe they speak the truth (so far as ratiocination can deduce logical results from known premises,—*humanum est errare*), have enfeebled its value—except for purposes of archæological restorations of the Hebrew text—to such degree that, in this discussion, the ablest theologians have advanced into the *positivist's* stage of philosophy. No scientific exegetist of the present generation—save for purposes aforesaid—perils his Continental reputation on the letter of any Greek version, unless *chronological* computations be the objects of his research. Another Essay (III.) of this book gives parallel tables wherein the *Septuagint* system is compared with others; but, to evince the numerical discrepancies between Text and versions, it suffices here to note, that, from the creation of Adam to the "Deluge," computations (based upon the Hebrew original, as now extant) generally yield 1656; upon the Samaritan Pentateuch, 1307; and upon the *Septuagint*, 2242 years.

The indefatigable labors of a profound Hellenist and Egyptological scholar, enable us to sweep away any *chronological* superstitions, yet in fashionable vogue, built upon the Septuagint: —

"The chief disagreement between the [Hebrew] original and the [Greek] translation is in the chronology, which the translators very improperly undertook to correct, in order to make it better agree with Egyptian history and the more advanced state of Alexandrian science. They only made the Exodus of Moses 40 years more modern; but they shortened

(170) BUNSEN: *Op. cit.*; p. 185.

(171) DE WETTE: *Note*, p. 150; — HENNELL: *Origin of Christianity*; pp. 454, 455, note.

(172) "Bear witness! God is *one*. He is the God eternal. He never has begotten, and was never begot" (*Kur'àn*; *Sura* cxii).

(173) *Recueil des Inscriptions*; Paris, 1843; Introd., i. p. xliii. We clip the following from the London *Inquirer*, 1853: "*The Cost of Converting a Jew.*—After some twenty years of labor — after the erection of a church on Mount Zion, at an enormous cost — after the expenditure of hundreds of thousands of pounds, the 'London Society for promoting Christianity among the Jews' (a mission presided over by a bishop and endowed by the joint efforts of the kingdoms of Prussia and England) produces as its fruits, according to its own statistics, a congregation of just *thirty-seven* Jewish converts. During the whole of last year, the result of its labors was the conversion of *one* Jew. The cost of this one convert was the annual outlay at Jerusalem alone, besides the bishop's stipend, of £1228 expended on the mission, £445 on the church, £1173 on the hospital, and £400 (we beg pardon, £399 19*s*. 11*d*.; see *Report*, p. 111) on the house of industry. The Jerusalem Mission, then, if we add to its cost the £1200 per annum paid to Bishop Gobat, arising from the endowment, has actually, in the past year, baptized converts at the moderate rate of only £4443 7*s*. 2*d*. per head."

the residence of the Jews in Egypt by 275 years, allowing to it only the more probable space of 155 years. But having thus made the great Jewish epoch, the migration of Abraham out of Chaldæa, 315 years more modern, they thought it equally necessary to make such a large addition to the age of the world as the history of science and civilization, and the state of Egypt at the time of Abraham, seemed to call for. Accordingly, they added to the genealogies of the patriarchs neither more nor less than a whole Egyptian cycle [*Sothic-period*] (174) of 1460 years; or 580 between Adam and Noah, and 880 between Noah and Abraham, though in so doing they carelessly made Methuselah *outlive the Flood.* (175)

This plain matter-of-fact solution of the reasons why the *Septuagint* chronology differs from that of the Hebrew — between Adam and the *Deluge* — upon popular computations only 586 years! — relieves us from the bootless trouble of attaching any importance to opinions current at Alexandria among those successors of the *Founder* of chronology; who, with the original copies of MANETHO (176) before them, paid homage to his accuracy in their endeavors to assimilate their own foreign estimates of time to his.

Archæological rules also permit two deductions to be drawn from these premises:—

1st. That the differences of numerical results among early Christian and Judaical computators of the *Septuagint* proceed less from wilful perversions of numbers (as heretofore attributed to Josephus and others), than from radical discrepancies then existing between the *manuscript* consulted by one computator, and those exemplars whose numeration was followed by his compeers. This becomes obvious by comparing the eras severally reached by modern computations upon manuscript and printed copies now extant.

	Creation B. C.	Deluge B. C.
HALES's *Septuagint* computation—edition to us unknown—	5586	3246
Alexandrinus MS.	5508	——
Vaticanus MS.	5270	——
JOSEPHUS, on some lost MS.—probably	5555	3146

2d. That already in the time of Josephus, during the first century after Christ, the manuscript he followed must have differed in numeration from the parental *exemplars* of those transcriptions that, under the modern names of various codices, *Cottonianus, Alexandrinus, Vaticanus, Bezæ,* &c. (none earlier than A. D. 500), have reached our day; and *ergo* there must have been many corruptions and variants among *Septuagint* MSS., about and prior to the Christian era.

Hence we conclude, that it is as vain a task for computators, now-a-days, to recover more than a vague approximation of chronological notions (deducible from the *Septuagint*) current at Alexandria before the Christian era, as, after the foregoing analysis of the natural origin, history, and manifold corruptions of Greek codices, it would be to insist upon Divine authenticity for king James's version; on the plea that, in the majority of cases, its forty-seven translators rendered from the *Greek* of editions, or manuscripts, so rotten in basis as those of the *Septuagint.*

We proceed to the *Hebrew Text;* with the remark that, although we now know that it could have had little to do with the formation of our "authorized version," we shall examine it under the hypothesis (customarily put forward) that it had a great deal.

In the year 1603, at the time when king James authorized a new English translation, there were numerous printed editions of the *Hebrew Text* familiar to biblical scholars. That of Soncino, 1488, the first printed; of Brescia, 1494, used by Luther for his translation; Bomberg's, 1518–45; Stephens's, 1544–46; Munster's, 1546; are the most prominent of the number. Whether the translators consulted any, or what, Hebrew *manuscripts,* does not appear from works within our present reach. We have shown how trivial was their acquaintance with the language of the editions, and may be persuaded that they did not

(174) CHAMPOLLION-FIGEAC: *Egypte Ancienne;* 1840; pp. 236–240; — GLIDDON: *Chapters on Early Egyptian History;* 1843; pp. 50, 51, 52, 61; — LEPSIUS: *Chronologie;* 1849; i. pp. 165–180.

(175) SHARPE: *Op. cit.;* p. 196.

(176) BUNSEN: *Op. cit.;* pp. 56–96.

greatly distress themselves about the latter; for, a century and a half elapsed before Kennicott proclaimed how — "the Hebrew Bible was *printed* from the *latest*, and consequently the *worst* manuscripts;" (177) thus corroborating his previous acknowledgment — "that the Sacred Books have not descended to us, for so many ages, *without some mistakes and errors of transcribers.*"(178) He enlarges upon the certainty of corruptions in the *printed* Hebrew Text, powerfully refuting those who claim textual unity; and then passes on to establish the absurdity of attributing perfection, either, to the *manuscripts.*(179)

Of all men down to his epoch, 1780, Kennicott had the best right to speak decisively; his conclusions being drawn from the collation of no less than 692 *manuscripts* of the Hebrew text; whereof about 250 were collated by himself personally, and the remainder by Mr. Bruns, under his direction. Of the most ancient relics, but *two* were assigned by him to the tenth century after Christ; to the eleventh or twelfth centuries, only *three;* while all the rest ranged between the years 1200 and 1500 A. D. (180) The bulk of his work, its costliness and comparative rarity, combine with its Latin idiom to render it inaccessible to ordinary readers, save at second-hand. But few of the facts established by this great and upright scholar are popularly known; or they have been misrepresented, more or less, by some of the ecclesiastical mediums (181) through which they have reached the public eye. Cardinal Wiseman, (182) for example, would lead his readers to infer, that the innumerable variants and corruptions of the Hebrew Text, verified by Kennicott, were of small importance; and even the Rev. Moses Stuart (183) slurs lightly over those depreciatory results which it will be archæology's duty presently to enumerate, in saying: —

> "Indeed, one may travel through the immense desert (so I can hardly help naming it) of Kennicott and De Rossi, and (if I may venture to speak in homely phrase) not find game enough to be worth the hunting." So again, "Have they (the Jews) added to, or diminished from, their Scriptures during all this period of 1800 years? Not the least. . . . *Their* Bible has remained inviolate."

Now, to continue the sagacious Professor's simile, the quantity of game to be found in a given wilderness frequently depends upon the keenness of the huntsman; its quality upon his individual tastes; some sportsmen being partial to *tomtits*, whilst others sigh that nothing fiercer than *grizzly-bears* encounters their ferine combativeness. And, with respect to the "inviolate" state of the Text, Kennicott shall speak for himself, after we have opened a volume of De Rossi.

G. Bernardo de Rossi, of Parma, was that august Italian critic who resumed investigation into the actual condition of the Hebrew Text at the point where his English predecessor had left off; recasting also (wherever the same MSS. could be reached by him) the work of the illustrious Oxonian. Written in Italian, and intended solely for the lettered, his books are not very familiar to the general reader. A quotation or two, therefore, may place matters in their proper light:

> "Here it suffices to observe, that the totality of *manuscripts collated* is 1418, of editions 374; that to the English 577, and 16 Samaritan, I have added 825; of which my cabinet alone furnished 691, and 333 editions; besides the ancient versions, the commentaries, the works of criticism and other sources that are also themselves in the greatest number." (184)

In another work he states: — "Of the manuscript codices most ancient of the sacred Text" . . . the *oldest*, that of Vienna, dates in A. D. 1019; the next is Reuchlin's, of Carlsruhe; its age being A. D. 1038. There is nothing in manuscript of the Hebrew Old Testa-

(177) *State of the printed Hebrew Text;* 2d Dissert.; Oxford, 1709; p. 470.

(178) *Ibid.;* 1st Dissert.; 1753; Introd.

(179) *Ibid.;* pp. 234, 263.

(180) *Dissertatio Generalis in Vetus Testamentum Hebraicum;* Oxford, 1780; in folio; pp. 110–113.

(181) "By 'ecclesiastical persons' are understood such as are indeed subjects, yet their office and works is [*sic!*] in matters of Religion; they act *between* GOD *and man*, as messengers, and mediators between them. They deliver GOD's *mind* to men; and offer men's prayers and *gifts* to GOD"; says the Rev. GEORGE LAWSON, *Protestant* Rector of More (*Politica Sacra et Civilis;* London, 1660; p. 230).

(182) *Connection between Science and Revealed Religion;* 1844; ii. pp. 168, 169.

(183) *Crit. Hist. and Defence of the O. T. Canon;* Andover, 1845; pp. 193, 239.

(184) *Compendio di Critica Sacra;* Parma, 1811; ii. p. 37.

ment now extant of an earlier date than the eleventh century after Christ. (185) And, "of the most ancient manuscripts of the Greek Text of the *New* Testament," . . . the *oldest* are the Alexandrian and Vatican, which may ascend to the fourth, but cannot be much later than the fifth century after Christ.

Considering such circumstances, our credulity is not strained by accepting what De Rossi asserts, as rather more authoritative than the fiats of some "teologini" we might name; for he, at least, had advanced by studious discipline to the *positive* stage of philosophy. These are his Italian views rendered into English:—under the head of "Premure degli Ebrei per loro Testo:"—

"It is known [?] with what carefulness Esdras, the most excellent critic they have had, had reformed [the Text] and corrected it, and restored it to its primary splendor. Of the many revisions undertaken after him none are more celebrated than that of the *Massoretes*, who came after the sixth century [ANNIS D.]; who, in order that the Text should not in after time become altered, and that it might be preserved in its integrity, numbered all the verses, the words, the letters of each book, together with their form and place. But their fatigues being well analyzed, one perceives that they had more in aim to fix the state of their Text, than to correct it; that, of infinite interesting and grave variants they do not speak; and that, ordinarily, they do not occupy themselves but with minutiæ of orthography of little or no weight: and all the most zealous adorers and defenders of the Massora, Christians and Jews, while rendering justice to the worthiest intentions and to the enormous fatigues of its first authors, ingenuously accord and confess that it [the Massoretic Text], such as it exists, is *deficient, imperfect, interpolated, full of errors; . . .* a most unsafe guide." (186)

Why, "the single Bible of *Soncino* [earliest printed Text] furnishes more than *twelve thousand* (variants)!" Which said, our authority continues through above eleven 8vo pages to deplore and make manifest "the *horrible state of the Text*," resulting from his own comparisons of 1418 Hebrew manuscripts, and 374 printed editions. Such being the truth, published a quarter-century before the Rev. Dr. Hales's "Analysis of Chronology," (187) the reader can qualify the following attestation of an ecclesiastic by what epithet he pleases:—

"It is not more certain that there are a *sun* and *moon* in the heavens, than it is, that not a single error of the press, or of a *Jewish* transcriber, has crept into the present copies of the *Masorete* Hebrew Text, to give the least interruption to its chronological series of years."

And yet, so devoid of consistency is this theologer, that he designates the *Hebrew* chronology as "spurious," and actually follows that of the *Septuagint!*

From the loud denunciations of one of the most learned Church-of-England Protestant divines, and the sterner sorrow of an Italian Catholic cenobite, turn we to the wild despair of the Hebrew Rabbis:—"Peruit consilium! Computruit sapientia nostra! Oblivioni traditæ sunt leges nostræ! Multæ etiam *corruptelæ*, et *errores*, ceciderunt in Legem nostram sanctam!" (188)

But Kennicott substantiates that the disorderly condition of the Hebrew Text, and its multitudinous vitiations, resile from the works, or are lamented in the language, of all claimants to biblical knowledge for 1700 years previously to the Rabbis and himslf; equivalent to 1730 prior to De Rossi. Here is a skeleton of his list, omitting citations:—"Justin Martyr, died A. D. 165—Tertullian, 220—Clemens Romanus, 102—Origen, 254—Eusebius Cæsarienensis, 340—Eusebius Emisenus, flourished 350—Ephraim Syrus, died 378—Hieronymus, 420." We pause to illustrate.

1st. King James's version.—PAUL, *Galatians*, iii. 13:—"for it is written, Cursed *is* every one that hangeth on a tree." [The English of the Greek passage in *Griesbach's text* is, apud Sharpe, "(for it is written; *cursed is every one that is hanged on a tree;*)"].

(185) *Introduzione alla Sacra Scrittura;* Parma, 1817; pp. 34, 47.

(186) *Compendio;* ch. iv. p. 7; and pp. 9–22. DE ROSSI furthermore proves these positions in his "Specimen Variarum Lectionum Sacri Textus"; Rome, 1782.

(187) *Analysis;* 2d edit.; 1830; i. p. 277.

(188) *Hebrew edition* of 1751; the preface, cited in *Dissert. Generalis;* p. 27.

2d. This is a quotation by the Apostle from *Deuteronomy* xxi. 23; which, in king James's version stands — "(for he that is hanged *is* accursed of God;)" [The French of Cahen reads — "car un pendu est une malédiction de Dieu" (v. pp. 93, 94); which conforms better to the context, and resembles current superstitious aversion to *gibbets.*]

Apart from illiteral citation, the New Testament, in this passage, leaves out the word ELoHIM, 'God.' Theologists who combat for "plenary inspiration" can doubtless answer the following interrogatories. If those words be Paul's (always provided for), did he quote from memory? then his recollection was faulty. If he copied the LXX, then, in his day, the Greek already differed from the Hebrew; and who can tell which of the two transcripts preserved the original reading?

The catalogue continues with—"Epiphanius, 403 — Augustine, 430"— but we abridge twenty-two folio pages of extracts from later Christian writers, who protest to the same effect, into a line; epitomizing the series by one name — Ludovicus Capellus, founder of sacred criticism in 1650.

All the subjoined commentators vouch for inaccuracies in the Text: viz.— "Raymond de Pennaforti, 1250—Nic. Lyranus, 1320—Rudolphus Armachanus, 1359—Tostatus, 1450—Jacob Perez de Valentia, 1450 — Marsilius Ficinus, 1450 — Baptista Mantuanus, 1516 — Zuinglius, 1528—Martin Luther, 1546—Bibliander, 1564," &c. The same corruptions are certified through the decrees of the *Council of Trent*, 1546; through the *Vulgate of Sixtus V.*, 1590; and through king James's version, 1604–1611: on which the Oxonian critic remarks (p. 50, § 108): — "To the AUTHORS of the *English version* that which is due: many examples prove that they did not always mind what they found in the Hebrew, but what they thought ought to be read therein: tantamount to that, in their opinion, the Hebrew Text was corrupt. This the reader evolves from twenty places:—*Gen.* xxv. 8: xxxv. 29: *Ex.* xx. 10: *Deut.* v. 14; xxvii. 26; xxxii. 43: *Jos.* xxii. 34: *Jud.* vii. 18—vid. com. 20—1 *Sam.* ii. 23: 2 *Sam.* iii. 7; v. 8; xxi. 19; xxiii. 8: 2 *Kings* xxv. 3: 1 *Chron.* vii. 6; ix. 41; xxiv. 23: *Ps.* xxxiv. 17; lxx. 1: *Isa.* xxviii. 12: *Ezech.* xxvi. 23."

After citing "Jos. Scaliger; the Buxtorfs, father and son, defenders of the purity of the text; Capellus; Glassius; Joseph Mede; Usher, Morinus, Beveridge, Walton, Hammond, Bochart, Hottinger, Huet, Pococke, Jablonski, Clericus, Opitius, Vetringa, Michaelis, Wolfius, Carpzovius, Joseph Hallet, Francis Hare" — Kennicott concludes (§ 132): —

"Id autem a me maximè propositum fuit, ut ostenderem — produci posse testimonia multa et insignia, per intervallum fere 2000 annorum, ad probandas *mutationes* in Hebraicum Textum *invectas:* quanquam in contrariam sententiam, annis abhinc triginta, docti fere omnes abierint." (189)

One would have thought (to return to Prof. Stuart's metaphor), that this "immense desert" contained "game enough," in all conscience! but, in some men, the love of chase is insatiable. "Defence," as he justly observes, "would seem to be needed. The contest has become one *pro aris et focis*" — "truly become one, as I have said, *pro aris et focis.*" (190)

"It has become plain," frankly declares this lamented Hebraist, "that the battle which has been going on over most European ground these forty or fifty years past, has at last come even to us [alluding to the exegetical works of his learned and reverend New England colleagues, Noyes, Palfrey, Norton, Parker, &c.], and we can no longer decline the contest. Unbelief in the Voltaire and the Thomas Paine style we have coped with, and in a measure gained the victory. But now it comes in the shape of philosophy, literature, criticism, philology, knowledge of antiquity, and the like.[!] Hume's arguments against miracles have been *exhumed*, clothed with a new and splendid costume, and commended to the world by many among the most learned men in Europe. Before them, all revelation falls alike, both Old Testament and New." (191)

And, considering who these "most learned men" veritably are, it is not for us to question the uprightness of his outspoken recognition, that—

(189) *Dissertatio Generalis;* 1780; pp. 7, 8, 33–43, 55, *seq.*
(190) *Op. cit.;* pp. 3, 422.
(191) *Op. cit.;* p. 420.

"The unbelief that consistently sets aside the whole, shows a *more manly and energetic attitude of mind;* and, in my opinion, it is much more likely to be convinced at last of error, than he is who thinks that he is already a believer and is safe, while he virtually rejects from the Gospel all which makes a Gospel, in distinction from the teachings of Socrates, of Plato, of Plutarch, of Cicero, and of Seneca." (192)

We have quoted the highest contemporary authority of the Calvinist school; and impartiality requires that a member of the "Chiesa Cattolica Apostolica Romana" should make up for the mild notice taken of Kennicott's and De Rossi's researches by His Eminence the Cardinal.

If the man of science mourns, with as much fervor as the most devout, over the irrecoverable loss of Hebrew *manuscripts* of the Bible—of those precious documents that would have linked the *Bodleian codex* (about 800 years old, said to be the most ancient) (193) with the transcripts of Ezra's copy; and filled up the frightful chasm that now divides, in Hebrew palæography, the tenth century after Christ from the fifth century before his advent — to whose acts is he indebted, and by whom are his sorrows caused? Lacour shall answer:—

"At the commencement of the thirteenth century, it was expressly forbidden to the laity to possess the books of the Old and New Testament. The Church permitted only the Psalter, the Breviary, or the Hours of the Sainted Mary; and these books were required not to be translated into the vulgar tongue. Decrees of Bishops interdicted the use of grammar." (194) Other sources confirm this assertion.

Gregory the Great, A. D. 590, censured Didier, Archbishop of Vienna, for suffering grammar to be taught in his diocese; "boasting that he (himself) scorned to conform his latinity to grammatical rules, lest thereby he should resemble the *heathen.*" (195) In the ninth century, Alfred the Great laments that there was not a priest in England who really understood Latin, and, for ages after, English Bishops were termed "marksmen," because they could not sign their names otherwise than by a *cross!*

"In 1490, the Inquisition caused the Hebrew Bibles to be burned, that is to say, the work in default of the author; in the absence of Moses, his *Pentateuch.*" At Salamanca, the fiendish Dominican, Torquemada, reduced some 6000 Hebrew volumes to ashes; and besides such as were ravished from libraries in Spain and Italy, about 12,000 Talmudic rolls perished, *circa* A. D. 1559, in Inquisitorial flames at Cremona. (196) These unnameable deeds were induced by orthodox doubts that, the Hebrew Text, as represented in the *square-letter* copies, was ever quoted by the Apostles; (196) but, in those ages of darkness, little respect could have been paid to MSS. even of the *New* Testament; for such ancient copies as had been preserved, down to A. D. 1749, at Alcala in Spain, were sold to one Toryo, a pyrotechnist, as materials for sky-rockets. (197) Quintillian (*Inst. Orat.* i. I), in the first century after Christ, complains that writing was neglected; but it was not until after the barbarian irruptions of the eighth century that "la crasse ignorance" prevailed in Western Europe. It is uncertain if even Charlemagne could write. The tenth to twelfth centuries exhibit Bishops, Abbots, Clerks, &c., incredibly ignorant: as even in earlier times, before the seventh century, at the Episcopal Conference of Carthage, the "brigandage" of Ephesus, and the Council of Chalcedon—at which last there were *forty* most incapable Bishops (Labbe, *Concil,* iv). Few Romish monks could read, in the eleventh; the laity began about the end of the thirteenth; but in the fourteenth, the number was small. (198)

From these fearful destructions (the Inquisitorial agents having acted in obedience to orders sent from Rome), Lacour draws a singular argument in behalf of his own free restorations of the Hebrew Text, maintaining:—

(192) *Op. cit.;* p. 320.

(193) Kennicott: 2d *Dissert.;* p. 317 — "*Laud,* A, No. 162," in catalogue Bodleian Library.

(194) Æloïm: Bordeaux, 1828; i. p. 28.

(195) Mandeville, apud Taylor; p. 34; — also, Righellini: *Examen;* iii. p. 537; — and Vico: *Scienza Nuova,* trad. Michelet; ii. p. 67; for other examples.

(196) Lacour: p. 29; — and Kennicott: *Dissert. Gen.;* p. 16.

(197) Marsh's *Michaelis;* ii. p. 44.

(198) Condensed from an excellent article on Alphabets, in vol. ix. pp. 727-739, of the great "Encyclopédie Catholique"; Paris, 1846: conducted by the Abbé Glaire and M. Walsh.

"That the Hebrew Text of the Bible, tried and condemned by the Holy Tribunal, burned as an act of faith at Seville, and in the Square of St. Stephen at Salamanca, proscribed during the sixteenth century, prohibited in the pulpits of Catholic preachers, declared dangerous, infected with Judaism, and causing those Christians who read it to Judaize likewise, finds itself—owing to this solemn condemnation from which it cannot be purged save through the adoption of a *new translation* — finds itself, I repeat, does this Text, to have lost the character and authority that, in the spirit of Christianity, the Fathers [only Origen and Jerome] attributed to it. One may, therefore, after all, study this Text in a new point of view, purely philosophical and philologic; and seek in it a new interpretation, without being scared at the sense which such interpretation may produce. The anathema with which it has been stricken has abandoned it to criticism and to the investigations of the world; *tradidit disputatione:* its testimony is no longer anything but mere human testimony, liable to error like all things that proceed from man." (199)

Conceding his premises, and allowing for his peculiarly catholic point of view, the deduction is logical; but they who deny Papal infallibility may continue to reverence the Hebrew Text just as if excommunication had never been pronounced upon it; notwithstanding the avowal of those manifold corruptions which, owing to these Inquisitorial holocausts of ancient *manuscripts*, it seems now humanly impossible to expunge. To persecutions and to the expulsion of the Jews from Spain, after 1491, the extinction of the most precious Hebrew exemplars may be, in part, attributed; for Muslim intolerance had never knowingly laid the hand of sacrilege upon documents which Christian charity has for ever destroyed. (200) MOHAMMED had built up his *Kur'àn* upon the monotheistic foundations of MOSES; (201) and his faithful disciples have been always too consistent, whatever barbarities they may have inflicted upon the Jews, to injure that chosen people's sacred *books*, and thereby stultify themselves. With reference to textual corruptions, says Kennicott (202 : —

"Hæc denique sunt verba eruditissimi Professoris J. A. Starck — 'cum negari prorsus nequeat (si quidem luminibus uti, et antiquos libros ab omnibus præjudicatis opinionibus liberi inter se conferre velimus) *multa et ingentia σφαλματα inisse sacris libris;* qualia sunt, gravissimi in chronologicis errores; in historicis manifestæ contradictiones; numerorum exaggerationes; literarum, nominarum, sententiarum, omissiones, additiones, transpositiones: quæstio jure orietur — Unde tot tamque graves immutationes originem suam habeant? Et si gravissimis argumentis, quibus solis permota ita sentio, fides habenda est; prorsus omni caret dubio, Judæorum imprimis fallaciam et malevolam mentem accusandam esse, post librariorum inertiam et negligentiam.' "

To avoid mistakes we have given the Latin text, and now offer its straightforward signification in English: —

"Since it cannot altogether be denied (if indeed we free ourselves from all prejudiced opinions, and wish to compare ancient books with each other and to avail ourselves of the instructions of the learned,) that *many and enormous σφαλματα* [*lapsi*, mistakes] *exist in the sacred books;* such as, most grave errors in chronological (matters); manifest contradictions in historical; exaggerations in numbers; omissions, additions, transpositions of letters, of names, of sentences: — the question will naturally arise, Whence have such and so many serious mutations their origin? And if faith is to be placed in most weighty arguments, by which alone I am influenced, every doubt is altogether wanting, (that) first one must accuse the fallacious and malevolent mind of the Jews, (and) afterwards the inertness and negligence of librarians."

Such are the published *facts*. Yet one marvels at the ways of theology; on seeing the Rev. Prof. Stuart skip nimbly over that "immense desert" with his "gun, man, and dog," (*Arma virumque cano*,) and the *dégagé* air of a juvenile Nimrod, without finding "*game* enough to be worth the hunting;" and then asserting with equal frivolity, that the Jewish "Bible has remained *inviolate*"! How can the unlettered distinguish truth from error, when their Teachers mystify the plainest results that scholarship the most exalted, honesty the most unbending, and science the most profound, have striven to make public to all men for the last hundred years?

---

(199) LACOUR: *Op. cit.;* i. p. 33.

(200) SISMONDI, not now before me, gives many other examples of literary destructions in Italy, Portugal, and Spain.

(201) Compare LANE: *Selections;* pp. 183–225, 270, 271.

(202) *Op. cit.;* p. 33; note to § 76.

Nevertheless, a time has come in which opinions, that ignorance had laid down as fundamental *principles*, begin to compromise those institutional structures beneath which they were placed. Enlightened manhood in a free Republic is fast approaching the hour when such opinions will be openly recognized as nothing more than *opinions of ignorance.* To attempt to impede reform, when it is necessary, is to jeopard the whole system. To refuse to repair foundations whose vetustity perils an edifice, is to desire that the downfall of such edifice shall prove that its foundations are rotten. "Creeds," says Sharpe, speaking of the decrees of the œcumenic Councils, "composed in the dark have now to be defended in the light, and those who profess them have the painful task of employing learning to justify ignorance." (203)

A point has been now attained in this exposition, when a brief recapitulation of the halts made during our journey will enable us to dismiss king James's version from further consideration. We opine that the foregoing pages have established, upon archæological principles and adequately for the demands of positive philosophy, —

1st — by *authority* of the highest Biblical critics;

2d — by *exegetical* exposure of some of its false-translations;

3d — by *historical* testimony, that all versions in English, (being mere popular accommodations of defective editions printed in the "Original Sacred tongues,") have only perpetuated or increased whatever errors their antecedent editions contain;

4th — that because the *Latin Vulgate*, printed or manuscript, abounds in mistakes;

5th — that because the *Greek Septuagint*, if ever a faithful representative of the Hebrew original, is so no longer, in any printed editions or manuscript copies now known; and that tradition, well authenticated, proves its vitiated state as far back as the first century of the Christian era;

6th — that because the only men, Protestant, Catholic, or Rabbinical, whose decisions (owing to their respectively minute collation of every printed edition or manuscript exemplar of the *Hebrew Text*) can be weighty in the premises, have pronounced the whole of them to be radically, enormously, and irretrievably *corrupt*; —

in view of all of the above facts, we have a right to conclude that, our English "authorized *Translation*," made 250 years ago under circumstances naturally adverse upon documents so faulty, can claim, in science, no higher respect than we should accord to a poor translation of mutilated copies of Homer; and finally, that those individuals who are most clamorous in its praises only bear witness that they possess the least acquaintance with its origin and history, however familiar they may be with its contents.

But, universal orthodoxy, regardless of the collective researches of three centuries, insists upon our credence that Moses *wrote the Pentateuch;* and still stigmatizes those who respectfully solicit some evidences of this alleged authorship (a little more conclusive than ecclesiastical tradition) with terms intended to be opprobrious; of which, perhaps, the most courteous form in vogue nowadays is "skeptic." (204) If by this harmless vocable nothing more is implied than that a "skeptic" has, by laborious study, attained to the positive stage of philosophy, while "orthodoxy" vegetates in a sub-metaphysical stratum, it should be cheerfully endured; if not with Christian fortitude, at least with gentlemanly equanimity.

The real question, however, posited in logical shape, is this: —

The Hebrew Moses wrote the Hebrew Pentateuch. Did the Hebrew Moses write the Hebrew Pentateuch? If the Hebrew Moses wrote the Hebrew Pentateuch, where is the Hebrew Pentateuch the Hebrew Moses wrote?

For ourselves, we do not perceive what essential difference it would make, in positive philosophy, supposing even that *he did:* but, inasmuch as we have embarked in an inquiry

(203) *History of Egypt;* p. 490.

(204) The Rev. Dr. SMYTHE of Charleston, S. C.: *Unity of the Human Races;* Index, p. 401

for the purpose of ascertaining the importance which progressive Ethnology must assign to *one document;* and this document happens to be the Xth *Chapter* of a Book called "Genesis," (which some vehemently protest is *Mosaic,* while others as flatly contradict them,) it behooves us to test certain points of these disputed allegations by archæological criteria; and, authority against authority, the citation of a few may help us in making ready for the voyage.

"And yet no one, I believe, has the pretension to understand perfectly the sense of *Genesis;* no one denies that the text of this book contains many parables, or Oriental allegories, of which the most skilful and the wisest of the Fathers of the Church have sought in vain for the meaning.—But, thanks to the massoretic points and to the susceptibilities of orthodoxy, things have come at the present day to such a pass, that if Moses himself arose from the tomb to cause all uncertainty to cease; if he interpreted his own book literally; if he expounded it as he had conceived it and reflected upon it; Jerusalem, Rome, Constantinople, and Geneva, [Great Britain, Germany, France, and the United States,] would convoke their Doctors of Divinity from all corners of the world, to prove to him—that he knows nothing about the genius of the Hebrew tongue—that his translation is contrary to the grammar and dictionary of Mr. *This* or Mr. *That*—that he does not possess even common sense—that he is an impious (fellow) whose book they had done perfectly right [*Rome's orders,* XIII–XVIth centuries] to burn; and that it is wonderful how he had not been served so himself in the other world." (205)

Having now fulfilled my published pledges to the reader, so far as relates to the exhibition of a few atoms of the vicissitudes through which the *Xth Chapter of Genesis* has travelled to reach our day, I am obliged to bring this "Archæological Introduction" to an abrupt close at this point. The reasons are these:—

When my colleague Dr. Nott, at Mobile (in April, 1852), agreed with me to erect a literary cenotaph "*To the* MEMORY *of* MORTON," it was mutually arranged that, in our division of labor, he would undertake the anatomical and physical department, embracing those subjects that belong to the *Natural Sciences;* while the execution of the archæological and biblical portions was to devolve upon myself.

No two men have ever worked together in the same harness with more perfect harmony of object. In the midst of professional engagements, whose onerous character none but the most laborious of the medical faculty can adequately appreciate, Dr. Nott, at the sacrifice of every instant of repose, succeeded in accomplishing, not merely all that appertains to his part of our enterprise as set forth in Part I., but also the revision of my studies as exhibited in Part II.: each of us, notwithstanding, being wholly responsible for whatever naturally falls within the specialities severally assumed, but neither of us being fairly amenable for mistakes in other than our own departments as above classified.

On the other hand—independently of three months, December 1852 to March 1853, spent by myself in travelling; and aside from all supervisions of the press since the 25th of August—I devoted nearly twelve months of day and night to the performance of my "spécialité" of our joint undertaking; some of the fruits of which have been already submitted to the reader's criticism.

Resolved, in my own mind, to pursue inquiries into biblical questions, once for all, *usque ad necem,* my manuscripts have, I think, completely answered the Aristotelian proposition above stated as concerns the *Pentateuch.* Nevertheless, I postpone their publication:—

1st. Because they do not directly concern *Ethnology,* and the main subjects of this work.

2d. Because the printers assure me that my "copy" could not be condensed, satisfactorily, within 300 more of these pages: thereby rendering it impossible to keep "Types of Mankind" within one volume.

Ample, however, and far more gratifying than a dry archæological disquisition can be to the general reader, are the compensations which displace my own performances: and it is with unfeigned pleasure that, in order to make room for the papers of our collaborators, I

(205) LACOUR: ÆLOÏM; i. p. 180.

mutilate my own essays in substituting theirs. Perhaps it is for the best; because the nature of this work may elicit some hostile comments; and he is the prudent soldier who "keeps his powder dry." In consequence, I suppress about 300 of these pages, after submitting an outline of the *Periods* of misfortune which the canonical Hebrew Text has, to a great measure, survived, down to CAHEN's *Bible*, A. D. 1831–1851.

Walton, Kennicott, and De Wette (to say nothing of other sources), the reader perceives are tolerably familiar to us. To extract from their works is merely mechanical; but the fear of tedium warns us to be eclectic. In these matters it is our private opinion that, if Titans were again to pile Ossa upon Pelion, after rolling upon "Ossa the leafy Olympus," (206) they would fail to startle, far less convince, those who lie below the *metaphysical* stratum of intellectual development; for, "as Jannes and Jambres withstood Moses, so do these men withstand the truth." (207) It will be more interesting to the enlightened reader to view a brief historical schedule of the *changes* which eighteen centuries have entailed upon the Hebrew Text — condensed principally from Kennicott's results in his *Dissertatio Generalis*: —

1st PERIOD, B. C. — "In most ancient times, the Hebrew Text was corrupt;" and the codex (say, "fragmentary books") used by the Greek interpreters of the Old Testament, at Alexandria, was undoubtedly Hebrew, but a copy not sufficiently emended. Even Buxtorf is obliged to admit — "Judæos a tempore Esdræ negligentiores fuisse circa textum Hebræum, et non curiosos circa lectionem veram."

The numerals were expressed by *letters:* the five *final* letters (*kaf, mim, nun, pay,* and *tsàde*) had not then been invented: the words were still *undivided.*

2d PERIOD, A. D. down to 500. — The texts were more corrupt in the time of Philo and Josephus. Neither in their day, nor in that of Origen, third century, were the *Commandments* (*Exod.* xx. 3–17) divided into ten, in the manner they are now. In Philo the division is *quinary*, after the fashion of Pythagoræans. About the latter epoch commences the Talmudic *Mishna;* and, in the fifth century, the *Gemara;* each of which books proves the increase of textual errors. So do the writings of the Fathers during all this age — notably St. Jerome; while the apostolic books demonstrate that the *Greek* differed, more or less, from the Hebrew original.

3d PERIOD, A. D. 500 to 1000.—Aside from the later and less reliable Fathers, two Hebraical works establish, that no expurgations of error had been made in the Text: viz., the *Robboth*, after A. D. 700, and the *Pirke Eliezar*, after 800. About the sixth century, the Rabbis of Tiberias commenced the "Masora": a labor that would not have been undertaken but for the reasons above given, and the wretched condition of the Text in their time; as proved by the multitudes of *Keri velo Kethib* (the read, but not the written) or *Kethib velo Keri* (the written, but not the read). (208)

4th PERIOD, A. D. 1000 to 1450.—The Jewish schools of Babylonia seek refuge in Spain about 1040; between which era and 1240 flourished the four great Rabbis. Their works prove not merely different readings, but absolute mistakes in copies of the Text: things then existing in manuscripts of the Old Testament now exist no longer, and *vice versa;* while the "Masora," itself, already in confusion inextricable, only rendered matters worse. It is of this age alone that we possess those Hebrew manuscripts by us called *ancient*—not one 900 years old!

5th PERIOD, A. D. 1450 to 1750.—Printing invented; the art was first applied to *Psalms* in the year 1477; and to the whole Hebrew Text in 1488; that entire edition, save one-third of a copy, being immediately burnt by Neapolitan Jews. But here, upon editions now following each other with rapid succession, the Rabbis begin their restorations and their lamentations. Continental scholars now set to work upon *Hebrew* in earnest, without professorships: whilst, in England, king James's version is a splendid

---

(206) VIRGIL: *Georg.;* i. 281.
(207) 2 *Tim.* iii. 8 — apud SHARPE.
(208) DE WETTE: i. pp. 345, 353–358.

record of Professors without Hebraism, during the years 1603–'11. Fifty years later, Walton redeems the shame of Oxford; and yet, one hundred years later still, Kennicott himself chronicles — "the reader will be *pleased* to observe, that as the study of the Hebrew language has only been *reviving* during the last one hundred years:" (209) to end which sentence logically, we ourselves consider that there could be no "revival" where, in 1600, there was scarcely a *beginning;* and, ergo, that the Doctor's attestation must refer to incipient efforts, in his century commencing, to resuscitate the *Hebrew tongue* after twenty centuries of burial.

6th and present PERIOD, A. D. 1750 to 1853.

Taking Eichhorn as the grand point of departure, we find, after the lapse of a century, how, through the operations of that "*rational* method" of which he and Richard Simon were, among Christians, the first qualified exponents, the Hebraical scholarship of our own generation (proud of its hundred champions) has truly kept pace, on the European continent, with the universal progress of knowledge.

Nevertheless, on every side, we still see and hear the crocodile whimper how "nobody undertakes a *new* translation (into English) of Holy Scripture" commensurate with the imperious demands of all the sciences at present advancing — news of the onward steps made by each being actually transmitted through magnetic telegraphs (210) — and yet, withal, few men in America so blind as not to perceive that, even in evangelized England, such pecuniary superfluities as those said to have been realized through a "WORLD'S *Exhibition,*" are expended (God alone knows how or why) upon anything, or everything, rather than in behalf of a conscientious *revisal* of our ENGLISH BIBLE.

G. R. G.

ESSAY II.

PALÆOGRAPHIC EXCURSUS ON THE ART OF WRITING.

THE same imperious necessity that has constrained us to suppress the continuation of Part III., Essay I. (*supra*, p. 626), renders it obligatory to curtail our History of the "*Art of Writing*, from the earliest antiquity to the present day." This subject, perhaps the most vital in any researches into the antiquity of the Hebrew *Pentateuch*, has never yet publicly received adequate attention from modern scholarship. With ourselves it has been a favorite pursuit ever since 1844; (211) nor, did space permit the insertion of what we had prepared in manuscript for the present volume, should we not have taken some pride in the presentation of a series of facts and arguments that would entirely justify every point set forth in the accompanying *Tableau* [*infra*, pp. 630, 631].

(209) 1st *Dissert.;* 1753; p. 307.

(210) Rev. JOHN BACHMAN, D. D.'s *Doctrine of the Unity of the Human Race;* Charleston, S. C., 1850; p. 288 — "And even telegraphing to America, through the convenient wires of Mr. Gliddon, the yet unpublished discoveries of Lepsius." These discoveries have since been published, and much JOHN BACHMAN knows about them! MORTON'S refutations, in the Charleston *Medical Journal*, 1850–'51, render it quite unnecessary for me to waste more ink upon the extinguished author of the above "Doctrine." — G. R. G.

(211) Vide GLIDDON, in LUKE BURKE'S *Ethnological Journal*, No. ix.; London, Feb. 1849; pp. 400–416: — republished in *Otia Ægyptiaca;* London, Madden, 1849; pp. 99–115: — and, without *text*, but with some improvement of the "Table," in *Hand-book to the Panorama of the Nile;* London, Madden, 1849; pp. 41–45; under the heading of "Philology." Of this pamphlet, rather more than 3000 copies have been distributed in the United States, from Maine to Louisiana, and, accompanied by my oral Lectures, have somewhat familiarized American auditors with themes but little known in Europe beyond collegiate precincts.

As it is, we can merely recommend the reader, after viewing the three distinct geographical origins and independent developments of the art of writing, to study well the place which palæography now assigns to the modern *square-letter* (ASʜURI) Hebrew alphabet of "22 letters;" while we discuss a few general principles, to be amply corroborated in detail on some future occasion.

Digressional Remarks on the ensuing Table.

I.—The *principle* followed (probably for the first time in palæographical disquisition) and exhibited through the annexed table, is a consequence of the work which it accompanies. As "Types of Mankind" tabulates the various species of the "genus homo" according to their several relations to the *Flora* and the *Fauna* of their respective centres of creation, the harmonious unison of all sciences,(112) when directed to the elucidation of a given fact, cannot be better exemplified than by cleaving into *three* well-ascertained masses the grand enigma of graphical *origines*.

We hold, without mental reservations, that history does not justify, archæology permit, or ethnology warrant, *any*, the slightest, intercourse, between Egypt and China prior to the days of Cyrus (as an extreme point); nor between either of these two primordial nations, and the Aborigines of that continent which, pronounced by Agassiz to be the *oldest* land, was unknown (from us trans-atlantically) to inhabitants of the Oriental hemisphere before Columbus. Some of the physical reasons are set forth in the present volume: and it is pleasing to find that *palæography* entirely corroborates results deduced from other investigations. To chivalrous opponents, "blanched under the harness" of scientific pursuits, we respectfully throw down our gauntlet upon three propositions:—

A — Prior to B. C. 500, Egypt had no intercourse with America or China.
B — " " America had no intercourse with China or Egypt.
C — " " China had no intercourse with Egypt or America.

Until some student, qualified through knowledge of the archæological actualities inherent in this triad of problemata (knowledge to be evinced by the weight in science of his demurrer), overthrows the *principle* upon which our table is erected, we shall not fear for its stability: nay, we offer to his use the weapons of our armory, by indicating the shortest path to verification of bibliothical accuracy.

II.—The researches of Gesenius (213) and of Champollion-Figeac (214) have been our points of departure in the construction of the *Table*. We have remodelled them by the lights which, in the former case fifteen, in the latter twelve, years of discovery demand; fusing the results of both authorities into one; and then separating the whole into *three* grand stems; 1st, HAMITIC, with its *Semitish* branches—2d, MONGOLIAN, with its offshoots—3d, AMERICAN, whose slender twigs were cut short, for ever, by Pizarro and by Cortez.

1st. The HAMITIC ORIGIN — start with Champollion le Jeune,(215) continue with Lepsius,(216) and close with Bunsen,(217) Birch,(218) Burgsch,(219) and De Saulcy.(220)

The *Semitic* streams have been followed in the subjoined order.

Aside from personal verification of the "old travellers" — Pietro della Valle, Chardin, Corneille le Brun, Kaempfer, Niebuhr, &c.; and of the later, Rich, Ouseley, Ker Porter, Kinnier, Morier, and Malcolm; the perusal of De Sacy, Tychsen, Münter, Grotefend, Saint

(212) Humboldt: *Cosmos;* Introduction to *French* edition; 1846; i. pp. 36–48.

(213) *Scrip. Ling. Phœn. Mon.;* 1837; pp. 62, 63, and Table of *Alphabets*, p. 64.

(214) *Paléographie Universelle;* 1841; i. p. 46 — "Tableau général pour servir à l'histoire de l'Écriture."

(215) *Grammaire Égyptienne;* 1836; — *Dictionnaire Égyptienne;* 1841.

(216) *Lettre à Rosellini* — Annali dell' Instituto di Corrispond. Archeol.; Roma, 1837; vol. ix.

(217) *Ægyptens Stelle in der Weltgeschichte;* 1845; vol. i. part 2d.

(218) In Bunsen's *Egypt's Place;* 1848; i. pp. 448–600; — and in Gliddon: *Otia Ægyptiaca;* 1849; pp. 113-115.

(219) Burgsch: *Scriptura Ægyptiorum demotica* ex papyris et inscriptionibus explanata; Berlin, 1848; — and *Numerorum apud veteres Ægyptios demoticorum doctrina;* Berlin, 1849.

(220) De Saulcy: *Lettre à M. Guigniaut;* Paris, 1843; — and *Analyse grammaticale du Texte Démotique du Décret de Rosette;* i., première partie, 1845.

THEORY OF THE ORDER OF DEVELOPMENT IN HUMAN WRITINGS.

"Primi per figuras animalium *Ægyptii* sensus mentis effingebant (ea antiquissima monumenta memoriæ humanæ impressa saxis cernuntur) et litterarum semel *inventores* perhibent; inde Phœnices, quia mari præpollebant, intulisse Græciæ, gloriamque adeptos, tanquam repererint, quæ acceperant."—*Tacitus*, Annal., lib. xi. c. 14.

PRIMORDIAL, or *ante-chronological* PERIODS—when incipient Humanity, requiring no records, had invented no form of *writing* (like the *Papuan, Eskimaux, Hottentot*, or *Patagonian* types at this day, A. D. 1853) .... B. C. 00,000

1st AGE; the PICTORIAL—current in the *ante-monumental* days of Egypt and China (as seen upon ancient *Peruvian* and *Mexican* monuments; and as practised now (1853) by *American Indians, Polynesians, African* and *Asiatic* barbarous tribes, &c.) 0,000

HAMITIC ORIGIN.

	B. C.
2d AGE; the Egyptian HIEROGLYPHICAL—earliest *monuments* extant, the pyramids and tombs of the IVth Memphite dynasty (say, with Lepsius)	3500
Secondary form—*Hieratic*—VIth dynasty	2800

SEMITIC STREAMS.

3d AGE; the Asiatic SEMI-ALPHABETIC—earliest *monuments* extant—*Cuneiform* Inscriptions: Assyro-Babylonian, partly ideographic, but chiefly phonetic	1000
4th AGE; the ALPHABETICAL—earliest Egyptian *monuments* extant (*i. e.* those indicating a close of transitions) are of the Tertiary form—*Demotic* papyri	600
Persepolitan *cuneiform*	530
Boustrophedòn inscriptions; Grecian	650
Himyaritic	?]
Phœnician	?

MONGOLIAN ORIGIN.

	B. C.
2d AGE; the Chinese IDEOGRAPHICAL—earliest *monuments* extant, the Inscription of YU, with later remains in the *Kou-wen* character,	2278

CHINESE STREAMS.

	A. D.
3d AGE; *Japanese*—	
First form	200
Second form	840

Infiltrations of the *Semitic* element by Jewish exiles, Christian Syriac missions, and Muslim extensions—with some slight *Arian* influences thro' the *Sanscrit*—produce changes upon Mongolian conditions and originate the

4th AGE; Alphabet *Mongol*	1260
Mandchou	1600

Siamese forms, &c., &c.; some inclining toward Chinese, others rather toward *Sanscrit* affinities.

AMERICAN ORIGIN.

	B. C.
2d AGE; *Peruvian monuments*	?
Mexican monuments	?

[Ante-Columbian—principles unknown.]

The post-Columbian influences break all links, palæographically, with the past. What atoms of extraneous civilizations reach the *American* type of humanity, begin with the sixteenth century after Christ, and are simply European. Even the *Cherokee* alphabet is the invention of GEORGE GUESS, *Sequoyah*—a half-breed Scotchman.

"*Phœnices* primi, famæ, *si* creditur, ausi
Mansuram rudibus vocem signare figuris."
Lucan, Pharsal. iii. 220, 221.

"Quidam Cecropem Atheniensem, vel Linum Thebanum, et temporibus Trojanis Palamedem, Argivum, memorant, *sexdecim* litterarum formas —— reperisse." . . .
"Quippe fama est, Cadmum, classe *Phœnicum* vectum, rudibus adhuc Græcorum populis artis ejus auctorem fuisse." — *Tacitus*, Annal. lib. xi. c. 14. [Vide also *Herodotus*, v. 57–8; *Diodorus*, iii. 66; *Pliny*, vii. 56; *Josephus*, Contr. Apion. i. 2.]

SEMITIC STREAMS. — *Continued.*

ASSYRO-PHŒNICIAN, B.C. 700.

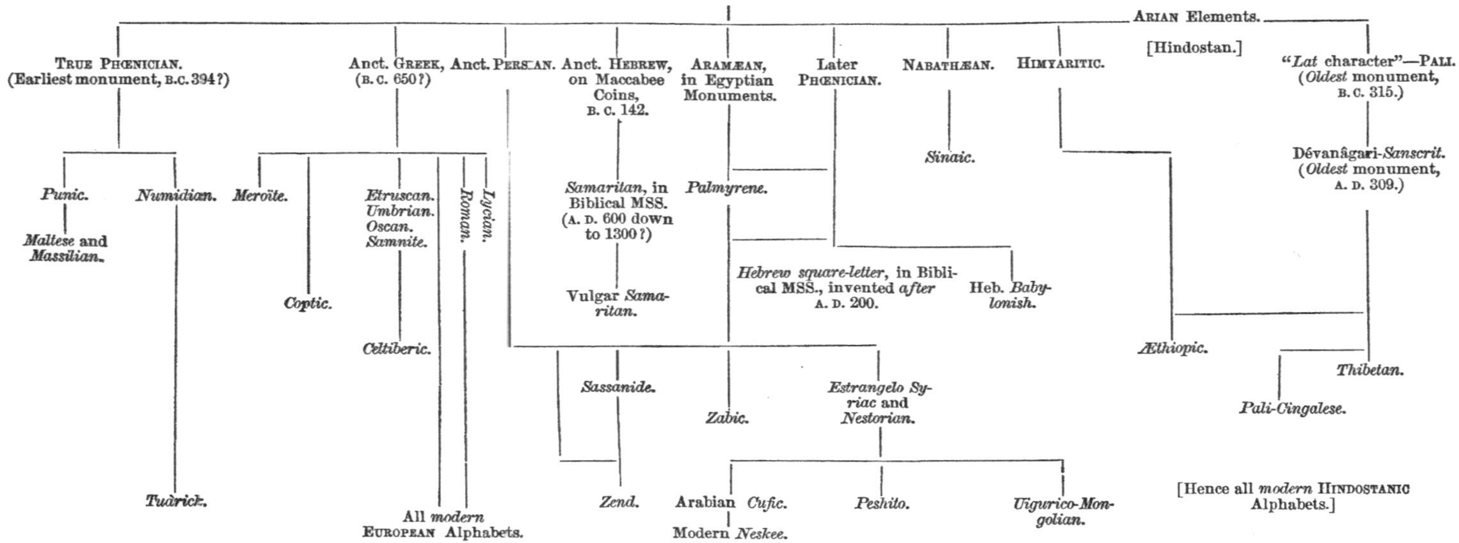

Martin, Rask, Burnouf, Lassen, and Westergaard; the possession of the major portion of the folio plates and texts of Botta, Flandin and Coste, Layard, Texier, &c.; and the inspection of what of Assyrian sculptures were in London and Paris during 1849: (221) — our views upon *Assyro-Babylonian* writings take their departure and are derived from the series at foot, appended in the order of our studies.(222)

Egyptian hieroglyphical discoveries had long ago revealed the fact that, as early at least as Thotmes III, of the XVIIIth dynasty, about the sixteenth century B. C., the Pharaohs had overrun "Naharina," or Mesopotamia, with their armies. Accepted, like all new truths, with hesitation, since Rosellini's promulgation of the data in 1832; or at first entirely denied by cuneatic discoverers, who claimed a *primeval* epoch for the sculptures of Nineveh and Babylon; nothing at this day is more positively fixed in historical science than these *Egyptian* conquests over "Nineveh" and "Babel," at least three centuries before Derceto (the earliest monarch recorded in *cuneiform* inscriptions) lived; assuming Layard's last view to be correct, (223) that he flourished about B. C. 1250. At foot we present the order in which an inquirer may investigate the discoveries that have finally set these questions at rest; (224) while the following extracts from Rawlinson will render further doubts irrelevant: —

> "That the employment of the Cuneiform character originated in Assyria, while the *system of writing* to which it was adapted was *borrowed from Egypt*, will hardly admit of question: . . . the whole structure of the Assyrian graphic system evidently betrays an *Egyptian* origin. . . . The whole system, indeed, of homophones is *essentially Egyptian*." (225)

It is upon such data that, without adducing other reasons derived from personal studies, we have made the earliest *Semitic* stream of our Table flow outwards from Egypt into ancient Mesopotamia — assigning the period of its Eastward flux, according to well-known conditions in *Egyptian* history, as bounded by the XIIth and XVIIIth dynasties: that is, between the twenty-second and sixteenth century B. C.; — which age, placed parallel with Archbishop Usher's scheme of biblical chronology, implies from a little before Abraham down to the birth of Moses. No Egyptologist will contest this view: the opinions of those who deny, without acquaintance with the works submitted, are "vox et præterea nihil."

---

(221) Three Archæological Lectures, on "Babylon, Nineveh, and Persepolis," delivered before the *Lyceum of the 2d Municipality* at New Orleans; 6th, 9th, 13th April, 1852; by G. R. G.

(222) BOTTA: *Lettres à M. Mohl;* Paris, 1845; — DE LONGPÉRIER and DE SAULCY, in *Rev. Archéol.;* 1844–1852; — LÖWENSTERN; *Essai de Déchiffrement de l'Écriture Assyrienne;* Paris, 1845; — BOTTA: *Sur l'Écriture Cunéiforme;* 1849; — RAWLINSON: *Tablet of Behistun;* 1846; — and *Commentary on Cuneiform Inscriptions;* 1850; — HINCKS: *On the Three kinds of Persepolitan Writing;* Trans. R. Irish Acad., 1847; — NORRIS: *Memoir on the Scythic Version of the Behistun Inscription;* and RAWLINSON'S communications; in Jour. R. Asiat. Soc., 1853; xv. part 1. Many other works upon this speciality, no less than upon the writings of every historical nation of antiquity, are cited in the manuscripts we suppress for lack of space. But, by anticipation of their future appearance, it would be injustice to an author "qui a puisé à des bonnes sources," not to recommend earnestly to the sincere inquirer after truth, a perusal of the first and only work in the English language which has grasped this vast subject in a manner commensurate with the progress of science. It arrived at the *Philadelphia Library,* and was kindly pointed out to us by our accomplished friend Mr. Lloyd P. Smith, after our own "Table" was already *stereotyped.* We have read it with admiration; and although upon *three* points, the hieroglyphical, the cuneiform, and especially the *Hebrew,* we might suggest a few critical — that is to say, more rigidly *chronological* — substitutions; yet, upon the whole performance we are happy to offer the warm commendations of a fellow-student. The reader will find it, in the meanwhile, an excellent adjunct to our "Table"; and the following extracts, with an interlineary commentary, suffice to indicate that Mr. Humphrey's views and our own differ upon but a single point: — "The world has now possessed a purely *alphabetic* system of writing for 3000 years or more [say rather, about 300 years *less*], and iconographic systems for more than 3000 years longer [say, considerably *more*]. . . . . There can be little doubt that the art of writing grew up independently in many countries having no communication with each other [entirely agreed]": (vide HENRY NOEL HUMPHREYS: *The Origin and Progress of the Art of Writing;* London, 1853; pp. 1, 3).

(223) *Babylon;* 2d *Ex.;* 1853; p. 623.

(224) LETRONNE: *La Civilisation Égyptienne;* pp. 1–55; Extrait de la Revue des Deux Mondes; Feb., April, 1845; — BIRCH; *Statistical Tablet of Karnac;* — *Obelisk of Thotmes* III.; and on *Two Cartouches found at Nimroud;* Trans. R. Soc. Lit., 1846–'48; — GLIDDON: *Otia;* p. 103; — LAYARD: *Nineveh;* 1848; ii. pp. 153–235; — SHARPE, in *Bonomi's Nineveh;* pp. ——; — LAYARD: *Babylon;* 1853; pp. 153–159, 186–196, 280–282, 630; — and, particularly, BIRCH: *Annals of Thotmes* III.; London *Archæologia,* xxxv., 1853; p. 160, &c.

(225) *Commentary;* 1850; pp. 4–6.

Scholars, guided by the books cited for justificatory details, will find little to alter in the general features of these several *alphabetical* streams as their respective *monumental* rocks first pierce through the mists of traditionary history: except in one direction; viz.: where we have made a *Semitic* rivulet (probably through Chaldæan channels) commingle with "ARIAN elements" in Hindostan. "Indology" will protest against profaning the sanctified soil of *Indra* and *Brahma* with the mere "tail-race" of a *Semitic* pond, originally filled by the Nile! Shades of Wilford, Faber, Hales, and spirit of Edgar Quinet! In Germany, appeal will at once be made to Von Bohlen! In Wales, to Arthur James Johnes, Esq.! (226) Does not every body know, it will be said, that primordial civilization (unceremoniously kicked out of Ethiopic *Meroë* by Lepsius,) first dawned upon the Ganges? that Memphis, (if not also Palenque, and Copan,) received her holiest Penates at the hands of *Siva*, *Vishnu*, *Bhairava*, *Crishna*, or any other Indian Deity a pundit may invent? (227)

With all deference, after the first horrors excited by our outrage shall have calmed down into philosophical contempt, we beg to offer a quotation:—

"The people of Hindostan and the ancient nations of Europe came in contact at a single point. The expedition of Alexander the Great begins, and in some sort ends, their connexion. Even of this event, so recent and remarkable, the *Hindus* have no record; they have not even a *tradition* that can with certainty be traced to it." (228)

Our author, who stands out in bold relief among the *Sanscrit* scholars of England, wonders at the credulity of those who reject Chaldæan and Egyptian antiquity to worship Hindostanic; administering stern rebukes to writers who trust in the "absurdity of *Hindu* statements,"—a people utterly "destitute of historical records."

The same historian, in Notes on the *Mudra Rákshana*, says:—

"It may not here be out of place to offer a few observations on the identification of Chandragupta and Sandracottus. *It is the only point on which we can rest with anything like confidence in the history of the Hindus*, and is therefore of vital importance in all our attempts to reduce the reigns of their kings to a rational and consistent chronology."

Turnour, (229) sums up his review of Hindoo literature with saying,—

"That there does not *now exist an authentic*, connected, and chronologically-correct Hindoo history; and that the absence of that history proceeds, not from original deficiency of historical data, but *from the systematic perversion* of those data adopted to work out the monstrous scheme upon which Hindoo faith is based."

The preceding extracts, we hope, may serve to break the fall of huge Indianist edifices from the highest peak of the Himalaya to a level but little expected by general readers. That we are not altogether freshmen in these Hindoo demolitions may be inferred from a passage, printed five years ago, which we now take the liberty of repeating, with its Italian preface:—

"Cadono le città, cadono i regni,
E l'uom d'esser mortal par che si sdegni!" (230)

"That the peninsula of Hindostan, thronged with varied populations, possessed great Empires and a high state of culture, in ages parallel with the earliest monuments of Egypt and China, upon whose civilizations India exerted, and from which she experienced influences, in the flux and reflux of Humanity's progressive development, no one, *nisi imperitus*,

(226) *Philological Proofs of the Original Unity and Recent Origin of the Human Race;* London, 1846; pp. 131–133. For "Celto-mania," this work out-Herods BETHAM'S! We can only observe with CHAMPOLLION (*L'Égypte sous les Pharaons*, 1814), of a *philologist* who derived the Greek name of Egypt from the *Gaelic* dialects of Lower Brittany—"Certainly, even admitting that the Greeks spoke *Bas-bréton*, there is some distance from AIGUPTOS to *Écou-é-vet*."

(227) PRICHARD: *Egyptian Mythology;* 1819; p. 35, *seq.*;—HEEREN: *Hist. Res., Indian Nations.*

(228) WILSON: *History of British India;* 1840; "Chronology and History of the Hindus;" i., book 2, ch. 1, pp. 163–169.

(229) Author of the "Buddhist Pali Historical Annals of Ceylon," called *Mahawanso*, "Royal Chronicles"; compiled from earlier sources in A. D. 302: if not *later*.

(230) METASTASIO: paraphrase of *S. Sulpicius's Letter to Cicero;* epist. v. lib. 4. The second line has been latterly rhymed—"E nel cader un c*****n par che si sdegni." The English is—"Cities fall, kingdoms fall; and (yet) man seems to scorn that he is mortal!"

will deny: but the hallucinations about early Brahmanical science in Astronomy, when their Zodiacs are *Greek*, their Eclipses calculated *backwards*, and their fabulous chronology is built upon *Chaldean* magianism, leave the historical antiquity of India prostrate beneath the axe of the *short*-chronologist. 'Un astronomo può, se vuole, far le tavole dell'ecclissi che avranno luogo di quì a cento-mila anni, se il mondo esisterà; e può ugualmente determinare lo stato, nel quale sarebbesi trovato il cielo centomil'anni fa, se il mondo esisteva:' (Testa, 'Dissertazione sopra due Zodiaci,' &c.; Roma, 1803, p. 23.) The Hindoos, in concocting their primeval chronology, merely added a *naught* to Babylonish cyclic reckonings;—4,320,000 years, instead of 432,000! (De Brotonne, 'Filiations des Peuples,' 1837; vol. i., pages 234 to 251, and 414.) See ample confirmations of the above view in the critical work of Wilson ('Ariana Antiqua,' 1841; pages 17, 21, 24, 419; 44, 45; and particularly page 439, wherein it is shown, that numismatic studies cease to throw light on Indian antiquities about the middle of the third century B. C.").

"When, therefore, the contenders for the ante-diluvian remoteness of the *forty-eight*-lettered *Sanscrit* Alphabet can produce any *stone*, or other record older than the 'column of *Allahabad* in honor of TCHANDRA-GOUPTA, *Sandracottus*,' cotemporary with SELEUCUS NICATOR, B. C. 315, it will be time enough for Hierologists, Sinologists, Hellenists and Hebraists, to take into account the pseudo-antiquity of *Sanscrit Alphabetical* literature." (231)

Our profession of faith in these matters, identical with the doctrines we hold at this day, shocked some literary prejudices. Nevertheless, it was based upon tolerably extensive perusal of works on Hindoo antiquities; and it is supported by the cuts and thrusts of a swordsman, whose trenchant blade, notched on the battle-fields of Hindostan, still preserves its keenness amid the bloodless strifes of archæological polemics — Lieut. Col. Sykes. (232)

From his matchless overthrow of European superstitions, in regard to Indian antiquity, we have already extracted two paragraphs containing the decisions of Wilson and Turnour. We now condense his own applications of cold steel to some of the vitalities of Hindostanic pretension.

There exists but *one* Sanscrit composition that can be called "history;" viz. the *Raja Taringini*, compiled A. D. 1148. It contains anachronisms of 796, and of 1048 years! Prior to the fifth century after C., "inscriptions in *pure Sanscrit* are entirely wanting" — the *earliest* Sanscrit inscription ascends to the fourth century, but it is impure in language and not orthographic. Between the tenth and seventeenth centuries of our era, Sanscrit inscriptions "roll in thousands!" The very Sanscrit *language*, in the polished form in which its literature reaches us, can no more be found *monumentally* in India, before the fifth century after C., than the English of Byron could appear in the days of Gower or Chaucer. In consequence, those Germanic writers who, in their assimilations (which are positive enough) of Greek, Latin, German, or other Indo-European idiom, forget that *Sanscrit* has undergone even greater transmutations than our Saxon vernacular has since the reign of Alfred, often commit philological oversights of sublime magnitude!

"Why are there not," asks Sykes, "the same tangible and irrefragable proofs extant of the Sanscrit as of the *Pali* language: the more particularly so as Brahmanism and Sanscrit have hitherto been believed to emanate from the fabled ages?"

Commencing his deep researches with the more recent Sanscrit inscriptions, and tracing them backwards as far as they recede, Prinsep (233) resolved the modern forty-eight *Deva-Nagari* characters absolutely into the primitive letters of the *old* inscriptions written in the "Lat" character and *Pali* language — the rencontre of graphical forms that approximated to the ancient *Pali* type increasing exactly in the ratio of the antiquity of each Sanscrit inscription. Of these last, the most *ancient* known dates A. D. 309; being just 624 years posterior to the *oldest Pali* inscription discovered throughout the Hindostanic peninsula!

Now, this oldest *Pali* inscription is found on the "column of *Allahabad*," whereupon it

---

(231) *Otia Æg.;* p. 110, and note.

(232) "Notes on the Religious, Moral, and Political State of Ancient India before the Mohammedan Invasion" —*Jour. R. Asiatic Soc.;* London, 1841; vol. vi. pp. 248–484.

(233) *Journal Asiatic Soc. of Bengal;* 1834–'41. Conf. *Jour. R. Asiatic Soc.*, 1853; xv. part i. p. xxv; for "Nassik Inscriptions," the date of the *cave* being only A. D. 338! Also, concerning *Arian* superpositions upon a dark autocthonous population of Hindostan, Gen. BRIGGS's Lecture "On the *Aboriginal* Race of India;" reported in London *Literary Gazette*, July 17, 1852.

was chiselled in the reign of Tchandra-Gupta, who is the *Sandracottus* of Greek history, coetaneous with Seleucus Nicator in the year B. C. 315. All India affords nothing, written *alphabetically*, more ancient; and this age is 220 years later than the alphabetic cuneiform of Persepolis; or above 300 years after the Greeks had already adopted the *Aleph* (alpha), *Beth* (beta), *Gimel* (gamma), *Daleth* (delta), of the anterior *Phœnician* alphabet! The identification of "Sandracottus" is moreover proved by the next early inscriptions known in the *Pali* tongue; viz.: two edicts of PISADASI-*Asoka*, a king of India in the year B. C. 247; who refers to his contemporary ANTIOCHUS *the Great;* just 62 years after the *oldest* inscription, whose epoch stands parallel with SELEUCUS. Thus, palæographically, the antiquity of India has fallen, never to rise again: and, inasmuch as the Brahmans certainly stole their *Zodiac* from the post-Macedonian Greeks; and probably some Levitical ceremonials of MANOU from Jewish exiles; there is no reason whatever, yet published, against our theory, that *alphabetic* writing also reached Hindostan, through *Arian* channels, from those Semitic streams the source of which is now irrevocably traced back to HAMITIC *origines* in Egypt.

"All those ancient systems of Persic writing with which we are acquainted, although applied to Arian dialects, are obviously formed on a *Semitic* model. I may notice, in chronological succession, the writing on the Cilician Darics; the Arianian alphabet (of which the earliest certain specimen is the transcript of the Edicts of Asoka), with its derivatives, the numismatic Bactrian, and the character of the Buddhist topes; the Zend; the Parthian; exhibiting in the inscriptions of Persia at least three varieties; and the Pehlevi, lapidary, numismatic and cursive. These several branches of Palæography are all more or less connected. (234)

Thus much to justify our table. But, "Titius or Sempronius" exclaims, have we not the *Sanscrit Vedas*, the Epics *Mahabharata* and *Ramayana*, the "Laws of MANOU," and the *Puranas?* Did not Sir William Jones fix the age of the Vedas in the fifteenth century B. C.; that of the "Institutes of Menu" in the twelfth? (235) Were not similar opinions held by Colebrooke and Schlegel; and are they not supported by great Indianists of our own time? Conceded, gentlemen. Knowing nothing of Sanscrit ourselves, we are as little able to speak decisively as those *littérateurs* who will be most startled at our audacities. Linguistically, there are not twenty-five men in the world whose judgment, matured by comparative archæology, is really authoritative in this discussion. In the meanwhile, *palæographical* facts speak intelligibly to all educated minds. We might add that Professor Wilson thinks the *Vedas* may, in part, ascend almost to the sixth century B. C.: but Sykes's sabre is not wanting in our defence; so let us continue.

In the first place, it is historical, that the Brahmans, in their efforts to destroy Buddhism, dealt, by the *ancient* texts of Hindoo treatises on religion or traditions, precisely as the Inquisition did with Hebrew Scriptures that existed before the tenth century of our era— *i. e.*, destroyed them. In the second, two Chinese travellers in India—Fa-hian, in the fourth century, and Hiuan-thsang, in the seventh after Christ — have (unfortunately for Brahmanical respectability) chronicled how, in this interval of three hundred years, the disciples of Brahma had expanded, from an incipient bud, into that detestable flower in which *Sanscrit* literature portrays them—ever noxious as *Upas* blossoms. (236) Their accounts are confirmed by the Chinese encyclopædist, Ma-touan-lin; (237) who registers that, about 502 A. D., the *Brahmans* were but a small sept among the Buddhists — "first among the tribes of *barbarians*." It may also be mentioned that, in the time of BUDDHA, sixth century B. C., the Hindoo population was classed already into those four grand divisions which attest, as

(234) RAWLINSON: *Behistun;* part i. pp. 43–44.

(235) We have recently re-read most of Sir W. JONES'S Papers with increased reverence: for his immense erudition qualifies all dogmatic opinions attributed to him with "*ifs*" of his own. Before us lie PAUTHIER'S *Livres Sacrés de l'Orient;* 1843: also MUNK: *Réflexions sur le Culte des Anciens Hébreux;* 1833; wherein the fifth book of MANOU is compared with *Leviticus;*—and other Sanscrit commentators "quos recensere supervacaneum esset." We have read BURNOUF: *Boudhisme*, and *Yaçna;* and nothing therein opposes, while much justifies, our view.

(236) REMUSAT; *Mélanges Asiatiques.*

(237) PAUTHIER: *Chine;* p. 381.

Pauthier well remarks, (238) "the diversity of races conquering and vanquished at a very early epoch;" viz: *Brahmans*, priests; *Kchatriyas*, soldiers; *Vaisyas*, tradesmen; and *Soudras*, serviles: (239) but the Chinese Fa-hian shows how, even in the fourth century after C., these divisions were merely *civil*, and not yet religious ordinances. In short, it is now certain that the "*caste*-system,"(240) which (it is likewise thoroughly established) was never known in Egypt, had not been invented in Hindostan until Brahmanical superstitions obtained predominance long after the Christian era. So again with respect to most of those prohibitions of animal sustenance, and other "unclean things," which some have supposed that Moses learned from primeval gymnosophists. Forbidden, for practical hygienic motives, among Pharaonic priests, Pythagoræan philosophers, and among Israelitish no less than Mohammedan Arabians, *pork* was equally proscribed by MANOU: (241) "The regenerate man who knowingly may have eaten mushroom, domestic *hog*, garlic, wild-cock, onion, or leek, shall be degraded." Now, as Sykes inquires, if the laws of MANOU had been in existence prior to the Christian era, how came it that BUDDHA died of dysentery from eating *pork*, and that *hog's* flesh should have been the aliment of early Brahmanical ascetics?

When enthusiastic Indologists shall have explained away the above palæographical and historical objections, they will be at leisure to defend the alleged antiquity of the *Sanscrit* books themselves. Here is a little thing calculated, as Lanci writes, to "scaponire i grattacapi." (242)

The "Puranas" claim for RAMA a date something like 867,102 years before their compilation. Bentley fixed the poem *Ramayana*, by its intrinsic evidences, at A. D. 291: and Wilson, together with the best Sanscrit critics, determines the age of the earliest "Puranas" between the eighth and ninth century after Christ. Such being the facts, Sykes educes as follows.

Sir W. Jones (Preface to the *Institutes of Menu*), assumed "that the *Vedas* must therefore have been written three hundred years before the Institutes of Menu, and these Institutes three hundred years before the Puranas." Then, Sykes's deadly sword gives point—as Wilson has proved, from internal evidence, that the "*Puranas* were written or compiled between the eighth and fourteenth centuries of the Christian era, it follows, according to Sir W. Jones's hypothesis, that the *Institutes of Menu* date from the fifth century (*Annis* D), and the *Vedas* from the second century." *Monumental* calligraphy supports this view; while the *Vishnu* Purana (dated by Wilson at A. D. 954) brings the polished Sanscrit language down as late as the tenth century. Analogy also, in adjacent countries, points to the same solution as to how Lamaism and Romanism present such striking identities. It is said by Father Georgi that "Writing, laws and religion were introduced into Thibet about the year 65 after Christ."(243) Thus, we learn that *Thibetan* pretensions, which have more affinity with those of Hindostan than of China, lend no support to Hindoo antiquity.

The geographical *names* in Hindoo literature wofully invalidate the antiquity of some books: because, if the mention of "Yavanas" (*Ionians*, IUNîm in Hebrew and in Assyrian cuneiform, *Yoonân* in Arabic, and YUNIN in old Egyptian), does not positively prove a writer posterior to ALEXANDER, B. C. 330; that of "Tchinas" (inasmuch as the Celestial Empire was not called *Thsin*, China, before the year 250 B. C.), at once knocks down a book to times *after* that era.(244) So again, as Indo-Scythians did not penetrate into India before B. C. 125, allusion to the *Sakas* must proceed from an author who lived subsequently. Now, the *Ramayana* and the *Mahabharata* both speak of "Yavanas, Tchinas, and Sakas;" and ergo, the latter cannot well be older (aside from other reasons) than the

---

(238) *Lois de Manou;* Introd.: p. 22.

(239) *Id.;* book i., sloka 31.

(240) GLIDDON: *Otia:* p. 90.

(241) Book v. 19:—The reason why neither Judaism nor Islàmism ever made progress in *China* is owing to its inhabitants' fondness for little *pigs*. The same tastes render either religion utterly impossible at *Cincinnati*.

(242) "*Remove the obstinacy of head-scratchers.*"

(243) *Alphabetum Tibetanum;* apud DE BROTONNE: *Filiations;* i. p. 445.

(244) The fleets of HOANG-TI first visited the ports of Bengal about the year 280 B. C. (*Chine*, p. 2).

second century after Christ, nor the former earlier than the fifth; in no case can either antedate B. C. 250. But, wildly shriek our Brahmanists — the grottos of *Ellora*, *Elephanta*, *Adjunta*, &c.? Alas, gentlemen — Sykes says, not one antedates the ninth century after Christ! Even Prichard, following Prinsep, does not consider these caves earlier than "a century or two prior to the Christian era, when Buddhism flourished in the height of its glory from Kashmir to Ceylon." (245)

We delude ourselves, probably, with the belief that our opponents in biblical studies will concede that, in our hands, the knife of criticism is double-edged; and that we apply it equally to the notions of Hindoo as well as of Judæan commentators. In the last century it was the fashion to exalt *Sanscrit* literature at the expense of Jewish; greatly to the discomfort of orthodoxy. The latter may now console itself with the assurance, that its Hindostanic apprehensions were puerile — for, beneath the most ruthless scalpel, a "Book of the *Law* of MOSES" stands erect with vitality, in the sixth century B. C.; that is, 200 years before the oldest *Pali* document of India was inscribed by CHANDRAGUPTA.

With the judicious reflections of another *Sanscrit* authority we take leave of Hindostan; merely mentioning that our own analysis of Xth Genesis has entirely confirmed the doctrine broached by the learned Col. Vans Kennedy. (246)

> "Although I do not derive all the nations of the earth from *Shem*, *Ham*, and *Japhet*, I still think that Babylonia [we read, ARIANA] was the original seat of the Sanscrit language and of Sanscrit literature. . . . But this error [i. e. the contrary hypothesis] necessarily proceeds from the assumption, that the first *eleven* chapters of Genesis give an authentic account of the creation and of the earlier ages of the world; which renders it necessary to insult common sense, and to disregard the plainest principles of evidence and reasoning, in order to prove that all the races of mankind and all systems of polytheism were derived from one and the same origin."

Those who have leaned upon Faber's broken reed would do well to peruse our author's *Appendix* — "Remarks on the Papers of Lieut. Col. Wilford contained in the Asiatic Researches." To others it may be satisfactory to know, that the earliest Greek mention of *India* (Sind) occurs in Æschylus, B. C. 525–456: while, about the same times (if Ezra compiled the "Book of Genesis," as patristic authority sustained), tradition — which, in our version (*Gen.* iv. 16), sends Cain into "the land of *Nod*, on the east of Eden" — probably consecrated some legendary rumor that the forlorn outcast had escaped to the *Hindus* — "*hi*NUD, towards the *East* of Eden," itself located in Mesopotamia; which Indian people are still called HINooD, by the Arabs. (247) *India* became known to Jews and Greeks after the former had been captive in Babylonia, and after the Persian invasions had given new ideas upon Asiatic geography to the latter.

FIG. 360.

Intending to publish other justifications of the correctness of our *Tableau* [*supra*, pp. 630, 631] on some future occasion, we suspend further discussion of the "*Semitic* streams," and merely submit specimens of that character upon which we have bestowed the name of "Assyro-Phœnician." If, as Dr. Layard states, some of these relics were positively found in the "chamber of records" opened by him at Kouyunjik, (248) and if, as he declares, they are really of the time of Sennacherib, B. C. 703 to 690, the reader beholds the very earliest known samples of *purely-alphabetic* writing hitherto discovered. They will become the more precious to his eyes, inasmuch as (in the contingency that Dr. Layard is certain that Fig. 360 belongs to Sennacherib's reign) here is the closest approximation to that (unknown) character in which the *oldest Hebrew books* of the Bible were originally written: which fact we shall demonstrate elsewhere. For

FIG. 361.

(245) *Researches;* 1844; iv. pp. 120, 121.

(246) *Researches into the Nature and Affinity of Ancient and Hindu Mythology;* 1831; pp. 368, 369; also pp. 406–422.

(247) MUNK: *Palestine;* p. 429.

(248) *Babylon;* 2d Exped., 1843; pp. 346, 591, 601, 606.

FIG. 362.

fear of misapprehensions, let us also note that the above *ancient* characters are entirely distinct in age from those on the *modern* and rabbinical "Bowls" (249) from Babylonia which Mr. Ellis's remarks might lead others than archæologists to invest with the halo of antiquity. They cannot attain even to the *third century after* C.; and, indeed, may descend to days after the Mohammedan conquests. Until we can resume the subject, the reader will find a place assigned to them in our Table under the heading of "Hebrew *Babylonish.*"

2d. MONGOLIAN ORIGIN. — We give this designation to a system of writings distinct organically, chronologically, linguistically, geographically, palæographically, ethnologically — in short, *aboriginally* — from any affinity with *Semitic* streams, or with the latter's common HAMITIC source. To comprehend us, the reader need but open the works of Pauthier; (250) without perplexing himself with other definitions, until he finds the former inconsistent with science, history, reason, and probability.

It is, however, from his *Sinico-Ægyptiaca* that the principles and examples of our author's critical results must be gathered; and, having advocated them on a former occasion, (251) we return to them with pleasure increased by subsequent verifications of their accuracy.

PAUTHIER'S THREE AGES OF WRITINGS.

"1st AGE. — The *figured* representation of objects and ideas; otherwise the *pictorial* age.

"Of this age we possess nothing that can be safely referred to primeval antiquity. All barbarous nations, like the tribes of North America, still strive to perpetuate their simple traditions by *pictures.*

"To this age, with a probable infusion of the *symbolical* element (although, as yet, whether of their lost languages, undeciphered writings, or chronology, it may be said that we literally *know* nothing), may perhaps be referred the *pictures* and so-called *hieroglyphs* of the ante-Columbian monuments of Mexico, Central America, and Peru.

"2d AGE.—The *altered* and *conventional* representation of objects; otherwise the *transition*-period; when the pictorial signs pass into the *symbolical,* and thence gradually into the syllabico-*phonetic.*

"To this age belong the *ideographic writings* of the Chinese secondary period, classified as follows: (252) 1st. — HIGH ANTIQUITY; B. C. 2637 to 3369 — according to the *Chinese* annalists, the KOU-WEN, or *antique writing.* 2d. — MEDIUM ANTIQUITY; B. C. 820 — the TA-TCHOUAN, or *altered image of objects.* 3d. — LOW ANTIQUITY; B. C. 227 — the SIAO-TCHOUAN, or *image still more altered of objects.* 4th.—MODERN TIMES; B. C. 200 to A. D. 1123, and still in use—*four* kinds of current writing and typography.

"The above are formed upon principles presenting some few analogies, but in the main remarkable differences, when compared with the Egyptian *phonetic* system.(253) Under the same age may be classed the *hieroglyphical* and *hieratic* system of Egypt, the latter being a tachygraphy or *short-hand* of the former.

"Albeit that we have but very vague data in this respect, it is exceedingly probable that all writings began by being *figurative* and *syllabic* before they became purely *alphabetical.* Many alphabets, such as the *Sanscrit* alphabet, the *Ethiopic* alphabet, the *Persepolitan* (without speaking of the *Japanese* and *Corœan* alphabets), are still almost completely *syllabic,* and bear evident traces of a *figurative* origin.(254)

"3d AGE.—The purely-*phonetic* expression of the articulations of the human voice: otherwise the strictly *alphabetical* age; to which belong all writings which represent no more than the vocal elements of human articulations, reduced to their simplest expression; i. e., A, B, C, D, &c.

---

(249) *Op. cit.;* pp. 509–526; figs. 1, 3, 5, 6.

(250) 1st. *Sinico-Ægyptiaca* — Essai sur l'Origine et la Formation *Similaire* des Écritures Figuratives *Chinoise* et Égyptienne; Paris, 1842. 2d. *Systèmes d'Écritures Orientales et Occidentales;* 1838. 3d. *Chine Ancienne,* d'après les documents *Chinois;* 1837. 4th. *Civilisation Chinoise* — containing the Chinese Books, CHOU-KING, Y-KING, TA-HIO, TCHOUNG-YOUNG, LUN-YU, and MENG-TSEU; 1843.

(251) *Otia;* pp. 100–102.

(252) PAUTHIER: *Sinico-Ægyp.;* p. 24.

(253) *Op. cit.:* pp. 98 to 110.

(254) *Op. cit.:* p. 34; and on each alphabet, consult his "Orig. des Alphabets," *passim.*

"To this belong the *Enchorial*, *Demotic*, or *Epistolographic* characters of Egypt, detached from occasional figurative and symbolical signs."

Nothing to the student of Pauthier's work can be more clear than that the primeval type of Mongol man, whose centre of creation lies along the banks of the *Hoang-ho*, and that other (organically distinct) Hamitic type whose centre is the *Nile*, after each one in its own region had passed through all preliminary phases of its individual development, reached, at an age on either side equally beyond *traditions*, the power of recording things by *pictures;* just as the American Indian around us, spurning every inducement to profit by our graphical art, still traces on the bark of trees, on rocks, on buffalo-robes, those rude designs whereby he hopes to annihilate space and time in the transmission of his thoughts.

If it be granted that an Egyptian, or a Chinese, could singly arrive at the discovery of this the humblest stage of letters for himself, why refuse the same capacities to the other? One nation of the two, at least, must have discovered this pictorial art for itself, most certainly: how then attribute tuition of another world of man to either, when the graphical systems of both are radically different?

Nearly a century ago, after applying vigorous strictures to the theories of Needham and De Guignes (we might add Kircher, De Pauw, Paravey, Wiseman, indeed orthodoxy generally), who claimed that either China taught Egypt, or Egypt China, Bishop Warburton thus emphatically placed the question in its only philosophical light:—

"To conclude, the learned world abounds with discoveries of this kind. They have all one common original; the old inveterate error; that a similitude of customs and manners, amongst the various tribes of mankind the most remote from one another, must needs arise from some communication. Whereas *human nature*, without any help, will, in the same circumstances, always exhibit the same appearances." (255)

How, it may be asked, do we know that the *pictorial* was the first, or rather the anterior, age of writing in Egypt, or in China? Aside from all arguments of analogy that *pictures* are the rudimental writings of semi-barbarism at this day—already a vast step higher than the savage *Bosjesman*, *Papuan*, or *Patagonian*, has ever attained—it is proved, in Egyptian hieroglyphics of the most ancient and pure style,(256) by their being, as far as perfection of sculpture and vivid coloring can make each thing, the exact representatives of natural and artificial objects, every one indigenous in nature to the *valley of the Nile:* and utterly foreign elsewhere. In China, the pictorial epoch is reached by tracing backwards each mutation of characters, age by age, to the primitive Kou-wen; which is a tachygraph, or abridgement, of natural or artificial productions, all autocthonous to the region of the *Hoang-ho.*

Of course, copies however rude of the same things must present certain identities, whether delineated in China, Egypt, or America; but just as a parent instinctively detects which of his children has scrawled a given form; or that a man betrays to others his individuality by his handwriting; so archæological practice enables an observer to point out the distinctive peculiarities of a given people's designs. The latter, moreover, tell whence they came by the very subjects figured. Thus, if, in a series of characters called "*Egyptian* of the IVth Memphite dynasty," a *camel*, a *horse*, a *cock*, were designed, the presence of either of these animals would prove the document to be a forgery; because camels, horses, and cocks, were unknown in the valley of the *Nile* for a thousand and more years later. In China, *cocks* and *horses* (257) were indigenous, like the silkworm, from the commencement of creation in this geological period; but, in her primitive pictures, there are no Egyptian *ibises*, nor *papyrus*-plants. No *rattlesnakes*, *magnolias*, or *bisons*, can be discovered in

(255) *The Divine Legation of Moses demonstrated;* 1766; 5th ed.; iii. p. 99.

(256) Lepsius: *Denkmäler;* for illustrations.

(257) There seems to be some doubt about the *horse* in China proper at an early period, because, about B. C. 900, this animal was imported from *Tartary* (*Chine*, p. 100). Nevertheless, Fo-hi is said to have taught his people to raise the six domestic animals—*horse, ox, fowl, pig, dog,* and *sheep:* and under the three mythical "Hoangs," his antecedents, there was a period of time called the *horse* (Pauthier: *Temps Antérieurs au Chou-king;* Liv. Sac.; pp. 20, 33). We cite the *pictorial* horse merely by way of popular illustration.

the pictures of China, or of Egypt, because these things are indigenous to the American continent — until COLUMBUS, segregated from the entire Old World: neither will the Grecian *acanthus*, the African *lion*, or the Asiatic *elephant*, appear in the sculptures of Yucatan or Guatemala; simply because, to American man, these objects were unknown. Each centre of creation furnished to the human being created for it the models of his incipient designs. It was materially impossible for him, without *intercourse* with other centres, to be acquainted with things alien to the horizon of his nativity. An *ornithorhynchus*, or a *kangaroo*, if found in a picture, would establish — 1st, that such picture could not be Egyptian, Chinese, or American; and 2d, that it was made within the last two centuries — that is, since the discovery of Australia by European navigators. Payne Knight laid down the rules:—

> "The similitude of these allegorical and symbolical fictions with each other, in every part of the world, is no proof of their having been derived, any more than the primitive notions which they signify, from any one particular people; for as the organs of sense and principles of intellect are the same in all mankind, they would all naturally form similar ideas from similar objects; and employ similar signs to express them, so long as natural and not conventional signs were used. . . . The only certain proof of plagiary or borrowing is where the animal or vegetable productions of one climate are employed as symbols by the inhabitants of another. . . . As commercial communication, however, became more free and intimate, particular symbols might have been adopted from one people by another without any common origin or even connexion of general principles." (258)

These few remarks suffice as suggestives, to the thoughtful and educated, of the radical distinctions which the first glance perceives when comparing the ancient sculptures of three aboriginal worlds of art, Egyptian, Chinese, or American. But, just as a physician's writings presuppose that his readers have passed beyond the elementary schoolroom, so it is not in "Types of Mankind" that any one need expect to find an archæological "Primer."

We return to the *ante-monumental* pictures of the Nile and the Hoang-ho — the former, long anterior to B. C. 3500; the latter, to B. C. 2300; being the *minimum* distance from our generation at which the graphical system of each river's denizens first dawns upon our view.

Impelled by the same human wants, though absolutely without inter-communication, the Mongol Chinese for his part, and the Hamitic Egyptian for his, attained, at periods unknown, the power of representing their several thoughts *pictorially*. Where they copied the same universal things — the *sun*, a *star*, a *goat*, a *pigeon*, a *snake*, a *tree* (though here even, in Flora and Fauna, already the two countries exhibit distinct "species"),— those copies necessarily resemble each other; although, in each, art betrays the individualities of a separate human type. Where the Chinaman, however, portrays a man, that man is a *Mongol:* where the Egyptian draws a human being, that being is an *Egyptian.*

No stronger exemplification of human inability to conceive that which is beyond the circumference of local experiences, can be met with, than in Squier's exhumations from the primeval mounds of the West. (259) Not merely is the *skull*, divested by time of its animal matter, osteologically identical with those of American Aborigines of this day; not only does every fragmentary relic which accompanies it limit that antique man's boundaries of knowledge to a space longitudinally between Lake Superior and the Gulf of Mexico, and laterally within the Alleghanian and the Rocky Mountains; — but, every *pipe-bowl*, or engraved article, that bears a human likeness, portrays an American *Indian*, and no other type: because man can imitate only what he knows. And finally, to bring the case home to our biblical researches, does not every line of the first *nine* chapters of Genesis prove that Hebrew writers never conceived, in speculation upon creative origines, anything alien to *themselves* and to their own restricted sphere of geography? At their point of view, the first *pair* of human beings conversed, at once, in pure *Hebrew:* — nay, the Talmudic books

---

(258) R. PAYNE KNIGHT: *Inquiry into the Symbolical Language of Ancient Art and Mythology;* Valpy's 8vo ed., 1818; par. 230, 231.

(259) *Ancient Monuments of the Mississippi Valley;* 1848: compare *wood-cuts*, pp. 194, 244–251.

show, that this divine tongue is to be the future language; the speech in which the "ultima ratio" will be meted out to all humanity in heaven!

"Concludam . . . verbis Rabbi Jehosuæ in *Talmud.* qui cuidam curiosè percontanti de statu resurgentium ad vitam æternam respondat, *Quando reviviscemus, cognoscemus qualis futurus sit eorum status.* Sic de futura lingua Beatorum in cœlis, quando reviviscemus, cognoscemus illam." (260)

Independently of one another, then, Mongolian man on the Hoang-ho, and Egyptian man on the Nile, each arrived for himself at *picture*-writing: yet, after casting a retrospective look at the relative epochas of both achievements, we behold that the difference between their chronological eras is almost as immense as when we, who in this day actually "*print* by lightning," see an Indian spend hours of lifetime in the effort to adorn a deer-skin with the uncouth record of his scalping exploits. At the time when Prince Mer-het (261) caused his sepulchre to be carved and painted with those exquisite hieroglyphs, that, through 16 *phonetic*, many *figurative*, and a few *symbolical* signs, relate his immediate descent from King Shoopho (262) builder of the mightiest mausoleum ever raised by human hand,— under the shadows of which great pyramid this (probably) son reposed: at that time, which, it is far more likely, ascends rather beyond than falls within the thirty-fifth century B. C., or 5400 years backward from our day — what was the state of civilization in China? Now, the most exacting of native *Chinese* archæologists will confess that their first Emperor Fo-hi (whose *name* emblematizes to the Chinese mind above 1000 years of meta-history, as that of Moses did to the Hebrew intellect in the age of *Hilkiah* the high-priest),(263) that this Fo-hi —*inventor* of writing,(264) through the legendary "8 *koua*" — scarcely floats upon the foam of tradition's loftiest surge: because, no Chinese scholar claims for Fo-hi's semi-mythical reign a date earlier than B. C. 3468; while conceding that perhaps it may have begun 600 years later.

And, if we compare *monuments*, then the *oldest* (265) written record of China claims no higher date than the "Inscription of YU," estimated at B. C. 2278 — being above 1000 years posterior to the Egyptian tomb of Mer-het, now in the Royal Museum of Berlin. All earlier Chinese documents being lost, the times anterior to Yu are, *palæographically*, blanks; but skepticism (scientific, not, the most obdurate, theological,) has no more reason to reject what of rational story pierces through the gloom of generations preceding, as concerns China, than we have to consider fabulous the British periods of the *Heptarchy*, although we cannot now individualize many events, and possess no Saxon "Saga" coeval with their occurrence.

A moment's pause will illustrate in what respect *Egypt's* monuments tower as loftily above Chinese antiquity, as *St. Peter's* at Rome above New York "Trinity Church." Our remarks are not directed to personages who, stifled beneath ante-metaphysical strata, read little and know less; but to readers who have perused, or will examine, the writings of at least Bunsen, Lepsius, Birch, and De Rougé; without disparagement of these scholars' ardent colleagues, too numerous for specification.

Whilst the pyramids and tombs of the IVth Memphite dynasty in Egypt stand, about B. C. 3500, at the uppermost terminus of that lengthy monumental chain — the coils of which, within a range of twenty miles, may still be unwound from Mohammed-Ali's mosque at Cairo, link by link, century by century, and stone by stone, back through all the vicissitudes of Nilotic annals, for 5400 years, till we touch the sepulchre of Prince Merhet — these pyramids, these tombs, themselves reveal infinite data upon ages to their construction long anterior; but, how long? Utterly unknown.

For instance, we here present the hieroglyphic for *scribe*, *writing*, or *to write*. It is compounded of the *reed*, calamus, or pen; the *ink*-bottle; and the scribe's *palette*, with two little cavities for his black and red inks. It may be seen

(260) WALTON: *Prolegomena*; ii. par. 25, p. 19.

(261) LEPSIUS: *Denkmäler*; and *supra*, p. 238; fig. 154.

(262) *Ibid.*; *Briefe aus Ægypten, Æthiopien*, &c.; Berlin, 1852; pp. 37, 38 — "Superintendent of all constructions of the king."

(263) About B. C. 625 — 2 *Kings* xxii. 8; 2 *Chron.* xxxiv. 14.

(264) PAUTHIER; *Chine*; pp. 24–26.

(265) *Ibid.*; p. 53.

81

on all monuments of the IVth Dynasty:(266) and its presence proves that *writing* must have been common enough in Egypt during ages antecedent. So again, here is  —a roll of *papyrus*-paper, a volume, tied with strings — meaning a "Book." Its presence upon the monuments, not merely of the XIIth, but of the VIth, and even of the same old IVth dynasty, establishes that the invention of *paper*, and the usage of written *volumes*, antedate the earliest hieroglyphics now extant.

It would require an especial treatise to convey to readers any adequate idea of the copiousness of ancient Egyptian documents written on *papyrus*-paper existing and deciphered at the present day. There are some of the IVth (B. C. 3400) and succeeding dynasties down to the XIIth B. C. 2200) in legible preservation; but the great "age of the Papyri" belongs to the XVIIth and following dynasties; (267) that is, from the 17th century B. C. downwards. Independently of the thousands of copies of the "Book of the Dead," there are *poems, account-books, contracts, decrees, chronological lists, histories, romances, scientific essays,* — in short, it is really more difficult now to define what there is not, than to catalogue the enormous collections of *Papyri*, some written ages before Moses's birth, existing in European cabinets. At foot we indicate where the curious inquirer may satisfy himself upon the accuracy of this statement. (268) And if he wishes to behold the *transitions* of Egyptian writing from the hieroglyphic into the hieratic, he need only open Lepsius's *Denkmäler.*(269) We have no space to enlarge upon these facts here, which the writer's *Lecture-rooms* have exhibited in most of the chief cities of the Union.

All which premised, as facts at this day open to everybody's verification, the reader comprehends that, if *picture*-writing, as well on the Nile as on the Hoang-ho, was the first stage towards *phonetic* orthography; nevertheless, according to *monumental* evidences, the Egyptians had already been inscribing their thoughts in perfect *hieroglyphics*, "sacred sculptured characters," a thousand years before the Chinese had perfected a system of *ideographics*, to us represented by their primitive character Kou-wen.

It is from Champollion's *Grammaire Egyptienne* (270) that the reader must draw clear definitions of Nilotic classifications into the *phonetic*, *figurative*, and *symbolical*, elements of calligraphy: and Mr. Birch's definition of Egypt's pristine 16 monosyllabic articulations— *a, b, f, g, h, i, k, m, n, p, r* × *l, s, t, sh, kh, u,* — is the most accessible to the English reader. (271) For *Chinese* analogies and discrepancies, as said before, there is no satisfactory work but the *Sinico-Ægyptiaca.*

Through their study the reader will glean how — starting both from the same springs, although chronologically and geographically distinct, viz., PICTURE-WRITING — the Egyptian rivulet, gushing forth naturally in one direction, formed the HIEROGLYPHICS; whence, in due time, through *Semitish* channels, streamed those mighty rivers that, from Chaldea, have watered Europe, Hindostan, Northern Asia, Africa, America, and Australia, with the refreshing rills of *Phœnicia's* alphabet: and how the Chinese fountain, its waters taking an opposite direction, created the IDEOGRAPHICS; which, cramped within gutters artificially if ingeniously conceived, have enabled the Chinamen to attain a system, it is true, essentially phonetic, and which, originating in a Mongolian brain, suffices for all the necessities of Mongol articulations: notwithstanding that A B C are as alien to its complex construction as our English language is remote from the agglutinations of an Indian, or the "gluckings" of a Hottentot. The Chinese never have had an *alphabet*. It is impossible, without organic changes which human history does not sanction, that the Sinico-Mongol ever can possess that, to us the simplest, method of chronicling our thoughts.

---

(266) Lepsius: *Chronologie;* i. p. 33; — *Todtenbuch;* 1842; Pref. p. 17; — Bunsen: *Eg's Pl.;* i. p. 8.

(267) Hincks: *Trans. R. Irish Acad.;* 1846.

(268) *Select Papyri;* published by the British Museum; — Lepsius: *Chronologie;* i. pp. 39, 40; — Prisse, De Rougé, and Champollion-Figeac's papers, in the *Revue Archéologique;* — and Birch's in *Trans. R. Soc. Lit.*, and in the *Archæologia;* &c.

(269) *Abth.;* ii. bl. 98, 99.

(270) A synoptical sketch is in Gliddon: *Chapters;* 1843.

(271) Gliddon: *Otia;* pp. 113-115; but better in Lepsius: *Vorläufige Nachricht;* 1849; p. 35.

In consequence of which reflections, fortified by the physical deductions elsewhere embodied in "Types of Mankind," we have assigned to MONGOL-*origins* a distinct column in our theoretical Tableau of human palæographic history.

For the objects of anthropology, the above explanatory remarks would be sufficient, were not notions current among those readers, who look to theology for biblical criteria, to metaphysics for archæological—1st., that the "Chinese" are recorded in Scripture; and ergo, that *Mongolian* races were familiar to Jewish writers; 2d., that "Chinese vases" have been found in tombs of the XVIIIth dynasty at Thebes; and ergo, that Egypt and China were in positive communication about the time of Moses. (272) So we digress.

Once upon a time an adage prevailed in literary controversies—*Cave hominem unius libri.* Through what impairing causes is to us unknown, but certain it is, that in proportion as one ascends in English theological literature to the Kennicotts, Warburtons, Lowths, Cudworths, Stillingfleets, Waltons, and other intellectual giants of that deceased school, so one's respect for divines and one's reverence for Scripture augment. They had *one* book to study professionally, and that book they knew well; because they actually read it.

It would appear that there are cycles of deterioration, as evident in theology as in the weather, to judge by what took place in China about A. D. 1368; and inasmuch as our inquiries first concern the Chinese, it is but fair that they should open proceedings.

The Emperor Houng-Wou, appalled at the degradation of scholarship consequent upon the tragic events that preceded him, one day convoked the "Tribunals of Literature" (equivalent to the French Ministère d'Instruction Publique), (273) and made to them a common sense speech, the pith of which is here in extract:

> "The ancients," said he, "the ancients used to write but few books, but they made them good. . . . Our modern *litterati* write a great deal, and upon subjects that cannot be of the slightest real utility. . . . The ancients wrote with perspicacity, and their writings were suited to the comprehension of everybody.
>
> . . . In former times their works were read with pleasure, and one reads them at this day [A. D. 1368, in China!] with the same.
>
> . . . You [addressing himself to the Censors of the Press], you, who stand at the head of literature, make all your efforts to restore good *sense:* you will never succeed but by imitating the ancients. (274)

In the days between Walton and Kennicott, a theological student who might have ventured to opine that the Chinese are mentioned in the Bible, would have been sent incontinently to read the Hebrew text of Isaiah. (275) When this task was executed (and, formerly, divinity students could read a little Hebrew), the young man would have found a place on the lowest form, by command of the Professor of History, for ignorance of the rudiments of his class. Shame would soon have impelled an ingenuous youth, of those days gone by, to cram his head with simple facts of which some of his elders in theology now seem unaware. (276)

Chinese history—in this question the most valid—proves that, until the year 102 after Christ, the Chinese never knew of the existence of any countries situate north and west of Persia. Between the years 89–106 A. D., in the reign of Ho-Ti, a vast Chinese army, under General Kan-Ying, detached by the Commander-in-Chief, Pan-tchao, halted on the shores of the Caspian Sea; (277) receiving the submission of the Tad-jiks (Persians) and

---

(272) Vide GLIDDON'S IVth *Lecture*—reported in "Daily Dispatch," March 18; and in "Richmond Examiner," March 21; Richmond, Va., 1851. Also, more extensively, in "The Union," Washington, D. C., April 25, 1851. The abusive writers alluded to in that discourse, as

"Mere youths in science, and to fame unknown,"

were the reverend authors of "Unity of the Human Races," 1850; of an article in the *Princeton Review*, 1851; and of a third article, the one prelauded [*supra*, p. 587], as emanating from an Ass. of Min. at Col., S. C.

(273) ED. BIOT: *Essai sur l'Instruction publique en Chine;* 1846.

(274) PAUTHIER: *Chine d'après les Documents Chinois;* pp. 393, 394.

(275) ISAIAH; xlix. 12.

(276) Rev. THOMAS SMYTHE, D. D.: *Unity of the Human Races;* 1850; p. 43;—Rev. Dr. HOWE: *Southern Presbyterian Review;* Columbia, S. C., No. 3, Jan. 1851; &c.

(277) REMUSAT: *Mém. sur l'Extension de l'Empire Chin. du coté de l'Occident;*—PAUTHIER, *Chine;* pp. 258-260

of the *Asi* [*supra*, MaGUG, p. 471]. A powerful interest, however, incited these last to withhold correct information on western countries from the Chinese officer; viz.: that, hitherto, they had held the monopoly of the raw *silk* trade, by caravan, between China and the West; which silk, dyed and woven into then-priceless raiments by the Parthians, found its way occasionally to the grandees of Europe; and, on the other hand, one of the practical motives which carried Roman eagles to the Tigris, was a hope to discover the unknown source whence the crude material of these exquisite fabrics had reached Persia. It was during this, the most distant military expedition ever undertaken before Genghis-Khàn, that the Chinese heard, for the first time, of the existence, far west from the *Asi*, of the Roman Empire. Deterred from advance for its conquest by the discouraging report of the Parthians that his commissariat ought to be supplied for three years, the Chinese General renounced the enterprise, and returned to headquarters at Khotàn.

From the opposite direction, the arms of Rome had not been turned towards Persia until, about B. C. 53, Pro-Consul Crassus perished by Parthian arrows on the western frontier of Persia; some 155 years before the Chinese had penetrated to its south-eastern provinces. Within four years after the retrograde march of the Chinese armies, Parthia was invaded by Trajan, A. D. 106; and it was about that generation, a few years more or less, that the Romans first heard, through the Persians, of the remote country whence the silk came. (278) In A. D. 166, Antoninus sent the first Roman embassy to China; the hospitable reception of which is chronicled, by contemporary Chinese annalists, in the reign of their Emperor Houan-Ti.

No nations, then, situated to the north-west of Persia, so far as history or monuments relate, had ever heard of China; nor had the Chinese known anything about such nations until after the Christian era. Surmises to the contrary require, nowadays, to be justified by something more substantial than the *ipse dixit* of moderns, however erudite, whose opinions were formed before geographical criticism had fixed the boundaries of antique intercommunicational possibilities.

With this historical basis, let us take up the only word in the entire canon of Scripture, upon which living theologists have erected a fable, that the Chinese are mentioned in the Old Testament. Even king James's version suffices for this discussion: — "Behold these [the Jewish Babylonian exiles] shall come from far; and, lo, these from the north and from the west; and these from the land of *Sinim*." (279) "Our modern *litterati*," says the Emperor Houng-Wou, "write a great deal;" and sustain that *Sinîm* means the Chinese; because, after stripping away the Hebrew plural IM, there remains the word SIN; and the native name of China is THSIN.

Now, the whole context of the prophet refers to the return of the Jews from bondage in Babylonia. It must, therefore, be in Mesopotamian vicinities that the SIN*s*—"inhabitants of SIN;" or, otherwise, "cities, districts, localities of" SIN—should be sought for, before traversing Central Asia, in such impassable ages, to recall from China unknown Jewish fugitives who might have escaped thither from Babylonia.

The root SIN of Isaiah is not SINI; (280) and, furthermore, that SIN*ian* was a Canaanite. Nor is it either of the "wildernesses of SIN" familiar to the Mosaic Israelites; because the first, (281) spelt with the letter *sameq*, lay close to Egypt: and the second (282) was T*s*iN, near the Dead Sea. Far less could it have meant the Egyptian city of *Pelusium*; called Sin, (283) or dialectically T*h*IN, anciently, as *Teen* now by the Arabs. Why travel to China, when Mesopotamia itself offers to every eye, in an excellent map, (284) at the

(278) On "Sérica," and the fact that little or nothing was known about it by writers antecedent to Claudius Ptolemy, in the second century after Christ; compare the excellent *critique* of ANTHON, *Class. Dict.*, voce "Seres."

(279) ISAIAH: xlix. 12.

(280) *Genesis*; x. 17; *supra*, p. 531.

(281) *Exodus*; xvi. 1; xvii. 1.

(282) *Numbers*; xiii. 21; — *Deuteronomy*; xxxii. 51; &c.

(283) EZEKIEL: xxx. 15, 16.

(284) FRASER: *Mesopotamia*; 1841; — XENOPHON: *Anab.*; lib. ii. 4.

mouth of the river Lycus, the vestiges of a city termed *Kainai* by Greeks, *Cænæ* by Romans, and *Senn* by Arabians? Or, if it be absolutely necessary to obtain SINIM (more SINs than one), add to the preceding *Senn* the site of *Sina*, (285) about fifty miles northeastward of Mosul; together with the "large mounds" called *Sen*, on the banks of the Euphrates, opposite Dair.

One, or two, or all of these localities, amply suffice for the extremest points whence the Jews were to be summoned from captivity; and, singly or collectively, they are comprehended in the LXX translation; where *Sinîm* is paraphrased by *εκ γης Περσων* — "from a land of the Persians."

Aside from the obvious adaptation of these places, near the Euphrates or the Tigris, to the natural sway of Nebuchadnezzar who captured the Jews, no less than of Cyrus and Artaxerxes who released them; it is physically impossible, as well as unhistorical, that ancient Jews should have been expatriated to China: a country none of their descendants ever reached until centuries after the Christian era. (286) It is equally out of the question that the Septuagint translators could have known anything of China — a land beyond the horizon of Alexandrian knowledge previously to the time of Trajan, about a century after C.; or some 230 years after the various Hellenistic-Jews, called the LXX [*ubi supra*], had completed their labors. Indeed, they pretend to nothing of the kind; for they well knew that the SINIM were in the "land of the Persians;" while Orientalists of the present day always understand, with the Chaldee paraphrast, "from the southern country" of Assyria, in that passage. (287)

We forbear from reagitating here the question elsewhere treated, whether there were really "twelve tribes" of Israel before the times of Sennacherib; nor what became of the ten said to have remained — where? Some moderns (288) claim that these Israelites marched round by Behring's Straits into America; and, after building the cities of ancient Mexico and Peru, have run wild in our woods—in short, unaccountably become our Indians. Others have sought for them in Affghanistan; (289) although the portraits of Dost-Mohammed, Shah-Soojah, and their fierce cavaliers, are as little Jewish in lineaments as are their speech, and still more their bellicose habits: for the Bible shows that the Jews of Palestine, except under supernatural circumstances, were beaten and enslaved by any adjacent tribe that happened to covet their persons or property. If ever supposititious offshoots of the "ten tribes" wandered as far as Cabul, Bokhara, Balkh, or Samarcand, they were Jews at their migration, and Jews they would have remained in type and in religion, if certainly not in language. Wolff found his compatriots everywhere. Indeed, we know, personally and positively, that had the reverend renegade not been a true Hebrew, he could never have traversed Central Asia in 1832–'5. But he narrates that the fathers of those who kindly welcomed him, on the score of his inextinguishable Judaism, had established themselves in Affghan provinces very long after the fall of Jerusalem. We also know that Arabs (to the Abrahamidæ closely allied) settled in Persia, Khorassan, Balkh, &c., ever since the Muslim invasion, one thousand years ago, having rarely intermarried with Tartars, remain physiologically distinct to this day. Yet while they have preserved the name, religion, and appearance of Arabs, they have lost their Arabian language. (290) So it is with the Hebrew nation in every clime—indelibility of physical type, coupled with a most pliant faculty for change of tongue. If, then, exactly "ten tribes" of Israel were swept away into Chaldea, they did but return to their aboriginal centre of creation; and (mixing voluntarily with no type of mankind but their own) they have naturally disappeared amid the

(285) LAYARD: *Second Expedition, Babylon;* 1853; *Map of Journeys;* and p. 297

(286) About 60,000 Jews are reputed to be there now; others reached Malabar about A.D. 490; — See NOTT: *Phys. Hist. of the Jewish Race;* 1850; pp. 12, 13; and *supra*, pp. 117–123.

(287) CAHEN: *Bible;* ix. p. 176, note 12.

(288) DELAFIELD: *American Antiquities.*

(289) DUBEUX: *Afghanistan;* pp. 65, 66.

(290) MALCOLM: *History of Persia;* 1815; p. 277; — MORIER: *Second Journey through Persia;* 1818; i. pp. 47 48; — PICKERING: *Races;* 1848; p. 240.

waves of a homogeneous population. These opinions, long avowed by the authors, are confirmed by the views and new facts of Layard.(291)

But we finish with orthodoxy's "Chinese":—

From a previously small feod of the Celestial Gates, called *Thsin*, given by Hiao-Wang, about B. C. 909, to one of his jockeys, issued a line of princes whose constant acquisitiveness had enabled them, by the year B. C. 249, to incorporate a fifth part of the Chinese realm, and to extend over it their patronymic title of *Thsin*. Out of this stock sprung Thsin-Chi-Hoang-Ti, at once the Augustus and the Napoleon of China—founder of the fourth or *Thsin* dynasty, whose name signifies "the first absolute sovereign of the dynasty of *Thsin*." About B. C. 221, all the principalities of China were consolidated under his supreme sway; and, as a consequence, the name *Thsin* became, in common parlance, synonymous with the whole empire. Proud of his mighty exploits, although detesting the individual, the Chinese, from and after his day, adopting the word *Thsin* as typical of China itself, originated the Hindoo appellative "Tchina," whence we inherit our corrupt designation "China." Under these circumstances we tender to future sustainers of Chinese in Scripture a many-horned dilemma:—

Either the Prophet Isaiah (whose meaning is so naturally explained above) by the word SINIM does not refer to the Chinese, or inasmuch as the Chinese empire was not called *Thsin* previously to B. C. 221—which is about 450 years after Isaiah wrote—the verse 12 of chapter xlix of the book called "Isaiah" cannot possibly have been penned by Isaiah, but is the addition of some nameless interpolator: who must have lived, too, later than the first century after Christ, when the existence of China first became known, under its recent name *Thsin*, to nations dwelling west of the Euphrates. The writers called the "Seventy" knew nothing of this absurd Chinese attribution, as their "Land of the Persians" attests.

Were it not for them who thus had paraphrased SINIM between B. C. 260 and 130, the interpolation of a mere verse, after the year A. D. 100, in a prophetic book wherein whole chapters had been previously interpolated, would excite small surprise among biblical exegetists. "If, for example," writes the great Hebraist of the "Bibliothèque Impériale," (292) "in a prophetic book, bearing the name of Isaiah, they speak to you of the return from Babylonish exile; if they go so far as even to name Cyrus, who is posterior to Isaiah by about two centuries, be assured that it is not Isaiah who speaks." And if that explanation does not satisfy theological exigencies, then let some people bear in mind that the word SINIM occurs in the forty-ninth chapter of Isaiah; and that, according to the highest biblical critics of Germany, whose mouth-piece is the eminent Professor of Theology at Basle,(293) "the whole of the second part of the collection of oracles under Isaiah's name (xl.—lxvi.) is spurious." But they say Chinese vases have been found in tombs of the Mosaic age in Egypt; and, *ergo*, that China was known some 3300 years ago to the ancient Egyptians. The archæological interest of this alleged fact has been revived in the present year by two new phases:—

First. The presence at New York, among a variety of Egyptian antiquities, less authentic, of—

"No. 626.—A Chinese vase, with 17 others of different forms. All found in tombs. Some from Thebes; others from Sakharah and Ghizeh.

"These vases are curious, inasmuch as they prove the early communication between Egypt and China. *Vide* Rosoleni [*sic* for Rosellini]; Sir Gardner Wilkinson's Manners and Customs; Sir John Davis's Sketches of China, p. 72, and Revue Archœologique, by Mr. E. Prisse.

"No. 627.—A Chinese padlock, found in the tombs at Sakharah." (294)

This last *bijou* is a confirmation of ancient intercourse between Pharaonic Egypt and

(291) *Op. cit.;* pp. 373, 383–386.

(292) MUNK: *Palestine;* p. 420.

(293) DE WETTE: Parker's transl. ii. p. 336; and also HENNELL: *Origin of Christianity;* 1845; pp. 354, 355

(294) "*Catalogue of a Collection of Egyptian Antiquities,* the property of Henry Abbott, M. D., now exhibiting at the Stuyvesant Institute, No. 659, Broadway, New York"; 1853; p. 44.

China, of which orthodox navigation may well be proud, especially now that two additional vases have been discovered since Joseph Bonomi, in his sly way, indicated the extreme rarity of such antiques at Cairo, 1843.

"No. 254.—Padlock, Chinese, said to be found at Sakhara.

"No. 255.—Thirteen Chinese bottles, of the usual form, and with the inscription in the Chinese characters; and three bottles of different shape, found in Egyptian tombs, both in Upper Egypt and Sakhara. The larger portion of this collection was found in Sakhara. Bottles exactly similar may be purchased in the perfume bazaar of Cairo; and in 1842 the Jannissary of the Prussian Mission purchased ten of them." (295)

Second. The deterration of two similar Chinese vases by Layard, one from the mound of Arban, and another from its vicinity. These are the more precious as they show the orthodox and primeval overland route of Egypto-Chinese intercourse by way of Assyria, in ages preceding the discovery of the monsoons, about A. D. 45, by the Greek pilot Hippalus. (296)

"In a trench on the south side of the ruin, was found a small green and white bottle, inscribed with Chinese characters. A similar relic was brought to me from a barrow in the neighbourhood. Such bottles have been discovered in Egyptian tombs, and considerable doubt [not the remotest] exists as to their antiquity, and as to the date and manner of their importation into Egypt. (*Note.* —Wilkinson, in his 'Ancient Egyptians,' vol. iii. p. 107, gives a drawing of a bottle precisely similar to that described in the text, and mentions one which, according to Rosellini, had been discovered in a previously unopened tomb, believed to be of the eighteenth dynasty. But there appears to be considerable doubt on the subject.) The best opinion now is, that they are comparatively modern, and that they were brought by the Arabs, in the eighth or ninth century, from the kingdoms of the far East, with which they had at that period extensive commercial intercourse. Bottles precisely similar are still offered for sale at Cairo, and are used to hold the kohl or powder for staining the eyes of the ladies." (297)

Since the conquest of Algeria, Parisian naturalists have been constantly employed by the French Government to collect every specimen of natural history that region affords. One of these enthusiastic savans, lamenting that his predecessors had exhausted the resources of the country, was supplied by the Zouaves with sundry live examples of a wild rat, the species of which was entirely unknown at the Jardins des Plantes. The soldiers called it *rat à trompe.* On arrival of these novelties at the Museum, (298) it was perceived that each rat was adorned by a flexible and hairy proboscis. In time these appendages happening to drop off, some assistant ascertained that the malicious Zouaves had inserted an amputated tail of one species of rat into the nasal cartilage of another! It behooves archæologists, therefore, to view any such marvels as Sinico-Nilotic "padlocks" with more than caution; for, as De Longpérier, the Conservator of the Louvre Museum, writes to De Saulcy, Director of the Musée d'Artillerie, "above all things, now-a-days, *gardons nous des rats à trompe.*"

Chinese vases, of the genus mentioned, having been familiar things to the writer ever since his boyhood's visit to Cairo in 1823, no less than during his official residence there from 1831 to 1841, it was against his wishes (while aiding his revered friend Morton with a few hieroglyphical indices in 1842–3) that the following passage ever saw the light without some qualifying reservation: "That the Chinese had commercial intercourse with the Egyptians in very early times, is beyond question; for vessels of Chinese porcelain, with inscriptions in that language, have been repeatedly found in the Theban catacombs. (Wilkinson's *Ancient Egyptians*, vol. iii. p. 108.)" (299) But Dr. Morton relied upon the accuracy of Wilkinson, and the latter upon that of Rosellini, (300) as to the matters of fact; at the

(295) BONOMI: *Catalogue* of ditto: Cairo, 1846; pp. 25, 26, 35. [Printed in London. We saw its proof-sheets there.]

(296) PLINY: lib. vi. p. 26.

(297) *Babylon:* p. 279.

(298) Vide *Histoire Naturelle de MM. les Professeurs aux Jardins des Plantes:* 12mo, Paris, 1847.

(299) *Crania Ægyptiaca:* 1844; p. 63.

(300) Compare CHAMPOLLION-FIGEAC: *Égypte Ancienne:* 1840; voce "Nechao," p. 369; and *Notice sur deux Grammaires de la Langue Copte:* June, 1842; pp. 7-10. The perusal of these two *critiques* might benefit the author of *Horæ Ægyptiacæ.*

same time that, in the United States, there was no sinologist to whom we could refer the inscriptions themselves. Nor, indeed, was it until the writer studied at Paris, (301) in the winter of 1845–6, that appeal had ever been made from the learned opinion of Davis. (302)

In the letter cited at foot, the Chinese scholar defends his view against the "Quarterly," (February, 1835); which maintained that these vases could not have been found in ancient Egyptian tombs — that the supposition of their being so found depended upon hearsay; neither Lord Prudhoe, Mr. Wilkinson, nor Mrs. Bowen (quondam Mrs. Col. Light), having seen those specimens they had purchased at Coptos and Thebes, extracted from any ancient tomb. To repel which attack, Davis exhibits a letter from Rosellini to the effect, that he saw one withdrawn from an ancient tomb during the Tuscan excavations at Thebes, in 1828–9. And thus, the only archæological process of determining the vastly important fact of Pharaonic intercourse with China, so far as depended upon these vases, stood over until, at the writer's suggestion, and in his presence, *four* specimens were submitted by his valued colleague, Prisse, at the latter's apartments, to their mutual friend, the high sinologue, Pauthier. It is also desirable to note, that the question of the authenticity of these vases arose amongst us at Paris, in consequence of their forming a prominent feature in the "Notice" which M. Prisse was at that time preparing of the identical "Collection of M. H. Abbott;" (303) — a collection that, rejected by Europe, has "fata profugus" since been transferred, with the augmentation of a Chinese padlock, in 1852, from Egypt to New York. "Iisdem in armis fui;" although M. Prisse's own doubts first prompted him to consult the opinion of so old an Egyptian fellow-sojourner as the writer.

M. Prisse had already projected the substance of the following in manuscript:

"It is pretended that these little flasks have been found in Egyptian tombs; but as the fact is contestable, I think it useful to discuss it. Whenever an error is met with in your path, says Bacon, fail not to eradicate it, as a traveller cuts down a bramble in passing. I ought to strain myself the more to destroy this error that I have aided in its propagation, by coöperating in the 'Collection of Dr. Abbott,' and by giving to N. L'Hôte two of those little flasks for the Royal Museum of the Louvre, where they figure under the title of 'Vases Chinois trouvés dans les tombeaux de l'Egypte par MM. Champollion et L'Hôte.' Champollion had bought one of these little vases at Thebes (*Monuments de l'Egypte et de la Nubie*, Pl. 424, No. 28.) N. L'Hôte received from me the two others; and none of them, to my knowledge, had been found in an Egyptian tomb. Rosellini, the only one who pretends to have found a similar one himself (*Monumenti Civili*, vol. iii. p. 397), in a tomb of which he makes the epoch ascend to the XVIIIth dynasty, is not an author very worthy of credit. Sir G. Wilkinson (*Man. and Cust.*, iii. p. 108) believes that these little flasks which held perfumes, had been brought into Egypt by the commerce of India, with which country the ancient Egyptians appear to have been in relation from a very remote epoch: but he does not discuss the authenticity of these vases. Upon the testimony of these two authors, and upon that of the Arabs, I had believed for a long time that these flasks issued from the excavations, and I bought many that I gave away. Soon after, a traveller having assured me that he had seen similar vases at some ports of the Red Sea, (304) I began to conceive doubts. Pressed by questions, the Arabs avowed to me that the greater number of these vases came from Qous, from Qeft and from Qosseyr, successive entrepôts of Indian commerce. This avowal seemed to me peremptory."

It was here that M. Pauthier's call with the writer led opportunely to the sequel.

"Nevertheless, the stability of the arts in China might have caused repetitions of the forms of these vases from early centuries; and the nature of the characters employed in the inscription could alone remove all objection. I consulted at Paris two learned sinologists, MM Stanislas Julien and Pauthier, who assured me that the characters *thsao*, painted upon these vases, dated solely from the second century of our era. M. Pauthier has been pleased to indite a note upon this subject, which I hasten to publish in order to terminate the discussion."

From Pauthier's "Note upon the Chinese vases found in Egypt," we have condensed the

---

(301) PRISSE: *Recherches sur les legendes de SCKAI:* Revue Archéol., 1845; pp. 457-475, note.

(302) *Lettre à M. Bunsen sur les Vases Chinois trouvés dans d'Anciens Tombeaux:* translated from the English in *Annali dell' Instituto di Corr. Archeol. di Roma*, 1836; p. 322, *seq.*, and plate G.

(303) *Notice sur le Musée du Kaire, et sur les Collections Égyptiennes de MM. Abbott, Clot Bey, et Harris:* Revue Archéol., 15 Mars, 1846; tirage à part, pp. 3–28, and wood-cuts, pp. 18, 19.

(304) Compare PICKERING: *Races of Men and their Geographical Distribution:* 1848; p. 400.

subjoined. In his work, "The Chinese," under the article "Porcelain," Gov. J. F. Davis, of Hong-kong, refers to the exceptions taken by the Quarterly Review, citing Wilkinson and Rosellini for the fact of the discovery of such vases in Egyptian catacombs.

"M. Letronne, when giving account, in the Journal des Savans, (Nov. 1844, p. 665,) of the work of Mr. Wilkinson, thus expresses himself: 'The author believes in the Chinese origin of certain porcelain vases, found in the tombs at Thebes, of which one is of the XVIIIth dynasty. He gives the figures of four of these vases, with Chinese inscriptions, which Mr. Davis flatters himself with having read. We know that other sinologues doubt this origin. The fact deserves to be cleared up by a contradictory discussion. . . . There is nothing in it impossible, but it seems *little verisimilar*. . . . Yet, if these inscriptions are really *Chinese*, the fact must be accepted. All lies in that.'"

It is merely justice to Morton's memory here to remark that his "Crania Ægyptiaca" had appeared in the spring of 1844, at Philadelphia. Nor is his discrimination amenable, on questions alien to his special studies, to the charge of hastily adopting, in good faith, that which Parisian science had not begun to ventilate for six months later.

After stating that no sinologist doubted that these vases "are really and purely *Chinese*," M. Pauthier holds that all the question does "not lie in that;" and then eliminates the facts as follows:—

1. The inscriptions upon these vases are in the cursive Chinese character called *thsao.*
2. This cursive character was not invented in China until the second century after Christ. Hence "it is materially impossible that vases, bearing inscriptions in that writing, could have been manufactured and transported to Egypt in the time of the XVIIIth dynasty; that is to say, about 1800 years before the said epoch!"

   Gov. Davis, "well versed in the study of the vulgar Chinese (language), seems, like some other sinologues, to have completely neglected the study of Chinese archæology." Nevertheless, on the vase published by him (No. 4 of Wilkinson, and of M. Prisse), one reads easily:—
3. "*Ming youé soung tchoung tchao:* 'the brilliant moon is resplendent through the pines.'"
4. This is a line from a "strophe composed by Wang-gan-chi, who lived under the Soung dynasty, in 1068 of our era; and corrected in the last syllable by Sou-toung-po, who flourished fifty years later."
5. The highest antiquity of the cursive character on these vases being 200 years after Christ, and the verse written upon them being from an author who lived early in the twelfth century of the same era — it follows that the vases in question have been transported into Egypt since the year 1100 A. D. M. Pauthier gives reasons, from Chinese history, why some of them may have been brought back from China by Arabian embassies in the fifteenth century after Christ; to which age probably belong the two specimens recently exhumed from the Khabour mounds by Dr. Layard.

But, as the writer, and Mr. Bonomi, and M. Prisse, and others, have known for these twenty years, such vases abound in Egypt; especially after the annual return of the *Hadj*, or Mecca pilgrims, to Qossèyr and Cairo. The Mosaic Theban tombs are supplied through the former; the ante-Abrahamic catacombs of Memphite Saccàra through the latter mercantile channels; while the drug bazaars of Cairo and of Qenneh have always a stock on hand — price fluctuating, according to the demands of antiquaries, between two and a half and three and a half cents apiece, retail. Arab curiosity-mongers are thus enabled to furnish imbecilities travelling along the Nile with Sinico-Ægyptian vases even of ante-diluvian antiquity, on application. In the meanwhile, archæologists are aware of the sort of proofs of "early communication between Egypt and China" the New York collection embraces.

To close the digression. The reader will duly take note that the New York catalogue, above cited, refers to the "Revue Archœologique, by Mr. E. Prisse." The proprietor of the invaluable "Revue *Archéologique*" is M. Leleux; but while the author of the "catalogue" aforesaid mentions both the work and the savant whose inquiries, seven years ago, demonstrated a "Chinese vase with 17 others" to be, as antiquities, spurious; readers of that document need not wonder at the appropriate association, in the same unique cabinet, of *similia similibus.*

All obstacles to the appreciation of what we mean by "Mongolian Origin," in the theory of human graphical development, being now removed, but a few paragraphs are necessary to elucidate that section of the General Table devoted to

3d. AMERICAN ORIGIN.—To another department of "Types of Mankind" belongs the argumentative exhibition of those data, whereby the aboriginal groups of American humanity are disconnected from other centres of creation [*supra*, Chap. IX]. The purposes of our tableau are served by reference to Morton for the *craniological*, to Gallatin for the *philological*, and to Squier for the *archæological* bases of discussion.

It is unnecessary to reiterate the emphatic disclaimers of Dr. Morton, concerning any recognition by himself of such notions as an exotic origin for *American Indians*. Dr. Patterson's Memoir [*supra*, pp. xlvi–xlix] and our various *Chapters* [VII. p. 232; IX. p. 275; X. pp. 305–307, 324–326] have removed from Morton's cherished memory any further attributions to him of these philosophical heresies. (305)

The total segregation of American aborigines from other types of man throughout the rest of our globe, deduced in the present volume from the former's osteological peculiarities, animal propensities, geographical constitution, and what of history has been made *for* Indian nations by post-Columbian foreigners, results equally from the matured philology of Gallatin.

"I beg leave once more to repeat that, unless we suppose that which we have no right to do, a second miraculous interposition of Providence in America, the prodigious number of American languages, totally dissimilar in their vocabularies, demonstrates not only that the first peopling of America took place at the earliest date which we are permitted to assume, but also that the great mass of existing Indian nations are the descendants of the first [imaginary] emigrants; since we must otherwise suppose that America was peopled by one hundred different tribes, speaking languages totally dissimilar in their nature." (306)

Dr. Young it was who first made languages the subject of mathematical calculation:—

"It appears, therefore, that nothing could be inferred with respect to the relation of two languages, from the coincidence of the sense of any given word in both of them; and that the odds would be three to one against the agreement of two words; but if three words appear to be identical, it would then be more than ten to one that they must be derived in both cases from some parent language, or introduced in some other manner; six words would give more than seventeen hundred chances to one, and eight near one hundred thousand; so that, in these cases, the evidence would be little short of absolute certainty." (307)

Comparative philology now recognizes the grammatical structure of tongues as the sole criterion, which point we have explained in its proper place; but those whose minds have been led astray by the plausible application of arithmetical formulæ to the chances of intercourse between ante-Columbian American nations and the aborigines of Europe, Asia, Africa or Australasia—based upon vocabularies said to be coincident in about one hundred and eighty words—would do well to ponder upon the fiat of the greatest archæologist of our generation, Letronne:—

"Profound mathematicians have essayed, principally since Condorcet, to apply the calculus of probabilities to questions of moral order, and above all to the divers degrees of certitude in historical facts. They have flattered themselves upon ability to calculate how much might be bet against one, that a given event had or had not happened. Unfortunately, they have not seen that such a probability can yield but a result chimerical and illusory. In no case could it replace that conviction, intimate, absolute, admitting neither more nor less, which the examination of the diversified circumstances accompanying a real event produces. To those who may yet preserve any confidence in this abusive employment of mathematical analysis, I would venture the counsel that they should undertake to find out, through calculation, what new chance of probability is added by the fortuitous discovery of all these contemporaneous testimonies [such as Squier has disinterred from the primeval mounds of the West] which seem to emerge from the earth expressly to con-

---

(305) The substance of our remarks appeared, under the heading of "The Progress of Knowledge *versus* the Increase of Crime," in the New Orleans *Picayune*, June 12 and 19, 1853; signed G. R. G.

(306) *American Civilization:* Trans. Amer. Amer. Ethnol. Soc.; 1845; i. p. 179.

(307) *Experiments on the Pendulum:* Philos. Trans.; London, 1819; p. 7.

firm history. They will feel, I think, the uselessness, the vanity of their efforts; because that which results naturally from this unexpected accord, is not one of those definite probabilities estimable in numbers and in ciphers; it is a complete certitude which, with irresistible force, takes possession of every mind that is honest and exempt from prejudice." (308)

Not a solitary point of identity which cannot, at a glance, be explained by the rule — that similar causes operating upon similar principles produce everywhere the same effects — exists between the sculptured and architectural monuments of the Old World and those of the New, as known in 1853 to archæologists: not a tongue, habit, custom, mythe or idea found among the aborigines of America by Columbus, can be traced back to any anterior communication with other inhabitants of our planet. The real differences, moreover, in the geological constituents, the fauna, the flora, and the entire range of physical nature whence American man drew his artistic models, preponderate infinitely over those partial resemblances which, when not caused by the circumscribed necessities of all human things, are simply accidental—if accidents can occur in the organic laws of creative power.

Take up the works of Squier. (309) What relic of art, what natural object, what human or non-human thing, unearthed from those forest-clad mounds, is not solely and exclusively American? Run your finger along the map from the sub-polar limit of the Esquimaux down to the Terra del Fuego, and where, in published designs, of respectable authenticity, can you point out a fact, in native human economy, anterior to the fifteenth century after Christ, that compels your reason to travel off the American continent for its origin? We cannot find, at this day, pretensions to any but one. There is nothing, earnestly insists Mr. Squier, (310) even in the most curious of all mythological coincidences yet discovered between the Old and New Hemispheres, viz: the "serpent worship," that necessarily drives an archæologist away from this continent for explanation: the very figurative expression of this American mythe is, "ab ovo," a rattlesnake! Mr. Squier's subsequent pursuits in Europe (311) have opened, he tells us personally, hopeful prospects of filling up some gaps between tribes of Indians still extant and the Azteq and Tolteq scribes of ancient Mexico. He is now in Central America exploring untrodden ground; and may he succeed in his indefatigable restorations.

The possibility of Malayan, Polynesian, Japanese, or other shipwreck on the American Pacific coasts, having been established by such accident within our generation, is not disputed; but there are three common-place reasons that militate against the probability that contingencies of this sporadic nature had any the slightest influence in stocking this continent with its groups of Indian aborigines: 1st. No memento of any similar event exists in the speech, semi-civilization, art, or mythe, of the American world to induce such hypothesis; which originates simply in evangelical cravings — European fathers "of that thought." Nor, were it proven, could such petty accident establish intercourse; because these ancient castaways never returned home again; and (still stranger to relate) there are no "Indians" in the countries whence originally they sailed. 2d. In the ratio that antiquity is claimed for such a supposititious chance, so, owing to proportionate diminution of human navigatory ability, the physical possibilities of its occurrence become "fine by degrees, and beautifully less." 3d. As Morton long ago declared, "If the Egyptians, Hindoos, or Gauls have ever, by accident or design, planted colonies in America, these must have been, sooner or later, dispersed and lost in the waves of a vast indigenous population;" so that, Indians existing before the arrival of such metaphorical colonists, the old difficulty remains.

Of Irish or Welsh "Indians" it will be time enough to speak, when their "coprolites" —we dare not say their historical vestiges — are found, not merely on this continent, but west of the European "Ultima Thule" of established Celto-maniac migrations.

(308) *Recueil des Inscriptions Grecques et Latines de l'Égypte:* 1842; i., Introd., p. 63.

(309) *Observations on the Aboriginal Monuments of the Mississippi Valley:* New York, 1847; — *Ancient Monuments of the Mississippi Valley:* 1848; and, besides fragmentary papers, *Nicaragua:* 1852.

(310) *American Archæology:* "The Serpent Symbol;" 1851; pp. 170, 171.

(311) Sketched in the New York *Tribune:* 24 Nov. 1852.

Far be it from us to disparage the Icelandic researches of the "Royal Society of Northern Antiquaries at Copenhagen;" nor their "Scriptores Septentrionales Rerum Ante-Columbiarum."(312) Most laudable are their national resuscitations of "Sagas" recounting the voyages of Eric-rufus, or of Thorfinn Karlsefne; particularly those affording *American* proofs of that genealogy of Thorvaldsen, the great sculptor, back to the eleventh century after Christ. In our humble opinion, however, Thor, with his hammer, is much older; but, unable to seize the exact threads of connection between the "Fornmanna Sögur" of Iceland and the autocthones of the American continent, we are fain to leave their unravelling to the incredulous author of the "Monumental Evidences of the Discovery of America by the Northmen critically examined."(313)

We have said that to the evidences of non-intercourse between Ancient America and the other hemisphere there was but one exception. Here it is: —

In the printed "Inquiries respecting the History, present Condition and future Prospects of the Indian Tribes of the United States," circulated gratuitously by the Department of the Interior, (314) contributions are solicited from "persons willing to communicate the results of their reading or reflection." Applauding most heartily any Government action in the rescue of some mementoes of national tribes whose span of life is but short, we deem it the part of good citizenship to coöperate. Our respectful mite is tendered gratis.

"*Appendix* (Inquiries, p. 560): — 305. Is the Inscription found on opening the Grave Creek Mound, in Western Virginia, in 1839, alphabetic or hieroglyphic?"

Neither the one nor the other.

Originally a forgery — its disappearance from the "Museum" at Grave Creek is accounted for in the discovery of an imposture; its sempiternal reappearance, in an unique series of works, is due to individual idiosyncracy.

An old acquaintance of ours is this inscription; which was first started, about A. D. 1838, by some "Grave Creek Flat." (315) Flat at its origin, the Ohio pebble has become flatter through scholastic abrasions; and so terribly worn away, that the United States Department, at no trivial expense, is doomed to advertise perpetually for its recovery through official inquiries.

Already, before our sojourn at Paris, 1845–'6, the vast palæographic erudition of this inscription's composer had been exemplified by the reduction of its twenty-two rudimental apices, into four Greek, four Etruscan, five Runic, six Gallic, seven Erse, ten Phœnician, fourteen British, and sixteen Celtiberic letters; being no less than sixty-six chances drawn from twenty-two, that an Ohio pebble had made, in primeval times, an outward voyage to Europe and the Levant; and, after receiving the engraved contributions of eight antique nations, had recrossed the Atlantic to its pristine geological habitat.

Unhappily, we were too late. Our venerable friend, M. Jomard (having accepted a copy of this inscription, for the "Bibliothèque Royale," in scientific good faith), had already printed the learned and skilful analogies deducible between the scratches on this pebble and the Numidian alphabet. Other scholars, native and foreign, were misled; and there really seemed no prospect that the bewilderments produced by this contemptible petroglyph of a "Grave Creek Flat" should not become universal, when Squier's sudden mallet flattened it out forever, in 1848.(316) The pebble vanished from the Grave Creek Mound; and while, at this day, there is but one man who yet slumbers in a fool's paradise concerning it, we may echo its annihilator's felicitous dictum — "sic transit gloria moundi."

We have seen how the fabled communications between the ancient denizens of the Nile and those of the Hoang-ho have reposed upon Sinico-Ægyptian "vases" — to which has recently been added a "padlock"; and we now know the archæological worthiness of the *only*

---

(312) *Antiquitates Americanæ:* opera et studio CAROLI C. RAFN; folio, Copenhagen, 1837.
(313) SQUIER: in LUKE BURKE'S London *Ethnological Journal*; Dec. 1848; especially p. 319.
(314) *Office of Indian Affairs:* 4to, Washington, 1851.
(315) *Trans. Amer. Ethnol. Soc.:* 1845; i. pp. 369–420.
(316) London *Ethnological Journal: loc. cit.*

proof yet standing to sustain idiocratical theories of ante-Columbian intercourse between the American continent and any other centres of human creation on our terraqueous planet. Until something very different in calibre be discovered by future explorers, the section of our General Table devoted to AMERICAN ORIGINS will survive, as the plain result of palæographic science in Anno Domini 1853.

G. R. G.

ESSAY III.

MANKIND'S CHRONOLOGY—INTRODUCTORY.

OUR brief inquiries into a subject which possesses such manifold ramifications may be conveniently heralded by an extract or two from the works of some learned contemporaries: —

"We must therefore acquiesce in the conclusion, that the *Hebrew* copies represent the original and authentic text of the book of Genesis. . . . On historical grounds, very formidable objections present themselves to the *Hebrew* Chronology. . . . The difficulties are still greater when the Mosaic chronology is applied as a measure to profane history. . . . It is not, however, in these difficulties alone that we find reason for doubting whether the genealogies of the book of Genesis, taken either according to the Hebrew or the Septuagint, furnish us with a real chronology and history. . . . No evidence, therefore, remains, by which we can fix the interval which elapsed between the origin of the human race and the commencement of the special history of each nation. . . . The consequence of the method which has been commonly adopted, of making the Jewish chronology the bed of Procrustes, to which every other must conform in length, has been, that credence has been refused to histories, such as that of Egypt, resting upon unquestionable documents; and we have voluntarily deprived ourselves of at least a thousand years, which had been redeemed for us from the darkness of ante-historical times." (317)

"From this discrepancy we may infer, securely as it seems to me, that the Biblical writers had no revelation on the subject of chronology, but computed the succession of times from such data as were accessible to them. The duration of time, unless in so far as the knowledge of it was requisite for understanding the Divine Dispensation, was not a matter on which supernatural light was afforded; nor was this more likely than that the facts connected with physical science should have been revealed. . . . The result of this part of our inquiry is, in the first place, that a much longer space of time must have elapsed than that allowed by modern chronologers between the age of Abraham and the Exode; (318) and, secondly, that generations have certainly been omitted in the early genealogies. . . . By some it will be objected to the conclusions at which I have arrived, that there exists, according to my hypothesis, *no chronology*, properly so termed, of the earliest ages, and that no means are to be found for ascertaining the real age of the world. This I am prepared to admit, and I observe that the ancient Hebrews seem to have been of the same opinion, since the Scriptural writers have always avoided the attempt to compute the period in question. They go back, as we have seen in the instance of St. Paul's computation, to the age of Abraham, at the same time using expressions plainly denoting that they make no pretension to accurate knowledge, and could only approximate to the true dates of events; but they have in no instance, as far as I remember, attempted to carry the computation of time further back, nor has any one writer alluded to the age of the world. . . . Beyond that event (the arrival of Abraham in Palestine) *we can never know how many centuries nor even how many chiliads of years* may have elapsed since the first man of clay received the image of God and the breath of life." (319)

---

(317) Rev. JOHN KENRICK: *Primæval History;* London, 1846; pp. 56, 57, 58, 61, 62.

(318) The contrary is now held by the highest Egyptologists: viz.—there being but ISAAC, JACOB, LEVI, KOHATH, and AMRAM—*five* generations, or about 165 years—between ABRAHAM and MOSES, this interval must be curtailed. Vide LEPSIUS: *Chronologie der Ægypter;* and *infra.*

(319) PRICHARD: *Researches into the Physical History of Mankind;* 1847; v., "Note on the Biblical Chronology," pp 557, 560, 569, 570.

"The Roman researches of Niebuhr had proved to me the uncertainty of the chronological system of the Greeks, beyond the Olympiads; and that even Eusebius's chronicle, as preserved in the Armenian translation, furnishes merely isolated, although important, data for the Assyrian and Babylonian chronology beyond the era of Nabonassar. Again, as regards the Jewish computation of time, the study of Scripture had long convinced me, that there is in the Old Testament *no connected chronology, prior to Solomon.* All that now passes for a system of ancient chronology beyond that fixed point, *is the melancholy legacy of the 17th and 18th centuries;* a compound of intentional deceit and utter misconception of the principles of historical research." (320)

With Germanic virility of diction, Bunsen further insists—

"This fact must be explained. To deny it, after investigation once incited and begun, would imply, on the part of such investigator, small knowledge and still smaller honesty." (321)

"But (il s'en faut) much is wanting, we are convinced of it, that religious truth should be thus tied to questions of literature or of chronology. Christian faith no more reposes upon the *chronology* of Genesis, than upon its *physics* and its *astronomy;* and besides, to restrain ourselves to the subject that occupies us, the career of examination has been largely opened to us by men who certainly were far from holding Christian orthodoxy cheap." (322)

Nor does our learned authority confine himself to mere assertion; because, within a year after the publication of the above passage, he illustrates the slight estimation in which he holds *Genesiacal* chronology in the following emphatic manner:—

"It must be known that I wish to make public a monument of which the interpretation, if this be admitted, will push back the bounds of historical certitude beyond everything that can have been imagined up to this day. . . . Because, one must not dissimulate, Manetho places king MENCHERES in the IVth dynasty; and the most moderate calculation, if one follows the ciphers of Manetho, makes the author of the third pyramid remount beyond the *fortieth* century before our era. A monument of six thousand years! And what a monument! . . . We obtain the sum of 63 years, which, joined to the 4073 years, result of the preceding calculations, would give, to the end of the reign of Mycerinus, *the date of* 4136 *before* J. C." (323)

That is, our author means, the third Pyramid was built in Egypt just 153 years before the world's *Creation,* and exactly 1809 years before the *Flood;* according to the "Petavian" chronology of that Catholic Church in which M. Lenormant is a most devout communicant.

We have thought it expedient to preface our chronological inquiries with the above four citations. Each of them will protect us, like an *Ægis* raised on the stalwart arm of Jove or of Pallas. We have selected, out of the multitude before us, the highest representatives of distinct schools; who, nevertheless, perfectly agree in rejecting Scriptural chronology:—

1st. The Rev. Dr. John Kenrick—author of many standard classical works, and of "Egypt under the Pharaohs," 1850,—one of the most brilliant Protestant scholars of England.

2d. James Cowles Prichard, M. D., F. R. S.—the noblest champion of the "Unity of the human species."

3d. Chev. Christian C. J. Bunsen—the successor of Niebuhr as Prussian Ambassador at the court of Rome, and of Wilhelm von Humboldt at that of St. James; the pupil of Schelling, and the friend of Lepsius. (324)

4th. Prof. Charles Lenormant—the companion and disciple of Champollion-le-Jeune; alike famed for Hellenic erudition, and for severe Catholicity; who now fills the chair of Egyptology, vacated by Letronne's demise, at the Collège de France. (325)

It will moreover be remarked that our quotations set up no claim, as yet, for the respect-

---

(320) BUNSEN: *Egypt's Place in Universal History;* London, 1848; i., Preface, pp. 1. 2.

(321) *Ibid.: Ægyptens Stelle in der Weltgeschichte;* Hamburg, 1845, i., Einleitung, pp. 6, 7—unaccountably omitted in *Egypt's Place* by the accomplished English translator.

(322) LENORMANT: *Cours d'Hist. Ancienne;* Paris, 1838; p. 122.

(323) LENORMANT: *Éclaircissements sur le Cercueil du Roi Memphite Mycerinus;* Paris, 1839; pp. 3, 6, 24.

(324) Read Dr. ARNOLD's eulogies of this illustrious gentleman.

(325) GLIDDON: *Otia Ægyptiaca;* 1849; pp. 91, 92.

ability of the chronological systems of other nations at the expense of Judaism. On the contrary, they bear with undivided force upon *Hebrew* computations, viewed for themselves alone.

Not less truthfully does the language of a profound thinker — expression of a fifth, and far more liberal philosophy, — set forth the effeteness of Jewish chronology. Luke Burke's writings are unmistakeable: his "Critical Analysis of the Hebrew Chronology" (326) is one of the most masterly productions our literature can boast. Curtailment is injustice to its author: to the reader garbled extracts would be unsatisfactory; and the sincere investigator knows where to peruse the whole. We content our present requirements with one specimen: —

"Such, then, is the character and importance of 'the most brilliant and important of Primate Usher's improvements in chronology!' [as Dr. Hales terms the fabulous notion that Abraham was *not* the eldest son of Terah!] It consists, first, of an argument that turns out to be groundless, in every one of its elements; and, which, if well founded, would prove the Old Testament to be one of the most absurdly written books in existence: and secondly, of an assumption which, apart from this argument, is wholly gratuitous and improbable; and which also, if admitted, would bear equally hard against the character of the very writings for the support of which it was invented. And it is by such arguments as these that grave and learned divines seek to ascertain the realities of ancient history, and endeavor to place chronology upon a rational and sure foundation! And it is to such as these that men of science are required to bow, at the risk of being deemed sceptical, dangerous, profane, &c., &c. For it must not be supposed that the present is an isolated or exceptional instance of theological argument. On the contrary, it is a rule. Volumes upon volumes have been written in precisely the same spirit — volumes numerous enough, and ponderous enough, to fill vast libraries. Until a comparatively late era, all historical criticism, on which Scriptural evidences could in any manner be brought to bear, was carried on in this spirit. Nothing else was thought of; nothing approaching to genuine independence would have been tolerated. And thus the human world rolled round, century after century; the brave trampled upon by slaves; the wise compelled to be silent in the presence of fools; the learned alternately serfs and tyrants, deluded and deluding, cheating themselves, and cheating others with sophistries which, upon any other subject, would disgrace even the mimic contests of schoolboys! For ourselves, we should feel a humiliation to contend with such sophistries seriously, and in detail, were we not firmly convinced that to do so is not merely the most legitimate, but also the only mode by which truth can be rendered permanently triumphant. Wit and sarcasm may obtain a temporary success, they may awaken minds otherwise prepared for freedom, but they are often unjust, usually unbenevolent, and consequently, in the majority of cases, they merely awaken antagonism, and cause men to cling with increased fondness to their opinions. Nothing but minute, searching, inexorable argument will ever obtain a speedy, or a permanent triumph over deep seated prejudices." (327)

"But, fortunately," winds up another and a sixth formidable adversary to Hebrew computation — no less an archéologue than the great Parisian architect, Lesueur — "fortunately, questions of ciphers have nothing in common with religion. What imports it to us, to us Christians, who date so to say from yesterday, that man should have been thrown upon our globe at an epoch more or less remote; that the world should have been created in six days, or that its birth should have consumed myriads of centuries? Can God, through it, become less grand, his work less admirable? We are, since the last eighteen hundred years, dupes of the besotted vanity of the Jews. It is time that this mystification should cease." (328)

Italian scholarship speaks for itself: — (329)

"The Bible is, certainly, as the most to be venerated, so the most authoritative fount of history; but, in so many varieties of chronological systems, which are all palmed off by their authors as based upon indications of time taken from the Bible; in the very notable difference of these indications between the Hebrew and the Samaritan text, and the Greek version, and between the books of the Old and of the New Testament: finally, in the indecision, in which *the* CHURCH has always left such controversy, that, I do not see any certain standard, by which the duration of the Egyptian nation has to be levelled, unless this

(326) London *Ethnological Journal;* June, July, November, December, 1848.

(327) *Op. cit.;* pp. 274, 275.

(328) *Chronologie des Rois d'Égypte* — ouvrage couronné par l'Académie: Paris, 1848; pp. 304, 305.

(329) BARUCCHI, Director of the Museum of Turin; *Discorsi Critici sopra la Cronologia Egizia;* Torino, 1844; pp. 29, 43, 44, 147.

become determined through an accurate examination of all its historic fountains. . . . Leaving therefore aside anysoever system of biblical chronology; because, of the quantity hitherto brought into the field by the erudite none are certain, nor exempt from difficulties the most grave; and, because *the* CHURCH, to whose supreme magistracy belongs the decision of controversies appertaining to dogma and to morals, has never intermeddled in pronouncing sentence upon any one of the systems aforesaid, of which but one can be true, while all peradventure may be erroneous. . . . I shall finish by repeating in this place that which already I declared elsewhere, viz.: it is not my intention to combat any systems regarding biblical chronology; but inasmuch as, of these, not one is propounded as true under *the* CHURCH's infallible authority; I have placed all these (systems) aside in the present examining, in order to treat Egyptian chronology through the sole data of history and of Egyptian monuments."

Finally, we quote Lepsius:—(330)

"The Jewish chronology differs in a most remarkable manner from every other; and even in times as modern as those of the Persian kings the difference amounts to no less than 160 years, from known dates. Its several sources present but little difference among themselves. They count according to *years of the world;* a calculation which, as also IDELER (*Hand. d. Chron.* I. pp. 569, 578, 580), considers most probable, was invented, *together with the whole present chronology of the Jews*, by the Rabbi HILLEL HANASSI, in the year 344 after Christ: and thenceforward gradually adopted. They fix the creation of the world 3671 B. C.; and all agree, even Josephus, in the usual calculation of the Hebrew text. They fix the deluge at 1656, the birth of Abraham at 1948, Isaac's 2048, Jacob's 2108, Joseph's 2199, Jacob's arrival in Egypt 2238, Joseph's death 2309, years after Adam." . . . "The question is now, how must we explain this obvious dislocation of facts as compared with the true dates. IDELER has demonstrated that the introduction of the *era of the world*, and consequently of the whole system of chronology, must be ascribed to the author of the *Moleds*, (or 'New Moons,') and in general of the whole later Jewish calendar, the Rabbi HILLEL who flourished in the first half of the IVth century."

Reserving further extracts until we take up the *Hebrew* chronology, it here suffices to notice that MOSES, who lived about the fourteenth century B. C., is not amenable for numerical additions made, to books that go by his venerable name, about 1800 years after his death, by a modern *Rabbi.*

The unanimity of science in the rejection of any system of biblical computation might be exemplified by many hundred citations: either, of savans who, establishing grander systems more in accordance with the present state of knowledge, pass over the rabbinical ciphers in contemptuous silence; or, of divines who, like the Rev. Dr. Hitchcock (President of Amherst College, and Professor of Natural Theology and Geology) strive, vainly we opine, to reconcile the crude cosmology of the infantine Hebrew mind with the terrestrial discoveries of matured intellects like Cuvier, De la Beche, Murchison, Owen, Lyell, or Agassiz. Nevertheless, Calvinism in the pages of Hitchcock begins to affect a more amiable disguise than was worn by the magnanimous slayer of SERVETUS, or by the iconoclastic John Knox; to judge by the following admissions:—

"If these positions be correct, it follows that, as we ought not to expect the doctrines of religion in treatises on science, so it is unreasonable to look for the principles of philosophy in the Bible. . . . But a still larger number of [clerical] authors, although men of talents, and familiar, it may be, with the Bible and theology, have no accurate knowledge of geology. The results have been, first, that, by resorting to denunciation and charges of infidelity, to answer arguments from geology, which they did not understand, they have excited unreasonable prejudices and alarm among *common Christians* respecting that science and its cultivators; secondly, they have awakened disgust, and even contempt, among scientific men, especially those of sceptical tendencies [!], who have inferred that a cause which resorts to such defences must be very weak. They have felt very much as a good Greek scholar would, who should read a severe critique upon the style of Isocrates, or Demosthenes, and, before he had finished the review, should discover internal evidence that the writer had never learned the Greek alphabet." (331)

How true the latter part of this paragraph is, the reader has convinced himself by the perusal of our Essay I. [*supra*]; where the Hebraical knowledge of Calvinistic divines in Ame-

---

(330) *Chronologie der Ægypter:* "Kritik der Quellen," i. pp. 259, 360, 361, 362.
(331) *The Religion of Geology;* Boston, 1852; p. 3, and Preface, p. 7.

rica has been compared with that of coetaneous Lutherans and Catholics in Europe. Contentions between scramblers for the loaves and fishes may, however, be left to the diverted contemplation of the gatherers of St. Peter's pence. None of them have real bearing upon the science of mundane *chronology*, to which our present investigations are confined.

Until very recent times, it was customary, among chronologers, to follow the Judaic and post-Christian system in assigning eras to events; viz.: by assuming that a given occurrence had taken place in such a year (*Anno Mundi*) of the Creation of the world. This arrangement would have been absolutely exact, if the precise moment of Creation, according to the "book of Genesis," had been previously settled, or even conventionally agreed upon: but, unhappily, no two men ever patiently reckoned up its numerals and exhibited the same sum total; as will be made apparent anon, in its place. Besides, this arrangement was found by experience to be theologically unsafe; because, on the one hand, the Christian Fathers, by assuming the *Septuagint* computation, demonstrated that Jesus, appearing exactly in Josephus's 5555th year of the world, could be no other than the Χριστος, "the *anointed;*" (332) whilst, on the other hand, the Jewish Doctors, proving through computation of the *Hebrew* Text that the birth of Jesus had occurred in the year of the world 3751, demonstrated that he could not possibly be their MeS*h*aiaH. (333)

"There was an old tradition," says the profound Kennicott, (334) "alike common among Judæans and Christians, sprung from the mystic interpretation of Creation in six days, that the duration of the world should be 6000 years: that the Messianic advent should be in the *sixth* millennium; because he would come in the *latter days.* The ancient Jews, therefore, their chronology having been previously contracted, made use of an argument sufficiently specious, through which they did not recognize Jesus: for the *Messiah was to come in the sixth millennium;* but *Jesus was born* (according to the computation of time by them received) in the latter part of the *fourth millennium*, about the year of the world 3760 (*Seder Olam*, edit. Meyer; pp. 95 and 111). The very celebrated [Muslim-Arab] Abul-Pharagius, who lived in the XIIIth century, in his history of Dynasties, thus proffers a sentence worthy of remembrance; by Pococke so rendered into Latin:—'A defective computation is ascribed by Doctors of the Jews—For, as it is pronounced, in the Law and the Prophets, about the Messiah, he was to be sent at the ultimate times: nor otherwise is the commentary of the more antique Rabbis, who reject Christ; as if the ages of men, by which the epoch of the world is made out, could change. They subtracted from the life of Adam, at the birth of Seth, one hundred years, and added them to the rest of the latter's life; and they did the same to the lives of the rest of the children of Adam, down to Abraham. And thus it was done, as their computation indicates, in order that Christ should be manifested in the *fifth* [fourth, K.] millennary through accident in the middle of the years of the world; which in all, according to them, will be 7000: and they said, *We are now in the middle of this time, and yet the time designated for the advent of the Messiah has not arrived.*' The computation of the LXX also indicates, that Christ should be manifested in the sixth millennary, and that this would be his time. . . . The old *Italic* version, which, according to St. Augustine, was 'verborum tenacior cum perspicuitate sententiæ,' is the foundation of the *chronologia major* of the Latin Church, to this day (1780); for, 'in the Roman Martyrology, which is publicly chanted in church, on the 8th Jan., the Nativity of the Lord is thus announced to the people from the ecclesiastical table: *Year from the creation* 5099 (5199 in Martyrol. Rom. Antwerp. 1678, p. 388): and *from the deluge year* 2957 (Hod., p. 447)."

A quotation from a Christian work next to canonical will establish the belief of those early communities who lived nearest to the apostles:—the 5500 years, be it noted, had been, by Nicodemus, "found in the first of the seventy books, where Michael the archangel" had mentioned them to "Adam, the first man."

"13 By these five cubits and a half for the building of the Ark of the Old Testament, we perceived and knew that in five thousand years and half (one thousand) years, Jesus Christ was to come in the ark or tabernacle of the body;

14 And so our Scriptures testify that he is the Son of God, and the Lord and King of Israel.

15 And because after his suffering, our chief priests were surprised at the signs which were wrought by his means, we opened that book to search all the generations down to the generation of Joseph and Mary the mother of Jesus, supposing him to be the seed of David;

---

(332) Hennell: *Christian Theism;* 1845; pp. 82, 83.

(333) *Seder Olam Rabba*, composed about A. D. 130; apud Hales.

(334) *Dissertatio Generalis;* § 75, pp. 32, 33, 76.

83

"16 And we found the account of the creation, and at what time he made the heaven and the earth, and the first man Adam, and that from thence to the flood, were two thousand, two hundred, and twelve years.

17 And from the flood to Abraham, nine hundred and twelve. And from Abraham to Moses, four hundred and thirty. And from Moses to David the king, five hundred and ten.

18 And from David to the Babylonish captivity, five hundred years. And from the Babylonish captivity to the incarnation of Christ, four hundred years.

19 The sum of all which amounts to five thousand and a half (a thousand.)

20 And so it appears, that Jesus, whom we crucified, is Jesus Christ the Son of God, the true and Almighty God. Amen." (335)

The conclusive logic of this passage derives support from another ancient Christian document, wherein is given the reason why the end of the world was expected some time ago:—

"Consider, my children, what that signifies, he [God] finished [creating] them in six days. The meaning of it is this; that in six thousand years the Lord God will bring all things to an end." (336)

Such being the whole story, the reader has now to make choice of whichever of the following dates may suit his views upon the

EPOCHAS OF CREATION.

Biblical Texts and Versions.	B. C.		B. C.
Septuagint computation	5586	Chinese Jews	4079
Septuagint Alexandrinus	5508	Some Talmudists	3761
Septuagint Vatican	5270	Vulgar Jewish computation	3760
Samaritan computation	4427	Seder Olam Rabba, great Chronicle of the World,	
Samaritan Text	4305	A. D. 130	3751
Hebrew Text	4161	Rabbi Lipman	3616
English Bible	4004	*Christian Divines.*	
Jewish Computations.		Clemens Alexandrinus, A. D. 194	5624
Josephus — Playfair	5555	Hales, Rev. Dr.	5411
Josephus — Jackson	5481	Origen, ——, A. D. 230	4830
Josephus — Hales	5402	Kennedy, Bedford, Ferguson	4007
Josephus — Universal History	4698	Usher, Lloyd, Calmet	4004
Talmudists	5344	Helvetius, Marsham	4000
Seder Olam Sutha	4339	Melancthon	3964
Jewish computation	4220	Luther	3961
" "	4184	Scaliger	3950

These are mere excerpts of 120 different opinions, on the date of *Creation*, tabulated by Hales. (337) This list can easily be swelled to above 300 distinct and contradictory hypotheses. Between the highest epoch, B. C. 6984 (the Alphonsine tables), and the lowest, B. C. 3616 (Rabbi Lipman), there is the trifling difference of 3268 years!

It is but fair to set off Catholic against Protestant authorities, so we cull a few more instances from the learned pages of De Brotonne (338).—"Among authors who deny the eternity of the world, not one, from its creation to the advent of Jesus Christ, counts more than 7000 years, nor less than 3700." He also supplies a schedule of 70 more disputants, ranging between B. C. 6984 and 3740, from Riccioli; (339) but the subjoined are some of his own, *extra.*

	B. C.		B. C.
Suidas	6000	*Hebrew Text*	3834
Nicephorus, Constantinopolitanus	5500	St. Isidore	5336
Eusebius Cæsariensis	5200	Montanus	3849
St. Jerome, and Beda	3952	Vossius	5590
Hilarion	5475	Petavius (Romanist authority)	3983
St. Julian, and the LXX	5205		

(335) *Gospel of Nicodemus;* chap. xxii.—Apochryphal New Testament, pp. 51, 52.

(336) *General Epistle of Barnabas;* xiii. 4: *op. cit.*; p. 101.

(337) *Analysis:* i. p. 212.

(338) *Filiations et Migrations des Peuples:* Paris, 1827; 428–436.

(339) *Chronologia reformata:* pp. 290–292, 293.

Riccioli shows that computations upon different exemplars of the LXX oscillate, also, between a maximum of 5904 years B. C., and a minimum of 5054, for the *Creation* alone! Nevertheless, "Cœlum ipsum petimus stultitiâ." Not satisfied with human inability to define, through biblical or anysoever methods of reckoning, the age when Creative Power first whirled our incandescent planet from the sun's fire-mist, some intelligences, at the supernatural stage of mental development, have actually fixed the *month*, *day*, and *hour!*

"And now hee that desireth to know the yeere of the world, which is now passing over us this yeere 1644, will find it to bee 5572 yeeres just now finished since the Creation; and the year 5573 of the world's age, now newly begunne this *September at the Æquinox.*" (340) *Anno Mundi* I; "VIth *day* of Creation, . . . his (Adam's) wife the weaker vessell: she not yet knowing that there were any Devils at all . . . sinned, and drew her husband into the same transgression with her; this was about *high noone*, the time of eating. And in this lost condition into which Adam and Eve had now brought themselves, did they lie comfortlesse till towards the cool of the day, or *three o'clock* afternoone. . . . (God) expelleth them out of Eden, and so fell Adam *on the day* that he was created." (341)

"We do not speak of the theory set forth in a work entitled *Nouveau Système des Temps*, by Gibert father and son. This system, which is not so new as its title seems to announce, gives to the world only 3600 years of duration down to the 1st July, 1834; and makes Adam's birth 1797 years before J. C., on the 1st *July*." (342)

"It is, besides, generally allowed by Chronologists, that the beginning of the patriarchal year was computed from the autumnal equinox, which fell on *October* 20th, B. C. 4005, the *year* of the creation." (343)

But the Promethean intrepidity of orthodoxy is not content with mathematical demonstrations of the year, the month, the day, nor the hour of Creation. It ascends, in some extatic cases, far beyond! Thus, Philomneste heads an especial chapter with

"*Antégénésie*—What God was about *before* the creation of the world." (344)

Albeit, none of these profanations of science contain one solitary element, in regard to Creation, that is strictly *chronological*. "Passons au Deluge" (345)—let us descend to the Flood; and see what resting-place a "dove" could find amid these wastes of waters and of time. For the

EPOCHAS OF THE DELUGE,

out of sixteen opinions published by Hales—maximum, B. C. 3246; minimum, 2104; difference 1142 years—the following are singularly in accordance:—

	B. C.		B. C.
Septuagint version	3246	Vulgar Jewish computation	2104
Samaritan Text	2998	Hales	3155
English Bible	2348	Usher	2348
Hebrew Text	2288	Calmet	2344
Josephus	3146		

So are also the intervals of time assigned, by the subjoined computators, to mundane existence, between the Creation and the Flood. We borrow them from De Brotonne.

CREATION TO DELUGE.

	YEARS.		YEARS.
Josephus	2256	Later Rabbis, St. Jerome, Beda, Montanus, Scaliger, Origanus, Emmius, Petavius, Gordonus, Salianus, Torniellus, Hervartus, Philippi, Tirinus, Riccioli	1656
Suidas, Nicephorus, Eusebius, St. Julian, St. Isidore	2242	St. Augustine—"From Adam to the Deluge, according to our sacred books (*i. e.*, the LXX), there have elapsed 2242 years, as per our exemplars; and 1656, according to the Hebrews."	
Clemens Alexandrinus	2148		
Hilarion	2257		
Vossius, Riccioli	2256		
Cornelius a Lapide	1657		

(340) Rev. Dr. LIGHTFOOT: *Harmony of the Foure Evangelistes;* London, 1644; 1st part, Proleg., last page.
(341) *Ibid.: Harmony, Chronicle, and Order of the Old Testament;* London, 1647; p. 5.
(342) DE BROTONNE; *op. cit.;* ii. p. 160.
(343) Rev. Dr. F. NOLAN: *The Egyptian Chronology Analysed:* London, 1848; p. 392.
(344) *Livre des Singularités:* Dyme, 1841.
(345) DANDIN, in *Les Plaideurs:* iii. 54.

But these discrepancies are increased by the computations made, since 1623 A. D., upon MSS. of the *Samaritan* Pentateuch, which generally yield an interval between the Creation and the Deluge of years 1307.

The basis of all these calculations lies in the hyperbolical lives of the *ten antediluvian Patriarchs.* It will be seen, through the skilful synopsis of a learned divine, how admirably the numerals of the Hebrew and Samaritan texts correspond, not merely with each other, but with those of the Septuagint version, and of Josephus: —

"The following tabular schemes exhibit the variations; the numbers expressing the parent's age at the son's birth, except in the cases of Noah and Shem.(346)

Ante-Diluvian Patriarchs.	Hebr.	Samr.	LXX.	Josep.	Post-Diluvian Patriarchs.	Hebr.	Samr.	LXX.	Josep.
1. *Adam*	130	130	230	230	11. *Shem* (aged 100 at the Flood)	2	2	2	12
2. *Seth*	105	105	205	205	12. *Arphaxad*	35	135	135	135
3. *Enos*	90	90	190	190	[Cainan spurious...	...	...	130	...]
4. *Cainan*	70	70	170	170	13. *Salah*	30	130	130	130
5. *Mahalaleel*	65	65	165	165	14. *Heber*	34	134	134	134
6. *Jared*	162	62	162	162	15. *Peleg*	30	130	130	130
7. *Enoch*	65	65	165	(1)65*	16. *Reu*	32	132	132	130
8. *Methuselah*	187	67	187	187	17. *Serug*	30	130	130	132
9. *Lamech*	182	53	188	182	18. *Nahor*	29	79	79	120
10. *Noah* (at the Flood)	600	600	600	600	19. *Terah* (Gen. xi. 32, xii. 4.)	130	130	130	130
*165 is doubtless the correct reading. Total	1656	1307	2262	2256	So to *Abraham*	352	1002	1002	1053

The above, like all other tables compiled by theological computators to illustrate so-called "Biblical chronology," assumes the *numerals* of current printed exemplars to be correct; but, if we set to work, archæologically, to verify the original Hebrew, Greek, and Samaritan *manuscripts*, we find even this apparent uniformity to be a delusion — indeed, another orthodox figment. A few instances pleasingly exhibit this fact (347): —

"In one of the manuscripts collated by Dr. Kennicott, and which is marked in his Bible, codex clvii., this century [in the Hebrew generation of Jared] is omitted, and there is much probability that it was also omitted in the copies used by the eastern Jews. According to the testimony of Ismael Sciahinshia, an eastern writer, all these copies reckon only 1556 years from Adam to the flood, instead of 1656. . . . According to the numbers still existing in the vast majority of [Greek] manuscripts, Methuselah dies 14 years *after* the deluge, and had not the fifty-three, of the generation of Lamech, been changed to eighty-eight, he would have died 49 years after the deluge. . . . The deluge occurred, according to the Septuagint, in the year of the world 2242, and by adding up the generations previous to his, we shall find that he was born in the year 1287. He lived 969 years, and therefore died in 2256. But this is 14 *years after the deluge!* . . . And had they [the theologers] not, by a previous system of changes, added a century [in Greek *MSS.*] to all the generations, he would have died 249 years after it. . . . Origen appears to have been the first who gave notoriety to the contradiction; and for a long time, the fact greatly disturbed theologians. The reader will be hardly surprised to learn that in a subsequent age some manuscripts were *found* with the error *corrected.* . . . Some [Greek *MSS.*] make the generation of Adam 330 years; one makes it 240. Another gives 180 to Canaan, a third 170 to Jared, while others allow 177 or 180 to Methuselah. . . . One [Hebrew] manuscript, codex lvii. of Holmes, makes the age of Methuselah 947: three or four other authorities make the generation of Lamech 180: the two corrections conjoined, bring the death of Methuselah to the year of the deluge. We also find three other authorities making the generation of Methuselah 180 years; this connected with the 188 of Lamech, places the death of Methuselah *only one year* after the deluge, even allowing him full age. Another manuscript makes his generation 177 years, three other authorities give the number 165, while one manuscript makes his total age 965. . . . Dr. Kennicott has given readings of 320 Hebrew manuscripts of the book of Genesis. 97 of these have been collated throughout, 223 in part only. . . . One manuscript (codex clvii.) omits the hundred years in his [Jared's] generation; two others (codices ci. and clxxvi.) omit it in that of Methuselah; and one (codex xviii.) in that of Lamech. Codex clxxvi. makes the generation of Lamech 172 and his total age 772, and codex xviii. makes his total age 909. . . . We also find that, in three

(346) Rev. E. B. Elliott, A. M.: *Horæ Apocalypticæ;* London, 1846; iv. p. 254, note. Compare "Tables of the discrepancies of the three Texts with regard to the Ante-diluvian Patriarchs" in Wallace: *Dissertation on the True Age of the World;* London, 1844, pp. 14–16.

(347) Burke: *Ethnological Journal;* 1848; pp. 27, 28, 82, 83, 84, 87, 78–91.

or four manuscripts, some of the numbers of Methuselah are written over *erasures*. This, of course, looks suspicious. One manuscript (codex clv.) makes Enoch live after the birth of Methuselah 'five and sixty and three hundred years' [i. e., the old 365 *days* of an Egyptian vague year !], instead of 300 years simply."

Thus far Luke Burke in his studies of the *Hebrew* variations exhibited by Kennicott. (348) The annexed Table shows how he found matters in the *Greek* of Holmes. (349)

"TABLE III.

NAMES.	BEFORE GENERATION.				AFTER GENERATION.				TOTAL AGES.			
	1.	2.	3.	4.	1.	2.	3.	4.	1.	2.	3.	4.
ADAM.........	330	330 240 130	4 1 2	MSS. 31,121, Ald., Theop. p. 13. MS. 77.................. Slav., Arm. Ed......	700	705 800	1 3	MS. 135 .... Slav., Ostrog., 127	930	1200	1	Corrected in the margin to 930, 300 having been accidentally put for 30. MS. 18
SETH..........	205	132 805	1 1	MS. 127 .... Coptic..................	707	807	1	MS. 127	912	910 902 772	1 1 1	MS. 19. MS. 18. Arab. 2.
ENOS..........	190	180 140 95	1 1 1	MS. 65.................. MS. 75.................. MS. 127	715	705 916 800	1 4 1	MS. 135 .... " 14, 78, 130, 133* MS 127............	905	915	1	MS. 79.
CAINAN......	170	180	1	MS. 106................	740	800	1	MS. 127.............	910			
MAHALALEEL	165	65	1	MS. 127	730	830	1	MS. 127.............	895	795	9	MS. 14, 25, 31, 38, 57, 73, 77, 78, 79.
JARED	162	170	1	MS. 75..................	800	...	...	....	962	847	1	MS. 127.
ENOCH.........	165	65	1	MS. X..................	200	...	...	....	365	465	1	MS. 71.
METHUSELAH	167	165 177 180 187	4 1 3 18 &c	MS. 106, 107, Compl., Georg. MS. 75.................. MS. 71, Slav., Theop. p. 133. MS. I, X, 15, 16, 55, 59, 64, 68, 83, 120, 121, 131, 135, 187, Ald., Alex., Chrysos. IV., Arm. Ed. and a few others.	802	782	32	MS. I., X., 14, 15, 20, 25, 55, 57, 59, 64, 68, 71, 73, 75, 77, 78, 79, 83, 121, 128, 130, 131, 133 135, Ald., Cat Nic., Arm. 1, Arm. Ed., Arab. 1, 2, Alex., Slav., & perhaps another examin'd by Vossius.	969	947 965	1 1	MS. 57. MS. 82.
LAMECH	188	180 182	3 2	MS. 75, 187, Chrysos. IV. Arab. 2. Chron. Orient.	565	595	1	Arab. 2.............	753	733 755 765 768 777	1 3 1 1 1	Arm. 1. MS. 19, 107, 107. MS. 25. Arab. 3. Arab. 2.

* In this case, nine hundred has been corrected by another hand into seven hundred. There are several minor remarks and explanations relative to this table, which we should have been glad to have afforded, were we not much pressed for time and space. These, however, would, after all, be of little interest to the general reader, and the learned reader will not need them.

. . . The first glance at this table will show the inquirer, that he has got into a region of various readings, very different from that presented to him by the Hebrew manuscripts. Instead of some *eight* or *nine* variations found in some *three hundred* manuscripts, he has about 118, found in a much smaller number of manuscripts ! . . . Are we to say, then, that the Christian scribes were, in general, so wretchedly careless, that they made twenty errors where a Jew made but one? . . . These things, therefore, evince design, not accident. We find one variation followed by more than 32 authorities, another by 18, a third by 9. There are three which are each copied by four manuscripts, four which are copied by three each, and two which have each two manuscripts agreeing in them: thirty-one only are single variations, and some of them, at least, are as clearly intentional as any of the others. As to the variation which makes Methuselah live 782 years after the birth of Lamech, instead of 802, no one can doubt of its being intentional. 788 is the Hebrew date, and it was here copied from the Hebrew for the same reason that the Hebrew was previously invented, viz.: for the purpose of bringing the death of Methuselah within the antediluvian period, instead of fourteen years after it. . . . Codex LVII. has the total age

(348) *Vetus Testamentum Hebraicum, cum variis lectionibus;* folio, Oxon. 1776–'80.

(349) *Vetus Testamentum Græcum, cum variis lectionibus;* folio, Oxon. 1798–1827.

of Methuselah 947, while four authorities have his generation 165. . . . The whole number of variations in the case of Methuselah is 60; more than half the number in the entire Antediluvian Chronology. Every one of them but four, or at the utmost five, viz., those making the generation 165, and codex LXXXII. making the total age 965, have reference to the error in the age of Methuselah. This fact is of course significant; and at once reduces, to nearly one-half, the number of variations that can be supposed accidental. This number is easily reduced still farther. Codex Arabicus II. has all the Hebrew numbers, in the case of Lamech. The Chronicon Orientalis has the generation like the Hebrew, and, for anything we know to the contrary, may have the other periods in harmony with this generation. Codex CXXVII. has the Samaritan numbers in five instances. The Sclavonic version gives us both the Hebrew numbers in the case of Adam, the Armenian edition gives one of them, and the Ostrogoth version the other. Thus we have 13 more intentional variations, making the whole number, thus far, 73 out of 118. Nine manuscripts make the total age of Mahalaleel 795, instead of 895; four make the generation of Adam 330 instead of 230; four others make the age of Enos after generation 915 instead of 715; and four make the generation of Lamech 180, instead of 188 or 182. Three make the total age of Lamech 755, while three others make it respectively 733, 765, and 768. These make 27 other cases in which the intention is apparent though less obviously than the former. So that we thus have 99 instances out of 118, which cannot be reasonably attributed to accident. And even of the remaining nineteen, there are not more than *two* that have any unequivocal indications of being accidental. The substitution of 300 for 30 in Codex XVIII., in the total age of Adam, is evidently accidental, as is the 805 for 205 in the Coptic version, of the generation of Seth. Accident may also have occasioned some of the other changes, but this is not probable. . . . When Origen, in the early part of the IIId century, began to collate these manuscripts and versions, he was confounded at the clashings which he discovered in them. Whole passages existed in some [Greek biblical MSS.] for which there was no counterpart in others, nor in the Hebrew, nor in the Samaritan. . . .

"The reader will here naturally ask, how is it that the commentators have managed to confront these hosts of difficulties, and yet avoid the inevitable inferences which a clear view of them discloses? The answer is simple. They never have fairly confronted them. They never have classified them, or analyzed them, in a manner likely to lead to the truth. They would not admit that any conclusion could be true which did not harmonize with their pre-conceived theory of the entire inspiration of every portion of the Scriptures — of every portion at least which they severally regarded as canonical. This with them was a settled point, from which they neither wished to recede, nor dared to recede. Their works therefore present us with little more than vain attempts to reconcile, to soften down, to slur over these contradictions.

"Thus, it is evident that this *antediluvian* chronology, as we now have it, is not the work of any one person, or of any one era. In its original form [not earlier than B. C. 130 to 420] it was not only contradictory to all human experience, and to the laws of organization, but also glaringly self-contradictory. It is plain, too, that it has been repeatedly altered, in various ages, and by various people, and that these alterations have been made in a perfectly arbitrary manner, and without any reference to facts or historical data bearing upon the subject. Who can say by whom, or when it was drawn up, or how many stages it has passed through previously to the changes we have spoken of? Is it not folly, then, to pretend to regulate history by a series of numbers thus tampered with, to say nothing of their scientific and historic impossibility?"

Folly! It is worse than folly: it is an absolute disregard of every principle of rectitude; an impudent mockery of educated reason; a perpetualized insult to honest understandings; and a perdurable dereliction, on the part of interested and self-conceited supernaturalists, of Almighty *truth*. Ignorance, abject ignorance, is the only plea through which future sustainers of genesiacal numerals can escape from the charge of knavery. Let imbecility impale itself, henceforward, on either horn of this dilemma for edification of the learned; and with the derisive jeers of men of science, who are now endeavoring to reconstruct a solid *chronology* out of the débris of universal and primeval humanity yet traceable, in their various centres of Creation, upon our planet's superficies.

The reader of Essay I. in the present work is aware of the conjectural hundreds of thousands of variants proceeding from what Kennicott, De Rossi, and the Rabbis, qualify as the "horrible state" of the *Manuscripts* of the Old Testament. He also may infer the historical metamorphoses of alphabets, and the alterations of *numbers* which, to suit different schools of theology, the Hebrew and Samaritan Texts, and Septuagint version, underwent between the third century before C. and the fourth century after. A pledge, too, has been incidentally made to him, that a future publication shall demonstrate why the "*ten* patriarchs," from A-DaM to NoaK*h*, were no more *human beings*, in the idea of their original writers, than are the ethno-geographical names catalogued in Xth *Genesis*. Abler hands, in another chapter [XI.] of this volume, have set forth what of geology and palæontology throws more or less light upon Types of Mankind.

Leaving the *Deluge*, its universality or its fabled reality, to professional reconcilers;(350) the chronological bearings of this hypothetical event compel us not to dodge, at the same time that it is far from our intention to dwell upon, its passing consideration. No *Hebraist* disputes that, according to the literal language of the Text, the flood was universal. To make the Hebrew Text read as if it spoke of a partial or local catastrophe may be very harmonizing, but it is false philology, and consequently looks very like an imposture.

"The waters *swelled up* (prevailed) infinitely over the earth; all the high mountains, beneath all the skies, were covered: fifteen cubits upward did the waters rise; the mountains were covered."(351)

The level of the flood was, therefore, 22½ *feet* above the Dhawalaghiri (28,074 feet) and over the Sorata (25,200 feet); according to Humboldt.(352) Equivalent to some *two miles* above the line of perpetual snow must, therefore, have been the level whereupon the Ark would have been frozen solid but for an universal thaw. This is what the Hebrew chronicler meant by KuL HaHeRIM, HaGiBuHIM — *all the high mountains;* even if Hindostan and America were as alien to his geography, as such an aqueous elevation is to the physicist.

"If there is any circumstance," declares Cuvier, "thoroughly established in geology, it is, that the crust of our globe has been subjected to a great and sudden revolution, the epoch of which cannot be dated much further back than *five or six thousand years ago;* that this revolution had buried all the countries which were before inhabited by men and by the other animals that are now best known."(353)

Science has found nothing to justify Cuvier's hypothesis, conceived in the infancy of geological studies; whether in *Egypt*, (354) in *Assyria*, (355) or on the *Mississippi:*(356) whilst, without delving into the wilderness of geological works for flat contradictions of this oft-quoted passage of the great Naturalist, here are three extracts by way of arrest of judgment: —

"Of the Mosaic Deluge I have no hesitation in saying, that it has never been proved to have produced a single existing appearance of any kind, and that it ought to be struck out of the list of geological causes." (357)

"There is, I think (says the President of the London Geological Society, 1831), one great negative fact now incontestably established; that the vast masses of Diluvial Gravel, scattered almost over the surface of the earth, *do not belong to one violent and transitory period*. . . . Our errors were, however, natural, and of the same kind which led many excellent observers of a former century to refer all secondary formations to the NOACHIAN DELUGE. Having been myself a believer, and, to the best of my power, a propagator of what I now regard as *philosophic heresy*, . . . I think it right, as one of my last acts before I quit this chair, thus publicly to read my recantation."

A later President of the same illustrious corps, 1834, uses similar language: —

"Some fourteen years ago I advanced an opinion . . . that the entire earth had . . . been covered by *one general but temporary deluge* . . . I also now read my recantation." (358)

Were it not for such denials of Cuvier's six-chiliad doctrine (to which hundreds might be added of the whole school of true geologists at the present day), then, it would be evident to archæologists that "geology" must be of necessity a *false science:* and for the following reason:—It has been shown [*supra*, p. 562], that the first chapter of the "book of Genesis" is an ancient cosmogenical *ode*, with a "chorus" like the plays of Grecian dramatists; — that its authorship, if entirely unknown, is not *Mosaic;* — that its age, the style being

(350) Such as, the Rev. Dr. PYE SMITH, the Rev. Dr. HITCHCOCK, or "The *Friend* of Moses."

(351) *Genesis;* vii. 18, 19; — CAHEN'S *Text;* i. p. 21.

(352) *Cosmos;* Otte's trans., 1850, i. p. 28, 31, 330-332.

(353) *Essay on the Theory of the Earth;* 1817; p. 171.

(354) GLIDDON; *Otia Ægyptiaca;* pp. 61-69.

(355) AINSWORTH: *Assyria, Babylonia, and Chaldæa;* London, 1838; pp. 101, 104-107.

(356) DOWLER: *Tableaux of New Orleans;* 1832; pp. 7-17.

(357) McCULLOCH: *System of Geology;* i. p. 445.

(358) Rev. Dr. J. PYE SMITH: *Relation*, &c.; 1841; pp. 138, 139, 141.

Elohistic and the writing *alphabetical*, cannot ascend even to the tenth century before C.; and that, being based upon the harmonic scale of 7 *notes*, in accordance with the erroneous planetary system of Chaldaic magianism (of 5 *planets*, and the *sun* and *moon*); it is an arbitrary human production, founded upon ignorance of the physical laws and phenomena of Nature — as this Nature is unfolded by science in the nineteenth century.

In consequence, did *geologists* pretend to arrange the dozen, or more, distinct creations manifested in the earth's crust through rocky stratifications and different fossil remains (divided from each other by immeasurable periods of interjected time), according to the "7 musical notes" of Genesis, they would perpetrate a caricature of God's works more gross, and less excusable, than that of COSMAS-*Indicopleustes:* at the same time that they would make parade of stolid ignorance of *philology* and biblical exegesis such as every Orientalist, versed in archæology, must laugh to scorn. On the other hand (whether practical "geology" be or be not a fiction), were a *philologist* at the present day to argue, that the writer of "Genesis i–ii. 3" possessed *more* knowledge between the fifth and tenth centuries before C., than Cosmas did in the sixth after that era, his logic would establish two things: 1st, his absolute ignorance of geology; 2d, of every principle of historical criticism.

Indifferent, ourselves, to the self-appropriation, by either side, of one or both of these branches of the alternative, we cannot leave the "Deluge" without one observation; the force of which theologers and geologists would do well to keep constantly in view. It is, that this genesiacal Flood is inseparable from NuK*h*'s *Ark*, or boat. Without the buoyant convenience of the latter, let ethnographers remember, the entire human race would have been drowned in the former.

We could quote a real historian, and living divine, who seriously speaks of Noah as "the great navigator." We have *seen* a wondrous plate of the "Ark,"(359) exhibiting the Noachic family pursuing their domestic and zoological avocations with the placidity of a Van Amburgh, and the luxuriousness of a Lucullus. We have read abundant descriptions of this diluvian packet-ship, in ecclesiastical and ponderous tomes, "usque ad nauseam." But, there is no work that does such pains-taking justice to the "Ark;" there is no man who has exhausted Noachian seamanship, antediluvian ship-building, cataclysmal proprieties, human and animal (from the "leopard *lying* down with the kid" in their berth, to the cheerful smartness of Ham the cabin-boy)—than Father Kircher,(360) almost two centuries ago. It is a shame that some great publisher does not reprint such a sterling good work, abounding in plates; as it might be a most useful field-manual to the orthodox geologist, and pleasing, at the same time, to children. Unable to do adequate honor to the Arkite researches of this Herculæan Jesuit, we must be content with the lucid description, in plain English, of the Rev. Dr. Lightfoot; who, living above two hundred years nearer to the Deluge than ourselves, no doubt knew considerably more than we do about the vessel that survived it. (361)

"The dimensions of the Arke were such, as that it had contained 450,000 square cubits within the walls of it, if it had risen in an exact square unto the top; but it sloping in the roofe, like the roofe of an house, till it came to be but a cubit broad in the ridge of it, did abate some good parcell of that summe, but how much is uncertain; should we allow 50,000 cubits in the abatement, yet will the space be sufficient enough of capacity, to receive all the creatures, and all their provisions that were laid in there. The building was three stories high, but of the staires that rose from story to story, the Text is silent; in every story were partitions, not so many, as to seclude one kinde of creature from another, for that was needlesse, there being no enmity between them, while they were there, and it would have been more troublesome to Noah to bring their provisions to them: but there were such partitions, as to divide betwixt beasts and their provisions in store: betwixt provisions and provisions, that by lying neer together might receive dammage. The doore was in the side of the lowest story, and so it was under water all the time of the flood; but God by so speciall a providence had shut them in, that it leaked not. In what story every kinde of creature had its lodging and habitation, is a matter undeterminable; how their excrements were conveyed out of the Arke, and water conveyed in, the Text hath con-

(359) YEATES: *Dissertation on the Antiquity, Origin, and Design of the principal Pyramids of Egypt;* London, 1833; pp. 9, 10, and pl. i.

(360) *De Arca Noë;* 1 vol. fol., Amsterdam, 1675.

(361) *The Harmony, Chronicle, and Order of the Old Testament;* London, 1647; ch. vi. pp. 8, 9.

cealed. All the creatures were so cicurated and of a tamed condition for this time, that they lived together, and dieted together without dissention: *The wolf dwelte with the lamb, and the leopard lay down with the kid, and the calf and the young lion together:* and Noah or any of his family might come among lions, dragons, serpents, and they had forgot the wildness and cruelty of their nature, and did not meddle with him."

Chronology, therefore, among men of science, possesses relation neither to the unknown epoch of the "Deluge," nor to that of the "Creation." These events, scientifically unseizable, are abandoned by positivists to theological tenacity.

Archæologists, in efforts to re-arrange the World's occurrences from the chaos into which ecclesiastical presumption had cast them, now pursue an altogether different process of inquiry. Beginning from *to-day*, as a fixed point in history if not in universal nature,(362) they retrograde, as closely as possible, year by year to the Christian era; said to be 1853 years backwards from the present year. From that assumed point, chronologers continue to retrocede, year by year, so long as history or monuments warrant such annual registration of events: but when, owing to absence of record or to confusion of accounts, the impossibility of identifying a given date for a given occurrence becomes manifest, they endeavor to define it approximately within a few years, more or less. In the ratio of their recession into the mists of antiquity, so does the possibility of fixing an approximate epoch diminish; and, therefore, it becomes necessary to group a given number of events into masses; which conventional masses become larger and less distinctly marked in proportion as they are remote from that era we call "the Christian."

The era of the miraculous birth of JESUS was the stand-point of chronologists; the pivot upon which every modern system turns. How minutely precise to the mathematician this era is, may be perceived, by archæologists, at a glance.

EPOCHS OF THE NATIVITY.

				Year of Rome.	Year before C.
According to	3	*authorities*	— Tillemont, Mann, Priestley	747	7
"	4	"	Kepler, Capellus, Dodwell, Pagi	748	6
"	5	"	Chrysostom, Petavius, Prideaux, Playfair, Hales	749	5
"	2	"	Sulpitius Severus, Usher	750	4
"	8	"	Irenæus, Tertullian, Clemens Alex., Eusebius, Syncellus, Baronius, Calvisius, Vossius	751	3
"	7	"	Epiphanius, Jerome, Orosius, Bede, Salian, Sigonius, Scaliger	752	2
"	3	"	Alexander Dionysius, Luther, Labbæus	753	1
			The moment of the Nativity is, consequently,	*zero*	0
					Year after C.
"	1	"	Herwart	754	1
"	1	"	Paul of Middleburgh	755	2
"	1	"	Lydiat	756	3

35 authorities, of the most orthodox schools, here differ among themselves *ten years* about the era of the grandest præternatural event in human annals; which event is itself dependent in epoch upon the implied accuracy of a date—*Anno Urbis Conditæ*, the "year of the building of Rome"—that, in his next pages, the Rev. Dr. Hales (363) shows to be fluctuating, according to *six* dates established by 34 chronologists, between the assumed year B. C. 753 and B. C. 627!

And this is what theologers term "chronology." In the American edition of Calmet,(364) the date of the Nativity appears thus (the reader being free to adopt, in a free country, whichever date he pleases)—the editor naively remarking, "It must, however, be borne in mind, that the particularity of the dates here assigned rests chiefly on mere conjecture":—

Year of World.	Before Christ.	Before A. D.	Year of Christ.
CALMET.	HALES.	CALMET.	CALMET.
4000	5	4	1

(362) HUMBOLDT: *Cosmos*; i. p. 178; note, on "The English Sunday"!
(363) *New Analysis of Chron.*; 1830; i. pp. 214, 217; GLIDDON: *Chapters*; 1843; p. 33; and *Otia*; 1849; p. 42.
(364) *Dictionary*; "Chronological Table;" 1832; pp. 947, 981.

However, avers the Rev. Dr. Horne,(365) "The true date of the birth of Christ is *four* years before the common æra, or A. D." This date we should not be unwilling to accept but for the Rev. Dr. Jarvis (366) — "The date being taken of December 25, by reckoning back thirty years from his baptism, we come to his birth, A. J. P. 4707, *six* years before the common æra." It would not be decorous in us to hold fast to such dogmatic extension by a Churchman who sacrilegiously derides a *mitre* — "Abp. Newcombe could say, 'Jesus was born, says Lardner, between the middle of August and the middle of November, A. U. C. 748 or 749. (Cred. I. 796, 9, 3d ed.) We will take the *mean* time, October 1.'!!!" The notes of admiration are the Rev. Dr. Jarvis's.

We have preferred quoting the latest authorities; but it need not be observed to the learned that this discussion has been revived periodically during the last ten centuries with no better result, than when agitated previously between the unbelieving Rabbis and the all-believing Fathers. *Ex. gr.*, John *of Spain* (367) sums up: —

"That there has been sought in what season of the year, in what month, and on what day our Saviour was born: some place this birth at the winter solstice; others, at the equinox of autumn or at the equinox of spring."

And again, Bossuet, one of the most enlightened men of his age, winds up his chronological investigations as follows: —

"Birth of Jesus, son of Joseph and Mary. — It is not agreed *as to the precise year* when he came into the world, but it is agreed that his *true birth precedes* by some years our vulgar era. Without disputing further upon the year of the birth of our Lord, it suffices that we know it happened in the *year* 4000 *of the world.*" [!] (368).

If we inquire the *age* of Jesus at his death, Bossuet tells us, that—"According to Matthew, he was 33 years old; to Pagan legend, 21; to Luke, 39; to Bossuet, 40,"

"Common Christians," as the Rev. Dr. Hitchcock designates them (ubi supra), may start back in amazement at these results upon the *year* of the Savior's *birth*, which the first slashes of an archæologic scalpel have now laid bare. Mystified by childlike or fraudulent authorities, they may or may not be grateful for the truth; but their conscientiousness will hereafter whisper to their minds that it is safest, perhaps, to become more charitable towards men of science; whose unwearied struggles to arrive at *a* chronology are superinduced by acquaintance with these facts. In the meanwhile, readers of Strauss and Hennell know why the settlement of the *year* of Jesus's nativity is one of those things not to be looked for; because, as Scaliger wrote — "to determine the day of Christ's birth belongs to God alone, not to man."

To "uncommon Christians," whose effrontery has led them to accuse Egyptologists of dissensions as to the epoch of the first Pharaoh, Menes, (by no thorough hierologist dogmatically fixed) we have merely to advise their prior determination of the year of Christ's nativity, before they henceforward venture into Egyptian polemics wherein they themselves are the only parties liable to "get hurt."

In a recent hieroglyphical work, to which allusion will be briefly made in its natural department, the Royal Astronomer, Professor Airy, (369) through profound mathematical calculations, obtains a celestial conjunction which he designates "2005 B. C.; April 8th." "B. C." implies *before Christ*. Now, as no human being can determine the *year* of Christ's advent; and inasmuch as the foregoing table exhibits a difference of opinion oscillating between ten years at least; we would respectfully solicit the astronomical era upon which the learned Professor founds his minute coincidence. Is it upon the "*star* of the east"(370) seen by the *Magi?* Or does he take the unknown moment of time "C." to be *zero?* Among archæologists, to say "B. C.," merely implies before an epoch conjectural for one or more

(365) *Introd. to the Crit. Study and Knowledge of the Holy Scriptures;* 8th ed., London, 1839; iii. pp. 527, 535.
(366) *Chronol. Introd. to the Hist. of the Church;* London ed., 1844; Preface, p. vii., and pp. 535, 563.
(367) *Quad. Istor. del. Lit. Arm.;* Venezia, 1829.
(368) BOSSUET: *Discours sur l'Hist. Univ.;* and *Art de vérif. les Dates*, par les Bénédictins de Saint-Maur.
(369) *Horæ Ægyptiacæ;* London, 1851; pp. 216 217.
(370) *Matthew;* ii. 1, 9, 10; omitted by MARK; called an "angel" in *Luke* ii. 9–15; and unmentioned by JOHN. Vide STRAUSS: *Vie de Jesus;* 1839; i. pp. 254–292.

years; but, without some more mathematical indication of the *astronomical* date of the birth of Jesus, those Egyptian calculations made at the Royal Observatory must be pregnant with error; and, at present, seem as valueless to chronological science, as are the *hieroglyphic* malinterpretations that originated such a waste of official labor and of nationally-important time.

To us, however, the forms "B. C." and "A. D." are merely conventional. No astronomical certitude is implied by their use. This year, which is the LXXVIIth *of the Independence of these United States*, may be, for aught we know, "A. D. 1850" or "A. D. 1860;" although vulgarly termed "the year 1853." When we use the customary era, chronologically, it simply means *one thousand eight hundred and fifty-three years* backwards from the present day; and "B. C." signifies whatever number of years the necessities of illustration compel us to place *before* the 1853d year thus specified. We leave Astronomy to astronomers.

With this proviso constantly present, the reader will understand that the only ancient chronological era, positively fixed, is the *Nabonassarian* — "February 26, B. C. 747." All other dates in ancient history are to this subordinate; although, for ordinary purposes, save when phenomena in the heavens can be historically connected with human events passing on the earth, "B. C." is both usual and adequate to the requirements of archæological science; still more of ethnological, wherein precision of specific eras is less imperative.

Our object, in this Essay (III), is to lay before the reader a general view of the relative positions which *Egypt*, *China*, *Assyria*, *Judæa*, and *India*, now occupy, in the eye of the monumental chronologist, on the tableau of different human origins. Like every other science that of chronology is progressive: in the cases of Egyptian and Assyrian time-registry essentially so; for, at the present year, 1853, the former study is immature, the latter scarcely commenced. That of China must be accepted upon the faith (which there is not the slightest reason to impugn) of what Chinese historians who, having no *theological* motives for unfair curtailment or for preposterous extension, have rebuilt from the archæology of their own country. There is but one nation of the five of which the utmost limit can, nowadays, be absolutely determined, and that is the Judæan; whose chronicles, in lieu of the first place still claimed for them by ignorance, now occupy, among archæologists, a *fourth* place in universal history. For Greece, Rome, and more recent populations, according to the criteria of their own annals, we refer the reader to well-known histories.

It will be remembered that, in "Types of Mankind," chronology is only one element out of many; and that we here profess merely to present the *results* of those chronological laborers who are now reputed to be the most scientific, and consequently the most accurate.

CHRONOLOGY—EGYPTIAN.

"Un certain public, ce public qui tour à tour admet sans preuve ce qui est absurde, et rejette sans motif ce qui est certain, satisfait dans les deux cas, parce qu'il se donne le plaisir de trancher les questions en s'épargnant la peine de les examiner; ce public qui croit aux Osages quand ils viennent de Saint Malo, mais qui ne croit pas aux Chinois, quand ils viennent de Pékin; qui est fermement convaincu de l'existence de Pharamond, et n'est pas bien sûr que le latin et l'allemand puissent être de la même famille que le sanscrit; ce public gobe-mouche quand il faut douter, esprit fort quand il faut croire, hochait et hoche encore la tête au nom de CHAMPOLLION, trouvant plus commode et plus court de nier sa découverte que d'ouvrir sa *grammaire*." (371)

"Quant aux hommes éminens qui ont conquis une belle place dans la carrière des études égyptiennes, il ne peut être question ici d'analyser leurs livres; il suffit que l'on sache bien que tous ont marché franchement dans la voie ouverte par CHAMPOLLION, et que la science qui a dû sa première illustration aux Young, aux Champollion, aux Humboldt, aux Salvolini, aux Nestor l'Hôte, et dont la réalité a été proclamée sans rétinence par les Sylvestre de Sacy et les Arago, compte aujourd'hui pour adeptes fervens et convaincus, des hommes tels que MM. Letronne, Ampère, Biot, Mérimée, Prisse, E. Burnouf, Lepsius, Bunsen, Peyron, Gazzera, Barucchi, Gliddon, Leemans, — [Abeken, Birch, Böckh, Bonomi, Brugsch, Brunet de Presle, De Saulcy, De Rougé, Harris, Hincks, Kenrick, Lanci, Lenormant, Lesueur, Mariette, Maury, Morton, Nott, Osburn, Perring, Pickering, Raoul-Rochette, Sharpe, Ungarelli, Wilkinson,] &c.—On connait maintenant les amis et les ennemis du système de CHAMPOLLION." (372)

"In short, the little spring of pure water which first bubbled from the Rosetta Stone, has, in twenty-three years, now swoln into a mighty flood; overwhelming all opposition;

(371) AMPÈRE: *Recherches en Égypte et en Nubie;* 1st art.; Revue des Deux Mondes, Aug. 1846; pp. 390, 391;—see also, *Ibid.: Promenade en Amérique;* Rev. des D. Mondes, June, 1853, pp. 1225, 1226.

(372) DE SAULCY: *De l'Étude des Hiéroglyphes;* Rev. d. D. Mondes, June, 1846; p. 983.

sweeping aside, or carrying in its surges, those whose inclination would induce them to stem its force; and, at the present hour, we know more of positive Egyptian history and of the ancient inhabitants of Egypt, ages previously to the patriarch Abraham, than on many subjects we can assert of our acquaintance with England before Alfred the Great, or with France before Charlemagne!" (373)

The work last cited, accessible to every reader of English at an insignificant cost, renders explanations on the incipient steps of hierological discovery herein superfluous. As a synoptical report of the progress of Egyptian studies it is correct enough, for general purposes, to the close of the year 1841. Our present point of departure is A. D. 1822.

"With Dr. Young's key, and Champollion's alphabet contained in his letter to M. Dacier, a group of scientific Englishmen, headed by Henry Salt, and subsequently aided by A. C. Harris, commenced in Egypt itself, about 1822, the scrutiny and examination of all the monuments of antiquity existing, from the Sea-beach to Upper Nubia, from the Oases to the peninsula of Mount Sinai, and in every direction through the Eastern and Western Deserts. These gentlemen, mutually aiding and co-operating with each other, were enabled to take instant advantage of the true method of interpretation. Egypt was then all virgin ground. Every temple, every tomb, contained something unknown before; and which these gentlemen were the *first* to date, and to describe with accurate details. A more intensely interesting field never opened to the explorer — every step being a discovery. Nobly did these learned and indefatigable travellers pioneer the way, and mighty have been the results of their arduous labors. They procured lithographic presses from England; and, at their individual expense, for private circulation, Messrs. Felix, Burton, and Wilkinson printed (at Cairo—1826 to 1829) and circulated a mass of hieroglyphical tablets, legends, genealogical tables, texts mythological and historical, with other subjects, which, under the modest titles of "Notes," (374) "Excerpta," (375) and "Materia Hieroglyphica," (376) were disseminated to learned societies in Europe. Lord Prudhoe's distant excursions and correct memoranda rendered the collections of antiquities, with which he enriched England, extremely valuable; and his labors were the more appreciated, as his lordship's liberal mind and generous patronage of science were above any sordid motives of acquisitiveness. Mr. Hay's own accurate pencil, aided by various talented artists whom his princely fortune enabled him to employ, amassed an amount of drawings that rendered his portfolios the largest then in the world. The researches of all these gentlemen have been of incalculable value to the cause. They have preserved accurate data on subjects,(377) that the destroying hand of Mohammed Ali has since irrevocably obliterated; and as they all pursued science for itself, they deserve and enjoy a full measure of respect. The rumor of their successes reached Europe; and Champollion, with reason, apprehended that, if he delayed his visit to Egypt any longer, the individual labors of English travellers would render that visit as unprofitable as unnecessary. National jealousy was excited; and, to preserve her position as the patroness of Egyptian literature, France determined not to be anticipated.

"In 1828, the French government sent a commission, consisting of Champollion le Jeune, and four French artists, well supplied with every necessary outfit, to Egypt, in order that the master might, for his own and his country's honor, and at her expense, reap the harvest for which his hand had sown the seed. A similar design having suggested itself to another patron of arts and sciences, the Grand Duke of Tuscany, the celebrated archæologist and oriental scholar, Professor Ippolito Rosellini, of the University of Pisa, and four Italian artists under his direction, were appointed a commission to proceed to Egypt, with the same intent as the French mission. It was amicably arranged by the respective governments, and between the chiefs of each expedition, that their labors should be united; and, in consequence, the French and Tuscan missions were blended into one, and both reached Alexandria in the same vessel, and prosecuted their labors hand in hand from Memphis to the second Cataract. They returned in 1829.

"It was amicably arranged, between Champollion and Rosellini, that they were to combine their labors in the works that were to be issued; each, however, taking separate branches—Champollion undertaking the illustration of the "Historical Monuments," and the grammar of the hieroglyphic language of Egypt — to Rosellini was assigned the task of elucidating, by the "Civil Monuments," the manners and customs of this ancient people, and the formation of a hieroglyphical dictionary. Each set to work by 1830; but Champollion, finding his end approaching, hastened the completion of his grammar. Intense application had prostrated the fragile frame which enveloped one of the most gifted mental

(373) GLIDDON: *Chapters on Early Egyptian History;* New York, 1843; p. 10: 15th ed., Philad., 1850.

(374) FELIX: republished in Italian, at Pisa; but now out of circulation.

(375) JAMES HALLIBURTON: out of print, and extremely rare.

(376) WILKINSON: like the preceding.

(377) GLIDDON: *Appeal to the Antiquaries of Europe on the Destruction of the Monuments of Egypt;* 1841; London, Madden.

capacities ever vouchsafed to man. The government gave him, in the Collège de France, a professor's chair, created for him alone; and his address to his pupils, at the first and only occasion accorded to him by Providence, is a marvel of eloquence, sublimity of thought, and classical diction.

"He finished his grammar on his death-bed, and summoning his friends around him, delivered the autograph into their custody, with the injunction 'to preserve it carefully, for I hope it will be my *visiting card* to posterity.' A few weeks after, Champollion le Jeune was followed to the grave by the noblest men of France; and the wreath of 'Immortelles' hung over his sepulchre (at his native town, *Figeac*), symbolized the imperishable fame of the resuscitator of the earliest records mankind has hitherto possessed."

His posthumous works were put to press at the expense of the nation, nor is their entire publication as yet complete. Death removed Rosellini (1841) before the *Monumenti dell' Egitto e della Nubia* received his final touches: and his worthy Italian colleague, Ungarelli, also died (1846) previously to the termination of the latter's *Interpretatio Obeliscorum Urbis.*

We may now proceed with a brief historical sketch of the steps through which Egyptian Chronology has become the criterion whereby the annals of all antique nations are now measured; subjoining references sufficient for the educated inquirer to verify bibliographical accuracy.

When Fourier, the polytechnic philosopher, in that masterpiece of eloquent erudition—the *Preface* to the "Description de l'Egypte"—claimed a period of *twenty-five hundred years* before the Christian era, (378) for the monuments which he, and the corps of illustrious Savans of whom Jomard is the surviving patriarch, had beheld in the valley of the Nile, his intuitive grasp of the amount of time adequate to the construction of then-unnumbered piles as gigantic in their architecture as diversified in their sculptures, obtained but little favor with the scholars, and none with the public of Europe, from 1810 to 1830. As when the immortal Harvey announced his discovery of the circulation of the blood, no surgeon, over forty years of age, but died an unbeliever in the theory; so forty years after the utterance of this chronological estimate by Fourier, and notwithstanding the victorious labors of the hierologists, do we still encounter cultivated minds unwilling to accept, or incapable of comprehending, the general truth of his proposition.

Equally unpalatable was this scale of 2500 *years*, at the time of its publication, to the representatives of two distinct schools; whom, for convenience sake, we will designate as the *long* and the *short* chronologists. On the one hand Dupuis and those astronomers who had claimed as much as 17,000 years B. C. for the erection of the temple of Dendera, and on the other, the followers of the *Petavian* and *Usherian* computations of the chronological element in Scripture, coincided in its rejection; the former deeming it too restricted, the latter too extensive for their respective cosmogenical theories. And, in a controversy in which the first principles of historical criticism, and a common basis of debate were alike wanting; before Young had deciphered the first letter in the hieroglyphical name of *Ptolemy;* before Champollion-le-jeune's "Précis" broke the spell in which the antique writings of the Egyptians had been bound for fifteen centuries: and at a day when absolutely *nothing* was known of the respective ages of Nilotic remains; the dogmatical assertions of the latter were infinitely preferable to the hallucinations of the former.

On his death-bed, in 1830, Fourier was solaced by the glimpse which Champollion, then just returned from his triumphant mission to Egypt, afforded him of the probable accuracy of his prospective vision: but, before the founder of Egyptological science could arrange the enormous materials collected for his *chronological* edifice, the 4th of March, 1832, overtook Champollion on his own death-bed, in the act of bequeathing the manuscript of his immortal *Grammar*, as "my visiting-card to posterity." (379)

In the same year, Rosellini commenced the publication of the "Monumenti dell' Egitto

(378) Champollion-Figeac: *Fourier et Napoleon — l'Égypte et les cent jours;* 1844; p. 61.

(379) *Grammaire Égyptienne;* 1835; Introduction. See also in Champollion-Figeac (*Notice sur les Manuscrits autographes de Champollion le Jeune*, perdus en l'année 1832, et retrouvés en 1840; Paris, 1842) the account of that wretched larceny which, while it accounts for the non-publication up to this hour of all the *Manuscripts* left by this indefatigable scholar, compels the historian to wipe his pen after writing the name—Salvolini. The example had, however, been previously set by the plagiarist of John Hunter's MSS.

e della Nubia;" in which, for the first time, an effort was made to embrace in one grand compendium all Egyptian documents in that day deciphered. Inheritor of the ideas, and associate in the labors of the great master, the Tuscan Professor's frame-work of chronology reflects Champollion's views on Pharaonic antiquity down to the close of 1830. The practical result of the erudite Italian's researches was the monumental restoration of the lost history of Egypt, back to the XVIIIth Dynasty, computed by him at B. C. 1822,—and the vindication of the general accuracy of Manetho, back to the XVIth dynasty, at B. C. 2272: (380) confirmed by Champollion-Figeac,(381) with many improvements and valuable suggestions; mainly drawn from "les papiers de mon Frère."

In 1835, Wilkinson's admirable work, "Topography of Thebes," presented a summary of the learned author's personal exploration of Egyptian monuments during some twelve years of travel in the valley of the Nile. The epoch of Menes, first Pharaoh of Egypt, was conjecturally assigned to the year B. C. 2201; but the accession of the XVIIIth dynasty placed at B. C. 1575, corroborated by the collation of hieroglyphical and Greek lists, evinced the critical author's appreciation of the solidity of Egypt's chronological edifice, and of Manethonian authority, at least up to the latter era.

We thus reach the year 1836; when B. C. 1822 as the *maximum*, and B. C. 1575 as the *minimum*, for the accession of Manetho's XVIIIth dynasty of Diospolitans, were already recognised by the world of science in general principle as established *facts:* and sixteen centuries of lost monumental history became resuscitated from the sepulchre of ages, through *hieroglyphical* researches that only commenced in A. D. 1822. (382)

But there had been, in Egypt, times *before!* there were still extant the *pyramids*, with the lengthy chain of tombs extending for above 20 miles along the Memphite necropolis, *unexplored;*—there were the "unplaced Kings" recorded in the "Materia Hieroglyphica" —the "Excerpta"—and the "Notes"—of Wilkinson, Burton, and Felix;—and there existed in the museums of Europe, as well as throughout the valley of the Nile, innumerable vestiges, recognised by every qualified student of Egyptology to belong to ages long anterior to the XVIIIth dynasty — immensely older than the year 1575—1822 B. C.; to say nothing of many biblical and classical texts that attested the necessity for more elbow-room in the chronology of the ancient Egyptians. Every one felt it: — every man who had *beheld* the storied ruins in Egypt itself asserted it, with more or less assurance according to the elasticity of the social atmosphere he breathed: — every hierologist *knew* it.

How was the conscientious discussion of these overwhelming questions avoided? Why were the countless monumental documents, that vindicated the claims of Manetho's first fourteen human dynasties to historical acceptance, left out of sight? Rosellini, while faithfully publishing all the materials in his possession, and throwing back pyramidal questions into the category of things anterior to the XVIth dynasty, having the fear of Petavius before his eyes, modestly declares —"Nè a me occorre indagare più addentro in *tanto bujo* di tempi." (383) Wilkinson, — in whose invaluable "Materia Hieroglyphica," among a host of "unplaced Kings," the names of *Shoopho*, *Shafra*, and *Menkera*, builders of the three great pyramids of Geezeh, had been published years before, and *two* of them at least read and identified, — Wilkinson, appalled perhaps at the authority of Usher, jumps at a bound, in his Plate I. of the "Dynasties of the Pharaohs," from MENaI, over SE-NEFER-KE-RA and RA-NEB-NAA, to RA-NUB-TER (which last he places in the XVth dynasty at B. C. 1830); omits every "unplaced King" published in his previous researches; ignores some *fifty Pharaohs* whose *monuments* prove they lived between Menes and the XVIIIth dynasty; and assigns only the year B. C. 2201 (!) to Menes, "for fear of *interfering* with the Deluge of Noah, which *is* 2348 B. C."

"I am aware," wrote, in 1835, the yet-unknighted Mr. Wilkinson, "that the era of Menes might be carried back to a much more remote period than the date I have assigned

(380) GLIDDON: *Chapters;* 1843; pp. 48, 49, and General Table, pp. 64, 65, 66.

(381) *Égypte Ancienne;* Univers Pittoresque, 1839.

(382) CHAMPOLLION: *Lettre à M. Dacier;* 1822.

(383) *Monumenti Storici;* 1832; vol. 1. p. 111

it; but as we have as yet no authority further than the uncertain accounts of Manetho's copyists to enable us to fix the time and the number of reigns intervening between his accession and that of Apappus, I have not placed him earlier, for fear of interfering with the date of the deluge of Noah, which *is* 2348 B. C." (384)

The inconsistencies inherent in this scheme of chronology were exposed in 1843; (385) nevertheless, in his most excellent later work, "Modern Egypt and Thebes," 1843, as well as in his "Hand-book," 1847, this erudite Egyptologist has left *chronological* disquisitions pretty much as he had defined them in 1835 — as if inquiry had been *stationary* in Europe during twelve years! — although, when treating *geologically* on the antiquity of the *Delta*, "il laisse percer le bout d'oreille" in the following scientific assertions: —

"We are led to the necessity of allowing an immeasurable time for the total formation of that space, which, to judge from the very little accumulation of its soil, and the small distance it has encroached on the sea, since the erection of the ancient cities within it, would require ages, and throw back its origin *far beyond the Deluge*, or even the *Mosaic era of the Creation.*" (386)

In consequence, Sir J. G. Wilkinson granted a reprieve of some few years to poor Menes; for (1837) in the same "Manners and Customs," this Pharaoh's accession is placed at B. C. 2320; or only 28 years after the Flood!

It is sufficient, herein, to point out to the reader, that the year 1836 closed with a mighty stride, already accomplished, into the "darkness of Egypt;" through which a *mass of time*, exceeding *fifteen centuries* in duration, was irrevocably restored to the world's history. The mutilated annals of the oft-maligned Priest of Sebennytus were vindicated by an unanswerable appeal to monuments contemporaneous with the Pharaohs recorded by him, back to his XVIIIth Theban dynasty. More than one-half of the *twenty-five hundred* years claimed by Fourier, and Napoleon's "Institut d'Égypte," was thenceforward restored to positive history by the *Hierologists*.

The years 1837 to 1839 witnessed the munificent expenditures, and fulfilment of the grand conception, of a Vyse; the self-sacrificing exertions of a Perring, but for whose fortitude, enthusiasm, and engineering skill, small, indeed, would have been the scientific results accruing from such immense undertakings; and the archæological acumen of a Birch, in deciphering and assigning an historical place to the fragmentary legends disenterred among some 39 pyramidal mausolea (387) of the *Memphite* and *Arsinoïte* nomes. (388) Simultaneously with these successes, the *Tablet of Abydos*, that most precious register of the genealogy of the RAMESSIDES, found its way to the British Museum. (389)

Lenormant, (390) we believe, was the first to apply the new discoveries to *chronology*; and Nestor L'Hôte (391) to retread the Memphite necropolis, and verify some of the data obtained by the English explorers.

The combined result of these researches, in the year 1840, was the recognition of the great principle, that the *pyramids*, without exception, antedated the XVIIIth dynasty, already established between the fifteenth and the eighteenth centuries B. C.: — that a mass of "unplaced Kings," and a vast field of unopened tombs in the burial-ground of Memphis; together with a prodigious variety of lesser monuments, stretching from the peninsula of Sinai to the temples of *Samneh* and *Soleb* in Upper Nubia; still preserved authentic records coetaneous with the first *twelve dynasties* of MANETHO: and that, from out of the chaos, the

(384) *Topography of Thebes;* 1835, pp. 506 and 509.

(385) GLIDDON: *Chapters;* pp. 51, 52.

(386) *Manners and Customs;* 1837–'41; i. pp. 5–11; ii. pp. 105–121;—compare *Otia Ægyptiaca;* pp. 61–69.

(387) *Operations carried on at the Pyramids of Geezeh*, from 1837 to 1839.

(388) SHARPE: *Chronology and Geography of Ancient Egypt;* 1849; pl. 11, Map, *Ancient Egypt under Ant. Pius.*

(389) LEPSIUS: *Auswahl;* 1842; pl. 11; — BIRCH: *Gallery of Antiquities;* part ii. pl. 29, and pp. 66–71; — LETRONNE: *Table d'Abydos*, imprimée en caractères mobiles; Paris, 1845; pp. 24–36; — BUNSEN: *Egypt's Place;* 1848; pp. 44–51; — DE ROUGÉ: *Examen de l'Ouvrage de M. Bunsen;* 1847 pp. 16, 17, *Extrait des Annales de Philosophie chrétiennes;* and *Ibid.: Deuxième Lettre à M. Alfred Maury, sur le Sesostris de la XIIme Dynastie;* Revue Archéologique, 15 Oct. 1847; pp. 479, 480; — LESUEUR: *Chronologie des Rois d'Égypte;* ouvrage couronné, Paris, 1848; pp. 260–263; — PRISSE: *Notice sur la Salle des Ancêtres de Thoutmes III.;* Rev. Archéol.; Paris, 1845.

(390) *Eclaircissemens sur le Cercueil de Mycerinus;* Paris, 1839.

(391) *Lettres d'Égypte:* Paris, 1840.

IVth Manethonian dynasty, cotemporary with the building of the *Geezeh* group of pyramids, loomed like a meteor in the night of time.

Some perceptions were entertained, about those days, even in America, of the probable extent to which *monumental* researches would eventually carry the epoch of MENES. In 1845, Bunsen's era for this monarch was B. C. 3643; and in 1849, Lepsius's is B. C. 3893. Our "*Chapters*" (1843) assert, that "if 1000 more years could be shown admissible by Scripture, there is nothing in Egypt that would not be found to agree with the extension." It is a happy coincidence, exhibiting how different minds, in countries widely apart, reasoning upon similar data, arrive at conclusions nearly the same, that, if the above "1000 years" be added to our former conjectural and *minimum* estimate, printed *ten* years ago, of the date of MENES, noted at *about* B. C. 2750,(392) the sum B. C. 3750 falls, almost equidistantly, between the eras assigned to this primordial Pharaoh by two of the *three* highest hierological chronographers: — the third, it need scarcely be observed, being Mr. Birch; who, whilst tabulating Egyptian events in the recognised order of Manethonian *dynasties*, (393) has never yet put forth an arithmetical *system* of hieroglyphical chronology. As remarked by us (*Otia*, p. 45):—

"We are dealing, in events so inconceivably remote, with *stratified masses of time*, and not with supposititious calculations of the *exact* day, week, month, or year; in futile attempts to ascertain which so many learned investigators "ne font qu'un trou dans l'eau."

Our sketch of the progressive conquests over the past, commenced by Champollion in 1822, through which a pathway has been hewn, inch by inch, by the axes of the Hierologists, far into the briery jungle of Pharaonic antiquity, has reached the year 1843; and already Fourier's "*twenty-five hundred* years B. C." for the monuments of the Nile, even to the uninformed eye, began to wear the garb of probability—to the hieroglyphical student, who had *actually beheld with his own eyes* these monuments *in Egypt itself*, they had assumed in that year the aspect of certainty.

It is a remarkable fact, that with the exception of Wilkinson, whose chronological consistency has been indicated (*supra*), not one of those Egyptologists of whom the critical opinion is now authoritative, and who, at this day, yet aspires to the name of a *short*-chronologist (that is, one to whom the *Usherian deluge*, at B. C. 2348, is a bed of Procrustes), has ever studied Egyptian monuments *in Egypt!* Much allowance, therefore, should be made for living English scholars who still, like the ostrich, bury their heads in sand; surrounded as they are, essentially, by the "intellectual *flunkeyism*" for which this age, in England, is eminently celebrated among scientific men on the Continent and in the United States. The ponderous weight of brains, congealed in the "cast-iron moulds" of Oxford and Cambridge, presses upon British intelligence and education with the numbing power of an incubus. Among recent vindicators of the claims of Egypt to the *longest* chronology is Ferguson ("True Principles of Beauty in Art," &c., London, 1849), to whose crushing pamphlet we must refer admirers of the educational "standard of a by-gone and semi-barbarous age," upheld in "the Sister Universities;" with which standard the citizens of republican America, of course, need have nothing to do, physically, morally, or intellectually.(394)

The discovery made by Lepsius, in 1840 (not publicly known for some years later), that the *Tablet of Abydos*, between *Cartouche* No. 40 and No. 39, omits the XIIIth, XIVth, XVth, XVIth, and XVIIth Manethonian dynasties, thus *jumping over the entire Hyksos-period*, (395)

(392) I am happy to find that this (by myself long ago abandoned — *Otia*, pp. 37–42) scheme of the possible epoch of Menes, approximates so nearly to the date adopted by Nolan; who places, according to the "Old Chronicle," Menes (whom he takes to be Noah!) at B. C. 2673; or only *ten* years difference from "my reduction of the *Old Chronicle*, B. C. 2683," five years previously—(compare *Egyptian Chronology analysed*; London, 1848; pp. 133, 156, 212, and 399, with *Chapters*, p. 51). Still less does it differ from the point at which a "great authority, whose permission I have not asked to give his name," fixes (*astronomically* speaking) the era of Egypt's first Pharaoh: viz., B. C. 2714–'15—the *very date* (B. C. 2715) to which I had reduced Manetho, in 1843. Compare *Literary Gazette;* London, 1849; pp. 485, 522, and 641; with *Chapters;* p. 51.)—G. R. G.

(393) "Relative Epochs of Mummies," in *Otia Ægyptiaca;* pp. 78–87; also, pp. 113–115.

(394) *Observations on the British Museum, National Gallery, and National Record Office;* London, 1849.

(395) BUNSEN: *Ægypten's Stelle;* 1845; ii. p. 277; and *Egypt's Place;* 1848; pp. 42, 49, 52. Compare HINCKS: *On the Egyptian Stele;* 1841; p. 68; and BARUCCHI: *Discorsi Critici sopra la Cronologia Egizia;* Torino, 1845; pp. 129–131.

had marked a new era in the chronological consideration to be awarded to some royal *genealogical* Tablets. This discovery was by far the most important feature of that day; but so varied and unforeseen were the victorious achievements effected, in the year 1843, by the Prussian Scientific Mission, among the *pyramids*, from Memphis to the Labyrinth; so completely have they revolutionized all preceding judgments upon Nilotic antiquity; that we must pause to indicate how they originated, and where they are to be found.

Chevalier Richard Lepsius, long celebrated as Corresponding Secretary of the *Institute of Archæological Correspondence* at Rome, directed his studies into Egyptology soon after the publication of a prize-essay, (396) that placed him in the front rank of linguistical scholarship in 1834. A *Lettre à M. le Prof. Hippolite Rosellini sur l'Alphabet Hiéroglyphique*, 1837, (397) next announced, to the world of science, that the loss of the illustrious Champollion had but momentarily arrested the onward march of his disciples. The return of Perring from Egypt after his indefatigable exploration of 39 pyramids, (398) rendered the fact generally known that, immense as had been his own successes, the necropolis of Memphis had, notwithstanding, scarcely begun to yield up its historical treasures. French and Tuscan national, with English private enterprise, had been rewarded, in the valley of the Nile, by victories over past time as noble as they were scientific. It remained for Frederic William IVth of Prussia to give full scope to the hitherto pent-up yearnings of Germany towards Egyptian discovery; and upon Lepsius, in 1842, naturally fell the mantles of his predecessors.

With eight coadjutors, the Chief of the Prussian Scientific Mission pitched his tents in the shadow of the great Pyramid on the 9th of November, 1842.

By May, 1843, he was enabled to announce that the Germans had gleaned the sites of "*thirty other pyramids*, entirely unknown to him (Mr. Perring), or to any preceding travellers. Of these, not a few are of very considerable extent, bearing evident traces of the mode in which they were raised, and surrounded by the ruins of temples, and extensive fields of tombs or burial-grounds. All these pyramids, without exception, belong to the *ancient kingdom* of Egypt *before* the irruption of the Hykshos, who invaded Lower Egypt about the year 2000 B. C., and the whole of them were erected (those at least between Abrorooàsh and Dashoor) by kings who reigned at Memphis. To the same period belong also the majority of the *effaced* tombs, of any importance, that surround them." (399)

After determination of the sites, and unfolding much of the history of "*sixty-seven* pyramids," sepulchres of ancient Egyptian sovereigns; together with "*one hundred and thirty* private tombs" of noble families, with these sovereigns coetaneous, back to the "*fourth thousand year* before Christ," the Prussians proceeded up the river; exploring every foot of ground, as far as *Soba* on the Blue Nile (*Bahr-el-Azrek*), and *Sennâr* to the 13th degree of N. latitude; returning to Thebes on the 2d November, 1844. While his able assistants prosecuted the necessary labors amid Theban ruins, Lepsius crossed the Red Sea and explored the Sinaic Peninsula; not only, thereby, rescuing from perdition hieroglyphical records of mining operations conducted between the IVth and the XIIth dynasty, 3400—2200 B. C., but also ascertaining that, if the *Gebel Serbâl* be not the Mount of Moses, of which there is little doubt, (400) the peaks above the Convent of St. Catherine most assuredly are not. Revisiting Thebes, Lepsius left it with his party on the 16th May, 1845: and after examining the land of Goshen, much of Palestine, and touching at Smyrna and Constantinople, landed at Trieste on the 5th January, 1846: having spent above *thirty-six months* in unparalleled monumental researches on the river, alluvium, and deserts of the Nile.

The reader will now perceive that we are dealing in realities; that our Egyptian deductions are based upon actual and positive researches, made by the "primi inter pares" of

---

(396) *Palæographie als Mittel für die Sprachforschung zunächst am Sanscrit nachgewiesen;* Berlin, 1835; 8vo

(397) *Annali dell' Instituto di Corrispondenza Archeologica;* vol. ix.; Roma, 1837.

(398) Vyse: *The Pyramids from Actual Survey;* iiird vol.; 1841.

(399) Lepsius; *Ueber den Bau der Pyramiden:* Berlin Academy, August, 1843; pp. 2, 3; — see the order of announcement of these discoveries in Gliddon: *Otia;* 1849; pp. 30–42.

(400) *Tour from Thebes to the Peninsula of Sinai, in March and April*, 1845; transl. Cottrell; London, 1846. We possess the *German* edition; with its tinted *map*, without which Lepsius's certain discovery is not so evident to the general reader.

living Archæologists, previously qualified by lengthened discipline, and furnished by munificent governments with facilities as unexampled as unbounded. We subjoin a list of the works (401) since published by Lepsius, that have been carefully consulted in the preparation of "Types of Mankind;" and may mention that, while one of its authors sojourned at Berlin in May, 1849, both are in frequent epistolary communication, on the themes this work discusses, with the esteemed Chevalier himself.

Consequently, whether the deductions drawn by the authors of the present volume be right or wrong, the *facts* upon which these are grounded are vouched for by the highest authorities. No attention is bestowed, in "Types of Mankind," to the puerilities of the ephemeral tourist, to the twaddling inanities of the unlettered missionary, or to the Egyptian hallucinations of the theological rhapsodist. At the present day (without disparagement to the less-known literary resources of other cities on our continent), (402) a qualified student, in this year A. D. 1853, can sit down quietly at *Mobile*, Alabama; and the books contained in four private libraries will enlighten him, upon almost every point our work discusses, with smaller trouble and greater economy of time, labor, and money, than if he resided for *years*, without previous knowledge of these works, in the valley of the Nile: or, should such student prefer Philadelphia, there, at her Library, his bibliothecal aspirations can be satisfied.

How utterly hopeless it is for any man (apart from erudition) unsupported by enormous pecuniary means, to advance Egyptian sciences, at the present day, by a steam-boat excursion up the Nile, may be inferred from three facts. In 1844–5, Ampère, one of the living luminaries of archæological knowledge, was sent out by the French Government expressly to make discoveries. His "Recherches en Egypte et en Nubie" in literary excellence are unsurpassable; yet, withal, his predecessors had left him so little to do, without a protracted sojourn, that he refers to Lepsius for every novelty discoverable:—

"Je n'ai pas touché, sans un certain respect, ce livre *des Rois*, commencé par lui avant son voyage d'Egypte, et qui contient une collection de noms royaux plus complète qu'aucune autre ne peut l'être, et un ensemble de chronologie Égyptienne depuis l'ancien roi Ménès jusqu'à Septime Sevère. Cette série va plus loin encore, car M. Lepsius ne s'arrête pas à ce nom, le dernier qu'eussent trouvé écrit en hiéroglyphes Champollion et ses autres successeurs. M. Lepsius a été assez heureux pour découvrir, dans un petit temple de Thèbes où Champollion avait trouvé le nom d'Othon, les noms de *Galba*, de *Pescennius Niger*, et, ce qui est plus important, de l'empereur *Dèce*. Par cette découverte, M. Lepsius prolonge la série hiéroglyphique d'un demi-siècle au déla de Septime Sevère, où elle s'arrêtait jusqu' ici. *On a donc une suite de monumens et d'inscriptions qui s'étendent depuis* 2500 *avant Abraham jusqu'à* 250 *ans après Jesus Christ.* Il n'y a rien de semblable dans les annales humaines." (403)

Two years previously, Prisse d'Avesnes had rescued the *Ancestral Chamber of Karnac*, the *Tablet of Ramses* XIV, (404) and other precious relics, from Turkish demolition. A residence of sixteen years in Egypt, of which about five in the Upper country among the monuments, had enabled this proficient Orientalist to fill his portfolios with every archæological item discovered, chiefly too by himself, between the departure of the French and Tuscan Scientific Commissions under Champollion and Rosellini, 1830, and the advent of the Prussians in 1842. So valuable were M. Prisse's self-sacrificing labors in Egyptology

(401) *Vorläufige Nachricht über die Expedition;* Berlin, 1849; — *Briefe aus Ægypten, Æthiopien, und der Halbinsel des Sinai;* Berlin, 1852; also, its excellent English translation, by Mr. KENNETH B. H. MACKENSIE: "Discoveries in Egypt," &c.; London, 1852; — *Einleitung zur Chronologie der Ægypter;* Berlin, 1848; vol. i.; — *Ueber der Ersten Ægytischen Götterkreis;* Berlin, 1851; — *Ueber den Apiskreis;* Leipzig, 1853; — *Ueber die Zwölfte Ægyptische Königsdynastie;* Berlin, 1853; — and, above all, the magnificent *Denkmäler aus Ægypten und Æthiopien;* Berlin, 1849; folio. Of this vast work, besides a series of the earlier ethnological plates kindly selected for him by Chev. LEPSIUS, and in his own possession, the writer has enjoyed the free use of two copies at Mobile, in the private libraries of Mr. A. STEIN and of the Rev. Dr. HAMILTON — to both of whom he here begs to reiterate his obligation — and of another in the Philadelphia Library. Altogether, he has seen the *plates* down to *Abth.* III., *Bl.* 172.

(402) I am speaking of public libraries. The private library of my honored friend, Mr. R. K. HAIGHT of New York, has been, from the commencement of my studies in 1842, the main source whence my individual facilities have been drawn.

(403) *Recherches en Égypte;* vii.; Thébes, 21 Jan. 1845; — *Revue des Deux Mondes;* 1842; p. 1035.

(404) *Salle des Ancêstres de Thoutmès III.;* Rev. Archéol.; 1845; pp. 1–23, tirage à part; — BIRCH: *Egyptian Inscription in the Bibliothèque Nationale;* Trans. R. Soc. Lit., new series, iv.; 1852.

deemed by Parisian science that, at national expense, he was appointed to continue the great folios of Champollion; (405) at the same time that his contributions to the *Revue Archéologique* are standard documents for posterity.

Last though not least, in Egypt itself resides a gentleman, affluent and influential, versed in many branches of ancient lore as thoroughly as 30 years of domicile have familiarized him with modern affairs, who never allows an opportunity of advancing archæological science to escape him; nor will any Egyptian student mistake our allusions to A. C. Harris. (406)

No clap-trap pretensions to acquaintance with hieroglyphical arcana recently made by theologers who speak not any continental tongue through which alone these subjects are accessible—no "ad captandum" figments of the possession of Oriental knowledge when men cannot spell a monosyllable written in the Hebrew alphabet — detract from the Memphite exhumations conducted at French ministerial expense by a Mariette; for whose enormous discoveries in the *Serapeum*, as yet confined to reports, we wait impatiently. 'T were well if, in view of the contemptuous silence with which Egyptologists treat their publications, some writers on these matters were to become *readers.*

Our part, however, is to indicate to the reader those sources upon which Egyptian chronology is dependent at the present day, in regard to the date of the first Pharaoh, Menes: a personage considered, in the subjoined works, to be historical; and neither connected with the mythical *Mestræans* invented by the Syncellus (407) in the seventh century after C.; nor, except nationally, with the MTsRIM (not *Mizraïm*) of the Hebrew Text, whom, in our examination of Xth Genesis, we have proved to be nothing more or less than the "Egyptians," inhabitants of MiZR, *Muss'r;* the Semitic name of "Merter," *Egypt* [*supra*, p. 494]: —

	Authorities.	*Dates of Menes.*
1839, Paris	LENORMANT: *Cercueil de Mycerinus*—	B. C.
	IVth Dyn. (p. 24) "Mycerinus, la date de 4136 avant J. C."	
	Add IIId " *Africanus* " 214 "	
	" IId " " " 302 "	
	" Ist " " " 263 "	
		4915
1840, Paris	CHAMPOLLION-FIGEAC: *L'Égypte Ancienne*	5867
1845, Berlin	BÖCKH: *Manetho und die Hundssternperiode*	5702
1845, Turin	BARUCCHI: *Discorsi Critici sopra la Cronologia Egizia*	4890
1845, Hamburg	BUNSEN: *Ægyptens Stelle in der Weltgeschichte*	3643
1846, Paris	HENRY: *L'Égypte Pharaonique*	5303
1818, Paris	LESUEUR: *Chronologie des Rois d'Égypte*	5773
1840, Berlin	LEPSIUS: *Chronologie der Ægypter*	3893
1851, Dublin	HINCKS: *Turin Papyrus*	3895
1851, London	KENRICK: *Egypt under the Pharaohs*	3892
1854, Philadelphia	PICKERING: *Geographical Distribution of Animals and Plants*	4400

The views of the authors of *Types of Mankind*, while with Humboldt, (408) for reasons to be given anon, they follow Lepsius, incline to the longer rather than to the shorter period. Ampère's opinion has been previously cited. The following is that of the first hierologist of France, Count Em. de Rougé, Conservator at the Louvre Museum: —

"Les efforts de M. de Bunsen seraient la meilleure preuve du contraire; après avoir, sans égard pour l'histoire et les monumens, supposé des *règnes constamment collatéraux*, trois dynasties à la fois et huit ou dix rois *simultanés* pendant la moitié des 12 premières dynasties, il n'en fixe pas moins le règne de *Ménès* à l'an 3643 av. J. C. L'obstiné fils de *Chanaan*, mutilé avec acharnement pendant 3 volumes, se relève enfin de ce lit de Procuste où l'avait étendu son critique impitoyable, et l'on s'apperçoit alors qu'il dépasse encore de plu-

(405) *Continuation des Monumens;* 100 plates; 1848; — *Papyrus Égyptien;* 1849.

(406) Mr. HARRIS's contributions, in the *Trans. of the R. Soc. of Literature*, the *Revue Archéologique*, and in the pages of several Egyptologists, are too numerous for specification here: but we may refer to his *papyrus*, "Fragments of an Oration against Demosthenes," London, 1848; also to the papyric fragments of "Books of Homer" (*Athenæum*, 8 Sept. 1849), and of the "Grammarian Tryphon" (*Athenæum*, 7 Dec. 1850): while of the very important work — "Hieroglyphical Standards representing Places in Egypt supposed to be Nomes and Toparchies, collected by A. C. HARRIS," M. R. S. L., 1852 — his kindness allows us to acknowledge receipt.

(407) LETRONNE: in BIOT's *Année Vague des Égyptiens;* p. 25: — *supra*, p. 494.

(408) *Cosmos;* ii. pp. 114, 115, 124: — *supra*, p. 245.

sieurs siècles les mesures qu'on lui avait imposées au nom des calculs que la chronologie ordinaire avait fondés sur la *genéalogie d'Abraham.*" (409)

We moreover coincide entirely in the same author's doctrine, when, after indicating the various chances of miscalculation inherent in Egyptian no less than in all other *chronologies,* he declares: —

"These causes of error, which cross each other in every direction, make up a large part of uncertainty, for any chronological sum that it may be wished to draw from the sole addition of reigns, after a number of centuries at all considerable. The chances of inexactitude augment with the number of partial sums; and I have always thought that an uncertitude of more than 200 years was very admissible, in the ciphers that result from monumental dates combined with the lists of Manetho, when one remounts to the XVIIIth dynasty, after the expulsion of the shepherds." (410)

Nor need any doubt be entertained upon De Rougé's adoption of the most lengthy chronology, when he declares elsewhere — "Were we to accept the data most clearly preserved in Manetho, the XIIth dynasty must have preceded the Christian era by *thirty-four centuries.*" (411)

We have already seen that, in England, the profoundest hieroglyphical scholar, Birch of the British Museum, tabulates Manethonian dynasties in their serial order, but without encumbering his monumental discoveries with any arithmetical *chronology.* Kenrick follows Lepsius. Hincks's former depression of the reign of Ramses II., in the XVIIIth dynasty, and of Thotmes III. to the year 1355 B. C., on the ground that Egyptian armies (born amidst solar calorics) avoided the heat of the weather, (412) was an argument too feeble to be seriously combated; but the matured judgment of this universal savant favors every scientifical extension demanded for Nilotic annals.

"A statement has been preserved, to which I am now inclined to attach more credit than I did formerly, that the Egyptians reckoned all the dynasties from Menes to Ochus as occupying 3555 years. If from this number we subtract 2291, which the Egyptians reckoned from Menes to the end of the XIIth dynasty, we have 1264 from the end of the XIIth dynasty to Ochus, or to 340 B. C. This would place the XIIth dynasty between the limits 1817 and 1604 B. C.; and I am disposed to accept these dates as the genuine Egyptian computation. Nor indeed do I see much reason to question their correctness."

Followers ourselves "of the German and French school," we pause not to debate the learned Irishman's deductions as to such an untenably modern date for the XIIth dynasty; but, adding his accepted 3555 years to the reign of Ochus, B. C. 340, we are gratified in finding that Dr. Hincks, (413) with several Germans and Frenchmen, places Menes at 3895 *years before* C.; and henceforward, therefore, can enrol, as we have already, his great name among the long chronologists.

On the opposite side, as representative of the shortest Egyptian computation, stands a gentleman, whose vast classical erudition, and keener criticism, we are always proud to acknowledge; and it is with pain that, having so often availed ourselves of his instructive pages, especially in regard to biblical history and exegesis, that, in Egyptian *chronology,* we must protest against the contracted system of a great Hellenist, Mr. Samuel Sharpe. With respectful deference we would, however, submit objections to his assumed dates for Osirtesen, whom he arbitrarily changes into an "*Amunmai Thor* I.;" (414) still more emphatically to his views upon Menes. Scientific criticism, to be practically useful, must be free; and pupils, often, of Mr. Sharpe in its application to the Greek *New Testament,* and to the theosophical notions of the Alexandria School, we feel persuaded that no writer of the day loves truth more than himself. We may therefore utter our mode of viewing it.

(409) *Examen de l'Ouvrage de M. Bunsen;* p. 82, Annales de Philosophie Chrétiennes, 1847.

(410) De Rougé: *Mémoire sur quelques Phenomènes Célestes;* Rev. Archéol., 183; p. 654; — Comp. *Otia,* p. 41.

(411) *Sur le Sesostris de la Douzième Dynastie;* Rev. Archéol., 1847; p. 482.

(412) Rev. Dr. Hincks: *On the Age of the XVIIIth Dynasty;* Trans. R. Irish Acad., 1846; xxi. pp. 5–9.

(413) *Observations of Dr. E. Hincks,* in Wilkinson's "Hieratic Papyrus of Kings at Turin," 1851; pp. 57, 58.

(414) *History of Egypt;* new edition; London, 1846; pp. 7, 9, 10; — *Chronology and Geography of Ancient Egypt;* 1849; pp. 4, 14, pl. 2, figs. 25, 32.

The *contemporaneousness* of Egyptian dynasties (415) we have always repudiated; (416) but, until the appearance of Lepsius's "Book of Kings," when our assent may possibly be yielded (if *monuments* to us now unknown establish it), in respect to the Ist and IId, VIth and VIIth (VIIIth), Xth and XIth, XIIIth and XIVth, and XVth and XVIth, Manethonian dynasties, we should commit the same fallacy, so frequently blamed in others, if we spoke dogmatically on that point without the new documents of the Prussian Mission. There is no more foundation, however, for Mr. Sharpe's dynastic arrangement than were we to make *Canute's* invasion of England coeval with WILLIAM *the Conqueror* in the reign of JAMES I., under the synthronic sway of GEORGE III and the Prince Regent. It is a favorite hypothesis of his own; in which not an Egyptologist coincides. But for the exposure of a radical error in Mr. Sharpe's system—root of all his deviations from hierological practice—our knife must be applied to one of its many vital spots. In his immensely-valuable folio *plates,* (417) through inadvertency, he had read

 *nfr,* (418) the "lute," *théorbe,* in lieu of  *tt,* (419) the "blade of an oar,"

as the sculpture stands. Through misapprehension of the groups (in *line* 9 compared with *line* 2, of the same inscription), Mr. Sharpe then deemed that this malcopied sign "*nfr*" was the homophone of  *b,* (420) the "human leg;"

and, in consequence, he always reads "*nfr*" as if it were the latter articulation—"That the arrow-shaped character is rightly sounded B or V is proved by its admitting that sound in the above four names, as also in No. 160 and No. 165." (421) The extraordinary metamorphoses of well-known royal names which this misconception, founded upon a *mistake,* has occasioned, are too evident to the hierologist to require comment. Unfortunately, through such concatenation of fallacies, Mr. Sharpe (422) transmutes the *prenomen* of Queen AMENSeT, (423) and the *nomen* of this queen's husband AMENEMHA, (424) and the oval of MENKERA, (425) into a fabulously bisexual "Mychera-Amun Neitchori"—rolls up the IVth, VIth, and XVIIIth dynasties into one—and thus makes the 3d pyramid of Geezeh (B. C. 3300) *contemporary* with the majestic obelisk (B. C. 1600) in the temple of Karnac! It is as if one were to call EDWARD *the Confessor* the same *personage* as "VICTORIA and ALBERT;" and then to insist that the former's *tomb* in Westminster Abbey must be *coeval* with the equestrian statue of WELLINGTON at Hyde Park corner! (426)

Mr. Sharpe's restricted system of Egyptian chronology, for times anterior to THOTHMOSIS III. (placed by him in the 14th century B. C.), may now be considered as "non-avenu." But, while compelled to shatter its superstructures down to his XVIIIth dynasty, let no one impute to us lack of respect for the profound author of the "History of Egypt"—a work that (from *page* 30 to 592) ever has our warmest admiration. Contenders for the longest

(415) SHARPE: *Chronology;* pp. 14, 15.

(416) GLIDDON: *Chapters;* p. 57;—*Otia;* pp. 39, 45.

(417) SHARPE: *Inscriptions in British Museum;* pl. cxvi., line 9, and line 2.

(418) BUNSEN: *Eg. Pl.,* i. p. 587, No. 31;—CHAMPOLLION: *Dictionnaire;* p. 293, No. 338—"NOFRE."

(419) BUNSEN: No. 30;—CHAMPOLLION: p. 378, No. 459—"TOUW."

(420) BUNSEN: p. 558, B, 1;—CHAMPOLLION: p. 100, No. 60—"B."

(421) *Chronology;* p. 4.

(422) *Op. cit.;* p. 6, Nos. 60, 61, 60; and plate ii., figs. 60, 61, 62.

(423) ROSELLINI: *Cartouche* No. 103.

(424) *Ibid.; Cartouche* No. 103 *f.*

(425) BUNSEN: *Ægyptens Stelle;* iii., pl. i.—*Men-ke-u-ra.*

(426) It is a year ago since this was written, and so reluctant do I feel to contradict a respected fellow-laborer, that I should have suppressed these comments but for a "rifacimento" of the same doctrines reported in the London *Athenæum,* Nov. 19, 1853. "The third aim of the paper was to show that the 3d and 4th pyramids were both made by Queen Nitocris, who governed Egypt during the minority of Thotmosis the IIId. The name of King Mycera has been found in both of these pyramids; Mycera is the first name of Queen Nitocris [!], and it was probably the name used in Memphis for Thothmosis the IIId." &c.—(*Syro-Egyptian Soc.,* Nov. 8.)

human chronology ourselves, it is imperative upon us to carry the outworks of truly-erudite short-chronologists before storming their last English citadel: a facile exploit now to be performed.

> "The thistle that *was* in Lebanon
> Sent to the cedar that *was* in Lebanon
> Saying, 'Give thy daughter to my son to wife':
> And there passed by a wild beast that *was* in Lebanon,
> And trode down the thistle." (2 *Kings* xiv. 9.)

On the part of one of the authors of "Types of Mankind," old Nilotic associations — on that of the other, convictions of the scientific worthlessness of HORÆ ÆGYPTIACÆ,(427) have, for two years, restrained both of them from printed notice of this production: and, if now they conjoin to chant its requiem, the necessity is superinduced, on one hand, by a desire to vindicate Egyptology; on another, the deed has been fastened upon the writer individually by the incessant officiousness of theologers in the United States, in local obtrusions uncalled-for, and in appeals continual to the illusory authority of an adolescent scholar.

It has been already shown [*supra*, p, 670] how Mr. Wilkinson, in 1835, had obliterated, with a dash of his pen, all the "unplaced kings" he had previously published; (428) and had cut down the *era of* MENES to the year B. C. 2201, "for fear of *interfering* with the deluge." During twelve years, Sir Gardner Wilkinson compassionately refrained from diluvial interference; but, from 1837 (429) to 1847, (430) he made a retrocession of MENES, on a sliding scale, to the year B. C. 2320; thereby placing this unfortunate king amid the paludic miasmata (he was killed by a *hippopotamus*) consequent upon that grand catastrophe — only *twenty-eight years* after Archbishop Usher's cataclysm, with which the gallant Knight scrupled to interfere.

The consequence was, that, for twelve years, no hierologist thought it incumbent upon him to quote Wilkinson in matters of chronology; even if scientific justice toward the latter's innumerable Egyptian discoveries occasionally induced Egyptologists to cite a most erudite author notoriously chary of mentioning the labors of continental contemporaries. (431)

Solitude, however, in time becomes tiresome even to an anchorite. Between the years 1835 and 1847, the bound made by Egyptian studies was enormous. Lepsius, followed by the whole school of *Champollionists*, had discovered the XIIth *dynasty of Manetho;* (432) and the XVI—XVIIth dynastic arrangement of Rosellini, abandoned by every other scholar, survived, in 1847, through Wilkinson's *Hand-book* alone. It became desirable, therefore, to "wear ship" in the smoke of Cairo, and to reappear to windward on the other tack; just as if the gallant Knight had been sailing in line with *Manetho's* XIIth dynasty all the time! A "cat's paw" of breeze, nevertheless, was requisite for these nautical evolutions, and *Horæ Ægyptiacæ* kindly wafted it over seas to the London "Literary Gazette."

"And I think this conjecture," wrote the author of *Horæ*, (433) "strengthened by the fact, that Sir G. Wilkinson has found with the name of Phiops (Pepi) a king's name, which I believe he agrees with me in considering as that of Othoes, the first king of the VIth dynasty."—"And this explanation is most strikingly confirmed by a fact [known 14 years previously (434) to every reader of Rosellini!], of which some very remarkable instances are found in some of the unpublished papers of Sir Gardner Wilkinson, which he has kindly shown me, as well as in some of his published works; that in numerous sculptures

(427) *Horæ Ægyptiacæ* — "or the Chronology of Ancient Egypt *discovered* from Astronomical and Hieroglyphical records upon its Monuments; including many dates found in coeval inscriptions from the period of the building of the Great Pyramid to the times of the Persians: and Illustrations of the History of the first nineteen Dynasties, showing the order of their succession, from the Monuments." London, Murray, 8vo, 1851.

(428) *Materia Hieroglyphica;* Cairo, 1827–'32; *Supplement*, and *Text*, Malta.

(429) *Manners and Customs;* 1837; i. p. 41.

(430) *Hand-book for Travellers in Egypt;* 1847; p. 17.

(431) GLIDDON: *Chapters;* p. 11, *a.*

(432) BUNSEN: *Ægyptens Stelle;* 1845; i., Vorrede, pp. 13, 19; ii. pp. 271–362; iii. pl. 3.

(433) *Literary Gazette;* 1849; p. 486; "Cairo, May, 1849."

(434) Compare also LEPSIUS — "Culte fréquent en Nubie de Sesertusen III." *Lettre*, 20 *Juin*, 1843; in Rev. Archéol., June, 1844, p. 208.

in Nubia, we find kings of the XVIIIth dynasty worshipping Sesertesen [Wilkinson always wrote "Osirtasen"] III. as a god."(435) — "I was unable to find it [*Hor-em-bai!*] during my last visit to Thebes, owing to its but once occurring, and to the great extent of the tomb; and I have to thank Sir Gardner Wilkinson in giving me a copy of it."(436) — "I must express my obligations to Sir Gardner Wilkinson, for his having greatly promoted these investigations, during his last visit to Egypt, *in discussing with me every point of importance* in the first four numbers (all I had then written), as well as for the kindness and liberality which he showed me in allowing me to examine and copy many of his unpublished transcripts from Egyptian monuments."(437)

These meritorious acknowledgments were due to the paternal solicitude with which the gallant Knight had watched at Cairo over *Horæ.* Nevertheless, expostulations were addressed from London to its author about the suppression of the names of so many other fellow-laborers; as well through private channels, as also hinted, in public session, before the "Syro-Egyptian Society."(438)

Years passed away. The 12 articles entitled *Horæ Ægyptiacæ*, originally published in the "Literary Gazette," having received unparalleled aid from the highest quarters, reappear, considerably altered, in a beautiful octavo.

We read first Sir J. Gardner Wilkinson's endorsement of *Horæ:* (439) —

"It is indeed the less necessary to enter into a detailed examination of the chronology, and the succession of the Pharaohs, as Mr. Stuart Poole's work on the subject will soon be published; and I have much pleasure in stating *how fully I agree with him* in the contemporaneousness of certain kings, and in the order of succession he gives to the early Pharaohs."

Secondly, we admire *Horæ's* re-endorsement of Sir Gardner Wilkinson: (440) —

"*I* have avoided, as much as possible, quoting or examining the works of others, *excepting* Sir Gardner Wilkinson. My object has been to explain what *I* learned from the monuments; not to combat the assertions of others. Sir Gardner Wilkinson stands in a position *different* from that of any others who have written on the subject; he has never written to support a chronological hypothesis ['in order not to interfere with the Deluge,' *supra*], and is entitled to the utmost confidence on account of his well-known accuracy, the many years which he has spent in the study of the monuments in Egypt, and the caution which he has shown in refraining from putting forth any complete system of Egyptian chronology: *I* am aware how greatly *I* disagree with all others who have written on this subject; but it is a sufficient consolation to me, since all differ, that it is little more to differ from *all others* than to differ from all of them but *one*."(441)

Thirdly, Sir Gardner Wilkinson again endorses *Horæ:* (442) —

"And the contemporaneousness of others [kings—*entirely arbitrary!*] have been very ingeniously and satisfactorily explained by Mr. Stuart Poole, in his *Horæ Egyptiacæ;* where he acknowledges that it was first suggested to him by Mr. Lane. That arrangement may be seen in the following table, which *he has obligingly communicated*, and which I have the more pleasure in inserting, as *I agree with him* in the contemporaneousness of the kings, and in the general mode of arranging those of the same line."

Fourthly, THE FRIEND OF MOSES endorses both: —

"So complete and satisfactory is the train of evidence adduced by Mr. Poole, that Sir J. G. Wilkinson, one of the most learned of living men, in all that relates to Egyptian archæology, has openly published in his last great work on the Architecture of Egypt, his entire concurrence in the views of Mr. Poole, and his conviction of the complete and satisfactory character of the evidence that gentleman has adduced from the monuments."(443)

Ever and anon, after reiterating this endorsement, the same FRIEND OF MOSES adds in Italics: —

"*Egypt, with all her splendid Monuments, is found a witness* [as much as and not less than Spitsbergen] *to the truth of the Bible, and to the correctness* ["credat Judæus Apella!"] *of the Mosaic chronology. . . .* These concessions of the Chevalier Bunsen prepare us to receive with greater confidence the statements of Mr. R. S. Poole, in his *Horæ Ægyptiacæ*, claiming to adduce proofs from the monuments themselves, that several of the dynasties which

(435) *Ibid.;* p. 552; "Cairo, June, 1849."

(436) *Ibid.:* p. 522.

(437) *Ibid.;* p. 910.

(438) London, 10th April, 1849; *Literary Gazette*, 28th April, 1849.

(439) *Horæ Ægyptiacæ;* Preface, p. 23 — citation from WILKINSON: *Architecture of Ancient Egypt.*

(440) *Horæ;* p. 23.

(441) *Horæ;* p. 23.

(442) *Hieratic Papyrus of Turin;* 1851; p. 29.

(443) "Mobile, Jan. 27, 1852" — *Southern Presbyterian;* Milledgeville, Ga., Feb. 19, 1852.

have been generally represented as successive were actually contemporaneous, as *e. g.* the twelfth and the fifth [!]; and that thus, the monumental history of Egypt covers not a period of duration beyond what may be readily reconciled with [poor Moses !] the Mosaic chronology as given in the Septuagint. A conclusion, to the accuracy of which, Sir J. G. Wilkinson has affixed the sanction of his great name in these matters." (444)

The FRIEND OF MOSES soon after becomes mystified : —

"I became acquainted with several gentlemen of distinction in the learned world. . . . Mr. R. S. Poole, a *bold* writer on Egyptian chronology." (445)

He next assures us : —

"I have carefully compared the copies taken by Champollion in *all these tombs and temples*, from the second Cataract to Thebes, and I have collated his hieroglyphics, *line by line* [this is the more miraculous, as it was performed between *Alexandria*, Nov. 12, and *Cairo*, Feb. 14—after going up the Nile, 1200 miles, to *Samneh;* and returning, 1050 miles, to Cairo !], and *character by character*, with the originals. . . . There is a magnificent error somewhere—though *I* am not prepared [!] to point out where ; nor how precisely it may be detected and exposed. Of one thing *I* am satisfied — that Sir J. G. Wilkinson, and *my kind young friend*, Mr. R. S. Poole, of the British Museum, are much nearer the truth, in their chronology, than is Dr. Lepsius, or the Chevalier Bunsen." (446)

The scientific reader now comprehends our local situation, and will compassionately forgive the inhumanities which such every-day offences compel us finally to perform. "Le jeu ne vaut pas la chandelle ;" else we would at once refute *Horæ Egyptiacæ*, page by page, and hieroglyphic by hieroglyphic ; in the interpretation of which last the juvenile author (or Sir G. Wilkinson) has committed blunders as egregious as they are multiform — altogether unpardonable in the actual state of hierology. For the present, our criticisms shall be chiefly confined to the publication of "three fragments," upon the principles of a world-renowned master, Letronne. (447) They are from the highest Egyptologists in Europe ; two of them in epistles to the authors ; one already in print.

First Extract. (448)

"I have nothing to say about the book of POOLE, if not that I regard it as a juvenile and sufficiently-pretentious essay, written without conscientiousness, and dangerous rather to the theologians than to science."

Second Extract. (449)

"Not one of its followers can read three lines of hieroglyphics correctly. The G. P. Y. (450) and G. P. M. (451) are only in the mind of the author. Examined by the microscope of philology, all vanishes into a few unimportant observations — for example ;

, , is not "the first month"—"the first half month," of the *Great Panegyrical Year ;* but merely

 = "monthly,"  = "half-monthly."

The consequence is that this expression does not fix the age of CHUFU [builder of the great pyramid]. The "7th ," (452) on the base of the Karnac obelisk, refers to the seven *smat*, or *periods-months*, I believe that the obelisk was in the quarry. Hence the whole cyclical part is a delusion ; and all the inferences are *nil*. The rest of the book is a string of hypotheses — where there are not actual misapprehensions."

Third Extract. (453)

"Mr. POOLE is of the number of those young workmen who deserve that one should tell them the whole truth. Either he has not read what recent archæologists have written

---

(444) *The Friend of Moses;* New York, 1852; pp. 376, 377, 514.

(445) *Mobile Daily Advertiser*, Oct. 9, 1852 — "Correspondence — Paris, Sept. 14, 1852."

(446) *Mobile Daily Register*, April 1, 1853 — "Letter from Egypt — Cairo, Feb. 14, 1853."

(447) *Trois Fragments* — Mémoires et Documents publiés dans la Revue Archéol.; Paris, 1849; pp. 100–119.

(448) *Letter to Mr. Gliddon.*

(449) *Letter to Dr. Nott.*

(450) *Horæ;* p. 59 — "Great Panegyrical Year."

(451) *Do.;* p. 56 — "Great Panegyrical Month."

(452) *Do.;* p. 66.

(453) DE ROUGÉ: *Phénomènes Célestes;* Rev. Archéol., 15 Feb. 1853; pp. 664, 665; and *note.*

upon this subject, which would be inexcusable; or he has read them and does not cite them, which would be still more grave. I have not read the name of Lepsius a single time in his book, in respect to all these questions so lengthily treated in the *Introduction to Chronology* [Berlin, 1848–9]. . . . Not content with this discovery [viz., the imaginary Panegyrical Months] M. Poole thinks also to find other new cycles, with the dates which refer to them. I confess that it has been impossible for me to comprehend how, in the presence of pretensions so important, Mr. Poole has not deemed himself obliged to prove the truth of his allegations, by minutely analyzing the inscriptions which he alleges. Far from that, he contents himself with indicating them, and sometimes even without producing their text in his plates. One cannot lean upon an Egyptian inscription, as upon a passage of Titus Livius, without new explanation, and I will frankly say that I believe in none of the cycles and in none of the dates of Mr. Poole. . . . It is evident that in thus handling the ciphers, without controlling their signification and the manner in which they are introduced into the inscriptions, one may end in imagining all the periods that one wishes, and in giving them a certain appearance of truth to the eyes of persons who can discuss but the results. A work thus based must pass for *non-avenu.*"

But, after all, *Horæ* has no "fear of interfering with the Deluge;" so the work becomes only another thorn in the side of orthodoxy. Mr. Wilkinson (1835, *supra*), devoutly following archbishop USHER and the margin of king *James's version*, says the date of the Flood "*is* 2348 B. C." In its author's first articles, *Horæ* had declared—

"The date of the accession of Menes, the first king of Egypt, is probably that of the commencement of the first great panegyrical year and first capital year. Eratosthenes and Josephus [say, modern *computators* on these ancient writers] place his accession somewhat later—namely, about 2300 years B. C., instead of 2715. The history of the 1st, 2d, 3d, 4th, and 5th dynasties [of the IV–Vth dynasties, LEPSIUS found the amplest details, while the author of *Horæ* dwelt only 15 miles off, at Cairo!] is but scantily furnished us by Manetho and the monuments, and the latter give us but one date [and that fabulous!], that of the commencement of what *I* have called the second great panegyrical year in the time of Suphis I., the builder of the great pyramid, and second king of Manetho's fourth dynasty, B. C. 2350." (454)

Horæ thus fixed the building of the great pyramid *two* years before Wilkinson's Deluge; and set MENES on the throne, in Egypt, 367 years before the same authority's catastrophe. But, it was promptly shown, that *Horæ*, in selecting the year B. C. 2715 for MENES, had merely stolen another man's thunder (455): wherefore, when its author came to reprint those twelve articles in an octavo volume, he so translated his hieroglyphics, astronomically, as to obtain *two* years' difference!—"The commencement of the great panegyrical year which preceded that of the Suphises, *I* have already shown to be in the year B. C. 2717" (456); and then he informs us that "the Septuagint chronology dates the *Dispersion* of Mankind about the year B. C. 2758; that is, about 41 years before the era of Menes"!

Computations upon the different copies of the LXX, every one of them as rotten as the MSS. themselves, cause the *Creation* to fluctuate between B. C. 5904, and B. C. 5054. (457) And the above sentence merely shows its penman's incompetency to discuss *Septuagint* questions. To the reader of our disquisition on Xth *Genesis* [PeLeG, *supra*, p. 545], the following specimens of *Horæ's* biblical knowledge will be amusing; as much as, to use its author's favorite adjective, the latter's credulity is "remarkable":—

"*I* therefore believe that the Vague year was instituted in the time of Noah; probably by Ham [!], not by Noah. . . . *I* have only to notice one other important epoch of Bible history—the dispersion of nations. The division [read "split"] of the earth is indicated as having occurred at the birth of Peleg [a "split"]; when we are told, (Gen. x, 25), 'unto Eber were born two sons; the name of the one (was) Peleg (or division); for in his days was the earth divided.' [*Vide supra*, what the *Hebrew* writer meant!] Now, it was a common custom of Hebrews to name their children from circumstances which occurred at their birth; and the custom of ancient Arabs was precisely the same, and has continued to the present day. We cannot reckon as exceptions to this the few cases where *God changed a name*, or imposed a new one; and in the latter case the old name was retained with the new one [!]. The *birth* of Peleg, according to Dr. Hales, happened B. C. 2754;

(454) *Art.* XII.; *Literary Gazette*, Dec. 15, 1849; p. 910;—compare *Art.* VII., p. 522.

(455) "By my reduction of 'Manetho'—2715" B. C.; GLIDDON, *Chap.*, 1843, p. 51:—and *Hand-book*, 1849, p. 41

(456) *Op. cit.:* p. 63, and p. 97.

(457) RICCIOLI: *Chronol. reformata;* p. 293.

but, calculated from *my* date of the Exodus, B. C. 2758."(458) — "*I* say that the Pharaoh of the Exodus reigned undoubtedly not more than about one year; for, although *his being drowned in the Red Sea* is not expressly mentioned by MOSES, it is so mentioned in the 136th Psalm [what a clinching argument!], and *I* hold all the books of the Bible to be equally true."(459).

It is to be deplored that, after being promoted for his Hebraism to a post in the British Museum, "my kind young friend," as *the Friend of Moses* affectionately terms him, should have expunged these delightful samples of pious feeling from the republication of *Horæ* in its octavo form. So imbued, we fear, is he likely to become in that enlightened institution with self-immolating principles, that it would not surprise us to learn through newspapers that *Horæ* likewise—as SCALIGER says, "ut signatius loquar"—for the sake of Oriental literature were to turn *Mohammedan.*

No inclination remains to follow *Horæ's* farthing-rush-light any further. We leave the pupil for the teacher, when we here exhibit on the margin a table printed by WILKINSON in the pamphlet-text accompanying the latter's truly-valuable contribution to archæological science — *The fragments of the Hieratic Papyrus at Turin: containing the names of Egyptian Kings, with the Hieratic inscription at the back.*

Here is that "magnificent error" which the FRIEND OF MOSES could not discover by going to Egypt: —

"Respecting the construction of the table, he observes: 'The relative positions and the lengths of most of these dynasties are founded *upon some kind* of monumental authority. The rest *I* have placed within approximative extremes. There are several points of *exact* [!] contemporaneousness, as in the 2nd and 4th and 5th dynasties, again in the 5th and 15th, and in the 9th and 11th; and these, with other evidence *of the same nature,* enable us to adjust the general scheme of all the dynasties.'" (460)

Reader! Suppose a Chinese archæologist, with a little red button on his cap, were to come all the way from Pe-kin to America, and tell us that good old king EGBERT was a

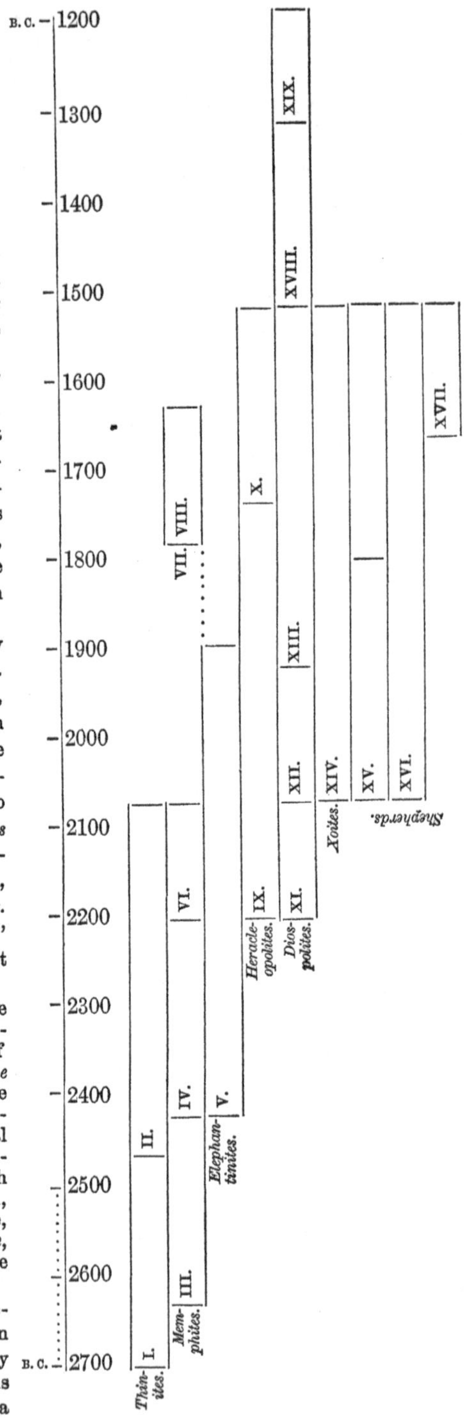

(458) *Art.* X.; *Lit. Gaz.;* p. 641. (459) *Art.* V.; *Lit. Gaz.;* p. 432. (460) *Hier. Papyr.;* pp. 30, 31, and *table,* p. 31.

mythe—that the consecutive dynasties of our common English father-land could fit no Hottentot's estimate of the chronology of John-Chinaman's sacred book, the *Chou-king;* unless, after rejecting Boudicea and Caractacus, we were to permit his reduction of *Danes, Saxons, Normans, Plantagenets, Lancastrians, Yorkites, Tudors, Stuarts, Orangites, Hanoverians,* &c.; together with all British, Scottish, and Irish, periods of anarchy; not forgetting Cromwell and the Commonwealth; into one century. Suppose that, after proving why every Anglo-Saxon had erroneously classified, as distinct, those personages, epochas, and historical events, which the "Tribunals of Literature" of China had pronounced to be identical, the said mandarin were to show us how beautifully the whole could be reduced, through electro-magnetic typography, into one line of a *table*, and expressed algebraically by an *x*, representing an infinitesimal fraction of a second of Creative time. What should we say to His Excellency "Uncle *Josh*"?

Now, whatever the American reader might be pleased to hint to such Chinese mandarin, would be uttered in demotic tongue with "brutale franchise" by old MANETHO (could his mummy arise) to Sir Gardner Wilkinson, at the first glance over the above table: where, in wilful disregard of Lenormant, Champollion, Böckh, Barucchi, Bunsen, Henry, Lesueur, Lepsius, Hincks, Kenrick, Pickering, Ampère, De Rougé, Birch, and of every hierologist past, present, and to come, the gallant Knight has made the IIId, IVth, VIth (VII), VIIIth Egyptian dynasties (consecutive in Manetho and, where mentioned, serial upon all monuments), *contemporaneous!*—has actually jammed *eleven* dynasties, VI, VII, VIII, IX, X, XI, XII, XIII, XIV, XV, XVI, into a space (2200 *a* 1700) of 500 years! And perpetrated, too, all these inexplicable vagaries with theological applause, when, by placing MENES (Ist dynasty, *Thinites*) at 2700 B. C., he shows that valiant knighthood, in A. D. 1851, no longer creeps all over "for fear of *interfering* with the Deluge of Noah; which (*was*) 2348 B. C." before an aspirant to ecclesiastical patronage had won his gilded spurs.

We dismiss, therefore, *Horæ Ægyptiacæ* as beneath scientific notice, reserving to ourselves the privilege of a reviewer's criticism, whenever circumstances may demand its annihilation. With it we snap off the last published peg upon which short-chronology can suspend its clerical hat; because Mr. Sharpe's arrangement of Egyptian dynasties anterior to the XVIIIth has been respectfully disposed of. When other writers, with hieroglyphical handles to their patronymes, adventure into the rude arena of archæology as champions of *short*-chronography, may their armor be well tempered and their lances tough!

The list of *long*-chronologists, above given, comprehends the "preux chevaliers" of archæological science at this day. The *minimum* of their respective dates for MENES is B. C. 3643; the *maximum* approaches the 6th chiliad B. C. By each authority all biblical computations, *Hebrew, Samaritan,* and *Septuagint,* are thrown aside among the rubbish of the things that were.

"The sum of all the dynasties varies according to our present sources from 4685 to 5049 years; the number of kings from 300 to 350, and even 500. It is evidently impossible to found a chronology on such a basis, but Syncellus tells us that the number of generations included in the 30 dynasties was, according to Manetho, 113; and the whole number of years, 3555. This number falls much short of what the summation of the reigns would furnish according to any reading of the numbers, but is nearly the same as 113 generations would produce, at any average of 32 years each." (461)

Fifteen years ago, the learned ethnographer, De Brotonne, reasoning upon this very number, "3555 de Manethon," obtained B. C. 3901 as "le chiffre le moins élevé" for MENES. (462)

To neither of the present writers have these results been unknown:—

"On my return to Cairo [April, 1840, from a voyage with Mr. Harris to the second cataract], I devoted a twelvemonth's leisure to the verification of the solidity of the basis upon which hieroglyphical revelations had placed Egyptian *monumental chronology*. The result was a conviction as profound then, as subsequent researches,—echoed by the voice of universal erudition, and embodied in the works of a host of savans whose names gild the

(461) KENRICK: *Ancient Egypt under the Pharaohs:* 1850; ii. p. 93.
(462) *Filiations et Migrations:* i. p. 203.

brightest page illuminated by science in the XIXth century,—have since demonstrated its accuracy, of the utter impossibility of reconciling *Egyptian facts*, geological, topographical, ethnological, hieroglyphical, and historical, with Archbishop Usher's system of patriarchal *chronology*.

"A *manuscript* compilation, over which an old and valued colleague, M. Prisse, and myself wiled away at Cairo many delightful weeks in reciprocal exchanges of our several gleanings, under the title of "Analecta Hieroglyphica," condensed every *cartouche*, with references to most of the historical monuments, known to hierologists up to April, 1841; and, as many personal friends are aware, this manuscript is still a most important *ground-text* and manual to those who, like myself, are anxious to ascertain the stability of prior investigations, before hazarding the erection of a theoretical superstructure." (463)

What, then, is the present state of scientific opinion on the era of MENES? The reader has it before him in the list on p. 682; and, without perplexing himself with vain speculations founded upon ignorance of the stupendous materials transferred from Egypt to Berlin by the Prussian Mission, let him do as we do, await patiently for the publication, hourly due, of Lepsius's "Book of Kings." The authors may be pardoned when stating that, in books, manuscript-notes, and epistolary communications from Egypt, Italy, France, Germany, and England, they probably possess as much specific and detailed information here at Mobile, on Egyptian monumental chronology, as most men in the world, less a dozen European hierologists — with whom they are in agreeable accord. When, therefore, they put forward no dogmatical system of their own, but wait for the "Book of Kings," they act themselves in accordance with the counsel offered to fellow-inquirers. Should Lepsius's work reach their hands before the issue of the present volume, a synopsis of its *chronology* will be appended to our essay. We may also look forward to Biot, the scholarlike astronomer of France, for a profound investigation of the *astronomical* data, revealed by Egyptian monuments, in their relations to mundane chronology; (464) which will supersede any future recurrence to the cyclic reveries of such youthful star-gazers as *Horæ*.

Should, however, a qualified student desire to prepare himself for thorough mastery of Lepsius's "Book of Kings," he should commence with Rosellini's *Monumenti Storici;* and, that being fundamentally acquired, his next guide is Bunsen, *Ægyptens Stelle in der Weltgeschichte;* wherein most of the royal Egyptian names, discovered up to 1845, are compared with the classical lists, and in which the grand alteration produced by Lepsius's resuscitation of the XIIth dynasty (unknown to the lamented Pisan Professor, or, in 1847, to Wilkinson), is abundantly set forth. "There is no royal road to the mathematics," nor is there a straighter path to the comprehension of Egyptian *chronology* than the one we indicate; but, after these two works, the study of Lepsius, *Chronologie der Ægypter*, "Einleitung, 1849," becomes imperative.

Such reader will appreciate the general correctness of the following method of verifying, archæologically, the progressive layers in which Egyptian history stretches backwards from the Christian era, assumed at 1853 years ago; until the unknown-commencements of Nilotic humanity merge into an undated, but ante-alluvial, period of geology. (465)

We gladly borrow the first points of departure, in our journey from the Christian era, backwards, from Sharpe (466): —

"The reigns of Ptolemy, of Darius, of Cambyses, and of Tirhakah are fixed by the Babylonian eclipses. Hophra and Shishank are fixed because they are mentioned in the Old Testament, since the length of the Jewish reigns, after Solomon, is well known, while those Jewish dates are themselves fixed by the earliest of the Babylonian eclipses in the reign of Tirhakah. Thus are fixed [by Mr. Sharpe] in the Table of Chronology the dynasties of Sais, Ethiopia, and Bubastis. Petubastes lived in the first Olympiad; this fixes the dynasties of Tanis."

Thus, king by king, and event by event, we ascend with precision back to Alexander the Great, B. C. 332; and thence, through the XXXIst, XXXth, XXIXth, XXVIIIth, XXVIIth,

---

(463) GLIDDON: *Hand-book;* London, Madden, 1849; p. 40;—conf. NOTT: *Biblical and Physical History of Man:* 1849; pp. 69–86; — also *Chronology, Ancient and Scriptural:* South. Quart. Rev., Nov. 1850.

(464) DE ROUGÉ · *Rev. Archéol.*, Feb. 1853; pp. 656, 686.

(465) GLIDDON: *Otia;* pp. 61–69.

(466) *Chronology and Geography;* 1849; p. 13, and table, pp. 14, 15.

XXVIth, XXVth, XXIVth, XXIIId Egyptian consecutive dynasties, back to S*h*eS*h*oNK, Shishak, founder of the XXIId dynasty; who, conquering Jerusalem "in the Vth year of king Rehoboam," (467) as is hieroglyphically recorded in Karnac, (468) enables us to establish a perfect synchronism, between Egyptian and Judaic history at B. C. 971–3.

Prior to this date, Egyptian monuments never once refer to the *Hebrews*, throw not a glimmer of light upon *Jewish* annals; and with Sheshonk also ceases the possibility of fixing any Pharaoh, to him anterior, within 5 or 10 years. *Chronology*, year by year, stops in fact at B. C. 972; as well in Israelitish as in Nilotic chronicles: although the foundation of Solomon's *temple* cannot be far removed from B. C. 1000.

Leaving Hebrew computation to ascend along its own stream, innumerable Egyptian documents — *tablets*, *papyri*, *genealogical lists*, public and private, together with an astounding mass of collateral and circumstantial evidence, — carry us upward, through the XXIst, XXth, XIXth, and XVIIIth dynasties, reign by reign, and monument by monument, to RAMSES I. (Ramesu); whose epoch belongs to the ——— *century* 15th–16th B. C.

Here intervenes a period, though for a few years only, of anarchy; represented in the *Disk heresy*, and by sundry royal claimants; at the head of whom stands ATENRA-BAKHAN, or *Beχ-en-aten*; (469) called by Lepsius "Amenophis IV." But upward from his *father's* reign, Amenoph III, every king is known, with many events of their respective reigns, through hieroglyphical sculptures and papyri, back to the beginning of the XVIIth Theban dynasty, in the reign of AAHMES, *Amosis*, I; computed, by Lepsius, to be about the year 1671 B. C. At this point, which begins the "Restoration," or "New Empire," after the expulsion of the *Hyksos*, we lose the thread of *annual* chronology, for times anterior to the 17th century, before C.

We refrain from discussion of the *Hyksos*, or shepherd kings. (470) They are supposed to occupy the XVIth and XVth dynasties; and, according to Manetho, their duration covered 511 years of time. The XIVth dynasty has not been disentangled clearly from the mutilated lists; and the hieroglyphical records have not yet spoken intelligibly, although they are numerous. We pause for Lepsius; and in the meanwhile refer the reader for a summary of the monumental edifices of the Old and the New Empires to his published travels. (471) To us at present this "middle Empire" is chaos; but, even supposing the XIVth, XVth, and XVIth dynasties could, by a *short*-chronologist, be expunged from Egyptian records, it must be remembered, by *long*-chronologists, that the XVIIth dynasty stands erect in the 17th century B. C. We leave the "middle Empire's" duration to be adjusted along a sliding scale from *zero* upward; and next proceed to show that we possess above 1500 years of positive monuments, behind this "middle Empire," by which all *Septuagint* computations of the Deluge, at B. C. 3246, or 3146, or 3155, encounter a "reductio ad absurdum."

The mists begin to clear off as we commence ascending to the latest representatives of the "Old Empire" in the land of K*h*aM, *Ham*, *Chemmis*: viz., the Sebakhetps and Nepherhetps of the XIIIth dynasty (472): but, at the XIIth dynasty, the glories of the olden time blaze forth again effulgently; (473) thanks to Lepsius's investigations of the *Genealogical Papyrus of Turin*. (474)

---

(467) 1 *Kings* xiv. 25; 2 *Chron.* xii. 2.

(468) GLIDDON: *Chapters*; p. 9.

(469) PRISSE: *Legendes de Sckai*; Rev. Archéol., 1845; pp. 472–474; also his arrangement of these kings, in WILKINSON, *Hand-book*, p. 393; — LEPSIUS: *Götterkreis*; 1851; pp. 40–43; — DE ROUGÉ: *Lettre à M. Alfred Maury*; Rev. Archéol., 1849; 120–124.

(470) GLIDDON: *Otia*; pp. 44, 45.

(471) *Briefe aus Ægypten*; pp. 364–369.

(472) BIRCH, in *Otia Ægyptiaca*; p. 82; and his *Historical Tablet of Ramses II.*; 1852; p. 19; — DE ROUGÉ. *Rochers de Semné*; Rev. Archéol., 1848; pp. 312, 313.

(473) BUNSEN: *Ægyptens Stelle*; ii. p. 271, *seq.*; — DE ROUGÉ: *Annales de Philosophie Chrétiennes*; xiv., xv., xvi., and HINCKS: *Turin Book of Kings*; R. Soc. of Lit.; iii., part i., pp. 128–150; but considerably emended in WILKINSON'S *Papyrus of Kings*; 1850; "Observations of Dr. E. HINCKS"; p. 55: — DE ROUGÉ: *Le Sesostris de la Douzième Dynastie*; Rev. Archéol., 1847; pp. 481–489.

(474) *Auswahl*; Taf. iii., iv., v., vi.: — most superbly recopied by Sir J. G. Wilkinson: *Fragments of the Hieratic Papyrus at Turin*; 1851: but consult also the critical history of this document as displayed by CHAMPOLLION-FIGEAC (Rev. Archéol.), with the caveat that the luckless disposal of these fragments is due to SEYFARTH alone.

The hieroglyphical names of some of these kings may be consulted in Bunsen; but we borrow from Lepsius this table of the XIIth dynasty; which cannot become more than slightly modified in his "Book of Kings." (475)

"THE XIITH MANETHONIAN DYNASTY.

		According to Manetho.			*According to the Turin Papyrus.*	*Highest year on the Monuments.*
1. Amenemhe I alone..................	9 y'rs	9 Am. I	[Afr. 16	Eus. 16]	9 y'rs	
2. Sesurtesen I and Amenemhe I.	7 "	} 46 Ses. I	[Afr. 46	Eus. 46]	45 "	8. of Am. and Ses. I.
Sesurtesen I alone..................	35 "					44. of Ses. I=2 of Am. II.
Sesurtesen I and Amenemhe II	4 "					
3. Amenemhe II alone................	28 "	} 38 Am. II	[Afr. 38	Eus. 38]	3(7) "	35. Am. II = 3. Ses. II.
4. Sesurtesen II & Amenemhe II.	10 "					
Sesurteseon II alone.................	28 "	28 Ses. II	[Afr. 48	Eus. 48]	(2)9 "	11. — —
5. Sesurtesen III........................	38 "	38 Ses. III	[Afr. 8	Eus. 8]	3(7) "	26. — —
6. Amenemhe III alone	41 "	} 42 Am. III	[Afr. 8	Eus. 42]	4(1) "	43. — —
Amenemhe III & Amenemhe IV	1 "					
7. Amenemhe IV alone........	8 "	8 Am. IV	[Afr. 8 }		9 y'rs 3 m. 27 d.	6. — —
8. Ra-Sebeknefru........................	4 "	4 Sebek.	[Afr. 4 }		3 " 10 " 24 "	
				Total	213 " 1 " 24 " "	

The XIIth dynasty ends, according to Lepsius, about B. C. 2124.

What relics are extant of XIth dynasty belong to the Enuantefs, (476) including perhaps Ra-nub-Cheper, discovered lately by Mr. Harris.

Little can here be related about the Xth, IXth, VIIIth, and VIIth dynasties, to be intelligible without a lengthy argument; but the duration of this last is felicitously suggested by Maury. (477) Solid as a rock, however, is the VIth dynasty; (478) so is the Vth on the *Turin Papyrus* and through the recovery of all its kings (but one?) from the tombs opened by the Prussian Commission at Memphis. (479) Of the IVth the vestiges surpass belief, to persons who have not opened the folio plates of Lepsius's *Denkmäler;* wherein the petroglyphs of these three dynasties, earliest and grandest relics of antique humanity, are now preserved for posterity, so long as the pyramids of *Geezeh* shall endure.

With the IIId dynasty Egyptian monuments cease. There is nothing extant of the IId, nor coeval with the Ist dynasty. Their existence is deduced from the high state of the arts, and the extensive knowledge possessed by the denizens of the Nile, as demonstrated by the *pyramids, sepulchres,* and *hieroglyphed records,* of the IVth dynasty, compared with the fragmentary catalogues of Manetho and Eratosthenes, and supported by Græco-Roman tradition.

MENES—Egypt's first Pharaoh—is recorded, in hieroglyphics carved, during the 14th century B. C. at the Theban Ramesium, by Ramses II. as his earliest ancestor; and, in hieratic, on the *Turin Papyrus,* a document written in the twelfth—fourteenth century B. C., "king MeN*ai*, of a firm life," is twice chronicled. (480)

By Lepsius, whose computations we adopt, Menes is estimated to have founded the Ist dynasty of *Thinites* about the year ———————————— B. C. 3893.

"There is nothing incredible in such an antiquity of the Egyptian monarchy." (481) Indeed, long before hieroglyphical discoveries had demonstrated its natural adaptation to all the circumstances of Egypt (when due allowance is made for *pre-Menaic* chiliads of years for alluvial existence), the researches of mathematicians had pointed to similar results.

"On supputing the 11340 years of Herodotus, taken for the Egyptian seasons of three months, we should have 2794 solar years, according to Freret, and 2835 years, according

(475) *Ueber die Zwölfte Ægyptische Königsdynastie;* 1853; p. 28.

(476) LEEMANS: *Lettre à Salvolini:* 1838; No. 22;—and *Lettre à M. De Witte:* Rev. Archéol., 1848, pp. 718–720;—BIRCH, in *Otia Ægyptiaca;* pp. 80, 81; and *Tablet of Ramses II.;* p. 18.

(477) *Chronologie des Dynasties Égyptiennes:* Rev. Archéol., 1851; pp. 166, 167.

(478) BUNSEN: *Ægyptens Stelle:* ii. p. 191, *seq.;*—MARIETTE: *Fragment du Papyrus Royal de Turin et la VIe Dynastie de Manethon;* Rev. Archéol., 1849; pp. 306–315;—HINCKS: Trans. R. Soc. Lit., Mar. 12, 1846; p. 137, and "Observations" in WILKINSON'S *Papyrus;* pp. 53, 54.

(479) GLIDDON: *Otia;* p. 38. For all details see authorities in the preceding note.

(480) *Column I., fragment 1, lines 11 and 12;* Sir G. WILKINSON'S copy.

(481) KENRICK: *Op. cit.;* p. 110.

to Bailly. These finished at the reign of Sethos and with the war of Sennacherib, in the year 710 before J. C. Following this hypothesis, the commencement of Menes fell about the year 3504 B. C., according to Freret; and in 3545 B. C., according to Bailly." (482)

Having thus indicated to junior students of Egyptian chronology the order in which they should read the works of our common seniors in this technical speciality of science, we will now reverse the process, and exhibit, from MENES downward, the stratifications in which Time's hour-glass has marked, historically, the consecutive events witnessed, during above forty-three centuries, by the Egyptian "Type of Mankind" down to the 4th century after the Christian era; assumed at 1853 years ago.

It is a convenient plan to group several portions of Egypt's history into the following separate masses, like the primary, secondary, and tertiary formations of our earth's crust; and to view the dynasties, in those masses included, as if they were so many distinct strata contained in such formations. We thereby divest the subject of the perplexities and dubiousness of arithmetical chronology; because, the viril existence of Menes, as an historical entity, is no more dependent upon *ciphers*, than Owen's *Dinornis giganteus* (in palæontology) hangs upon a "B. C. 2320" of a Knight's, or upon a "B. C. 2348" of an Archbishop's diluvian phantasms.

I. — The ANTE-MONUMENTAL period. This of course is an utter *blank* in chronology. Science knows not where geology ends, nor when humanity begins; and the definitive, or artificial systems, current on the subject, are of modern adoption and spurious derivation.

At what era of the world's geological history the River *Nile*, the *Bâhr-el-abiad* in particular, first descended from palustrine localities in Central Africa, along the successive levels of Nubian plateaux, through its Egyptian channel to the Mediterranean (beyond the indisputable fact that its descent took effect *after* the deposition of the so-termed DILUVIAL DRIFT upon the subjacent *limestone*) is a problem yet unsolved. But were proper investigations, such as those commenced in 1799 by Girard, (483) and cut short by European belligerent interference, entered upon, in the valley of the Nile itself, by competent geologists, the *alluvial* antiquity of the "Land of Khem" could be approximately reached. (484) The very rough estimates heretofore made by geologists yield a *minimum* of 7000 years for the depositions of the present alluvium by the river Nile. The *maximum* remains utterly indefinite; but, nevertheless, we are enabled to draw, from the data already known, the following among other deductions, of primary importance to Nilotic *chronology*: —

1st. — Previously to the advent of the "Sacred River" no deposition of *alluvium* having taken place upon the limestone, Egypt was uninhabitable by man.

2d. — Since the deposition of this alluvium, there has been no *Deluge*, in the literal Hebrew and genesiacal sense of the term, whether in Egypt, or in Asiatic and African countries to the Nile adjacent.

3d. — Humanity must have commenced in the valley of the Nile, under conditions such as exist at this day, *after* a sufficiency of alluvium had been deposited for the production of vegetable aliment, but at a time when the depth of this alluvium was at least twenty (fifty, or *more*, for aught we can assert to the contrary) feet below the level of the highest portion of the Nile's bed at this hour; but how much soil had been previously deposited — that is, what its *thickness* was over the limestone when humanity first developed itself in Egypt — it is yet impossible to define.

4th. — Many centuries (in number utterly unknown) must be allowed for the multiplication of a human Type *in Egypt*, from a handful of rovers to a mighty nation; and for the acquirement, by self-tuition, of arts and sciences adequate to the conception and execution of a pyramid: thus yielding us a *blank* amount of chronological interval, bounded on the one hand by the unknown depth and surface of the Nilotic alluvial,

(482) DE BROTONNE: *Filiations et Migrations;* i. p. 198, 199.

(483) *Description de l'Égypte:* tom. xx. p. 33, *seq.*

(484) GLIDDON: *Otia;* pp. 62–69; and "Geological Sections." For the *botanical* argument, vide PICKERING.

sufficient for the growth of human food, at the time of man's introduction; and on the other (after this nomad had been transmuted by time and circumstance into a farmer and then into a monument-building citizen) by the *pyramids* and *tombs* of the IVth Memphite dynasty; placed by Lepsius's discoveries in the thirty-fifth century B.C.

II.—The PYRAMIDAL period, or *Old* Empire.—Occupying, according to late scientific views, about fifteen centuries; probably beginning with Manetho's *first* dynasty (king OUENEPHIS); and ending with the XIIth or XIIIth, about twenty-two centuries prior to the Christian era. The XIIth dynasty is marked architecturally by the employment of *obelisks*.

III.—The period of the HYKSOS, or *Middle* Empire.—There being few *monuments* for this period extant, we are dependent, apart from Greek lists, upon the *Turin Papyrus*, and on the names chronicled long after on the "Chamber of Karnac" &c. Here is the *grand* difficulty in Egyptian chronology; it having been hitherto impossible to determine its duration; which is now generally considered to be far shorter than is estimated in Bunsen's "Ægyptens Stelle in der Weltgeschichte," and perhaps to embrace all *Scriptural* connexions with Egypt from Abraham to the *Exodus* inclusive; on every one of which the *hieroglyphics* are utterly silent. It includes, however, the XIVth, XVth, and XVIth dynasties.

IV.—The positive HISTORICAL period, or *New* Empire.—Commencing about 1600 to 1800 years B. C., with the *Restoration* (after the expulsion of the Hyksos tribes), under AAHMES, the founder of the XVIIth dynasty. It may be called the *Temple*-period; because, although temples existed in the Old Empire, all the grand sanctuaries standing at present upon the alluvia belong to the XVIIth dynasty downward.

Dated *hieroglyphical* records descend to the third century after Christ, with the name of the Emperor DECIUS: (485) but *demotic papyri* and *mummies* are extant as recent as the 4th century of the same era.(486) Greek inscriptions at Philæ corroborate Priscianus, who relates how, about A. D. 451, a treaty, between the Christian Emperor of Constantinople and the heathen Blemmyes, stipulated that—"every year, according to ancient customs, the Ethiopians were to take the statue of *Isis* from Philæ to Ethiopia;"(487) and a Grecian traveller bears witness, in an inscription, that he was once present at the temple when the goddess returned. In fact, history proves that *ISIS* was yet worshipped at Philæ, if not throughout Egypt, even in the year A. D. 486: and the pagan emblem of "eternal life," *Ankh*, continued still to be inscribed, in lieu of the Christian *cross*, over orthodox churches; as in the following instance discovered by the accurate Sir J. Gardner Wilkinson (488):—

"ΚΑΘΟ☥ΛΙΚΗ + ΕΚΚΛΗ☥CΙΑ"

Catho☥lic + Chu☥rch.

Finally, to enable the reader to classify, chronologically, the Egyptian data comprised in "Types of Mankind," a table is subjoined which the forthcoming "Book of Kings" will show to be in the main correct. It is made up, in part from the first volume of the *Chronologie der Ægypter*, and in part from Chevalier Lepsius's oral communications to the writer at Berlin, in May, 1849.(489) To it are added such excerpts of the Chevalier's subsequent epistolary correspondence with the authors as may give a general idea of his system, and a precise one of his scientific liberality.

(485) LEPSIUS: *Vorläufige Nachricht*, 1849; pp. 17, 29.
(486) BIRCH, in *Otia Ægyptiaca*, p. 87.
(487) LETRONNE: *Matériaux pour servir à l'Histoire du Christianisme.*
(488) LETRONNE: *Examen Archéologique*, "Croix Ansée Égyptienne," 1846; p. 23.
(489) GLIDDON: *Hand-book to the Nile:* London, Madden, 1849; pp. 20-2, 51.

MANETHO'S SYSTEM OF EGYPTIAN CHRONOLOGY, AS RESTORED BY LEPSIUS.

EPOCHAS anterior to MENES — Cyclic Periods : —

Divine dynasties :— 19 *gods* ...... reigned 13,870 Julian years = 19 Sothic *demi*-periods.
30 *demi-gods* " 3,650 " = 30 *twelfths* of a Sothic-period.
17,520 " = 12 Sothic-periods of 1460 years.
Ante-historical dyn.: 10 *Manes*, Thinites, 350 " — commencement of a *new* Sothic-period.

EPOCH OF MENES — commencement of *historical* period ; *thirty* dynasties : —

Old Empire:—	1st dynasty — Accession of Menes	3893 B. C.
	Commencement of *monumental* period; third dynasty.	
	4th dynasty — Pyramids and tombs extant — began	3426 "
Subdivisions : —		
	5th dynasty — Began about	3100 "
	7th " "	2900 "
	10th " "	2500 "
	12th " Ends about	2124 "
	13th " "	2100 "
Invasion of the Hyksos — comprising the		
	14th, 15th, and 16th dynasties — from about B. C. 2101 to about	1590 "
New Empire — *Restoration*:—		
	17th dynasty — Began	1671 "
	30th " Ending on the *second* Persian Invasion	340 "
Conquest of Egypt by Alexander the Great		332 "
Ptolemaic dynasty began B. C. 323 — ends		44 "
Roman dominion began		30 "
Hieroglyphical records of the Emperor Decius		250 A. D.

Thus, from an indefinite period prior to the year B. C. 3893, down to 250 years after the Christian era, the hieroglyphical character is proved to have been in uninterrupted use; while, from the year B. C. 3893, modern hierology has determined the chronologic order of Egyptian dynasties, through present archæological re-construction of the Nile's monuments.

The Romans held Egypt from the 27th year B. C. until 395 A. D.; when the sons of Theodosius divided the Empire. Egypt lingered under the sovereignty of the Eastern Emperors until A. D. 640–1; when, subjected by AAMER-EBN-EL-AS, she became a province of OMAR's Saracenic caliphate. In the year A. D. 1517—*Hedjra* 953—her valley was over-run by the Ottoman hordes of SOOLTAN SELEEM; and has ever since been the spoil of the Turk:—

O! Egypte, Egypte! . . . Solæ supererunt fabulæ et æque incredibiles posteris . . . sola supererunt verba lapidibus incisa. Et inhabitabit Ægyptum Scythus aut (ANGLO·) *Indus, aut aliquis talis.* (490)

CHRONOLOGY—CHINESE.

"The Philosopher said: SAN! (name of his disciple THSENG-TSEU) *my doctrine is simple and easy to be understood.* Thseng-tseu replied: 'that is certain.' The Philosopher having gone out, his disciples asked what their master had meant to say. Thseng-tseu responded: 'The doctrine of our master consists uniquely in possessing rectitude of heart, and in loving one's neighbor as oneself.'" (491)

Such were the ethics put forth in China by that "pure Sage" whom three hundred and seventy millions of humanity still commemorate, after the lapse of 2330 years, as the "most saintly, the most wise, and the most virtuous of human legislators:" this was Chinese "positive philosophy" in the VIth century before Christ; already at the *second* period of its historical development. (492)

About a century later, in a distinct Asiatic world, the school of EZRA at Jerusalem embodied a similar conception in the compilation termed *Deuteronomy*, or "secondary law:" (493)

(490) *Books of Hermes* — MERCURIUS TRISMEGISTUS's dialogue with *Asclepius*; — GLIDDON: *Appeal to the Antiquaries*: London, Madden, 1841, *passim*.

(491) The LUN-YU, or *The Philosophical Conversations*, of KHOUNG-TSEU (Confucius); ch. iv. v. 15; Livres Sacrés de l'Orient, p. 183.

(492) PAUTHIER: *Histoire de la Philosophie Chinoise*; Revue Indépendante, Aug. 1844; tirage à part, p. 9.

(493) N. B. My justification of this date is contained in the suppressed portions of our vol.; *supra*, pp. 626–7

"But if any man hate his neighbor. &c. . . . then shall ye do unto him, as he had thought to have done unto his brother." (494) At an epoch approximate, this idea became simplified into a maxim: "Better is a neighbor that is near, than a brother far off:" (495) and it is still more concisely expressed in *Leviticus*: "Thou shalt love thy neighbor as thyself." (496)

During the same fifth century B. C., the simultaneousness of moral as well as of other developments among Types of Mankind radically distinct, and remote from each other's influences, encounters a parallelism in the beautiful dictum of a Grecian Isocrates — "Do unto others as ye would they should do unto you."

About three generations earlier there flourished in Persia the philosopher Zoroaster; some of whose elevated doctrines have reached our day, although through turgid Grecian, Jewish, and Persic streams. "*Gate* the 71st" of his *Sadder* contains the following: —

"Offer up thy grateful prayers to the Lord, the most just and pure ORMUZD, the supreme and adorable God, who thus declared to his prophet Zardusht (Zoroaster): '*Hold it not meet to do unto others what thou wouldst not have done to thyself:* do that unto the people which, when done to thyself, proves not disagreeable to thyself.'" (497)

Five hundred years afterwards, the writer of *Matthew* (498) reported — "Ye have heard that it was said: *Thou shalt love thy neighbor and hate thine enemy;* but I say unto you, love your enemies." The writer of *Luke* (499) considerably extends the idea in language and contextual circumstances — "And he answering said: '*Thou shalt love the Lord thy God* [Hebraicè, IeHOuaH ELoHeK] *with all thy heart, and with all thy soul, and with all thy strength, and with all thy mind, and thy neighbor as thyself:*" thus combining, into one discourse, two citations from the Old Testament (500) slightly varied; owing probably to the evangelists' habit of following the Greek LXX in lieu of the Hebrew Text.

But, among the more exalted of the Hebrew nation, in the schools of Babylon and Jerusalem, such pure ethics had been taught long previously. Thus (as our learned friend, Dr. J. J. Cohen of Baltimore, opportunely reminds us while writing): —

"Let us recall the celebrated reply made by the Pharisee Hillel to a pagan who came declaring to him that he was ready to embrace Judaism, if the Doctor could make known to him in a few words the *résumé* of all the law of Moses: — '*That which thou likest not* [done] *to thyself*,' said Hillel, '*do it not unto thy neighbor;* therein is all the law, the rest is nothing but the commentary upon it.'" (501)

These comparisons made, we can revert with more pleasure to China and to CONFUCIUS.

"The lessons of KHOUNG-TSEU were often less indirect. His moral [doctrine] is summed up in the following lines: 'Nothing more natural, nothing more simple, than the principles of that morality which I endeavor to inculcate in you through salutary maxims. . . . 1st.— It is *humanity;* which is to say, that universal charity amongst all of our species, without distinction.'"

Father Amiot, the great Sinicized Jesuit, commenting upon this passage, observed — "Because it is *humanity*, and that *humanity* is nothing else than *man* himself." Which Pauthier explains: —

"In Chinese, JIN TCHE: JIN YE: word for word; *humanitas quæ, homo quidem. . . .* To render comprehensible how much *humanity*, or benevolence, universal charity, was recommended by KHOUNG-TSEU, it suffices to say that the word which expresses it is repeated above a hundred times in one of his works, the *Lun-yu*. And it is pretended, with as much levity as ignorance, that this grand principle of *universal charity* for mankind had only been revealed to the world five hundred years after the Chinese philosopher, in a little corner of Asia! *Quelle pitié!*" (502)

We have deemed it expedient to preface an inquiry into the archæological bases of

---

(494) *Deuteronomy*, xix. 11, 19.

(495) *Proverbs*, xxvii. 10.

(496) *Leviticus*, xix. 18.

(497) *Dabistan*, i. 338: and see the same quotation in HYDE, *De Relig. Vet. Persarum*, p. 471.

(498) *Good Tidings*, v. 43. SHARPE'S *N. T.*, p. 9.

(499) *Good Tidings*, x. 27, 27 — *Ibid.*, p. 132.

(500) *Deuteronomy*, vi. 5, with *Leviticus*, xix. 18.

(501) MUNK: *Palestine;* p. 565; from Babylonian Talmud (*Shabbath*, ch. 2). *Ibid.: Réflexions* in Appendix to CAHEN'S *Bible;* 1833; iv. p. 20.

(502) *Chine;* pp. 146, 147, and note.

Chinese *chronology* with the above extracts. They will furnish at once to the reader a very different idea of the teachings of Confucius (five hundred years before any Greco-Judæan writers of the Gospels lived) than he can gather from Macao supercargoes, Hong-kong opium-smugglers, or Canton missionaries. Whatever practical developments the latter may diurnally give to the sublime principle of "universal charity;" whatever merit may be due to the first human being who enunciated this exalted sentiment; or whatever thorough knowledge of humanity's best and loftiest interests such sentiments may imply; all these ascriptions, history attests, equally belong to a Sinico-mongol, Confucius; who died B. C.479, or about 2332 years ago. [See his portrait; *supra*, Fig. 330, p. 449.]

Whether among the *Hong* merchants "universal charity" (and there are noble instances) be unexceptionably practised, any more than in *Wall* street, *Lombard* street, or in the *Place de la Bourse*, concerns us not. These commercial princes are taught to reverence its principles as much as the DORIAS or the MEDICIS of Christendom; and they are exposed to infinitely greater temptations toward its violation, than are those Chinese archæologists, who, scattered throughout the empire, pursue, at national expense, their historical studies of their own monuments; in lettered seclusion, but with every honorable recompense scholarship may aspire to. (503) For above twenty-three centuries, moreover, the 4th and 5th maxims of Khoung-tseu have been instilled into each generation of them from earliest infancy.

"It is *uprightness;* that is, that rectitude of spirit and of heart, which makes one seek for truth in everything and to desire it, without deceiving oneself or deceiving others: it is finally *sincerity* or *good faith;* which is to say, that frankness, that openness of heart, tempered by self-reliance, which excludes all feints and all disguising, as much in speech as in action."

That the moral influence of such principles has not perished, even through the transitory irruption of the present and expiring dynasty of *Mantchou* Tartars, is testified by Sir Henry Pottinger in the eulogiums pronounced by him, at London, upon the high Chinese diplomatists with whom he concluded the Treaty of 1844. Nor should Americans forget the excellent conduct which such principles have already exhibited among thousands of our Chinese fellow-citizens in the State of California.

We have not the slightest right to doubt, therefore, whatever reasonable account Chinese scholars may furnish us of their nation's indigenous history; of which, otherwise, not a syllable is known to us prior to the fourteenth century after Christ; and, where not irrational, such annals, from such sources, may be received in the more good faith, that the *Chinese* archéologue, having none of our hagiographers' motives for chronological curtailment or extension, cares nothing about "outside barbarians," their alien history or superstitions, and did not compose his national chronicles with a view to such foreigners' edification.

The day is evermore passed that modern science should strive to reduce Chinese chronology, for the mere whim of adapting it to the spurious computations on a Hebrew Text, and Samaritan, Septuagint, or Vulgate version; as was the case before *Egyptian* monumental annals were proved to ascend, at least, to the thirty-fifth century B. C. (504) And we shall presently show (sketched also in our table of *Alphabetical origins, supra*, p. 638), how the highest point claimed by Chinese historians, for their nation's antiquity, falls centuries below that which hierologists now insist upon for Egypt: so that, if *Egypt* and *Egyptians* were a civilized country and populous people in the thirty-fifth century, B. C., it would be preposterous not to feel assured that Sinico-mongols (indeed every human type of Mongolia) were already in existence, in and around China, their own centre of creation, during the same parallel ages. What is the objection to believing that China was populated, by her Mongolian autocthones, chiliads of years previously? Reader! "one blushes" redder than St. Jerome to mention, that, now-a-days, the acceptance of this fact is questioned by the Rev. Dr. *This*, or the Rev. Mr. *That:* neither of whom, perhaps, has ever studied Sinology — never even opened a Sinological work!

---

(503) *Chine;* pp. 194, 218, 228, 236, 248, 286, 308, 336, 352, 359, 388, 397, &c.: also, BIOT, *Sur la Constitution Politique de la Chine au 12ème siècle avant notre ère;* 1845; pp. 3, 9, &c.

(504) DE BROTONNE: *Filiations et Migrations des Peuples;* ii. pp. 1–43.

The reveries of Fortia D'Urban (505) are now superannuated; the monstrous extravaganzas of a Paravey are preserved as ceaseless sources of merriment. (506) To refute either, seriously, would be sheer waste of time. The inundations of the river *Hoang-ho*, overcome by the engineer Yu, (507) lie parallel with the Egyptian XIIth dynasty; when, in the 23d century B. C., similar causes induced smaller constructions along the Nubian Nile: (508) and a reader of Pauthier will as soon associate those local dikings, buttresses, dams, and sluices, in China or Egypt, with Usher's universal Flood, as by anybody else the Noachian deluge might be proposed in explanation of the *levees* along our Louisianian Mississippi. It would be an equal outlay of labor to discuss Hales's views upon Chinese subjects; (509) after his Hebraical knowledge has been so repeatedly shaken throughout these pages: nor need we perplex the reader with other works whose authors, like ourselves, are not Sinologists; but who, in this respect unlike ourselves, do not seek for information at its only clear fountains.

It will be now plain that "Types of Mankind" recognizes for Chinese history none but Chinese historians. The chances of error lie uniquely in the channels through which its authors receive their accounts: and these, to our view, are completely guarded against when we accept Rémusat and Pauthier, as, above all Europeans at this day, qualified to be their interpreters. Furthermore, every relevant passage from the Jesuit missionaries is embraced within Pauthier's volumes.

Under the caption of *Mongolian Origin* and ideographic writings, we have displayed the argumentative process through which it becomes certain, that Europe knew naught about China, nor China aught about Europe, until the end of the 1st century after C.: but modern acquaintance with *Cathay* dates from the Venetian Marco Polo, who resided in China about A. D. 1275; followed by the first Jesuit missionary, Father Michæl Rogerius, who penetrated thither about A. D. 1581; and the second, Father Matthæus Riccius, in 1601. From that time, during more than a century, many accomplished Europeans *è Societate Jesu* flocked into the Celestial Empire; and to their vast labors are we indebted for complete reports upon China, derived by them from the highest scholastic and official sources of the realm — which narratives, now collated by Sinologists in Europe with the immense literary treasures accessible, in Chinese, to students at Paris and Rome, prove to have been conscientiously executed. No Europeans, before or since, have possessed such opportunities for acquiring thorough knowledge of everything Chinese as these lowly preachers of the Gospel. Indeed, the official report made, in 1692, by the "President of the Supreme Court of Rites" to the Emperor Khang-hi, and by him approved, alone suffices to show their powerful claims upon *Mantchou-Tartar* affections: —

"We have found that these Europeans have traversed vast seas, and have come from the extremities of the earth. . . . They have at present the supervision of astronomy and of the board of mathematics. They have applied themselves with great pains to making war-like machines, and to casting cannon; of which use has been made in the last civil troubles [that is, the missionary ordnance had been found effective in quelling *Chinese* revolts against the Tartar dynasty]. When sent to Nip-chou with our ambassadors [the reverend Fathers Pereyra and Gerbillon, *è Soc. Jesu,*] to treat about peace with the Muscovites, they caused those negotiations to succeed: in short, they have rendered great services to the [*Mantchou*] empire. . . . The doctrine which they teach is not bad, nor capable of seducing the [Chinese] people, or of causing any troubles. It is permitted to every body to go into the temples of the *Lamas*, of the *Ho-chang*, of the *Tao-ssé;* and it is forbidden to go into the churches of these Europeans, who do nothing contrary to the laws: this does not seem reasonable." (510)

The emperor himself had been previously instructed by the scientific Father Verbiest, "chief of the bureau of astronomers"; whose evangelical virtues comprised gnomonics,

(505) *Histoire Anté-diluvienne de la Chine.*

(506) *Documents sur le Déluge de Noë:* Paris, 1838.

(507) PAUTHIER: *Chine;* pp. 12–4; and his *Chou-king;* pp. 49–56.

(508) LEPSIUS: *Nachricht;* p. 11: — *Briefe aus Ægypten;* pp. 259, 260: — DE ROUGÉ: *Phénom. Célestes;* Rev. Archéol., Feb. 1853.

(509) *Analysis:* i. pp. 199–203.

(510) *Chine:* pp. 435, 440, 445–449.

geometry, land-surveying, and music. The reverend Fathers Bouvet, Regis, Jartoux, Fridelli, Cardoso, de Tartre, de Mailla, and Bonjour, at government expense, made official maps of the different provinces of China, after European methods; and, at the same time that such labors familiarized the whole of these Propagandic missionaries with Chinese literature, Fathers Amiot, Gaubil, and Du Halde, devoted their leisure more especially to minute study of Chinese archæology. In one word, the admiration avowed by the Jesuits for Chinese civilization on the one hand, and the influence which Chinese philosophy possessed over their intellects on the other, had led to such a fusion at Pe-kin, during the 17th century, that one is at a loss to decide whether the Chinese were becoming converts to spiritual Christianity, or whether the disciples of Loyola were adopting the materialistic "doctrine of the Lettered."

Unhappily for our desires to solve this curious problem, certain puritanic *Dominicans* arrived from Rome; and, Pandora-like, let loose fanatic ills heretofore preserved hermetically. It was they who started that everlasting question whether the Chinese word *chang-ti* be a synonyme for "God" or the "sky." Pig-tailed converts to Christianity *à la Jésuite* were incontinently bambooed by hog-tails *à la Dominicain;* for heretical notions upon an equivocal point by aliens indicated for Mongol salvatory "credo." Khoung-tseu's "universal charity" being interrupted by swinish brawls at which the writers of *Leviticus* (511) would have shuddered, policemen duly reported their real causes to mandarin magistracy: which reports, in official course, reached a new embodiment of the Sun upon earth, Young-tching. This unsophisticated Tartar at once relieved himself, and his successors for more than a century, of these foreign theologers, by shipment of a live cargo, including missionaries Jesuit and Dominican, consigned to Macao under judiciary "bill of lading," about the years A. D. 1721–'25.

It is to the *Jesuits*, nevertheless, that impartial science looks back, gratefully, for throwing the portals of Chinese history widely open to European Sinology: and it is especially to the late Rémusat, Klaproth, and Ed. Biot, as to MM. Stanislas Julien and Pauthier, that our generation owes the reappearance of Chinese studies on the continent, since the demise of the famed historian of the *Huns*, Deguignes. At Paris, the Chinese department of the Bibliothèque Impériale comprehends quantities stupendous of that country's literature.

Every element for our purposes being in consequence accessible, we proceed, Pauthier's works in hand, to sketch 1st, — the mode through which archæologists in China have definitely tabulated, in precise stratifications, the relative order of national events; and 2d, — to present a chronological table of Chinese dynasties, from such tabulations accruing.

It is as certain as any other fact in history (512) that about 1000 years B. C., parallel with the reign of Solomon, books existed in China with such titles as these: — "Laws of the administration of ancient kings;" and that recurrence was common to "ancient documents." It is also certain that arts and sciences continued to prosper down to the year 484 B. C., (513) when Confucius compiled the *Chou-king*, sacred book of the Chinese, from anterior documents. Literature was immensely diffused among the "Lettered" in China; when, B. C. 213, Chi-hoang-ti burned all the books which torture could extort, together with multitudes of their readers; (514) because the latter quoted the former against his imperial innovations. Nevertheless, this splendid miscreant served practical objects, not altogether indefensible, when he relieved the empire of its "old-fogiedom;" to judge by the withering oration of his prime-minister, Li-sse: —

> "Prejudiced in favor of antiquity, of which they admire even the stupidities, they are full of disdain for every thing which is not exactly chalked after models that time has nearly effaced from the memory of man. Incessantly they have in their mouths, or at the tips of their pencils, the *three* Ho-ang [the Chinese august triad], and the *five* Ti [the Chinese pentateuch]."

Nearly 2000 years previously, disputes among religious sects in China had risen to such

---

(511) XI. 7.
(512) *Chine;* pp. 59, 194, 200.
(513) *Chou-king, Préface du Père Gaubil;* PAUTHIER'S "Liv. Sac. de l'Orient," Paris, 1843; pp. 1, 2.
(514) *Chine;* pp. 222–228.

an intolerable pitch, that the pious Emperor Mou-wang, about B. C. 950, records how Yao, in B. C. 2337, in order to suppress false prophecies, miracles, magic, and revelations, —

"Commanded the two Ministers of Astronomy and Religion to cut asunder all communication between 'sky' and earth; and thus (says Mou-wang) there was no more of what is called this *lifting-up* and *coming-down.*"

And, so inveterate, in sporadic instances of the Chinese mind, was this childish reliance upon invisible powers, that fifteen centuries after the burning of the books, the Minister Tchang-kouei, about A. D. 1321, during a period of great physical calamities, pestilence, inundations, &c., felt it incumbent upon his office to include the subjoined remarks in a long and manly expostulation: —

"A prince must not think to govern his country save as the father of his subjects; and it is not through *Bonzes* [Budhist priests] that he must seek felicity. Ever since the *Bonzes*, the *Lamas*, and the *Tao-ssé*, make so many prayers and sacrifices to their idol, 'Heaven' has given constant signs of its indignation; and until such time as one sees the worship of Fo [Budha] abolished, and all these *priests* driven away, one must expect to be unhappy."

Such political necessities may palliate some of Chi-Hoang-ti's deeds; which obliterated so much of earlier literature extant down to the Chinese "era of the martyrs" for science, B. C. 213.

Upon accession of the famous *Han* dynasty, B. C. 202, a reaction in favor of letters immediately commenced; and from this period of "renaissance" downwards no nation upon earth possessed, till recently, annals comparable to the Chinese. About B. C. 176, the *Chou-king* of Khoung-tseu was recovered, partly, by taking down the recitations of a nonogenarian savant, Fou-cheng, who had been president of literature prior to the conflagration of libraries. Through this venerable scholar (who is to the Chinese what Ezra was to the Jews) and the fortüitous discovery, B. C. 140, of a copy of the *Chou-king* with other books in the ruined house of Confucius, the more important documents of Chinese antiquarian lore were restored.

European authors, who claim that we possess the plenary words if not the autograph of Moses, have doubted this account. We accept it, notwithstanding, in good faith; because neither the books themselves nor their transcribers pretend to supernaturalism in any shape; whilst the nature of the local researches subsequently undertaken renders nugatory such unwarrantable European objections.

"But the man who has thrown the grandest éclat over the reign of the Emperor Wou-ti, is Sse-ma-thsian, whom M. Abel Rémusat has called the *Herodotus of China.*"(515) His portrait is given under our Fig. 331 [*supra*, p. 349]. About B. C. 104 he commenced his *Historical Memoirs;* which, in 130 books (extant in European libraries, and consulted by the Sinologists we quote), furnish a vast encyclopædia of Chinese annals, of every kind, from the reign of the old Hoang-ti, 2697 years before C., down to B. C. 140.

"Sse-ma-thsian made good use of all that remained of the *Classical Books;* of those of the *Ancestral Temple* of the Tcheou-dynasty; the *Secret Memoirs of the House of Stone*, and of the *Golden Coffer;* and of the registers called *Plates of Jasper*. It is added that he stript the *Liu-ling*, for what concerns the laws; the *Tactics of Han-sin*, for what regards military affairs; the *Tchang-tching*, for what relates to general literature; and the *Li-yi* for every thing that is relative to usages and ceremonies."

There are no further breaks in Chinese archæological labors down to our time; which researches, for care and magnitude, may challenge the universe. We mention, however, only the *Researches profound of the Monuments left by Savans*, published at royal expense, in 348 books, by Matouan-lin, in A. D. 1321; which covers history from the twenty-fourth century B. C. down to the twelfth after C. Copies exist in European libraries. After the death of Chi-Hoang-ti: —

"The tombs, the ruins of cities, the canals and rivers, saved some moneys, some bronze vases, some urns and other objects of his proscription. A certain number of these has been found since the fall of the Thsin-dynasty. They have been carefully collected and preserved in museums or in private cabinets; descriptions have been made

(515) *Chine;* pp. 246–248.

of them, accompanied by figured designs that faithfully reproduce them with their ancient inscriptions. The emperor Kien-loung, who reigned from A. D. 1736 to 1796, caused to be published, in 42 *Chinese folio volumes*, a description and engraving of all the antique vases deposited at the Imperial Museum. An exemplar of this magnificent work, which has no rival in Europe, being at the Bibliothèque Royale of Paris."

Pauthier has selected, out of 1444 *vases* of different species contained in these "Memoirs of the Antiquities of Occidental Purity," those beautiful specimens we behold, reduced in size, in his work. (516)

The earliest originals, now extant in China, go back in date to the *Chang*-dynasty, B. C. 1766: — an epoch when Abraham, according to Lepsius's computation of biblical chronology, was yet unborn. One more ancient inscription, upon a rock of Mount *Heng-chan*, yet remains to vindicate the engineering ability of YU. It dates about the year B. C. 2278; (517) and is therefore parallel in age with the thousand records we possess of Egypt's XIIth dynasty. Its translation, given by Pauthier, disconnects it from any diluvial hypotheses; with which, moreover, no geologist or archæologist need distress himself further.

We trust the reader has now attained to our point of view, and perhaps perceives three things — 1st, the historical meritoriousness of Chinese literature; 2d, the nature of the materials examined by Jesuits whose evangelical prepossessions were essentially hostile to the literature they laud; and 3d, that there are Sinologists living in the world competent to liberate historical truth from chances of error. We now proceed to lay before him a brief summary of Chinese time-registry; commending to his perusal the "Researches upon times anterior to those of which the *Chou-king* speaks, and upon Chinese mythology," by Father de Prémare, together with an old rule of Vico's. (518)

"We have heard Diodorus Siculus declare, in respect to the *pride of nations*, that these, 'whether they may have been Greek or barbarian, have pretended, each one, to have been the first to discover all the comforts of life, and to have preserved their own history since the commencement of the world.'" (519)

Greece, Rome, and Judæa, possess first their fabulous and then their semi-historical periods. Tradition alone pierces through the gloom of the latter, in the ratio of approximation to the several epochas at which given nations first began to chronicle their events. In later days, progressive science invests such fables and faintly-shadowed incidents of a nation's childhood with the garb of mythico-astronomical sanctity. Thus does the founder of chronology, Manetho, preface his historical dynasties with cycles of *Gods*, *Demigods*, and *Manes;* thus do the compilers of *Genesis* antecede Abraham with symbolical names of mythic patriarchs gifted with impossible longevity; and so do the Chinese place mythology before history. The sole difference being that neither did Manetho nor the *Chinese* archéologues ever believe their respective mythologies to be otherwise than unhistorical: at the same time that the whole of these antique systems represent that instinctive consciousness of nations who feel that an unrecorded national infancy must have preceded a recorded national adolescence.

CHINESE ANTE-HISTORICAL PERIODS. (520)

PAN-KOU — first symbolical man — followed by the *three* HOANG, viz.: —

1st.—Reign of the *Sky*.
2d.— " " *Earth*.
3d.— " " *Man*.

They are comprehended in a grand cyclic period of 129,600 years; composed of twelve parts called *conjunctions*, each of 10,800 years.

(516) *Chine;* p. 201; Plates 38–44.
(517) *Ibid.;* pp. 53–54.
(518) *Liv. Sac. de l'Orient;* pp. 13–42.
(519) VICO: *Scienza Nuova;* Principles, axiom iii.
(520) *Chine;* pp. 22–24; — *Livres Sacrés*, pp. 16, 19.

Meta-historical Period.

Fou-hi—first Emperor—estimated at.. B. C. 3468
Several of his descendants are named, with traditionary discoveries in arts affixed to each personage.

Fou-hi, however, is a collective name under which the Chinese figure many centuries of national existence coupled with progressive developments in civilization, marked by consecutive artistic inventions: just as the Hebrews ascribe all legislation to their noun of multitude, Moses. This traditionary and semi-mythical *first* Emperor stands parallel with the Egyptian IVth *dynasty,* during the thirty-fifth century B. C. The latter is positively historical: to reject the former, on the imaginary ground of recent mundane antiquity, is renderëd futile by existing pyramids at Memphis. Fou-hi, Menes, and Abraham, to us appear equally historical, as human individuals who once lived; although of none of the three are contemporaneous monuments, carved by their respective people, now extant.

Historical Period.

Chronological Table.—We condense into dynasties that chronology of *all the Sovereigns* who have reigned in China, (from B. C. 2637 down to A. D. 1821), which Father Amiot transmitted from Pe-kin to Paris in 1769; and which is printed "in extenso" at the end of Pauthier's *Chine,* after collation with the learned Jesuit's manuscript notes, and with parts of the 100 *volumes* of the Chinese chronographic work *Li-tai-ki-sse.*

The 61st year of the Chinese emperor Hoang-ti, corresponding to our B. C. 2637, falls, according to Lepsius's computation, within Egypt's "Old Empire," and between the VIIth and Xth dynasties of Manetho, in any case during the pyramidal period.

Ist	*Dynasty*—	1st King, Hoang-ti, "Yellow Emperor," 61st year...	2637	B. C.
		Five successors down to Yao, B. C. 2337.		
	"	6th " Yao, 81st year	2277	"
	"	8th " Chun, ...9th of his synthronism	2277	"
		[*Monuments* commence—"Inscription of YU," B. C. 2278.]		
IId	"	"Hia"—1st King, Yu, 10th year of his synthronism	2205	"
	"	" 4th " Tchoung-kang 5th year of his reign, *eclipse of the Sun,* B. C. 2155 (521)	2155	"
IIId	"	"Chang"	1783	"
		[Contemporary *vases* exist, dating from B. C. 1766.]		
IVth	"	"Tcheou"	1134	"
Vth	"	"Thsin" [whence the name of "China"]	255	"
VIth	"	"Han"	202	"
		King Youan-ti, of the "Wei," A. D. 292.		
VIIth	"	"Tçin"	265	A. D
VIIIth	"	"Northern Soung"	420	"
IXth	"	"Tsi"	479	"
Xth	"	"Liang"	502	"
XIth	"	"Tchin"	557	"
XIIth	"	"Souï"	581	"
XIIIth	"	"Thang"	618	"
		The *Five Little Dynasties.*		
XIVth	"	1st, "Posterior *Liang*"	907	"
XVth	"	2d, "Posterior *Thang*"	923	"
XVIth	"	3d, "Posterior *Tsin*"	936	"
XVIIth	"	4th, "Posterior *Han*"	947	"
XVIIIth	"	5th, "Posterior *Tcheou*"	951	"
XIXth	"	"Soung"	960	"
XXth	"	"*Kin,* simultaneously with *Soung*"	1123	"
XXIst	"	Commencement of "Youan," *Mongols*	1260	"
XXIId	"	*Mongols*	1295	"
XXIIId	"	"Ming"	1368	"
XXIVth	"	"Taï-thsing," *Mantchou*-Tartars	1616	"
		Now reigning—and down to	1821	"

24 *Dynasties,* whose consecutive rule covers *years* 4458.

(521) *Chine,* p. 58; and *Chou-king,* p. 47:—but, compare Biot, *Syzigies,* 1848, for astronomical doubts.

Egyptian priests had told Herodotus, (522) that lengthened experience and observation of their own history enabled them to predicate the future through the cyclic recurrence of the past. In no chronicles do similar causes oftener reproduce similar events, through perpetual cycles, than the reader of Pauthier will recognize among the *Chinese.* No political acumen is required by historians to foretell the inevitable downfall of the present alien *Mantchou*-Tartar dynasty. Its doom is sealed; its knell is ringing. One fact will illustrate its Tartarian despotism, and explain the repugnance to prolongation of its hateful rule nurtured in the bosom of every true *Chinaman;* precisely paralleled by *Arab* hatred to the cognate Tartar-*Turks.*

In the same manner that the radical poverty of the Ottoman speech compels the Turk to draw all his polite terms from the *Persian,* his scientific from the *Arabic,* so, in China, the uncouth and slender vocabulary of the *Mantchou*-Tartars became enriched, after their conquest, with Chinese words of civilization. This gave offence to the Tartar emperor, Kien-loung; who, anxious to preserve the *Mantchou* idiom in its natural if barbaric "purity," appointed an Imperial Commission, to compose, from *Mantchou* radicals, 5000 *new words,* to stand in place of those which his courtiers had borrowed from the Chinese tongue. This new nomenclature, printed and proclaimed, was imposed upon all high government functionaries; who had thus to learn 5000 unknown words by heart, under severe penalties! Truly, as Champollion-Figeac remarks — "Il n'y a qu'un *Tartare* régnant sur des *Chinois* qui soit assez puissant pour introduire d'emblée et par ordonnance cinq mille mots dans une langue!" (523)

CHRONOLOGY—ASSYRIAN.

"The spider weaves his web in the palace of Cæsar;
The owl stands sentinel upon the watch-tower of Afrasiab!"
(FIRDOOSEE — *Shah Nameh.*)

THE eighteenth century, fecund precursor of those conquests in historical science that have immortalized the nineteenth, passed away, without permitting its contemporaries to illumine the gloom which, since the decline of the Alexandria School at the Christian era, for 2000 years had enveloped with equal obscurity the pyramids and temples of the Nile, the lightning-fused towers and crumbling brick mounds on the Euphrates and Tigris, or the rock-hewn sepulchres and thousand-pillared fanes of "lorn Persepolis."

In the year 1800, absolutely nothing was known about these huge colossi of the past beyond the fact of their existence!

A wondrous change has been wrought, by half a century of research, in historical knowledge: almost inconceivable when we reflect that, upon the Assyrian theme before us, modern science knew nothing in 1843 — only ten years ago. "Palpitants d'actualités," Lamartine would say, are these glorious discoveries — still damp from the press are the volumes that unfold them.

Antithesis serves to place past ignorance and present information in the strongest light. Persepolis and her arrow-headed inscriptions suffice by way of illustration.

The German Witte ascribed these ruins, not to human agency, but to an "eruption of the earth." De Rœsch deemed them the work of an antediluvian *Lamech,* "whose exploits are exhibited in these sculptures." Discarding *Homer's Iliad* in the sense vulgarly understood of its glowing heroics, De Rœsch believes Persia to be figured by Troy, Media by Europe, and Assyria by Asia. According to this logopœist, or compiler of invented facts, the Grecian siege of Ilium was but a war between Medes and Persians: and the cuneatic letters of Persepolis "record a series of kings from Cain to Lamech."

Chardin, in 1673, pronounced these remains to be about "4000 years old;" a limit too restricted for the astronomer Bailly: who attributes the foundation of Persepolis to the

(522) Aptly cited by HENRY, *L'Égypte Pharaonique,* ii. pp 27, 28.
(523) *Paléographie Universelle;* 1841; Introduction, p. 48.

Persian hero, *Djemshid*, (524) whose fabulous because mythic epoch he fixed at 3209 B. C. To the same Iranian demigod are these edifices assigned by Sir W. Jones, estimating their age at about 800 years before Christ.

Semitic historians without exception, as Sheridan neatly observed, "draw upon memory for their wit, and upon imagination for their facts:" wherefore slim clews to a reality could be obtained through them. Like the libraries of Alexandria, of Jerusalem, of China, of Budhic Hindostan, and of Hebraical Christendom, those of ante-Mohammedan Persia perished, from similar fanatical causes, in Saracenic flames with the dynasty of Chosroes, about A. D. 637. Such fitful traditions as survived the wreck of Persic literature became invested (after Bédawee destructiveness had become altered into caliphate restorations) with the hyperbolic extravagancies of Eastern poetry and romance.

One immortal epic, Firdoosee's *Shah Nameh*, or "Book of Kings," composed in the eleventh century, purports, indeed, to cover 3600 years of his country's annals, from the taurokephalic Kaiomurs down to the Arab invasion. Persepolis, under its local name of *Istakhàr*, is mentioned in twenty-eight passages, and its existence is referred to as coeval with Kai-kobad; whose apochryphal era, under Sir W. Jones's hypothesis, falls about B. C. 610: but, neither from the "History of the early kings of Persia" by Mirkavend, in the fifteenth century, nor from the "Dabistàn," was archæological acumen able to disentangle a solitary thread indicative of the age, the builders, or the writings, of Persepolis.

As in Egypt the present *fellàh*, or peasant, ascribes the pyramids to "Pharaòon" (525) or Pharaoh — a name to him the synonyme for Satan — so in Persia, the illiterate native is content that an ancient edifice should be the work of Suleymàn; at once the archimagus of Oriental necromancy and the sage monarch of Israel: for at Murghàb, *Pasargadæ*, the mausoleum whence we have drawn the portrait of that great man [*supra*, p. 138, Fig. 43] whose sculptured epitaph is simply "I am Cyrus, the king, the Achæmenian," is called *Takhti Suleymàn*, or "Solomon's throne." Like Jephtha's, who was buried "in the *cities* of Gilead," (526) Solomon's tomb is shown at Shiràz and again on the road to Kashgàr! Nimrod is even still more ubiquitous.

Equally futile were attempts to rescue history applicable to Persia's monuments from the *Zend-Avesta* of Zoroastric attribution, or from the later *Boundehesh-Pehlvi:* sacred books containing the rituals and theosophy of the Guebres, or Persian expatriated ignicolists of Guzerat, now called Parsees. From Greek writers alone (Herodotus, Xenophon, Ctesias, &c.) were such elements of early Persian history derived as have stood the test of monumental investigation: but the science of the last century had ransacked all these sources without obtaining a glimmer of light as to the nature of Persepolitan wedge-shaped characters. Like the once-mysterious hieroglyphs of Egypt, as interpreted by Father Kircher, the inscriptions of Persia were supposed to veil occult and awful things, black arts of magic, or diabolic talismans. With naught to guide them but the more or less faithless copies printed by De la Valle, Le Brun, Kaemfer, and other old travellers, how could the opinion of a student be other than a conjecture more or less rational according to the mental calibre of each critic?

Thus, by Leibnitz and by Cuper, these inscriptions were reasonably conjectured to contain the letters and elements of "some very ancient writing." Lacroze, the great Coptologist, conceived them to be hieroglyphical inscriptions similar to those of Egypt (at that day undeciphered) and of China, which last are not "sacred sculptured characters" at all.

---

(524) DJEMSHID is the Persic, as SAMSON is the Hebrew, *Hercules*. The former we opine to be DJoM, the Egyptian *Hercules*, coupled with ShaDI, the *strong:* the latter is simply SheMS-on, the *Sun*, with its Arabian euphonizing suffix. *Hercules* is but HaR-GoL, "revolution of heat." Compare LANCI, *Paralipomeni;* and RAOUL-ROCHETTE, *Archéologie Comparée;* with DUPUIS in *Anthon's Class. Dic.*, "Hercules."

(525) "*Yà Pharaòon ebn Pharaòon*" is generally rendered "Thou Pharaoh son of a Pharaoh"! Why not "Thou *crocodile* son of a *crocodile*"? Conf. ROSENMULLERI *Instit. Ling. Arabicæ;* 1818; p. 211.

(526) Text. *Judges* xii. 7. The *sacrifice* of Jephtha's daughter is beautifully told by EURIPIDES; for *Iphigenia*, in its Greek sense of *Ιφιγένεα*, is only a "daughter of Jephtha."

Chardin opined them to be a "veritable writing like our own;" and Le Brun happily describes these ruins as covered with "ancient Persian characters."

In the face of sensible speculations on matters then entirely inexplicable, the intrepidity of ignorance is exemplified from a quarter whence it would have been least expected; viz., in Hyde's *History of the Religion of the Old Persians* (Oxon. 1760). Not only does he deny that these Persepolitan inscriptions are "old Persian writings," but the author backs assertion with professions of faith:—"I am of opinion that they are neither letters nor intended for letters; but a mere playful *jeu d'esprit* of the chief architect; who, to adorn the walls of Persepolis, imagined a trial of how many divers forms a single elementary stroke (the *wedge*) could be produced combined with itself"! This is as pitiable for such a scholar, as the unfortunate Seetzen's mistake, when he took the *sunken spaces* between each Himyaritic letter for the characters themselves. In the same manner, one of Hyde's contemporaries (the Abbé Tandeau, 1762) stoutly maintained that Egyptian "hieroglyphics were mere arbitrary signs, only employed to serve as ornaments to the edifices on which they were engraven, and that they were never invented to picture ideas."

These arrow-headed sculptures, like the still-unintelligible carvings on aboriginal monuments of Mexico, Central America, and Peru, seemed so enigmatical even to the great explorer of Babylon in 1816, that J. Claudius Rich disconsolately embodies the sum total of knowledge in these words:—

"Their real meaning, or that of the Persepolitan obeliscal character, and the still more complicated hieroglyphics of Egypt, however partially deciphered by the labors of the learned, will now, perhaps, never be fathomed, to their full extent, by the utmost ingenuity of man."

By strange coincidence (serving to add another example of the simultaneousness of discovery, at every age of human development), while Rich penned the above lament, Grotefend in Germany communicated to Heeren, 1815, those successful decipherings of Persepolitan cuneiform inscriptions he had commenced in 1802; which is the identical year of the arrival in England of that *Rosetta Stone;* whence, about 1816, Young's deduction of the letter L in the name "Ptolemy" originated those astounding revelations from Egyptian sculptures which are now so familiar in the archæological world as no longer to require notes of admiration.

Egyptologists, by rough and ready processes, have so completely vanquished opposition, that, at this day, disbelievers in Champollion confine their lugubrious chants to hearers illiterate and inarticulate: but, to judge by the pertinacity with which one, who is no mean scholar, (527) insists that Moses wrote—"The Tigris flows *to the east* of Assyria;" (528) and, therefore, that Botta and Layard have discovered Nineveh on the *wrong* side of the river—the battles of cuneiformists have only commenced! Happily, the Louvre boasts of an Orientalist (529) who can always quote to M. Hoefer the Muslim poet's mnemonic to St. Louis:—

"(O king of the Franks!) if thou preservest the hope of avenging thy defeat, if any temerarious design should bring thee back to our country, forget not that the house of Ebn-Lokmàn, that served thee for a prison, is still ready to receive thee. Remember that the chains which thou hast worn, and the eunuch Sabèeh who guarded thee, are ever there and waiting for thee." (530)

Such was the picture on the obverse page of Assyrian archæology in the year 1843. Before contrasting which with its illuminated face in 1853, it is due to the memory of that master, whose teaching of the methods for deciphering the meaning of all antique records has been the true cause as well of Champollion's as of Grotefend's successes—and hence of the whole of our present Egyptian and Assyrian knowledge—to name Silvestre de Sacy.

(527) Hoefer: *La Chaldée,* &c.; 1852; p. 146.

(528) *Genesis;* ii. 14.

(529) De Longpérier: *Antiquités Assyriennes;* Rev. Archéol., 1850; pp. 429–432: who reads, most triumphantly, "Le Tigre coule *en avant vers* Assour."

(530) Michaud: *Hist. des Croisades;* iv. p. 274.

In that part of our work discussing *Alphabetic Origins*, the student will find a sufficiency of authorities cited to verify the accuracy of those results to which this volume is confined. Recapitulation here is needless: but, should ever such inquirer follow the developments of palæographical discovery, book by book, backwards from to-day, his bark will not ground until he reaches the year A. D. 1797, and touches the *Mémoire sur les antiquités de la Perse, et sur les médailles des Rois Sassanides*. Its author, De Sacy, is to palæography that which his colleague Cuvier is to palæontology: each being the inventor of the only true method of ratiocination in either science. From the former's Memoir we have borrowed many of the citations above presented; and, our remarks being but introductory to Assyrian chronology, a reference to the excellent compendium of Vaux (531) indicates the shortest road to summary annals of cuneiform investigation; no less than corroborates our assertion that *monumental* Assyria was a blank down to 1843.

Paul-Emile Botta (whose surname is dear to all American readers of his uncle's *Storia dell' Independenza*), appointed French Consul at Mosul in 1842, was the first to resuscitate *Nineveh* since her fall in B. C. 606. Proficient as an Orientalist and Eastern traveller, through residence in Syria, Egypt, Ethiopia, and Arabia, since 1829–30, none possessed higher qualifications for the task; yet, with rare modesty, he attributes his own discoveries (as Newton to an apple his finding the laws of gravitation) to an accident; viz., to a couple of bricks, brought to him by a Nestorian dyer, who unearthed them whilst digging a foundation for stoves and boilers on the mound of *Khorsabàd*. (532) But, these two forlorn bricks were impressed with arrow-heads — things which Botta's education at once permitted him to appreciate. Ten years have since elapsed. The Louvre proudly displays his sculptured deterrations — national typography splendidly perpetuates his unaffected narrative — and, those who weigh science by "dollars and cents" may sneer at legislative munificence on learning that France, in 1849, had already voted $150,000 to eternalize Botta's Assyrian deeds; without either forgetting an individual's future, or considering the balance of an account-current between a man and his country thereby stricken. His consulate is now at Jerusalem.

An intimate friend, and enthusiastic spectator of the French Consul's achievements, commenced operations where the latter relinquished them. Henry Austen Layard — of noble Huguenot extraction — born at Ceylon, and brought up at Florence, is essentially a man of the East. Leaving England in 1839, he reached Mosul, 1842, by way of Germany, Russia, Dalmatia, the Bosphorus, Asia Minor, Persia, and Kusistàn. His performances are familiar to all readers of *Nineveh and its Remains*, 1849; and *Babylon and Nineveh*, 2d Exped., 1853. The letters LL.D. and M. P., and the office of Under Secretary of Foreign Affairs, tell how a nation can reward living merit: at the same time that "Eastern questions" point to eventualities not less nationally important. The British Museum consecrates for science the innumerable exhumations of Layard.

Great as have been, however, the exploits of these discoverers, they must not dazzle our vision from beholding the less ostentatious if archæologically superior researches of Rawlinson and of Hincks; but for whom, the cuneiform records of Nineveh and Babylon might have yet remained sealed books: although, so closely followed have these savants been by a Löwenstern, a De Longpérier and a De Saulcy; so materially aided by Birch, Norris, and other skilful palæographers; that by grouping them all into a "Cuneiform School" the invidious task of assigning a place to any one is cheerfully avoided. Our inquiry simply is, what have they all done in Assyrian *chronology?*

Let it first be observed "en passant," that the long lists of Chaldæan, Arab, Assyrian, and Babylonish sovereigns, preserved by Ctesias, Ptolemy, and the Hebrews; (533) coupled with the pseudo-antiquity popularly assigned to the Xth *Chapter of Genesis;* had occasioned the most exaggerated notions, about 1844–50, of the epochas to which these sculptures of

---

(531) *Nineveh and Persepolis;* London, ed., 1852.

(532) *Lettres à M. Mohl;* Découvertes à Khorsabad, 1845, p. 2: — *Monument de Ninive*, chap. ii., p. 23.

(533) FRASER's excellent *Mesopotamia*, pp. 47 – 50; and CORY's *Ancient Fragments;* supply the classical authorities.

Assyria should be attributed. Nowhere was this sentimentality exhibited more strongly than at the British Museum. Ninevite bas-reliefs of the 7th century B. C. were reverenced by pious crowds who looked upon them as if their carving had actually been coeval with the "Tower of Babel"; at the same time that Egyptian relics of the IVth Memphite dynasty, belonging to the 4th chiliad before C., and those stupendous granites of the XVIIth–XVIIIth dynasties, positively dating in the 16th–13th centuries prior to the same era, were passed over in contemptuous silence; although displayed in gigantic halls, whilst Assyria (for want of room) lay in an underground cellar! And yet, withal, the only monumental proof of the existence of either BaBeL, or NINWE, 1500 years B. C., depended then, as it does now, upon Thotmes IIId's "Statistical Tablet" of Karnac! (534) Nor, excited by the magnificence of their monumental resurrections, can we be surprised that the two explorers somewhat participated, at that time, in the general feeling.

But, the habit of dispassionate comparison of art (upon itself alone) among sculptured antiquities of every period and region collected in European Museums, had instinctively led thorough archæologists to pronounce the word "modern," over every fragment brought to London and Paris from Nimroud or Khorsabâd; and this before a single Assyro-cuneatic inscription had been deciphered. First to undertake this thankless office was De Longpérier; (535) who proclaimed, to shocked orthodoxy, that nothing found or published of Assyrian bas-reliefs could possibly ascend beyond the 9th century; at the same time that Khorsabâd had then not yielded anything older than the 7th–8th century B. C.

Nevertheless, it was published —

"On the most moderate calculation, we may assign a date of 1100 or 1200 before Christ, to the erection of the most ancient [palace]; but the probability is, that it is much more ancient:" (536) and maintained — "There is no reason why we should not assign to Assyria the same remote antiquity we claim for Egypt" [B. C. 3500?].

Col. Rawlinson too, whilst conceding that "the whole structure of the Assyrian graphic system evidently betrays an Egyptian origin: first organized upon an Egyptian model,"(537) formerly considered the *Obelisk of Nimroud* to date about the 12th–13th century B. C.

Now, this age for Assyrian monumental commencements harmonizes perfectly with Egyptian conquests and dominion over much of that country, during the XVIIth dynasty, 15th–16th centuries B. C. It is merely the archæological attribution of any sculptures, yet found and published, to such an epoch that we contest. We are the last to curtail any nation's chronography; but, misled so often by hypotheses, we cease to depend any further upon arithmetic where not supported by positively archæological stratifications. Lepsius, it seems to us, has fairly stated the possibilities of *Chaldaic chronology*; (538) and future researches by cuneiform scholars will doubtless determine the relative position of each historical stratum as firmly for Assyria as has been already done for Egypt.

With these provisoes, we may safely present a synopsis of the last chronological results put forth by Layard. Possessing all the resources at present attainable, and profoundly versed himself in Assyrian studies, his tabulation of the monumental series of reigns inspires full confidence, at the same time that his results accord naturally with the histories of adjacent countries and people. (539)

ANTE-MONUMENTAL PERIOD.

Into this category are cast the vague and semi-mythical traditions of Nimrod, Ninus, Belus, and their several lines; which, according to classical writers, may ascend to 1903 years before Alexander, equivalent to 2234 B. C. (540)

---

(534) BIRCH: *Op. cit.;* 1846; p. 37: — *Two Egyptian Cartouches found at Nimroud;* 1848; pp. 161–177: — GLIDDON: *Otia;* p. 103. Vide also BIRCH, *Annals of Thotmes III.;* Archæologia, 1853, xxxv. p. 160.

(535) *Revue Archéologique,* Oct. 1847: — *Galerie Assyrienne,* Musée du Louvre, 1849; p. 16; — *Revue Archéol.* Oct. 1850.

(536) LAYARD: *Nineveh and its Remains;* Am. ed., 1849; pp. 176, 179, 185.

(537) *Commentary on the Cuneiform Inscriptions,* &c.; 1850; pp. 4, 7, 21, 71, 73, 74.

(538) *Chronologie der Ægypter;* i. pp. 6–12.

(539) *Babylon;* pp. 611–625: — already RAWLINSON extends Assyrian antiquity to the 14th century B. C.; *Jour. R. Asiat. Soc.,* 1853, p. xviii., note.

(540) LEPSIUS: i. p. 10.

Genealogical Period.

This class embraces those *Assyrian Kings*, of whose reigns no *contemporaneous* monuments have been discovered, but who are recorded in the pedigrees or archives of their successors: distinguishing Rawlinson's reading by R, and Hincks' by H.

	King (conjectural reading).	*About* B. C.
I.	Derceto (R.)	1250
II.	Divanukha (R.), Divanurish (H.)	1200
III.	Anakbar-beth-hira (R.), Shimish-bal-Bithkhira (H.)	1130
IV.	Mardokempad? } (R.)	—
V.	Mesessimordacus? } (R.)	—
VI.	Adrammelech I. (R.)	1000
VII.	Anaku Merodak (R.), Shimish Bar (H.)	960

Monumental Period.

VIII.	Sardanapalus I. (R.), Ashurakhbal (H.) — North-west Palace, Nimroud	930
IX.	Divanubara (R.), Divanubar (H.)—Obelisk; cotemporary with Jehu	900
X.	Shamas Adar (R.), Shamsiyav (H.)	870
XI.	Adrammelech II. (R.)	840
XII.	Baldasi? (H.)	—
XIII.	Ashurkish? (H.)	—
XIV.	? Pul, or Tiglath-Pileser	750
XV.	Sargon	722
XVI.	Sennacherib	703
XVII.	Essarhaddon	690
XVIII.	Sardanapalus III. (R.), Ashurakhbal (H.)	—
XIX.	(Son of preceding)	—
XX.	Shamishakhadon? (H.)	—
	Fall of Nineveh	606

The chronological approximations of our sketch hinge upon the name of Jehu, king of Israel, who, on the *Obelisk of Nimroud*, is made tributary to Divanubar; thus establishing a synchronism about the year 885 B. C.

Everything yet discovered on the site of *Babel* seems to belong to the reign of "Nabukudurruchur (i. e., *Nebuchadnezzar*), king of Babylon, son of Nabubaluchun, king of Babylon" — not earlier than about B. C. 604.

Time, the performer of so many marvels in archæology, will assuredly enable us soon to attain greater Assyrian precision; already foreshadowed through the pending excavations of M. Place, and the personal studies of M. Fulgence Fresnel and of Col. Rawlinson, on the sites of Mesopotamian antiquity.

CHRONOLOGY—HEBREW.

"For a thousand years in thy sight are but as yesterday when it is past."—(*Psalms* xc. 4.)

"One day is with the Lord [IeHOuaH] as a thousand years, and a thousand years as one day." (2 *Peter* iii. 8.)

It would be affectation if not duplicity, on the part of the authors of "Types of Mankind," after the variety of shocks which the plenary exactitude of Hebrew chronicles has received at their hands, not to place everything Israelitish on precisely the same human footing as has been assigned to the more ancient time-registers of Egypt and of China, and to the more solid restorations of Assyria.

The reader of our Essay I, in the present volume, can form his own estimate of the historical weight that Hebraical literature may possess hereafter in scientific ethnography.

Monumental history the Hebrews have none. Even their so-called "Tombs of kings," owing to the absence of *inscriptions*, have recently occasioned a discussion among such deep archæologists as De Saulcy, Quatremère, and Raoul-Rochette, (541) that shows upon how tremulous a foundation their attribution rests. The "arch" and massive basements of Jerusalem's temples (discovered by Catherwood, Arundale, and Bonomi, 1832–3) may

(541) *Revue Archéologique;* 1851–'52. Also, De Saulcy: *Journey round the Dead Sea;* 1853; ii. p. 131.

belong to Zerubbabel's or to Solomon's edifices; or, in part, to the anterior *Jebusites*, for anything by tourists imagined to the contrary. In the absence of monumental criteria, we are compelled to give the Hebrews but a *fourth* place in the world's history; at the same time that justice to a people whose strenuous efforts to preserve their records has encountered more terrible obstacles and more frequent effacements than any other nationality, demands the amplest recognition.

The numerous citations and tables with which the subject of chronology has been already ushered, spare us from recapitulation of the manifold instances whereby the Text contradicts the versions; the numerical designations of a given manuscript, those of another; and the modern computations of one individual, the estimates of almost every other individual; whensoever the date of any Jewish event, anterior to Solomon's semi-pagan temple, is the object sought after.

In fact, we may now realize with Lepsius, that the strictly-chronological element was wanting in the organism of Hebrew, as of other Semitish, minds; until MANETHO *the Sebennyte*, about B. C. 260, first established the principles of chronology through Egyptian indigenous records; and, by publishing his results, in Greek, for the instruction of the Alexandria School, first planted the idea of human "chronology" upon a scientific basis. All systems of computation (heretofore followed by Christendom) take their departure, historically, from Manetho.

It is deeply to be lamented, for the sake of education, that no qualified translator has yet honored Anglo-Saxon literature with an English version of Lepsius's "Introduction" to his *Chronology of the Egyptians;* of which the writers, through the Chevalier's complaisance, have possessed the *first-half* since December, 1848, and the second since May, 1849. Impossible, we fear, until such translation be accessible, is it to convey to the majority of our readers, the *entirely-new* principles of chronological investigation this wonderful grasp (of a mind at the pinnacle of the culture of our time) has condensed into 554 pages quarto. Erudition stands humbled at the aspect of this volume's conscientious and universal probity of citation; at the same time that its perspicacity of arrangement is such, that those who, like ourselves, possess no acquaintance with German, can track the footsteps of its author almost paragraph by paragraph. Through the kindness of many Allemanic friends, the writers have been enabled to annotate their copies of the *Chronologie der Ægypter* with marginal and other notes that justify whatever assertions they respectively make upon an authority otherwise to them Germanically concealed: and, in consequence, with reference to Rabbi Hillel and many of the facts subjoined, they may confidently refer the reader of "Types of Mankind" to Lepsius's compendium; (542) as a ground-text which the writers' comparative studies of works in other tongues, more or less familiar, have resulted in deeming the highest, in these peculiar branches, of our common generation. In any case, a German scholar can easily verify our desired accuracy by opening a *printed* book; four copies, at least, of which are now even at Mobile, Alabama.

We have said that Manetho is the founder of the science called "chronology." We mean that he is the first writer who developed through the Greek tongue, at his era the language of Occidental science, those methods of computation in vogue from very ancient times among the sacerdotal colleges of the Egyptians. He is the exponent, not the inventor of his country's system: Eratosthenes, Apollodorus, &c., are his successors; together with Josephus, Africanus, Eusebius, and the Syncellus; whose Judaico-christian theories have been the sources of that fabric of superstition heretofore reputed to inform us concerning the epoch of God's *Creation.*

No doubt remains any longer that, centuries prior to Manetho, the Egyptian priesthood did possess chronological registers; because, aside from inferences patent in his predecessor Herodotus's "Euterpe," we have before our eyes in the *Turin hieratic papyrus* (dating in the 12th–14th century B. C., or 1000 years before Manetho) the same system, often with the same numerals, of reigns of *Gods*, *Demi-Gods*, and *Men*, that this chronographer subsequently expounded to the Alexandrian schools. Alas! Manetho's mutilators, not his

(542) *Einleitung;* 1849; pp. 14–20, 359–404, 405–410.

own imaginary inaccuracies, are the cause of that confusion of personages and dates, from out of which modern archæology is now beginning, through hieroglyphical collations, to emerge.

Of course, *Chinese* computations are distinct: being the production of other lands, other races, other histories, other worlds of thought and action. So, likewise, may be the lost Chaldæan systems, of which fragments survive through scanty extracts of Sanconiatho and of Berosus; or, as we shall see, through the more recent Sanscrit astrologico-cyclic fables of the Hindoos: but, with the above exceptions, and (if you please) of Mexico and Peru, there is no system of what we call "chronology" but is historically posterior to Manetho, whose era stands at the middle of the 3d century B. C.

This is facile of comprehension to the reader of our Essay I. He therein perceives that the oldest computatory data based upon Judaic traditions are found in the Greek *Septuagint;* being itself a collection of translations manufactured at Alexandria after B. C. 250, and before B. C. 130; in which, Alexandrian Greek dialects and Alexandro-Egyptian "sothic periods" of 1460 years, betray a people, an age, and a fusion of philosophical notions, such as could have been produced, through natural causes, in no locality upon earth but Alexandria; and that too during Ptolemaic generations subsequent to Manetho.

The next in order is the Hebrew Text. Its canonical antiquity, in its oldest and last form, cannot reach up to Ezra in the 5th century, and descends unto the Maccabee princes in the 2d century B. C., *i. e.* after the writer of the book called "Daniel." But, our *Introductory* has effaced the validity of textual numeration in any Hebrew codex (no MSS. being 900 years old); because, while on the one hand its radically discordant numbers show that, when the *Septuagint* was translated, the original Hebrew exemplar in its patriarchal enumeration either did not then exist, or must have been identical with its copied Greek version; on the other, the *Hebrew square-letter* character, of this Text's present form, not having been invented until the *3d century after* C., the chronological elements now in the Text must originate from manipulations made above 400 years after Manetho.

Thirdly, and lastly, there is the *Samaritan* Pentateuch. Its numerical system altogether departs, for patriarchal ages, from both the Septuagint and the Hebrew Text. The age of its compilation is utterly unknown; but the palæographic shape of its alphabetic letters bring such MSS. as exist now to an epoch below that of our Hebrew Text itself. Supposing the rumored estimate of one *Nabloosian* codex did make that unique copy attain to the 6th century after C., such fact would merely prove our view to be correct; but, in Europe, no *Samaritan* MS. is older than the 13th century. In consequence, we cannot accept, in scientific chronology, any more than Siracides, the modern hypotheses of that "stultus populus qui habitat in Sicimis."

These facts being posited, one can understand the apparatus and the efforts made upon them by the learned Rabbi Hillel, about the year 344 after C., to place *Jewish* chronology upon a scientific basis that it never possessed before his labors. He was acquainted with Grecian calendrical computations; probably with the cycles of Meton and Callippus, the mathematical formulæ of Theon of Alexandria, and with the chronography of Africanus, perpetuator of Manetho.

A quotation from Lepsius has been submitted on a preceding page. Another extract will illustrate his views (543): —

"But then it is very improbable that Hillel went to work in the manner that Ideler believes. 'Evidently,' says Ideler, 'he started from the then-still-generally used (by the Jews) *Seleucidan era*, viz.: the autumn of the year 312 B. C. Calculating backwards, his next epoch was the destruction of the second Temple. This epoch he fixed at only 112 years (before); thus counting more than 150 years too little, and making Nebuchadnezzar contemporary with Artaxerxes I. Going back to the Building of the first Temple, the Exodus, the Deluge and the Creation, partly according to the express dates of the Bible, partly according to his explanation of those dates, he found, as the epoch of the *Minjan Shtaroth* beginning of the year 3450 of the World.' So gross and inconsistent an error of 160 years in so modern a time was *impossible* to a *savant* of the 4th century. But there is not much difficulty in explaining it, if we suppose, that the Rabbis, after the great *hiatus* in Jewish literature

(543) *Chronologie*—"Kritik der Quellen"; i. pp. 363, 364.

(which began with the conclusion of the Talmud, 500 A. D. to the 8th century,) did receive the few general points, which Hillel had connected with his universal calendar, from him, and that then, only then, they began to fill up their universal history of 5000 years according to the records of the Old Testament. Indeed, we find neither in the Talmud nor even in the ante-Talmudic writings,—ex. gr. in the *Seder Olam Rabba*, one of the most ancient of these writings—the whole chronological fillings up. This seems to have taken place in the 12th century; consequently at the epoch of a long-previously commenced scientifico-literary barbarism. From the Creation to the Deluge, and the Exodus, they had only to follow the numbers of the Pentateuch to attain the given date (A. M.) 2448 = 1314 (B. C.). But thenceforward they based themselves upon the convenient number of 480 years to the Building of the Temple (in the 1st Book of *Kings*), and according to this they arranged the chronology of the time of the Judges. By this, then, was the real link of chronology dislocated for 160–170 years, which occasioned the displacement of all the succeeding members. Only when arrived at the next fixed point, in the year (A. M.) 3450 = 312 (B. C.), was it found, that the chain of events, for the given space from the Building of the first to that of the second Temple, was much too long. The history of the second Temple, built under Darius Hystaspis, down to Alexander, from whom the Greek era took its name, shrunk then at once from 184 to 34 years. At first this created little sensation, but afterwards the difficulties becoming greater, they were removed by the simple means of adopting Darius II. and (Darius) III,, as one and the same person. In this manner alone can we explain the singular phenomenon of an entirely dislocated and mutilated chronology, which notwithstanding possesses two firm and only-sure points; and at the same time offers us the most important and probably most accurate determination of the epoch of the Exodus by a really learned chronologist."

It is from the original that the reader must gather, what our space and objects permit us not to transcribe, the citations, &c., through which the author establishes his view conclusively. To us the important facts are these—1st, that the Jews had made no attempts at scientific chronology prior to the 4th century after C.; nor did they complete such as their later schools adopt until the 12th.—2dly, that, through their childlike prepossessions, and owing to their superstitious notions that the era of "Creation" could be humanly attained, they ciphered out a fabulous number, equivalent to "B. C. 3762," for a divine act, which their ignorance of the phenomena of astronomical and geological unceasing progression, led them to imagine *instantaneous*—"Fiat lux!"—and 3dly, that, having blundered by 160–170 years, only between the *Exodus* and Solomon's temple, they sank deeper into the mud when, in efforts to account for their own imbecilities, they made one man of *two* Dariuses in order to rob the world's history (184 *minus* 34) of 150 years! And it is such wretched stuff as this rabbinical arithmetic which is to be set up, forsooth, against the *stone-books* of Egypt and Assyria, the records of China, the annals of Greece and Rome at the age of Alexander the Great, and every fact in terrestrial history! Well might Lesueur indite the passage above quoted—"Nous sommes, depuis dix-huits cents ans, dupes de la sotte vanité des Juifs:" and justifiably may archæological science hold cheaply the acumen of the whole series of those who, amid other conceits, have adopted 480 years between Solomon's temple and the *Exodus*.

Before examining which fact, it may be expedient that we should set forth our own point of view, founded upon the same principles hitherto pursued, viz., that our process is always retrogressive; ever starting from *to-day*, as the known, and going backwards, in all questions of human registration of events.

The *era of Nabonassar*, if astronomy be certainty, is a point fixed, by eclipses, &c., in the year B. C. 747. Thence, backwards to the "5th year of Rehoboam," when Jerusalem was plundered by the Egyptian Sheshonk (of which event the hieroglyphical register stands at Thebes), we have a positive synchronism *about* the years 971–3, "B. C.;" for, in ancient chronology, asserted precision to a year or so is next to imposition. Thence, taking Solomon with his "chariots dedicated to the sun," and his Masonico-zodiacal *Temple*, for granted, we accept the era "1000 years B. C.," as an assumed fixed point when that temple was already completed. We say "assumed," because Calmet's date for the completion of this edifice is B. C. 1000; whilst Hales's is B. C. 1020: and, rather than trouble ourselves with ascertaining which of these computations may be the least wrong, we would greatly prefer discussing whether Solomon ever built a *Temple* at all. Why, if for the second, or

Zerubbabel's Temple, we have to choose among 19 biblical chronologers, whose *maximum* is B. C. 741, and *minimum* 479—if, for a Jewish event of scarcely 2400 years ago, we cannot through Judaic books get nearer the truth, according to "chronological" arithmetic, than 262 years, up or down — how much nearer are we likely to get to another Jewish event (itself fraught with preternatural dilemmas), supposed to have happened somewhere about 2853 years ago, when the epoch of the building of the first Temple depends upon what computation we may elect to adopt out of 19 different orthodox authorities for the age of the second?

Thus much for the sake of furnishing our colleagues with practical means of rendering ecclesiastical opposers of "Types of Mankind," if not less supercilious, at least more malleable; whenever these may be pleased to obtrude Jewish "chronography" — or, as it is fashionably termed, "the received chronology"—into the rugged amphitheatre of Egyptian time-measurement.

Archæologically speaking (not "chronologically"), there is no material objection to such assumption as Solomon's *Temple* at (*circa*) B. C. 1000; a few years more or less. Under this historical view, apart from episodic circumstances (to be discussed hereafter), archæology may rationally concede that Hebrew tradition, through alphabetic facilities developed not much less than *three* centuries posterior, does really contain chronological elements back to about 2853 years ago — say to B. C. 1000.

We continue with Lepsius —

"The question is now whether we must give up, for lost, the number 480 (to which we cannot attach greater importance than to the numerous simple "*Arbaïnât*," or *forties* [40*s*], in the same parts of Israelitish history); and with it, also, every chronological helm for events anterior to the Exode? But such is not the case, because we find, in the [so-called] Mosaic writings themselves, a true chronological standard, by which we can compute [the chronological weight of] the views hitherto held, and confirm anew the truthfulness of Egyptian record. Such a standard I conceive to be the *Registers of generations*."

Allusion has been made, in other parts of this volume, to the Nos. 7, 12, 70 or 72, as mystic in original association; and how the latter always, the former two frequently, are unhistorical wherever found. To these numbers (of cabalistic employment since the days of Jeremiah), we may now add, as equally vague in Hebrew chronography, all the "*arbaïnât*" or "forties." By opening Cruden's *Concordance* the reader can see a list of above 50, out of many more instances, where the presence of "forty" renders the narrative, in this respect at least, unsafe. Here is a schedule of some that are positively apocryphal; especially when, through a conventional *No.* 40, an event, in itself præternatural, is rendered still more impossible by the numerals that accompany it.

APOCRYPHAL FORTIES.

Old Testament.

1. *Gen.* vii. 4........... "40 days and 40 nights."
2. *Exod.* xxiv. 18...... "40 days and 40 nights."
3. *Numb.* xiii. 25...... "40 days."
4. *Deut.* ix. 25........ "40 days."
5. *Josh.* v. 6............ "40 years."
6. *Jud.* iii. 11........... "40 years."
7. 1 *Sam.* iv. 18........ "40 years."
8. 2 *Sam.* v. 4......... "40 years."
9. 1 *Kings* xix. 8...... "40 days and 40 nights."
10. 2 *Kings* xii. 1....... "40 years."
11. 1 *Chron.* xxvi. 31.. "40th year."
12. 2 *Chron.* xxiv. 1... "40 years."
13. *Ezra* ii. 24........... "40 and *two*."
14. *Nehem.* v. 15........ "40 shekels."
15. *Job* xlii. 16......... "*hundred* and 40 years."
16. *Psalms* xcv. 10..... "40 years."
17. *Ezek.* iv. 6............ "40 days."
18. *Amos* ii. 10......... "40 years."
19. *Jon.* ii. 4............. "40 days."

New Testament.

20. *Matt.* iv. 2........... "40 days and 40 nights."
21. *Mark* i. 13........... "40 days."
22. *John* ii. 30........... "40 *six* years."
23. *Acts* i. 3................ "40 days."
24. *Heb.* iii. 9............ "40 years."
25. *Rev.* vii. 4, xiv. 1, 3 "*hundred* and 40 *four* thousand."

"It is evident from the narratives in the Pentateuch, as well as in other books of the Holy Scriptures, that in ancient times the number 40 was considered not merely as a round number, but even as one totally vague and undetermined, designating an uncertain quantity. The Israelites remained in the desert during 40 years; the judges, Athniel, Ehud (*Septuag.*), Debora and Gideon, governed each 40 years. The same did Eli, after the Philistines had ravaged the country during 40 years. The 40 days of the increasing and the 40 days of decreasing of the waters of the Deluge are well known. But one of the most striking instances of this use of the number 40 is 2 *Sam.* xv. 7, where, during the 40 years of David's reign it is said: 'And after 40 years it happened that Absalom went to the king and said, Let me go to Hebron, that I may fulfil the vow which I have made to Jehovah.'

"The Apocryphic books go still farther. According to them, Adam entered the Paradise when he was 40 days old—Eve 40 days later. Seth was carried away by angels at the age of 40 years, and was not seen during the same number of days. Joseph was 40 years old when Jacob came to Egypt; Moses had the same age when he went to Midian, where he remained during 40 years. The same use of this number is also made by the Phœnicians and Arabs. [See *Dissertatio Bredovii de Georgii Syncelli Chronographia* (second part of the edition of Bonn) Syncellus, p. 33, *seq.*] We must not forget hereby the *Arbaïnât* (the forties) in Arabian literature; a sort of books which relate none but stories of 40 years, or give a series of 40, or 4 times 40 traditions. They have a similar kind of books, which they call *Sebayât* (sevens). Their calendar has 40 rainy and 40 windy days. Also in their laws the numbers of 4, 40, 44, occur very often. In Syria the graves of Seth, Noah and Abel are still shown. They are built in the usual Arabian style. Their length is recorded to be 40 ells, and thus I have found them by my own measuring. This may also account for the tradition that the antediluvian men were 40 ells high, that is, not 'about 40 ells,' but '*very tall.*' Only afterwards was this expression so naively misunderstood. The Arabs give, in the conversational language, the same sense to *sîttîn*, 60, and *mïeh*, 100. I have already observed, in an earlier writing [*Zwei Sprachergleichende Abhandlungen* (Two lectures upon the Analogy of Languages), Berlin, 1836, pp. 104, 139], that of all the Semitic numerical words, *arbă*, 4, is the sole one which has no connexion whatever with the Indo-Germanic, and seems rather to be derived from *rab*, רב, 'much,' ארבה, 'the locust.' This would account for its undetermined use.'(544)

The historical spuriousness of the numeral 40, in its application to human chronology, may be illustrated by another example out of many. It is said, "Israel walked 40 *years* in the wilderness,"(545) after the Exode. On which Cahen: —

"It is probable that this itinerary contains but the principal stations: they are in number 42. In the first year they count 14 stations; in the last, or 40th, they count 8 stations; thus the 20 other stations occupied 38 years (*Jar'hi*, in the name of Moses the preacher). According to the ingenious remark of St. Jerome, the number 40 seems to be consecrated to tribulation: the Hebrew people sojourned in Egypt 10 times 40 years; Moses, Elias, and Jesus, fasted 40 days; the Hebrew people remained 40 years in the desert; the prophet Ezekiel lay for 40 days on his right side. This accordance shows us that Goëthe had some reasons for conjecturing that the 40 years in the desert might very well possess no historical certitude." (546)

Again — "Thus, during these 40 years, notwithstanding the miserable life which the Israelites had led in the desert, maugre the plagues, the maladies, and the wars, there was but a diminution of 1820 Israelites and an augmentation of [just!] 1000 Levites. Such results exist not within the domain of natural things, and consequently possess nothing historical." . . . "Savage tribes sing of their petty quarrels, their conquests and their disasters, upon the lofty tone of, and even loftier tone than, the greatest nations. Thus the septs along the river Jordan had their poets, their national ballads; these songs, there, as everywhere else, have preceded history. We have just read extracts from these productions, perhaps the most ancient that have reached us. It is probable that to them were afterwards added some events of a date much later than the political existence of Moabites, Edomites, &c." (547)

Finally, speaking of the "40 years" in the Sinaic desert, Cahen observes: —

"One finds in the Pentateuch only those events that occurred during the first *two* and the last or *fortieth* year. The history of the intermediary 37 years is totally unknown to us." (548)

All theological conjectures about this unhistoric interval are merely conjectures theological; because the Jews used the expression "forty," as we do "a hundred," for a vague number of anything uncounted. To Lepsius's numerous illustrations of the utter impossibility that uneducated nations or individuals can possess any clear ideas about *dates* for circumstances that may have happened during their respective lifetimes, we might add two parallels — the first (or Oriental) is that, in Egypt, if you ask an intelligent but illiterate

(544) Lepsius: *Chronologie der Ægypter:* i. pp. 15, 16, note.
(545) *Josh.* v. 6.
(546) Cahen: iv. p. 158; note on *Numb.* xxiii. 1.
(547) Cahen: *Op. cit.;* p. 134; note on the *two* censuses in the Desert: and p. 124, on Bilam and Balak.
(548) *Op. cit.;* p. 96.

native his age, he cannot express it by *years;* but replies, that his stature was about so high (holding out his hand at the elevation required), *fee ayàm en-Nussàra* — "in the days of the Christians;" alluding to Napoleon's conquest of Egypt, 1798–1802: or else tells you that he had not a white hair in his beard, *fee hurrèekut el-Qalàa,* "at the fire of the citadel" of Cairo, 1825. The second (or Occidental) is, that no *Indian,* or *Negro,* in the United States (save among the paucity that have been educated), can tell you his own age, by *years;* but the one dates either from such a time when "he and Col. —— shot that *bar;*" or the other from when he butted for cheeses against another negro-kephalus at such a local election.

This introduces a question upon which European biblical commentators, ignorant of living Oriental customs, have gone sadly astray. Whenever the number of personages, in a given Hebrew pedigree, has been found insufficient to occupy (that is, to fill up naturally, without improbable longevity) the length of time required to suit the chronological scale a given commentator may have elected to invent or follow, it has been incontinently assumed, that the Hebrew *numerals* were right; and that the anomaly proceeds from the accidental loss of one, or more, intermediary ancestors, in the genealogical list. Thus, says the learned Dr. Prichard, (549) adopting the suggestions of the great Michælis:—

"The result is that the difficulty which seems to have induced some of the ancients to alter the text requires a different explanation. It can only be solved, as it would seem, by allowing an *omission* of several generations in the genealogies of the Israelites. At present only two generations are interposed between Levi and Moses. It is probable that several are *omitted.*"

So again the Abbé Glaire, (550) in respect to the two genealogies of *Joseph:*—

"The first (method) is to suppose that these names (*Ochosias, Joas, Amasias*) were wanting in the genealogical tables the evangelist made use of; an hypothesis the more probable that the names of intermediary persons are often missing in many genealogies of the Old Testament. . . . Esdras, in his genealogy, omits seven of his ancestors, by jumping from Amarias to Achitob II, father of Sadoc II. . . . The genealogy of Saul, for a space of 800 years, names but seven persons. . . . From Mardocheus to Jemini or Benjamin, who lived 1200 years before, but four are named. . . . From Reuben to Beera, who was carried captive by Tiglath-pilesar, they give us but 12 generations to fill a space of more than 1000 years. In the genealogy of Judith, for a space nearly equal, there are but 16 generations. By fixing, as is commonly done, the generation at 33 years, one perceives that there are a good many degrees omitted in these genealogies. . . . Grotius, upon whose acquirements one may confide without difficulty, assumes that this happens frequently, as may be seen in genealogical trees. *Sœpe eodem temporis spatio familias inter se comparatas generationes habere unam aut alteram plures et pauciores; quod in omnibus stemmatibus videre est.* 'Veut-on un example d'une grande inégalité de générations dans les différentes branches d'une même souche? Scripture affords one very striking. The children of Jacob (*Numb.* i. 3) each formed a branch or tribe. When, a year after their issue from Egypt, Moses, by the order of God, caused the numbering of these tribes, there was found among them a prodigious inequality; but the most surprising is that which was beheld between the tribe of Levi and that of Judah: the latter comprised 74,000 males above the age of 20 years, and the former 22,300 counting (even) those above one month.'"

One would suppose, so naïvely does the Abbé accept all these *numerals* as historical, that he was actually present! But these violent statistics are susceptible of more rational solution. Such attempts at reconcilement have their unique origin in the uncritical ideas of eminent scholars upon the true ages of the composition of the fragments extant of Jerusalem literature; which the perusal of our *suppressed* pages might supersede: and similar weak explanations would not have been thought of by any Orientalist (Fresnel, Lane, or Layard, for instance) who had actually resided among Semitic populations. Lepsius (551) is the first, that we are aware of, to have placed the matter in its true light.

We know that unlettered Arabian Bédawees do preserve, for centuries, orally from father to son, their individual and clannish genealogies; and this too for an almost infinite number of generations. They even thus consecrate, legally, the pedigrees of their blood-

(549) *Researches;* 1847; v. p. 559.

(550) *Livres Saints Vengés;* ii. pp. 284–285, 201–202; quoted chiefly from Bullet: *Réponses Critiques.*

(551) *Op. cit.;* pp. 365, 366.

horses. (552) But, as for defining the *length of time* each tribe, man, or horse, may have lived, that the Bédawee has no means of doing beyond his own grandfather's lifetime; and for which he has no *annual* calendar. Thus, in ante-Mohammedan history, "the battle of Khazaz," fought by the *Mâadd* tribes under Koulayb-Waïl against the Yemenite confederacy, is the earliest stand-point of Arabian historical tradition; (553) but the era before *Islàm* — 250 — to which such battle is assigned, has been computed, *for* these wild children of the desert, by later and highly-cultivated Arab historians, and at best conjecturally.

It would be foolish to deny to the sedentary and somewhat educated Hebrews, of days anterior to the Captivity, equal faculties of preserving their own *genealogies*, that we recognize among cognate Semitish and still more barbarous tribes of Arabia: nor is there any reason to doubt the existence of *genealogical lists*, stretching backwards for many generations, from the days of Ezra. (554) These may even have ascended, ancestor by ancestor, to the times of Abraham. (555) But it was one thing to preserve, through saga, rythme, song, or oral legend, the *names* of predecessors in their natural order; and quite another to guess at the duration of these ancestors' respective lifetimes, or to infer, through traditionary events with any of the earlier ancestors coetaneous, the chronological remoteness of the age during which they lived, excepting approximately. In consequence, Lepsius (and we entirely agree with him) sustains, that the *genealogies* of the Hebrews are probably right; but that the chronological computations accompanying these lists are certainly wrong. Indeed, of this last fact there can be no doubt, when we remember that Rabbi Hillel, in the fourth century after Christ, was the first to regulate Jewish chronology by the verbal literalness of the Hebrew Text; independently of fabulous numeration such as that borrowed by Josephus from an Alexandrian Greek system adopted by the writers of the *Septuagint*. The manifest interpolation of an Egyptian "Sothic-period" of 1460–'61 years (so felicitously discovered Mr. Sharpe, *supra*, pp. 618, 619), obviates further necessity for recurrence to the spurious chronology of the Greek version.

These numerical estimates, we now see, are both modern and erroneous. But, to convince the reader of the fact; and to prove that the 480 years between the first Temple and the Exodus are erroneous; we copy Lepsius's synopsis, after remarking that, just as in all ancient pictures the artist gave colossal proportions to the figures of gods, or heroes, while the plebeian classes receive pigmaic stature, so among the antique Israelites, in their organic absence of "art," it was customary to assign to the royal line, or High-Priest pedigree, the attributes of longevity together with extensively-procreating capabilities; and to measure such exalted patricians by generations of 40 *years;* at the same time that to the vulgar herd were ascribed generations of only 30!

"I give here a Table of the principal genealogies, in which the Levitish generations follow in the same order as they are recorded in 1 *Chron.* chap. 7 (according to the LXX; in the Hebrew Text, ch. v. and vi.). These are preceded by the genealogical chain from Levi to Zadok according to Josephus, and also his list of the *High-Priests* from Aaron to Zadok. Lastly comes a genealogical table of Judah. Albeit I have excluded some other genealogies, *ex. gr.*, the three of Ephraim (*Numb.* xxvi. 35 — 1 *Chron.* viii. 20; xxi.24-27), because they were in evident confusion and led to no result.

"The first column," says LEPSIUS, (556) "contains the patriarchs from Abraham to Amram; next, 12 leaders (chiefs) of the people, beginning with Moses, who seem to have been regarded as representatives of the 12 *generations* of 40 *years* each; and thus to have occasioned the calculation of 480 years [as the chronological interval between the *Temple* and the *Exode*]. EWALD and also BERTHEAU give another list—for the subject, in general, admits of no precision; albeit, for us, the recognition of the *division into* 12 *parts* of this period is important. But one, likewise, (VIII.) of the aforesaid genealogies (1 *Chron.* vii. 39–43) contains 12 *generations of one and the same family.* It might therefore be possible that this last list, and not the other, had originated the calculation of 480 years. This list has the peculiarity of beginning with GERSOM, the *first-born* of LEVI. But the most noble line of the Levites was that of the *High-Priests*, who descended from AARON and Kahath (I.): this list, as well as that of MUSI (IX.), contains only 11 *generations.* This may be the reason why the LXX count but 440 *years.*"

---

(552) LAYARD: *Babylon:* pp. 220, 221, 250, 326–331.

(553) FRESNEL: *Arabes avant l'Islamisme;* Ist Letter; 1836; p. 16.

(554) *Ezra;* ii. 59–62; *Nehem.* vii. 61–64.

(555) *Numb.* i. 5–18, 26.

(556) *Chronologie;* pp. 367–371.

THE JUDAIC LINEAGES

Generations from ABRAHAM to DAVID.	Years.	*Plebeian generation.*	Series of *High-Priests* to ZADOK. [Josephus, A. J., 5, 11, 5.]		ZADOK'S Parentage. [Josephus, A. J., 8, 1, 3].		I. AARON'S Generations. 1 *Chron.* vii. 1–9, 50–53; *Ezra* vii. 2–5.		II. Generations Gersom-LIBNI. 1*Chron.* vii. 20,21. (=VIII.)	III. Generations Kahat-AMINADAB 1*Chron.*vii. 22–24. (=VI.)	
1. Abraham	100 *or*	30									
2. Isaac	100	30	—		—		|		—	—	
	200										
3. Jacob	100	30									
		90									
1. Levi	100	30			1. Levi		1. Levi		1. Levi	1. [Levi]	
2. Kahath	100	30			2. Kaathos		2. Kahath		2. Gersom	2. Kahath	
3. Amram	100	30			3. Amarames		3. Amram		3. LIBNI	3. AMINADAB	
	400	90									
1. Moses	40		1. Aaron	30	1. AARON	30	1. AARON	30	1. (Jahath)	1. Korah	30
2. Joshua	40		2. Eleazares	30	2. ELEAZARES	30	2. *Eleasar*	30	2. Simma	2. Assir	30
3. Othniel	40		3. Phineeses	30	3. PHINEESES	30	3. *Pinehas*	30	3. Joah	3. Elkana	30
4. Ehud	40		4. Abiezeres	30	4. Iosepos	30	4. *Abisua*	30	4. Iddo	4. Ebjassaph	30
5. Samgar	40		5. Bouki	30	5. Bokkias	30	5. *Buki*	30	5. Serah	5. Assir	30
6. Barak	40		6. Ozis	30	6. Iothamos	30	6. *Usi*	30	6. Jeathrai	6. Thahath	30
7. Gideon	40		7. Ilei	30	7. Maraïothos	30	7. Serahja	30	—	7. Uriel	30
8. Jephtha	40		8. (Phineeses)	30	8. Arophaios	30	8. Merajoth	30	—	8. Usija	30
9. Simson	40		9. Iokabes	30	9. Akitobos	30	9. Amarja	30	—	9. Saul	30
10. Eli	40		10. Akimelekos = Akias	30	10. Zadokos	30	10. Ahitub	30	—	10. [Jonathan]	30
11. Samuel, Saul	40		11. Abiatharos with Zadok	30			11. *Zadok*	30	—		
12. David	40										
	480			330		300		330			300

The practical result of which is, that *all* chronologers, by not perceiving the surplusage due to these absurd generations of 40 *years*, have assigned about 160–170 years too much between SOLOMON and MOSES; and *ergo*, the Exodus must descend from B. C. 1491, its date in the English version, to B. C. 1314–'22, *circa*.

After studying the above Table, the reader may perhaps perceive with us several things not generally known: —

1st. — That the whole of this Jewish chronology is unhistorical; because it is not based upon positive records of the number of years each personage lived, but it was fabricated, long after their times, by semi-scientific, semi-literary, computators; whose process was to assign impossible generations of 40 *years* to their country's pre-historic heroes; and then, having obtained a maximum-period in which the lives of such worthies were thereby inclosed, these modern computators (probably about the 3d century after C., when the Books were re-transcribed into the *square*-letter alphabet) apportioned to each hero, in the anew-manipulated Hebrew Text, those irreconcileable *numerals* that have come down to our time.

2d. — That, whether the *genealogical* catalogues be right or not, the chronology is a later intercalation.

FROM ABRAHAM TO DAVID.

IV. Generations Elkana-AMASAI. 1 *Chron.* vii. 25–28. (=VII.)		V. Generations Merari-MAHELI. 1 *Chron.* vii. 29, 30.	VI. HEMAN's Parentage to JEZEHAR. 1 *Chron.* vii. 36–38. (=III,)		VII. Parentage to AMASAI. 1 *Chron.* vii. 33–36. (=IV.)		VIII. ASSAPH's Parentage to JAHATH. 1 *Chron.* vii. 39–43. (=II.)		IX. ETHAN's Parentage to MUSI. 1 *Chron.* vii. 44–47.		DAVID's Parentage to JUDAH. *Ruth* iv. 18; 1 *Chron.* ii. 4–13; *Matth.* i. 3–6; *Luke* iii. 32, 33.	
—		—	—		—		—		|		—	
1. [Levi] 2. Elkana 3. AMASAI (and)		1. Levi 2. Merari 3. MAHELI	1. Levi 2. Kahath 3. JEZEHAR		1. [Levi] 2. Elkana 3. AMASAI		1. Levi 2. Gersom 3. (JAHATH)		1. Levi 2. Merari 3. MUSI		1. Judah 2. Perez	
1. Ahimoth	30	1. Libni	1. Korah	30	1. Mahath	30	1. Simei	30	1. Maheli	30	1. Hezrom	30
2. Elkana	30	2. Simei	2. [Assir]	30	2. Elkana	30	2. Sima	30	2. Samer	30	2. Ram	30
3. Elk.Zophai	30	3. Usa	3. [Elkana]	30	3. Zuph	30	3. Ethan	30	3. Bani	30	3. Aminadab	39
4. Nahath	30	4. Simea	4. Ebjassaph	30	4. Thoah (Thohu)	30	4. Adaja	30	4. Amzi	30	4. Nahesson	30
5. Eliab	30	5. Hagija	5. Assir	39	5. Eliel (Elihu)	30	5. Serah	30	5. Hilkia	30	5. Salma	30
6. Joram	30	6. Asaja	6. Thahath	30	6. Jeroham	30	6. Ethni	30	6. Amazia	30	6. Boas	30
7. Elkana	30	——	7. Zephanja	30	7. Elkana	30	7. Malchija	30	7. Hasabja	30	7. Obed	30
8. Samuel	30	——	8. Asarja	30	8. Samuel	30	8. Baesaja	30	8. Maluch	30	8. Isai	30
9. Vasni	30	——	9. Joel	30	9. Joel	30	9. Michael	39	9. Abdi	30	9. DAVID	30
10. ——	30	——	10. [HEMAN]	30	10. HEMAN	30	10. Simea	30	10. Kisi	30		
							11. Berechja	30	11. ETHAN	30		
							12. ASSAPH	30				
	300			300		300		360		330		270

3d.— That, as said before, there are no recorded *dates* in the Jewish Scriptures that are trustworthy; that, it is we moderns who must make Hebrew chronology for the antique Jews — who, until Rabbi Hillel, had not thought of doing it themselves; — and that, in these restorations, we cease to tread upon historical ground so soon as we retrograde to Solomon's era, said to correspond to B. C. 1000. Beyond that cipher, Jewish chronology is all conjecture, within a few approximate limitations.

Moses, or the *Hebrews*, being unmentioned upon Egyptian monuments of the 12th–17th centuries B. C., and never alluded to by any extant writer who lived prior to the *Septuagint* translation at Alexandria (commencing in the 3d century B. C.), there are no extraneous aids, from sources alien to the Jewish books, through which any information, worthy of historical acceptance, can be gathered elsewhere about him or them.

With these emphatic reservations, we are quite willing to consider Lepsius's computative synchronisms as not merely the most scientific but the only probable. His estimates place the Jewish *Exodus* in the reign of Pharaoh Menephthes, of the XIXth dynasty, about the year 1318 B. C.; (557) or rather *between* the years 1314 and 1322 B. C.: if we have understood our authority correctly: (558) to which we add the following comparative view

(557) *Chronologie;* p. 379, compared with pp. 335–337. (558) Vide GLIDDON: *Hand-book* · 1849; p. 44.

of *dates* for the Mosaic Exodus, as computed by Usher from the *Hebrew Text*, and generally appended to the English translation authorized since the reign of king James, A. D. 1611; and by Hales from the *Greek Septuagint version.* The new synchronisms between Hebrew and Egyptian events, put forward by Lepsius, may assist the hierological student in authenticating monumental history through what are still called the *established* dates of Scripture. It will be remarked that, while Hales extends, Lepsius reduces the antiquity assigned to each Israelitish era by archbishop Usher.

BIBLICAL SYNCHRONISMS.

Epoch of	*Pharaonic Contemporaries.*	A. D. 1660. USHER.	A. D. 1830. HALES.	A. D. 1849. LEPSIUS.
ABRAHAM	AMUNOPH III. (*Memnon*)	B. C. 1920	2077	about 1500
JOSEPH	SETI I. (*Sethos*)	" 1706	1863	" 1400
MOSES	RAMSES II. (Jewish oppression	" 1491	1648	1394—1328
Exodus (B. C. 1322?)	MENEPTHA			1328—1309

Jewish computation by "forties" ceases so soon as we ascend beyond Moses; who was 40 *years* old when he fled from Egypt; 40 *years older* when, after dwelling with Jethro, he returned to liberate his people; and *oldest by* 40 more years when he died at the age of 120 — "but no man knoweth of his sepulchre *unto this day*."(559) Vico supplies a formulary:

I. — The *indefinite nature of the human mind* is the cause that man, plunged in ignorance, *makes of himself* the rule of the Universe.

It is from this truth that are derived the two human tendencies thus expressed: *Fama crescit eundo et minuit præsentia famam.* Fame has travelled, since the world's *Creation*, a very long road; and it is during the voyage that she has collected *opinions so magnificent*, and so exaggerated, upon *epochas* which to us are but imperfectly known. This disposition of the human intellect is indicated to us by Tacitus, in his 'Life of Agricola,' where he tells us: — *Omne ignotum pro magnifico est.*" (560)

From Moses backwards to Abraham, post-Christian Jewish computation assumed 100 years for each generation; but every dozen MSS. of the Text or versions differ; and the general principle followed seems to have been, to make generations the longer, in the ratio that the lifetime of a given hero was more and more distant from each Judæan writer's day. The model copied was a Grecian theogonic idea, because the Esdraic Jews proceeded by the *four Hesiodic ages;* considering their own period to be the *Iron;* the Davidic the *Brazen;* the Mosaic the *Silver;* and that from the Abrahamic to the Adamic, to have been the *Golden* age of Hebrew humanity. To Moses, in consequence, they assigned only 120 years of longevity; but his worthier antecedents had their holier lives extended along a sliding scale, of which the numbers 240, 480, and 960, are the simple arithmetical proportion: their divisor being "40."

Here, then, we have finally arrived at the great fact; which, in different or less outspoken words, all the scientific authors we have quoted are at this day agreed upon: viz.: that the *Jews knew not an atom more of* "Humanity's Origins" *than we do now;* and that, as they really had no human historical ancestor before Abraham (whose epoch floats between Lepsius's parallel at 1500, and Hales's at 2077, B. C.), there is no *chronology*, strictly so-called, in the Bible, anteriorly to the Mosaic age; itself vague for one or more generations.

This posited, we shall close further argument with a Table of *Hebrew Origins;* conformably to the same principles upon which we have already tabulated the distinct histories of Egypt, China, and Assyria. Each of these nationalities possesses its *historical, semi-historical*, and *mythical* times. And, inasmuch as it is conceded by every true historian that the Israelites (under the literary aspect in which they first present themselves to the gentile world), had been previously educated in *Chaldæa;* it will be interesting to place the ante-diluvian "patriarchs" of the preceptors alongside those of the pupils. Berosus, Philo Byblius, Julius Africanus, Alexander Polyhistor, Eusebius, and the Syncellus, have preserved for us transcripts of the original Chaldæan catalogues: the whole texts of which are accessible in Cory's *Ancient Fragments*, or in Bunsen. (561)

(559) *Deut.* xxxiv. 6. (560) VICO: *Scienza Nuova;* 1720; "Elemento 1mo." (561) *Egypt's Place;* i. pp. 704–719.

Mythological Periods.

Symbolical Ante-Diluvian Patriarchs.

Græco-Chaldæan Decade.		*Hebræo-Chaldæan Decade.*	*Phœnico-Chaldæan Decade.*	
1. Alorus..........	years 36,000	ADaM	Protogonos	1. = First-born.
2. Alaparus........	" 10,800	SeT*t*	Genos, Genea	2. = Genus, family.
3. Almelon.........	" 46,800	ANoS*h*	Phos, pur, phlox	3. = Fire, light, flame.
4. Ammenon......	" 43,200	KINaN	Cassios, Libanos	4. = Cassius, Libanus (*mount's*).
5. Amelegarus.....	" 64,800	MaHaLaLeL	Memrounos, ousoos	5. = Celsus, "par cœlo," wood.
6. Daonus..........	" 36,000	IRaD	Agrios, alieus	6. = Peasant, hunter, fisher.
7. Edoranchus....	" 64,800	K*h*eNUK	Chrusor, hephaistos, artifex, geinos	7. = Vulcan, fire, artificer, earth-worker.
8. Amempsinus...	" 36,000	MeT*t*USeLaK*h*		
9. Otiartes..........	" 28,800	LaMeK	Agros, agroueros	8. = Rustic, agriculturist.
10. Xisuthrus......	" 64,800	NuK*h*	Amunos, magos	9. = Warrior, magician.
			Misor (Sydyc, Saduc)	10. = Egypt, and the "just" *king*, Melchisedek.
	Years 432,000			

CHALDÆAN DELUGE.

1st *Note.*—The 36 *Decans* of the Zodiac, (562) multiplied by the 12 months of the year, give the mystic number 432. The "grand year" of Astronomy—or the time anciently supposed to be required for the sun, planets, and fixed stars, to return to the same celestial starting-point—was at first 25,000, then 36,000, and lastly 432,000 years; being the supposed duration of the *ten* Græco-Chaldæan generations. A *Deluge* terminated the cycle. (563)

2d *Note.*—The *Phœnico-Chaldæan* list, derived from Sanconiatho, presents us with the Greek *translations*, not with the real names of its lost Oriental original. The Phœnicians had originally crossed from the Persian Gulf to the Mediterranean, and their intercourse with Chaldæa was incessant; while the two people spoke *Semitic* dialects. More saliently than the other two forms of the same theogony, this Phœnician stream exhibits the rationale of its "ex post facto" construction. According to it, we have the stages of *family, hunter, fisherman, artizan, husbandman, soldier, priest,* and *king,* through which antique humanity developed itself. A parallelism seems to be preserved in the offshoots of the Adamic stem in *Genesis*, where Abel *the wandering shepherd* is hateful to Cain *the sedentary peasant.*

Chaldaic Ethnological Division—[contained in Xth Genesis.]

Theoretical Post-Diluvian Commencements.

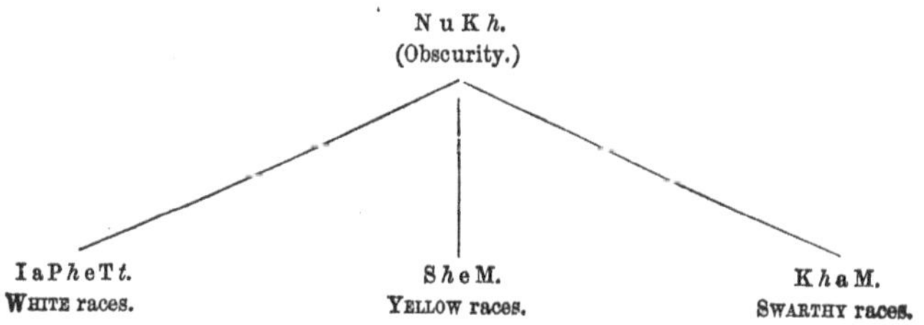

Babylonish Theory for Diversity of Tongues.

"City and Tower of BaByL"-on = *confusion* = "BaBeL-babblings."

Hebrew Geographical Origins.

ARP*ha*-KaSD = Orfa-the-*Chaldæan* (District).
SaLaK*h* = Salacha (City).
AĕBeR = the-*yonderer* (Tribe).
PeLeG = a-*split* (Earthquake?).

Earliest Legendary Ancestors.

RĕU.
SeRUG.
NaK*h*UR.
T*t*eRaK*h*.

---

(562) Lepsius: *Chronologie;* i. pp. 66–76.

(563) De Brotonne: *op. cit.;* pp. 234–246.

JUDAIC META-HISTORICAL PERIOD.

"Thou shalt no more be called AB-RaM (FATHER of the HIGH-land = *Aramæa*)—
Thy name shall be.............. AB-RaHaM" (FATHER of a MULTITUDE). (564)

Abrahamidæ.

ITsKhaK = "laughter."
IâKoB, surnamed ISRAEL.
(12 Signs of the Zodiac, 12 *Sons*, 12 Tribes of Israel.)
Levi.
Kohath.
Amram.
MOSES.

JUDAIC HISTORICAL PERIOD.

MOSES — assumed epoch .... 14th century B. C.

[Interval between *Exodus* and the *first Temple*, about 314–322 years.]

SOLOMON — (Chronological times begin) .... about B. C. 1000
First monumental synchronism, REHOBOAM and SHESHONK.......... " 971–3

[*Alphabetic-writing* does not begin until the 9th–8th century B. C.]

HILKIAH — "found a book of the Law" .... " 620
Jerusalem burnt, and *Captivity* commenced .... " 586

EZRA — *Second Temple* — "VIIth year of Artaxerxes" .... " 457
Esdraic School— "Renaissance" begins .... " 400

ALEXANDER — visits Jerusalem.......... " 332

Alexandria School:

MANETHO — the earliest known *chronologist*.......... " 260
Septuagint translations commence.......... " 250

ANTIOCHUS-*Epiphanes* — plunders Jerusalem, and burns the books .... " 164
DANIEL, the Satirist, wrote.......... " 160

JUDAS, the Hammerer — restores the books.......... " 150
Maccabee coin-letters extant — SIMEON.......... " 142
Septuagint translations finished.......... " 130

SIRACIDES, *Canon* closes.......... " 130

(Roman dominion — B. C. 49.)

CHRISTIAN ERA.

BETWEEN B. C. 7 and A. D. 3; but *assumed at* 1853 *years ago.*

HEROD — decorates the *Third Temple* with pagan Hellenic architecture .... A. D. 15

Fall of Jerusalem:

TITUS razes the Temple to its foundations .... " 74

JOSEPHUS — receives the *Templar*-copy of the Hebrew Text, as a present from VESPASIAN at Rome, about .... " 75

(Earliest citation of "Gospels" — JUSTIN MARTYR, died about 166.)
Controversies between the *Fathers* and the *Rabbis* here commence.
The *Oriental* Jews transcribe the *Text* into the *square-letter* alphabet, during the 3d century after C.

HILLEL HANASSI — computes *Jewish chronology*.......... " 344
The *Masoretic points* begun by Rabbis of Tiberias .... " 506
Oldest Manuscripts of Greek LXX extant, 5th century after C.
Oldest Manuscripts of *Hebrew* Text extant, 10th century after C.
King James's English Version, printed A. D. 1611.

---

(564) *Genesis;* xvii. 5; — CAHEN: i. p. 42, note 5.

CHRONOLOGY—HINDOO.

"Originally this [Universe] was naught but SOUL: nothing else existed active [or passive]. HE had this thought—*I will create worlds.* It is thus that He created these [divers] worlds, the water, the light, the mortals, and the waters. This water is the [region] above the sky, (365) which the sky supports; the atmosphere contains the light; the earth is mortal; and the regions beneath are the *waters.*"—(*Vedas,* "Aitarêya A'ran'ya"—PAUTHIER: *Liv. Sac.*, p. 318.)

Although, in our Table of *Alphabetical origins*, we have dealt as sternly with unhistorical Indian documents, as with the metaphysical fables of all other nations, it may be well to say a few passing words upon Hindoo *chronologies;* lest it be supposed that we are not prepared to reagitate that which, to us, is no longer a "vexata quæstio." Referring the reader to the citations from Wilson, Turnour, and Sykes, therein adduced, we repeat, that there is no connected chronology, to be settled archæologically by existing monuments, throughout the whole Peninsula of Hindostan, of a date anterior to the fifth century B. C.

That vast centre of creation swarmed with varied indigenous and exotic populations, from epochas coeval with the earliest historical nations; but, if any of these Indian philosophers ever composed a rigidly-chronological list of events, we have lost the record; or, what is more probable, the chronological element was wanting in the organism of Hindoo minds, until the latter received instruction (from Chaldæan magi scattered by Darius) through the Persians;—tuition greatly improved after contact with the Bactrian Greeks during the third century B. C.

In any case, the extract subjoined will show that the antiquarian dreams of Sir W. Jones and of Colebrooke are now fleeting away.

"Whether safe historic ground is to be found in India earlier than 1200 B. C., according to the chronicles of Kashmere (*Radjtarangini*, trad. par Troyer), is a question involved in obscurity; while Megasthenes (*Indica*, ed. Schwanbeck, 1846, p. 50) reckons for 153 kings of the dynasty of Magadha, from Manu to Kandragupta, from 60 to 64 centuries; and the astronomer Aryababhatta places the beginning of his chronology 3102 B. C. (Lassen, *Ind. Alterthumsk.*, bd. I., s. 473–505, 507, and 510)."

From Humboldt (566) we pass on to Prichard; whose Hindoo prepossessions of 1819 (567) have not only been nullified by Egyptian discoveries, but, with the learned ethnographer's usual candor, have become greatly modified by his own later reflections. (568) The inquirer can judge from the perusal of the passages referred to whether he can make out a fixed chronological idea, in India, prior to the age of *Budha* in the sixth century B. C.

Lepsius (569) contents his objects (confined to a general review of the world's chronological elements) by mentioning, that the Hindoo astronomical cycle *kali yuga* falls on the 18th Feb. 3102 B. C.; that the Cashmeerian king Gonarda I. is supposed to have reigned about B. C. 2448; and that king Vikramaditya's era is fixed at B. C. 58. But he also shows that the 4th–5th centuries B. C. comprise all we can depend upon, archæologically, in Hindoo history.

However, by opening the excellent work of De Brotonne, (570) the reader will easily perceive how the Chaldæan astrological cycle of 432,000 years became extended by later Brahmanical pundits to one, equally fabulous, of 4,320,000 years: and inasmuch as this fact merely invalidates *Sanscrit* hallucinations the more, we are fain to leave Hindoo chronology in the same "slough of despond" in which we found it.

Reader!—the task proposed to myself in the preparation of these three *supplementary* Essays here ends. It was assumed under the following circumstances:—

(565) This is the same cosmogony as that of COSMAS-Indicopleustes, herein-before described. Indeed, the notion was universal; and, in theography, is so still.

(566) *Cosmos;* transl. Otté; 1850; ii. p. 115.

(567) *Analysis of Mythology.*

(568) *Researches into the Physical History of Mankind;* 1844: iv. pp. 98–139.

(569) *Chronologie;* i. pp. 4–5.

(570) *Filiations;* i. pp. 238, 239, 414–433.

Within the past five years, various sectaries (momentarily suspending polemics amongst one another) had entered into a sort of tacit combination to assail those who, like Morton, Nott, Van Amringe, Agassiz, and others, were devoting themselves to anthropological researches. Each of the above-named gentlemen has successfully repelled the intrusions of dogmatism into his especial scientific domain.

In these literary "mêlées," it has so happened that my surname has been frequently made the target for indiscreet allusions on the part of certain *teologastri;* without any provocation having been given on my side, through a single personality, in the course of ten years' lectureship upon Oriental archæology in the United States. To treat such in any other manner than with silent indifference would have been unbecoming, as well as, at the moment of each offence, unavailing. I preferred abiding my own convenience; and, in the foregoing Part III., have indicated an easy method of carrying "the war into Africa."

I believe that, thereby, good service is done in the general cause of the advancement of knowledge, and in the special one of my favorite study, *Archæology.* Geologists, Naturalists, and Ethnologists (absorbed in the promotion of positive science through the discovery of new facts), have rarely devoted time adequate to the mastery of Hebraical literature; and, in consequence, they are continually laying themselves open to chagrin and defeat in the arena of theological wranglings. My former pursuits (in Muslim lands) were remote from Natural Science, and as they disqualify me from sharing the labors of its votaries, I have thought that a contribution like the present, to the biblical armory of scientific men, might be of utility; even if it should merely spare them the trouble of ransacking for authorities generally beyond the circumference of their higher sphere of research: at the same time that a work such as "Types of Mankind" would be deficient unless the Hebrew department of its themes were to some extent complete. To future publication [*supra*, pp. 626, 627], I reserve further analyses which, without these preliminary Essays, would be unintelligible to ordinary scriptural readers. Confident of her own strength, Archæology (let *one* of this science's thousand followers hint to her opponents) neither courts nor deprecates biblical or any other agitation, and will prosecute her investigations peaceably while she can, otherwise when she must.

Repeating the direct and manly language of Luke Burke — to whose conception of a real "Ethnological Journal" scientific minds will some day accord the homage that is its due:—

"For all our arguments, there is the ready answer that our statements directly contradict the express words of Scripture, and must therefore be false, however plausible they may appear. We may reply that the word of God cannot be in opposition to genuine history, any more than it can oppose any other truth, and that therefore the passages in question cannot be a portion of this word, or if so, that they cannot have hitherto been properly understood. But experience has abundantly proved that such answers as these give satisfaction to very few, until facts have become so numerous and unequivocal that further opposition is madness. In the meantime, a war of opinion rages, embittered by all the virulence of sectarian partisanship, and the credulous and simple-minded are taught to look upon the advocates of the new doctrines as the enemies of morality, religion, and the best interests of man. For ourselves, we have no ambition to appear in any such light, nor shall we quietly submit to be placed in such a position." (571)

And for myself — whilst thoroughly endorsing the sentiments of a valued friend and colleague — I cannot better express the feelings with which I close my individual portion of an undertaking that has occupied the thoughts and hands of some men not unknown in the world of science, than by applying to our antagonists the last words ever written by me at the dictation of him to whom, with being itself, I owe all that mind and heart still hold to be priceless after more than forty years' experience of a wanderer's life: —

"*La medicina diventa amara. Spero che sarà salutifera. Intanto, si prenderà.*"(572)

G. R. G.

(HOWARD'S — MOBILE BAY, 20th July, 1853.)

---

(571) "Critical Analysis of the Hebrew Chronology" — *Ethn. Jour.;* London; No. I., June, 1848; pp. 9, 10.

(572) JOHN GLIDDON, United States' Consul for Egypt (1832–'44): *Letter to H. Ex.* BOGHOS YOUSSOUF *Bey* — MOHAMMED ALI'S *Prime Minister* — "Cairo, li 5 Febbrajo, 1841."

APPENDIX I.

REFERENCES AND NOTES.

No. (*of Notes, &c.*)

1 Ethnological Journal, London, 1848; June 1, No. I.

2 Op. cit., pp. 1, 2. An excellent précis of the meaning and scientific attributes of "Ethnology" has long been published by the venerable Jomard, in Mengin, Histoire d'Égypte, 1839, iii. p. 403.

3 Nat. Hist. of Man, London, 1848. p. 6.

4 Varieties of Man, London, 1851.

5 North British Review, Aug., 1849.

6 Op. cit., p. 6.

7. Knox, Races of Man, Philadelphia ed., 1850.

8 Burke, op. cit., p. 30.

9 Researches, v. p. 564.

10 Jacquinot, Considerations générales sur l'Anthropologie (Voyage au Pole Sud), Zoologie, 1846, ii. p. 36.

11 Nott, Two Lectures on the Biblical and Physical Hist. of Man; New York, 1849, p. 64.

12 The Friend of Moses, New York, 1852; Preface viii, and Text, pp. 442, 446, 449-51, 492-7.

13 Briefe aus Ægypten und Æthiopien, Berlin, 1852, p. 35.

14 Genesis, vii., 19-23. We quote the Hebrew Text; referring the reader to Cahen, La Bible, Traduction Nouvelle, Paris, 1831; Tom. i. p. 21.

15 Cf. Jacquinot, op. cit., chap. i. From this remarkably scientific work we have borrowed freely in this chapter, and elsewhere.

16 We ought to mention that Dr. Pickering favored us with the sight of his pages while they were yet in "proofs."

17 Op. cit., pp. 161, 163.

18 Op. cit., p. 41.

19 Races of Men, pp. 75-99.

20 Des Races Humaines, p. 169.

21 Christian Examiner, Boston, July, 1850.

22 Nott, Two Lectures, 1849.

23 Researches, ii. p. 105.

24 Proceed. Acad. Nat. Sciences; Philadelphia, 10 Sept., 1850, p. 82 — Additional Observations on Hybridity in Animals, "Reply to the Rev. John Bachman, D.D.," Charleston Medical Journal, 1850, p. 8.

25 Bodichon, Études sur l'Algérie, Alger, 1847, p. 135.

26 Jacquinot, op. cit., p. 173.

27 Wood-cut, fig. 1. L'Égypte Ancienne, 1840, Pl. I., and Champollion-le-Jeune's description in pp. 29-31.

28 Rosellini, Mon. dell'Egitto, M. R. clvii., clvi., lx., &c. Mon. Stor., iv. pp. 238-44; iii. pp. 1, 433, seq. Lepsius, Denkmäler, Abth. iii, Bl. 136.

29 See the discussion in Bishop Warburton's Divine Legation of Moses; and in Munk, Palestine, pp. 146-150.

30 Hennell, Origin of Christianity, 1845, pp. 8-21.

31 Amédée Thierry, Histoire des Gaulois, Paris, 1844.

32 Strabo, lib. iv. p. 176—Fr. ed.

33 Thierry, p. xxxv., Introd. W. de Humboldt held the same opinion.

34 Hist. de la Filiation et des Migrations des Peuples, Paris, 1837; i. pp. 294-336.

35 British Association for the advancement of Science, 1850; reported in London Literary Gazette.

36 Antiquités Celtiques Antédiluviennes.

37 Retzius, cited in Morton's MSS.

38 Schmerling, Recherches sur les Ossemens Fossiles, Liège, 1833, i. pp. 59-66: referred to in our Chapter XI.

39 Vide infra, Part II., pp. 469, 470.

40 Edwards, Des Caractères Physiologiques des Races Humaines, &c., Paris, 1839.

41 Op. cit., p. 22.

42 Paulmier, Aperçus généalogiques sur les descendants de Guillaume, Rev. Archéol., 1845, p. 794, seq.

43 Virey, Hist. Nat. du Genre Humain, Disc. Prélim., i. pp. 14, 15.

44 On the question of hair, consult the microscopic experiments of Mr. Peter A. Browne, in Proceed. Academy Natural Sciences, Philadelphia, Jan. and Feb., 1851; also Ibid., in Morton's Notes on Hybridity, second Letter to Editors "Charleston Med. Jour.," 1851, p. 6.

45 Wood-cut, fig. 2. Italie, Didot's Univers Pittoresque.

46. August, 1849; American ed.

47 Edwards, op. cit.

48 Wood-cut, fig. 3. Pouqueville, Grèce, Pl. 9.

49 Wood-cut, fig. 4. Op. cit., Pl. 84.

50 Wood-cut, fig. 5. Bunsen, Ægyptens Stelle, ii., frontispiece.

51 Wood-cut, fig. 6. Pouqueville, op. cit., Pl. 85.

52 Wood-cut, fig. 7. Rosellini, M.R., Pl. xx., fig. 66.

53 Wood-cut, fig. 8. Ibid., Pl. xxii, fig. 82. N.B. The profiles are reduced with exactitude; but we have altered the eyes from the Egyptian canon of art to ours.

54 Edwards, op. cit. Mr. Gliddon's two years' residence in various parts of

No. (of Notes, &c.)

Greece led him, he tells me, to observe the same fact: particularly among the Speziotes; whence also sprung Canaris, the bravest Greek Admiral of the Revolution. — J. C. N.

55 Études, pp. 153, seq.

56 Wood-cut, fig. 9. Crania Æg. p. 54; from Rosellini, M. R. 161; M. S. iv. 53, 62, 250. Compare Wilkinson, Manners and Cust., i. pl. 62, fig. 2, *a*, *b*; and p. 367; with Osburn, Testimony, p. 137.

57 Morton's inedited Letter to myself, "Philadelphia, 23 Nov. 1842." — G. R. G.

58 Layard, Babylon, 1853, pp. 144, 231. We attribute differences of physiognomy chiefly to the ethnographic inferiority of Assyrian artists.

59 Phys. Hist. 1841, iii. pp. 24-5.

60 Varieties of Man, 1851, pp. 551-2.

61 De Brotonne, Filiations et Migrationes des Peuples, Paris, 1837.

62 In order that we may not be suspected of considering Plato's ethical romance about the "Atalantic Isles" to be historical, we refer the reader to Martin, Études sur le Timée de Platon, cited hereinafter.

63 The Archæology and Pre-historic Annals of Scotland, Edinburgh, 1851, pp. 700-1.

64 Genesis xi. 31; xii. 1, 2, 5 — Cahen, i. p. 31.

65 Genesis xvii. 5; Ib., p. 42.

66 Genesis xvii. 15; — Lanci, Paralipomeni, 1845. Travellers have not only hunted for, but narrate how they have actually found the "double cave" they call Machphelah! (Vide report of Syro-Egypt. Soc., Nov. 8—in London Athenæum, Nov. 19, 1853; p. 1391.)

67 Genesis xxiv. 3, 4; — Cahen, pp. 65-6.

68 Genesis xli. 45; — Lanci, Paral., i. p. 26.

69 Genesis xxxviii. 2.

70 Exodus ii. 19.

71 Exodus ii. 21.

72 Exodus xii. 38;—Cahen, Text, ii. p. 50.

73 Leviticus xxiv. 10.

74 1 Kings xi. 1, 2.

75 Crania Æg., pl. xi. fig. 2; p. 47.

76 Birch, Criteria, in Otia, p. 84.

77 Layard, Babylon, p. 610.

78 History of the Jews.

79 The Asmonean, New York, 27 March, 1850, contains a confirmatory article on the Jews of Malabar, translated from the Parisian "Archives Iraelites."

80 Missionary Researches, p. 308.

81 Remarks on the Mats'Hafar Tomar, or "Book of the Letter," an Ethiopic Manuscript: Syro-Egypt. Soc., London, 1848.

82 Encyclopædia Britannica.

83 Phys. Hist., 1844, iv. pp. 82, 83.

84 Wood-cut, fig. 13—Dubeux, Tartarie.

85 Borrow, Gipsies in Spain.

86 Lest our positions should be questioned, we refer to Prichard for Continental instances, to Wilson for the Pre-Celtic in Scotland and Scandinavia, to Logan, Crawfurd, and Earl, for those among islanders of the Indian Archipelago.

87 Races of Men; vol. ix. U. S. Exploring Exped., 1848, p. 305.

88 Wood-cut, fig. 14—Layard, Babylon, pp. 152, 153

No. (of Notes, &c.)

89 Wood-cut, fig. 15—op. cit., pp. 582-584.

90 Wood-cut, fig 16—op. cit., p. 105.

91 Wood-cut, fig. 17—op. cit., p. 583.

92 Wood-cut, fig. 18—op. cit., p. 538.

93 Wood-cut, fig. 19—Wilkinson, Man. and Cust., i. p. 384, pl. 69, fig. 8.

94 Lepsius, Auswahl, Leipsig, 1840, "Canon der Proportionen"; — ibid., Briefe aus Ægypten, Berlin, 1852, pp. 105, 106; —and Birch, Gallery of Antiquities, Br Museum, pl. 33, fig. 147.

95 Rev. Archéol., 1844, p. 213, seq.; 1847, p. 296, seq.:—Commentary on the Cuneiform Inscrip., 1850, pp. 4-7.

96 Wood-cut, fig. 20—Botta, Mon. de Ninive, pl. 36.

97 Wood-cut, fig. 21—ibid., pl. 68 bis.

98 Polyhym., lxxvii.; Bonomi, Nineveh, pp. 182, 301.

99 Wood-cuts, figs. 22, 23 — Botta, op. cit., pl. 14.

100 Wood-cut, fig. 24 — Lettres de M. Botta sur ses découvertes à Khorsabad, 1845, pl. xxii., and p. 28.

101 Essai de déchiffrement de l'Écriture Assyrienne, 1845, pp. 22-25.

102 De Longpérier, Galerie Assyrienne, 1850, p. 16; and Nos. 1, 12, 27, 33.

103 Gliddon, "Hist. Sketches of Egypt," No. 5, New York Sun, Jan. 14, 1850.

104 Wood-cut, fig. 25 — Botta, Mon. de Ninive, pl. 45.

105 Wood-cut, fig. 26 — Layard, Monuments of Nineveh, folio pl. 42.

106 Wood-cut, fig. 27—Layard, Babylon, pp. 150, 143-4.

107 2 Kings xviii.; Isaiah xxxvi.

108 Wood-cut, fig. 28—Layard, Babylon, pp. 617-9.

109 2 Kings xv. 19-21.

110 Wood-cut, fig. 29—Layard, op. cit., p. 361.

111 Vide infra, Part III., p. 714.

112 Deuteron. xxiii. 8, 9; Cahen, v. p. 99.

113 Egyptian Cartouches found at Nimroud, R. Soc. Lit., Jan. 1848, p. pp. 164-71

114 Mr. Birch's translation—Private letter to G. R. G.

115 Wood-cut, fig. 31 — Rosellini, M. R., pl. xii. fig. 46; — Conf. Bunsen, Ægyptens Stelle, iii. p. 133.

116 Bonomi, Nineveh and its Palaces, 1852, pp. 77, 78.

117 Babylon, pp. 153-9, 280-2, 630-1.

118 Egypt. Inscrip. in Bibliothèque Nationale, 1852, p. 17.

119 Wood-cut, fig. 32 — Layard, Babylon, p. 630: — Lepsius, Denkmäler, Abth. iii. Bl. 88.

120 Babylon, 623.

121 Birch, Stat. Tablet of Karnac, 1846, pp. 29, 37: — Gliddon, Otia Ægyptiaca, p. 103.

122 Birch, in Layard's Babylon, p. 630: — or Lepsius, Auswahl, Taf. xii. line 21.

123 Wood-cut, fig. 33—Rosellini, M. R., pl. i. fig. 2: — Conferre Lepsius, Denkmäler, Abth. iii. Bl. i., at Berlin. Lepsius (Letters, pp. 278, 381) calls her Amunoph's "mother, Aahmes-nufre-Ari"—"Amenophis I. and the black Queen Aahmes-nefruari." That she is painted black, as well as red, no one disputes; but did the Negro-black pigment ever accompany such osteological structure?

124 Crania Ægypt. p. 47.

No. (of Notes, &c.)

125 Wood-cuts, figs. 34, 35—Lepsius, Denkmäler, Altes Reich, Dyn. IV., Grab 75, Abth. ii. Bl. 8, 10.
126 Wood-cut, fig. 36—Bunsen, op. cit. ii. Frontispiece.
127 Wood-cut, fig. 37—Afrique Ancienne, Carthage, Univ. Pittor., from a coin.
128 Wood-cut, fig. 38—idem.
129 Wood-cut, fig. 39—Rosellini, M. R. pl. 157; M. S. iv. p. 237:—Osburn, Egypt's Testimony, pp. 114–6, fig. 1.
130 Wood-cut, fig. 40—M. R. 151, M. S. iv. p. 82:—Wilkinson, Man. and Cust. i. pl. 69, fig. 7:—Birch, Stat. Tablet, p. 34.
131 Wood-cut, fig. 41—M. R. 161, fig. 1; 159, fig. 3; M. S. iv. p. 1[illegible]:—Morton, pl. xiv. fig. 20, p. 48.
132 Rawlinson, Persian Cuneiform Inscrip. of Behistun, 1847, p. 270.
133 Wood-cut, fig. 43—Vaux, Nineveh and Persepolis, 1851, pp. 350–1.
134 Letronne, Civilisation Égyptienne, 1845, pp. 30–43.
135 Rawlinson, op. cit. p. xxviii.
136 Wood-cut, fig. 44—Coste et Flandin, Perse Ancienne, pl. 18.
137 Rawlinson, op. cit. p. 323.
138 Wood-cut, fig. 45—Perse Ancienne, pl. 154.
139 De Sacy, Antiquités de la Perse, et médailles des rois Sassanides, Paris, 1793; pp. 12, 64; A, No. 3—recopied in Perse Ancienne.
140 Woodcut, fig. 46—Perse Ancienne, pl. 185
141 Perse Ancienne, pl. 49, bas-relief A.
142 Woodcut, fig. 47—Perse Ancienne, pl. 51, bas-relief D.
143 Layard, Monuments of Nineveh, 1849, folio plate; Nineveh and its Remains, ii. pp. 329–31:—well described by Bonomi, op. cit. pp. 287–95.
144 Wood-cut, fig. 50—Rosellini, M. R. pl. 103, and 87; M. S. iii. part 2, p. 157:—Morton, Crania Ægypt. p. 63.
145 Pauthier, Chine, pp. 417, 427, 429. According to Callery and Yvan (L'Insurrection en Chine, depuis son origine jusqu'à la prise de Nankin, Paris, 1853) the present Chinese insurgents let all their hair grow, as their ancestry did under the Mings, to distinguish themselves from the Tartar usurpers.
146 Lepsius, Chronologie, i. p. 379. Ibid., Discoveries, transl. Mackenzie, p. 381.
147 De Sola, Lindenthal, and Raphall; New Transl. of the Scriptures, London, pp. 46–7:—Genesis xi. 10–26.
148 Monumenti Storici, ii. p. 461, seq.
149 Apochrypha, xiv. 17.
150 Wood-cuts, figs. 44 to 71—Rosellini, Monumenti Reali, pl. i. to xxiii.; and Mon. Storici, ii., "Iconografia de' Faraoni." Our selections are arranged in accordance with the more recent improvements of Egyptian chronology.
151 Prisse, Suite des Monumens de Champollion, 1848, pl. x.:—but compare Lepsius, Denkmäler, Abth. iii. Bl. 100. Ibid., Ægyptischen Götterkreis, 1851, pp. 40–5. Ibid., Briefe aus Ægypten, 1852, pp. 89, 362.
152 Morton, Cr. Æg. p. 44, pl. xiv. 3; from Rosellini.
153 Colossus at Aboosimbel; M. R. pl. vi. fig 22.
154 Chron. der Ægypter, i. pp. 321–2, 358, 379.
155 Notes upon an Inscription in the Bibliothèque Nationale of Paris, Trans. R. Soc. Lit. 1852, iv. pp. 16, 17, 21.
156 Gliddon, Chapters, p. 22; and Otia, p. 134.
157 Wood-cuts, fig. 71, bis—Rosellini, M. R. pl. 79.
158 Ibid., M. R. pl. clx. lxxx.; M. S. iii. pp. 2, 95, seq.; iv. pp. 245–9:—Morton, Cr. Æg. p. 55:—Osburn, Test., p. 121:—Birch, Tabl. of Karnac, pp. 14, 15–35.
159 Morton's inedited MSS.—Letter to Mr. Gliddon, entitled, "Reflections on Mr. G.'s Ethnological Charts," 1842; corrected by Dr. Morton's autographic notes, Philadelphia, 23d March, 1843. We shall refer to it as "Morton's MS. Letter."
160 Wood-cut, fig. 74—Rosellini, M. R. clvi. and lx; M. S. iii. pp. 1, 433, seq.; iv. pp. 228–44:—Lenormant, Cours d'Histoire Ancienne, 1838, pp. 322–36:—Champollion-le-Jeune, Lettr. d'Egypte, p. 250, seq.:—Champollion-Figeac, Ég. Anc. pp. 29–31, pl. i.;—Wilkinson, Topog. Thebes, 1835, pp. 106–7:—Man. and Cust. i. pp. 364, 371, pl. 62, No. 4, fig. *a*:—Mod. Egypt. ii. p. 105:—Osburn, Testimony, pp. 22–7, 114, 143:—Birch, Stat. Tab. Kar. p. 20.
161 Wood-cut, fig. 75—Lepsius, Denkmäler, Abth. iii. Bl. 136, fig. 37 *a*.
162 Woodcut, fig. 76—Rosellini, M. R. clxi. fig. 1; clix. fig. 3; M. S. iv. p. 150:—Morton, Cr. Æg. p. 48, pl. xiv. 20.
163 Denkmäler, Abth. iii. Bl. 136, fig. *d*.
164 Woodcut, fig. 78—Rosellini, M. R. clxi; M. S. iv. pp. 91, 251:—De Saulcy, Recherches, Inscrip. de Van, 1848, p. 26.
165 Wood-cut, fig. 80—Rosellini, M. R. lxix.; M. S. iii. part. 2, p. 29:—Birch, Gallery, pp. 93, 97, pl. 38:—Morton, p. 46, pl. xiv. 24. It is moulded in colors at the British Museum.
166 Wood-cut, fig. 81—M. R. cli.; M. S. iv. p. 82, seq.:—Wilkinson, M. and C. i. p. 384, pl. 69, fig. 7;—Osburn, p. 53;—Birch, Stat. Tab. p. 34.
167 Wood-cut, fig. 82—Rosellini, M. R. clix..—Champollion-Figeac, pp. 208–9, pl. 62:—Hoskins, Ethiopia, p. 329, pl. i. ii.:—Morton, p. 41, pl. xiv. 22;—Wilkinson, M. and C. i. pl. iv. p. 379:—Birch, Gallery, p. 80; and Stat. Tab. p. 61:—Prisse, Salle des Ancêtres, Rev. Archéol. 1845, p. 11, and note. N. B. After this page was stereotyped, we received Mr. Birch's freshest paper (Annals of Thotmes III., 1853) wherein he assigns these KeFa to the Island of Cyprus. Vide infra, pp. 479–480, voce "KTtIM."
168 Wood-cut, fig. 83—Rosellini, M. R. clix. M. S. iii. p. 435; iv. p. 234:—Birch, Gallery, pp. 88–9, 97, pl. 38:—Stat Tab. pp. 13–14.
169 Woodcuts, figs. 84, 85—Rosellini, M. C. xxii.:—Wilkinson, i. pl. iv.:—Champollion-Figeac, pp. 376–8:—Morton, p. 50; pl. xiv. 21:—Osburn, Testimony, p. 52:—Hoskins, Ethiopia, plates, part

No (of Notes, &c.)

iii. first line, p. 332:—Birch, Stat. Tab., pp. 18–9: — Pickering, Races, p. 372; also, Geog. Distribution, 1854.

170 References as above.

171 Wood-cut, fig. 86—Rosellini, M. C., xlix; M. C., ii. pp. 254–70: — Wilkinson, M. and C.. ii. p. 99: — Mod. Egypt, 1843, ii. p. 237:—Osburn, Antiquities, Relig. Tract Soc., 1841, pp. 220–1: — Keith, Demonstrations of Christianity:—Taylor, Illustrations of the Bible, 1838, pp. 79–84:—Kitto, Cyclopædia, i. pp. 353–4: — Morton, Cr. Æg., p. 47: — Lepsius, Denkmäler, Abth. iii. Bl. 40: compare ibid., Dyn. IV., Grab I., Abth. ii. Bl. 96 for "chin sprouts."

172 See references under Nos. 144, 145.

173 Wood-cut, fig. 88—Rosellini, M.R., lxiii.; M. S., iii. part ii. p. 12:—Morton, p. 48, pl. xiv. 19.

174 Wood-cut, fig. 89—Rosellini, M.R., clvii.; M. S., iv. p. 237; — Osburn, Test., pp. 114–6, plate, fig. 1.

175 Wood-cut, fig. 90—Lepsius, Denkmäler, Abth. iii. Bl. 116, fig. *a*.

176 Wood-cut, fig. 91 — Rosellini, M. R., lxxxiii; M. S., iii. part ii. p. 103: — Champollion-Figeac, pl. 79:—Morton's MS. letter.

177 Wood-cut, fig. 92 — Rosellini, M. R., cxlxxx. fig. 7; M. S., iv. pp. 91–4.

178 Wood-cut, fig. 93 — Rosellini, clviii; M. S., pp. 234, 239: — Birch, Gallery, pp. 89, 104:—Osburn, p. 27:—Morton, p. 46, pl. xiv. 23:—Layard, Babylon, pp. 142, 146, 628.

179 Lepsius, Denkmäler, Dyn. XIX. *a*, Abth. iii. Bl. 136; compared with Rosellini, M. R., pl. clv.; M. S., iv. pt. i. pp. 228–43. In common with Morton we were always at a loss to account for the presence of *two* white races in Rosellini's copy of this tableau. It turns out that an error of *coloring* on the part of the Tuscan artists was the unique cause of such perplexities; because they have tinted this figure *light flesh-color*, instead of tawny yellow.

180 Wood-cuts, figs. 97, 98—Rosellini, M. R., lxvii.; M. S., iii. part ii. p. 126:—Birch, Gallery, p. 99, pl. 38: — Osburn, pp. 77, 124.

181 Wood-cuts, figs. 99, 100— Rosellini, M. R., clx.; M. S., iv. p. 235: — Champollion-Figeac, pp. 30–1, pl. i. fig. 4:—Osburn, pp. 114, 142–3.

182 Wood-cut, fig. 101 — Rosellini, M. R., cxliii. fig. 9.

183 Wood-cut, fig. 102 — Rosellini, M. R., cxliii. fig. 5.

184 Wood-cut, fig. 103 — Rosellini, M. R., cxliii. fig. 10.

185 Wood-cut, fig. 104 — Rosellini, M. R., cxliii. fig. 3.

186 Wood-cut, fig. 105 — Rosellini, M. R., cxliii. fig. 8.

187 Wood-cut, fig. 106 — Rosellini, M. R., lxv. and Morton, p. 47. Compare with these heads, and with that one in M.R., cxliii. fig. 11; M. S., iv. p. 96 (also Wilkinson, M. and C., i. pp. 370–1; pl. 62, fig. 3, *a*, *b*, *c*:) what Layard (Babylon, p. 355) has written about the *Shairetana* of hieroglyphics contrasted with the *Sharutinian* in the cuneiform sculptures.

88 Researches, ii., chap. x., xi., pp. 193–205.

No. (of Notes, &c.)

189 Ibid., op. cit., p. 220. How is it possible that Dr. Prichard, in 1837, could have known nothing of the triumphant missions of France and Tuscany to Egypt of 1828–30—when all Europe rang with applause?

190 Appendix to first edition to the Natural History of Man, London, 1845, pp. 570–583; quoted in Dr. Patterson's Memoir of Morton, ubi supra.

191 Sopra i Popoli Stranieri introdotti nelle Rappresentanze Storiche de' Monumenti Egiziani — Annali dell' Instit. di Corr. Archeol., Roma, 1836, pp. 333–50.

192 Égypte Pharaonique, Paris, 1846, ii. pp. 352–4.

193 Prisse, Trans. R. Soc. Lit., 1841:—Gliddon, Appeal to the Antiquaries, London, 1841, p. 53:—Wilkinson, Materia Hieroglyphica, 1824, part ii. pl. 2; and Text, p. 118;—Top. of Thebes, 1835, p. 420, &c.:—Mod. Eg., 1843, ii. pp. 223–6:—Hand-book, 1857, pp. 306–7, 392–3: — Leemans, Lettre à M. Salvolini, 1840, pp. 149–51: — L'Hôte, Lettres, 1840, pp. 27, 93, 99, 131, 185, 198: —Perring, Trans. R. Soc. Lit.; followed by Morton, Cr. Æg., p. 54:— Hincks, On the Egyptian Stele, 1842, pp. 1, 18–9; Age of the XVIIIth Dynasty, 1843, p. 5:— Bunsen, Ægyptens Stelle, iii. p. 58. The Revue Archéologique contains the following—1845, Prisse, Legendes Royales, pp. 457–74; Lettre à M. Champollion-Figeac, p. 730; 1847, Antiquités Égyptiennes, pp. 693–723:—Leemans, Lettre à M. Witte, pp. 531–41:—1849, De Rougé, Lettre à M. A. Maury, pp. 120–3;—1851, Maury, Dynasties Égyptiennes, pp. 180–2: — Rosellini, Cartouches, Nos. 69, 69 bis: — For. Quart. Review, "Egyptian Hieroglyphics," Jan. 1842, p. 157: — Pauthier, Sinico-Ægypt., 1842, Frontispiece: — Prisse, Suite des Monumens, 1847, Preface:— Birch, Tablet of Ramses II.. p. 24: — Ampère, Recherches, Rev. des Deux Mondes, 1846–7: — Lepsius, Ægyptischen Götterkreis, 1851, pp. 37–46: — Briefe, 1852, p. 368: — Denkmäler, iii. 111.

194 Denkmäler, Abth. iii. Bl. 111. Even Lepsius's copies slightly differ among themselves — compare Bl. 99 with 100, 103, and 109.

195 Crania Ægyptiaca, p. 54—from Perring's paper in Trans. R. Soc. Lit., London, 1843, i. p. 140.

196 Letters, transl. Mackenzie, p. 297. Conf. Denkmäler, Abth. iii. Bl. 113.

197 Rosellini, M. R., xv. fig. 63.

198 Lepsius, Auswahl; and Wilkinson's Turin Papyrus.

199 Wood-cut, fig. 110—Dyn. XII., Abth. ii. Bl. 141.

200 Wood-cut, fig. 108 — Rosellini, M. R., xxvi. xxvii. xxviii.; M. S., i. p. 18[illegible]; iii. p. 48, seq.; M. C., i. p. 56: — Denkmäler, Altes Reich, Dyn. XII., Abth. ii. Bl. 31.

201 Stat. Tab. Karnac, p. 5.

202 Hist. Tab. of Ramses II., p. 28.

203 Letter to M. Humboldt, "Korusko, Nov. 20, 1843," London Athenæum, 2 March, 1844. Compare Briefe, 1852, p. 97–100.

No. (of Notes, &c.)

204 Discoveries in Egypt, Ethiopia, and the Peninsula of Sinai, in the years 1842–1845; London, 1852, pp. 108–10.
205 Denkmäler, Abth. ii. Bl. 123–33.
206 Geognostische Karte von Ægypten, Wien, 1842.
207 Wood-cut, fig. 111—Abth. ii. Bl. 107, Grab 2.
208 Wood-cut, fig. 112—Abth. ii. Bl. 109, Grab 2.
209 and 210 Wood-cuts, figs. 113, 114—Abth. ii. Bl. 73, Grab 26.
211 and 212 Wood-cuts, figs. 115, 116—Abth. ii. Bl. 10, "Pyr. v. Giseh," Grab 78.
213 Wood-cut, fig. 117—Abth. ii. Bl. 8, "Pyr. v. Giseh," Grab 75.
214 Woodcut, fig. 118—Abth. ii. Bl. 20, 22, "Pyr. v. Giseh," Grab 24; Briefe, pp. 36–8.
215 Wood-cut, fig. 119—Abth. ii. Bl. 2, "Wadi Maghara."
216 Abth. ii. Bl. 39 *f*; and Briefe, p. 336.
217 Researches, ii. p. 44. Where not referred to others, our citations are also taken from Prichard.
218 Beke, Journal. R. Geog. Soc., xvii.; and in Gliddon, Hand-book, 1849, pp. 26–33.
219 Ritter, Géog., transl. Buret, 1836, i.; and Jomard, Notes pour un Voyage dans l'Afrique Centrale, 1849, pp. 19–20.
220 This fact is established by D'Eichthal (Hist. et Origine des Foulahs), by Hodgson (Notes on the Sahara and Soudàn), by Perron (Transl. of Voyage du Cheykh Mohammed-el-Tounsy), by Jomard (Observations sur le Voyage au Darfour, &c.), and by Ritter, i. pp. 432–7.
221 Gliddon, Hand-book, p. 35.
222 Beke, Sections, in Map of Journey; Jour. R. Geog. Soc., xvii.
223 See all authorities in D'Eichthal.
224 Researches, ii. p. 97.
225 Op. cit., ii. p. 343.
226 Op. cit.
227 Prichard, ii. p. 129: — Beke, Jour. R. Geog. Soc.
228 Op. cit., ii. p. 132:—Harris, Highlands of Ethiopia, 1843:—Fresnel, Mém. sur le Waday, 1848: — Beke, Essay on the Sources of the Nile, 1848: — Origin of the Gallas, 1848:—Observations sur la communication supposée entre le Niger et le Nil, 1850:—Jomard, Sur la pente du Nil Supérieur, 1848.
229 Beke; and Newman; Trans. Philological Soc., London, 1843–5, i. and ii.
230 Larrey, Notice sur la conformation physique des Égyptiens; Descrip. de l'Égypte, ii.
231 Essai sur les Mœurs des habitants modernes de l'Égypte—id., ii. part 2, p. 361.
242 Prisse, Oriental Album, Madden, London, 1846, pl. 28, 29:—Pickering, Races, pl. xii. pp. 221–4.
233 Cherubini, Nubie, pp. 50, 51.
234 Gliddon, "Excursus on the Berbers," Otia, pp. 117–46.
235 "Et-Tullàk b'-et tellàteh," or "triple divorce."—G. R. G.
236 Cr. Æg., pp. 58–9: Gliddon, Otia, p. 119.
237 Tablet of Ramses II., 1852, p. 21.
238 Prichard, ii. p. 135.
239 Travels in Nubia, p. 439.
240 2 Chron. xii. 3.
241 Wiseman, Lectures, p. 136.

No. (of Notes, &c.)

242 Nott, Unity of the Human Race (Reply to "C."), Southern Quart. Rev., Jan. 1846, p. 24.
243 Champollion, L'Égypte sous les Pharaons, 1814, i. p. 255—"Coptic MS.":—Wilkinson, Mod. Eg. and Thebes, 1843, ii. p. 312—"Inscription of King Silco."
244 Tribus des Ababdeh et des Bicharis, Magazin Pittoresque, Paris, Nov. 1845, pp. 371–3.
245 Gliddon, Otia, pp. 134–5.
246 Compare Briefe aus Ægypten. pp. 220, 251, 263.
247 Gräberg de Hemso, Specchio geogranco e statistico dell' Impero di Marocco, Genova, 1834, pp. 251–6.
248 Notes on Northern Africa, the Sahara, and Soudan, New York, 1844, pp. 22-32: — also, Daumas, "Les Tuareg du Saharah," Revue d'Orient, Paris, Fev. 1846, pp. 168–171.
250 A Series of Chapters on Early Egyptian History, Archæology, and other subjects connected with Hieroglyphical Literature; New York, 1843, p. 58. Conf. Jomard, Études sur l'Arabie, in Mengin's Hist. d'Égypte sous Mohammed Ali; vol. iii., Paris, 1839: — Champollion-Figeac, Égypte Ancienne, Paris, 1840, pp. 28, 34, 417: — Champollion, Grammaire Égyptienne, p. xix.
251 Burke's Ethnological Jour., London, 1848, pp. 367, 368; and Otia Ægyptiaca, 1849, pp. 77–79.
252 Pettigrew, Encyc. Ægyp., 1841, pp. 2, 3.
253 Filiations, &c., 1837, i. pp. 210–17.
254 Asie Moyenne, 1839, i. p. 155.
255 Voyage en Syrie, i. p. 75.
256 Réflexions sur l'Origine, &c., des Anciens Peuples, 1747, pp. 303, 383.
257 Herodotus, lib. ii. § 105.
258 Trans. R. Soc. Lit., iii. part i.; 1836, pp. 345–6.
259 Gen. xlii. 23, 30, 33.
260 Deut. xxiii. 7, 8.
261 Gen. xli. 50–2.
262 Crania Ægyp., pp. 28–9: — Young, Discoveries in Hieroglyphical Literature, 1823, p. 63, &c.:—Champollion-Figeac, Contrat de Ptolémaïs, p. 43: — and John Pickering, Egyptian Jurisprudence, Boston, 1840, p. 313.
263 Wood-cuts, figs. 121, 122—Champollion, Monumens, ii. pl. 160, fig. 3.
264 Wood-cut, fig. 123—Rosellini, M. C., pl. 133, fig. 3.
265 Wood-cut, fig. 125—Hoskins, Ethiopia, pl. xi.
266 Cailliaud, Meroë, pls. xvi–xx.
267 Wood-cut, fig. 126 — Rosellini, M. C., pl. 133.
268 Champollion-Figeac, Égypte Anc., p. 356
269 Wood-cut, fig. 128 — Rosellini, M. C., pl. 97.
270 Wood-cuts, figs. 129, 130, 131, 132—ibid., M. C., 126.
271 Wood-cut, fig. 133—ibid., M. C., pl. 37.
272 Wood-cut, fig. 134—ibid., vol. i. pl. 4.
273 Wood-cut, fig. 135—ibid., M. C., pl. 86.
274 Wood-cut, fig. 136—ibid., M. C., pl. 41.
275 Wood-cut, fig. 137—ibid., M. C., pl. 29.
276 Wood-cuts, figs. 138, 139 — ibid., M. C., pl. 132.
277 Morton, p. 37: — Trans. R. Soc. Lit., 1794, pl. 16, fig. 4:—Gliddon, Chs., p. 23.

No. (of Notes, &c.)

278 Rosellini, M. S., parte 1ma, ii. 1833, pp. 476–521; Portraits, M. R., pl. i.–vii.
279 Vide infra, p. 688, "Chronology."
280 These *drawings* were our "stamps"; lithographed, infra, pls. i.–iv.
281 Humboldt, Cosmos, French ed. 1846, i. pp. 430, 579: on which see Dr. Patterson's commentary, supra, "Memoir." The heretical author of Vestiges of Creation (first Amer. ed., New York, 1845, pp. 209–242), however inaccurate in other theories—and the very orthodox Guyot (Earth and Man, Boston, 1851, p. 253, seq.), however exact in other data—owing to similar philanthropic sentimentalities, also break down when they discuss the Natural History of mankind.
282 Vansleb. in Quatremère, Recherches sur la langue Copte.
283 Manetho, apud Syncell. Chron., p. 40:—Lepsius, "Lettre à M. le Prof. Hippolyte Rosellini," Annali dell' Instituto di Corrispondenza Archeologica, Roma, ix. 1837, p. 18.
284 Kenrick, Ancient Egypt under Pharaohs, London, 1850, i. p. 99.
285 Op. cit., pp. 107–8.
286 Op. cit., p. 131.
287 Wood-cut, fig. 152—Rosellini, M. R., 155; M. S., iv. pp. 230, 241–2: — Osburn, Testimony, pp. 23–4.
288 Lepsius, Denkmäler, Abth. ii. Bl. 19.
289 Rosellini, M. R., 101, and 87.
290 Wilkinson, Man. and Cus., i. p. 285; iii. pp. 141, 346:—Henry, Égypte Pharaonique, ii. pp. 274–89: — Birch, Lettre à Letronne, Rev. Archéol.; and De Saulcy, Note, Rev. Archéol., 1847, p. 430.
291 Testimony, pp. 23–4.
292 Wood-cut, fig. 156—Rosell., M. R., pl. 96.
293 Wood-cut, fig. 157—ibid., M. C., pl. 13.
294 Wood-cut, fig. 158—ibid.
295 Wood-cuts, figs. 159, 160—Morton's MSS. for 2d ed. of Cr. Ægyp.
296 Wood-cut, fig. 161—ibid.
297 Ampère, Revue des Deux Mondes, Aug. 1846, p. 391.
298 Gliddon, Hand-book, pp. 20—22.
299 Denkm., Dyn. IV.–VI., Tombs at Berlin.
300 Crania Ægyptiaca, pp. 26, 27.
301 I was present in Dr. M's office when he opened it; and so vivid is my remembrance of the conversation its joint perusal superinduced, that, although I had never seen the letter from 1844 to this Sept. 1853, I sought for and found it among my deceased friend's papers.—G. R. G.
302 Pickering, Races of Men, 1848, p. 10.
303 Grammaire Égyptienne, Introd., p. xix.
304 Cosmos, ii. p. 147, French ed.
305 Jerem. xiii. 23:—Morton's notes for 2d ed. Crania Æg.; but vide infra, pp. 487-8.
306 Institutiones ad Fundamenta Linguæ Arabicæ, Lipsiæ, 1818, pp. 38–9.
307 Dubois, Voyage autour du Caucase, &c.; cited hereinafter.
308 Wood-cut, fig. 166 — Rosellini, M. R., 142; M. S., iv. p. 292.
309 Wood-cut, fig. 167—Nubie, p. 8:—Ros., M. R., 85; M. S., iii. part ii. p. 114: — Osburn, Testimony, p. 32:—Champollion, Monuments, pl. xvi.
310 Wood-cuts, figs. 168–170— Rosellini, M. R., pl. lxxxv.

No. (of Notes, &c.)

311 Birch, Gallery, pp. 68, 86, 104:—Gliddon, Otia, p. 119.
312 Madden's Oriental Album, pl. 25; "Nubian Females, Kenoosee Tribe, Philæ."
313 Wood-cut, fig. 171 — Rosellini, M. R., 156, 160; M. S., iv. pp. 231, 250.
314 Wood-cut, fig. 172—Rosellini, M. R., 60; M. S., iii. part i. p. 407.
315 Wood-cut, fig. 173—Wilkinson, Man. and Cust., p. 404, No. 73.
316 Otia, pp. 147–8.
317 Nott, Bibl. and Phys. Hist., pp. 138–146: — Gliddon, Otia, p. 147. James Cam, 1480, was the first who sailed along Africa to a little beyond the river Congo. Hottentot tribes were altogether unknown until after the voyage of Bartholomew Diaz in A. D. 1486 (Churchill's Collection of Voyages).
318 Anthon, Class. Dict., voce "Hanno." We have re-examined Heeren (Reflections on the Ancient Nations of Africa, i., chaps. ii., v., vi. — particularly pp. 214–241), and can find nothing but *hypotheses* to support Carthaginian possession of *Negro* slaves. The account of Hanno's voyage, &c., is given (op. cit., pp. 492–501).
319 L'Arménie, la Perse, et la Mesopotamie, Paris, folio, 1842, pl. 113: — compare pl. 126.
320 Botta et Flandin, Mon. de Ninive, folio, 1847–50, pl. 88.
321 Virgile, Moretum, "The Salad," Nisard's ed., Paris, 1843, p. 463.
322 Wood-cuts, figs. 177, 178—Rosellini, M R., xliv. bis, quater.
343 Abth. iii. Bl. 120.
324 Archæologia, xxxiv. pp. 18–22.
325 Compare Gliddon's assertions of the same fact in 1843, Chapters, pp. 47, 59; in 1849, Otia, pp. 78–81; and Hand-book, p. 35.
326 Hist. Tablet of Ramses II., London, 1852, pp. 1822.
327 Hincks, Hieroglyphical Alphabet, p. 16; pl. i. figs. 23, 26, 27: — Gliddon, Otia, p. 133.
328 Wood-cut, fig. 181—Mon. Civ., pl. xxii.
329 Travels, plate, part i. line 3.
330 Man. and Cust., i. pl. iv. line 3.
331 Egypte Ancienne, pl. 55.
332 Wood-cut, fig. 182 — Rosellini, Hoskins, Wilkinson, and Champollion-Figeac, supra No. 331.
333 Races, 1848, p. 224 — compare "Abyssinian," in plate xii.
334 Gallery, pp. 94, 97; pl. 38.
335 Topog. of Thebes, 1835, pp. 135, seq.: — Man. and Cust., i. pp. 58, 404; iii. 179: —Champollion, Monuments, pl. 158.
336 Gliddon, Otia, p. 148.
337 Gliddon's MS. Diary, "Thebes, February, 1840":—Wilk., Materia Hieroglyphica, "Amuntuonch": — Rosellini, Appendice, Oval No. 13: — Leemans, Lettre à Salvolini, p. 75. Compare Birch, Tablet of Ramses II., Tomb of Hui, p. 24.
338 Wood-cuts, figs. 183, 184 — Denkmäler, "Neues Reich," Dyn. XVIII., Abth. iii. Bl. 117. — N. B. The children sometimes are *red* — see the same paternity exemplified in Hoskins, Ethiop., "Grand Procession," lowest line.

No. (of Notes, &c.)

339 As among the "wrestlers" at Benihassan (Cailleaud, Arts et Métiers, pl. 39): —the "wine-pressers" at Thebes (ibid. pl. 34)—and other scenes.
340 Wilkinson, Man. and Customs, ii. p. 265.
341 Chev. Lepsius's private letters to Morton and to Gliddon.—Vide Chapters, 15th ed., Peterson, Phila., 1850, p. 68.
342 Crania Ægyptiaca, p. 41.
343 Wood-cut, fig. 187—Hoskins, pl. x.
344 Wood-cut, fig. 188—ibid.
345 Hanbury and Waddington, Travels in Ethiopia, pl. xiv.—compare Cailleaud, Voyage à Meroë; and Hoskins, pl. xxix.
346 Syncell. Chronograph., p. 120, ed. Venet.
347 Crania Ægyptiaca, pp. 49–50:—Rosellini, M. S., ii. pp. 174, 238.
348 Wood-cut, fig. 193, Crania Ægyptiaca, pl. xii., fig. 7; and p. 18:—Catalogue, 1849, No. 823.
349 Letronne, Matériaux pour servir à l'histoire du Christianisme en Égypte.
350 Crania Ægyp. p. 44:—Champ. Mons., I., pl. 1; Rosellini, pl. xxv. (eye wanting) —Cherubini, Nubie, pl. 10. p. 33.
351 Gliddon's Otia, p. 144.
352 Lepsius, Denkmäler, Part II., pl. 136; *i*, lines 1 and 2.
353 Mémoire sur quelques Phénomènes Célestes; Revue Archéol., 1853, p. 674, note 34.
354 Arundale, Bonomi and Birch's Gallery of Antiquities, selected from Brit. Mus.—before cited.
355 Champ. Mons. I., pl. lxxi, lxxii; Rosellini, M. R., lxxv.
356 Crania Ægyptiaca, pp. 61–2: corrected by "standing," for "seated," in MSS. for 2d ed.
357 "Parable"—It is well known that the earlier colonists of Barbadoes, Montserrat, and some other W. Indian islands, were Irish exiles. Odd to relate, while a few of their Negro slaves actually speak *Gaelic*, many have acquired the "brogue!" An Hibernian, fresh from the green isle, arrived one day at the port of Bridgetown, and was hailed by two Negro boatmen who offered to take him ashore. Observing that their names were "Pat" and "Murphy," and that their brogue was uncommonly rich, the stranger (taking them to be Irishmen) asked—"and how long have ye been from the ould counthree?" Misunderstanding him, one of the darkies replied, "sex months, y're honor." "Sex months! . . . . . only sex months, and turned as black as me hat!! J—!!! what a climate! Row me back to the ship. I'm from Cork last—and I'll soon be from here!"

Every one laughs at the verdant ignorance which believed that a Celt could be transmuted by climate into a Negro in 6 months. All would smile at the notion of such a possibility within 6, or even 60 years. Most readers will hesitate over 600 years. Anatomy, history, and the monuments prove that 6000 years have never metamorphosed one type of man into another.

No. (of Notes, &c.)

358 Second Visit to the United States, Part II., p. 188.
359 Tableaux of New Orleans, 1852, pp. 8–17:—also, Dickeson and Brown, Cypress Timber of the Mississippi, 1848, p. 3.
360 Scottish Archæologists, Dr. Wilson tells me, have found similar indications of early human existence in the Shetland Isles; and he considers this criterion very valuable.—G. R. G.
361 Morton, Crania Americana, p. 260.
362 "Information respecting the History, Condition and Prospects of the Indian Tribes of the United States," vol. I.
363 As Morton happily wrote—"The works of giants and the stature of pigmies"—MSS. for 2d ed. Cr. Ægyp.
364 The Serpent Symbol, &c., in America, 1851, pp. 26–7.
365 Westminster Review—"The Greek of Homer a Living Language." So true is this, that one word will illustrate the fact: *e. g.*, *νερο* is now the name for *water* in ordinary Grecian parlance, just as it was in Homeric days, to the exclusion of *υδωρ* which belongs to the classical ages intervening.—G. R. G.
366 Christian Examiner, Boston, July, 1850, p. 31.
367 Trans. Am. Ethnol. Soc., II.
368 Bunsen, Life and Letters of B. S. Niebuhr, New York ed., 1852.
369 Connection between Science and Revealed Religion.
370 Ancient Monuments of the Mississippi Valley, 1848, p. 304.
371 Wilson, Archæology of Scotland.
372 Op. cit., p. 168.
373 Layard's Babylon abundantly establishes this fact; but vide infra, p. 427, figs. 263, 264.
374 Morton, Cr. Ægyp. pp. 5, 7, pl. i.
375 Wood-cut, fig. 200—Martin, Man and Monkeys, p. 298, "Bushman."
376 Wood-cuts, figs. 201, 202—Wilson's Archæology—vide infra, pp. 369–70.
377 Hamilton Smith, Natural History of the Human Species, Edinb. ed., 1848, p. 93.
378 Trans. Am. Ethnol. Soc., New York, i. p. 192.
379 Rev. Dr. John Bachman, of Charleston, S. C., in a book on the Unity of the Races, did raise a question as to the American origin of maize, but Humboldt, Parmentier, Linnæus, and the best botanists are against him.
380 Gallatin, Notes, op. cit., p. 57.
381 Chronologie der Ægypter, i. pp. 131–3.
382 Pauthier, Chine, p. 180.
383 Gallatin, p. 58.
384 Vetruvius, lib. vi., cap. 1.
385 Kaimes, Sketches of the History of Man, 2d ed., Edinb., 1778; i. pp. 50, 75–7.
386 Layard, 2d Exped. Babylon, pp. 531–2.
387 Morton was here somewhat misled by a hastily written passage in my Otia. (Burke's Ethnol. Journal, p. 310.)—G. R. G.
388 This is by far too high a date for "castes"—see further on, pp. 635–6.
389 Also, and more probably, Petubastes; but the hieroglyphics reveal nothing for or against either supposition.—G. R. G.
390 They came from the old Jewish burial-

No. (*of Notes, &c.*)

ground, behind Muss'r-el-Ateeka, on the desert toward Bussateen: and no Muslim is interred near a Jew.—G.R.G.

391 Travels in Kordofan, London, 1844.

392 Proceed. Acad. Nat. Sciences, Philada., September, 1850, p. 82.

393 Canidæ, i. p. 104.

394 Want of space alone prevents the apposite citation of the corroborative statements of M. Hombron, "De l'Homme dans ses rapports avec la Création;" Voyage au Pole Sud; Zoologie, i. pp. 80–92, 110–7.

395 This is what the Halicarnassian states—"I am surprised (for my narrative has from the commencement sought for digressions), that in the whole territory of Elis no *mules* are able to breed, though neither is the climate cold, nor is there any other visible cause. The Eleans themselves say, that *mules* do not breed with them in consequence of a curse; therefore, when the mares' breeding approaches, they lead them to the neighboring districts, and there put the he-asses with them until they are in foal; then they drive them home again." (Melpomene, iv. 30—"A new and Literal Version, from the Text of Baehr"—by Henry Cary, M. A., Oxford—London, 1849, p. 247.)

396 Columella, p. 135.

397 Ham. Smith—Nat. Hist. of the Equidæ, p. 154.

398 Leidy; in Proceed. Acad. Nat. Sciences, Phila., Sept., 1847.

399 Equidæ, p. 183.

400 Ibid., p. 120.

401 Morton's posthumous papers.

402 Ibid.—Replies to the Rev. J. Bachman, &c., 1850–51.

403 Buffon, Quadrupèdes, xxii. p. 400; xxx. p. 230.

404 Chevreul, in Journal des Savans, Juin, 1846; p. 357. It was my good fortune to have marked, for Dr. Morton, that passage in Chevreul's skilful paper which Dr. Bachman so queerly ascribed to "old and musty" authorities.—G. R. G.

405 Karl Ritter's Geography of Asia; viii. Division 1st.—pp. 655, 659. Compare Frazer, Mesopotamia and Assyria, pp. 366–7; for "Turkoman Camel."

406 Canidæ, p. 19.

407 Sonnini's Buffon, Quad. xxxiii. p. 321, supp.

408 Pennant's Arctic Zoology, i. p. 42.

409 Fauna Boreale-Americana, Mamm., p. 61.

410 First Voyage, Supp., p. 186.

411 Fauna, p. 65.

412 Idem, pp. 74, 79.

413 American Edition, p. 365.

414 Martin, Nat. Hist. of the Dog, p. 30.

415 Hamilton Smith, Canidæ, ii. p. 123.

416 Nat. Hist. of Paraguay, p. 151.

417 Rural Sports, p. 16.

418 Lyell, Principles, ch. 38.

419 Wood-cut, fig. 235—Champollion, Grammaire, pp. 51, 173; Dictionnaire, pp. 117, 127:—Bunsen, Egypt's Place, i. p. 514, figs. 248, 249:—Wilkinson M. and C., iii. p. 32:—Lepsius, Denkmäler, IVth, Vth, and VIth, dynasty, passim.

420 Wood-cut, fig. 237—Denkmäler, Abth. ii. Bl. 9.

421 Wood-cut, fig. 238—Denkmäler, Abth ii. Bl. 96.

422 Wood-cut, fig. 239—Denkmäler, Abth. ii. Bl. 11:—See varieties in Cailleaud, Arts et Métiers des Anc, Eg., pl. 37.

423 Wood-cut, fig. 240—Denkmäler, Abth. ii. Bl. 20.

424 Wood-cut, fig. 241—Rosellini, M. C., xvii., fig. 3.

425 Wood-cut, fig. 242—Martin, Nat. Hist. of the Dog, p. 138.

426 Oriental Album, pl. 41.

427 Martin, op. cit., p. 53.

428 Wood-cut, fig. 243—Ibid., p. 50:—Denkmäler, Abth. ii. Bl. 132.

439 Wood-cut, fig. 244—Denkmäler, Abth. ii. Bl. 131.

430 Wood-cut, fig. 245—Rosellini, M. C., No. 5.

431 Wood-cut, fig. 246—Wilkinson, M. and C. iii. p. 13.

432 Wood-cut, fig. 247—Ibid., op. cit., p. 32.

433 Hoskins, Ethiopia, Plate i., line 3.

434 Bennett, Tower Menagerie, p. 83.

435 Wood-cut, fig. 248—Wilkinson, M. and C. iii. p. 12:—Lepsius, Denkmäler, ii. 131.

436 Wood-cut, fig. 249—Denkmäler, ii. 134.

437 The head resembles the skulls of Egyptian mummied-dogs now in the Academy, Philadelphia.

438 Wood-cut, fig. 250—Denkmäler, ii. 96.

439, and 440 Wood-cut, fig. 251—Layard, Babylon, p. 526:—Vaux, Nineveh, p. 198; discovered by Rawlinson. "Ctesias (says Photius in his Excerpta), in his description of India, speaks of the gigantic dogs of that country."—Indica, cap. 5; apud Heeren, Hist. Res.; London, 1846; i. p. 35.

441 Morton, Additional Observations on Hybridity, Oct., 1850, p. 26.

442 Lepsius, Denkmäler, Abth. ii. Bl. 131, and Passalacqua, Catalogue, 1826, pp. 231–3.

443 Zoologie, ii. p. 79:—Another, not less curious, arrived too late for us to use in our studies; viz: Courtet de l'Isle, "Tableau Ethnographique du Genre Humain," Paris, 1849. We shall revert to it elsewhere.

444 October, 1849:—Amer. Jour. of Med. Sciences, Jan., 1850.

445 Thoughts on the Original Unity of the Human Races, New York, 1830.

446 Zoologie, ii. p. 109.

447 Op. cit., p. 107.

448 Lyell, Principles, chap. xxxvii.

449 South. Quar. Rev., Charleston, S. C., Jan., 1846.

450 Second Visit to the United States, i. p. 105.

451 Hist. of Napoleon Buonaparte.

452 Notes to Azara's Quadrupeds, i. p. 24.

453 Amer. ed., No. ccciv, July, 1853. p. 55.

454 Genesis v. 4.

455 Études sur l'Algérie, p. 148.

456 Cahen's Hebrew Text, i. p. 8: Genesis ii. 20.

457 Layard, Babylon, p. 623.

458 Pauthier, Chine, p. 24:—Livres Sacrés de l'Orient, "Temps antérieures au Chou-king," p. 33.

459 De la Domestication du Llama et de la Vigogne; "Projêt d'une Ménagerie Nationale d'Acclimatation," 1848.

No. (of Notes, &c.)

460 The Black Man, "Comparative Anatomy and Psychology of the African Negro"—transl. Friedlander and Tomes, New York, 1853, pp. 11-12.
461 Crania Ægyptiaca, 1844. p. 1.
462 Observations on a Second Series of Ancient Egyptian Crania; Proceed. Acad. Nat. Sc., Phila., Oct. 1844, pp. 8-10.
463 Catalogue of Skulls, 3d ed., 1849: to which ought to be added those crania presented to him in 1851 by Mr. Gliddon; and, in 1851-2, the two shipments received from Mr. A. C. Harris of Alexandria, Egypt.
464 Cr. Ægyp., p. 3.
465 Gliddon's Otia, pp. 74-5, 80.
466 Ægyptens Stelle in der Weltgeschichte, ii. pp. 166-70.
467 Crania Ægyp., p. 19.
468 Observations, &c. Proceed. Acad. Nat. Sciences, Phila., Oct. 1844:—Lepsius, Briefe, p. 33.
469 Crania Ægypt., p. 20.
470 Exodus xii. 38; Cahen's Hebrew Text, ii. p. 50.
471 Champollion, L'Égypte sous les Pharaons, 1814, ii. p. 5. seq.: and Quatremère, Recherches sur la Langue et la Littérature des Coptes.
472 Abeken, Rapport à la Société Égyptienne du Kaire; in Bulletin de la Soc. de Géog., Paris, Sept., 1845; pp. 171-2.
473 Lepsius, Auswahl, pl. xx.; as well as in Briefe, pp. 105-6.
474 Cr. Ægyp., pl. ii. fig. 1.
475 Cr. Ægyp., pl. ii. fig. 2.
476 Cr. Ægyp., pl. ii. fig. 3.
477 Cr. Ægyp., pl. x. fig. 8.
478 Cr. Ægyp., pl. viii. fig. 1.
479 Cr. Ægyp., pl. xi. fig. 1
480 Cr. Ægyp., pl. x. fig. 1.
481 Cr. Ægyp., pl. x. fig. 4.
482 Cr. Ægyp., pl. x. fig. 5. Note to Woodcuts, figs. 263, 264; "Ancient Assyrian" (supra, pp. 426-7). After my remarks were stereotyped, I had the pleasure to receive another letter from Mr. J. B. Davis (dated, Shelton, Nov. 15, 1853), which affords the following, among other particulars, corroborative of the authenticity of this cranium:— * * "The skull is the veritable skull of an ancient Assyrian. It was found with the fragments of others, and a great many other bones and armor, in a chamber of the North-west palace at Nimroud, to which there was an entrance but no exit. This is marked in Mr. Layard's Nineveh, Vol. I., p. 62; Plan III., Chamber I. It was supposed to be the one to which the defenders of the palace had retreated. * * * * * The skull is undoubtedly allied to Morton's Pelasgic group, but, yet, I think possesses a distinct character which at once strikes my eye, as belonging to the people of the sculptures. The full, rounded, equable form like the ancient Greek, only decidedly larger and fuller, is striking."—J. C. N.
483 Égypte Ancienne, pl. 2. p. 261.
484 Gliddon, Appeal to the Antiquaries of Europe on the destruction of the Monuments of Egypt, 1841; pp. 125-129.

No. (of Notes, &c.)

485 Proceed. Acad. Nat. Sciences, Philadel., Dec. 24, 1850. On the "leathern straps," cf. Birch in Gliddon's Otia, p. 85; and Osburn's paper on the Leed's Mummy, 1828, pp. 4, 33-4, pl. ii.
486 Promenade en Amérique, Revue des Deux Mondes, Juin, 1853.
487 Martin, Man and Monks., p. 298, fig. 233.
488 Op. cit., p. 298.
489 Prichard, Phys. Hist. i. p. 297.
490 Ibid., op. cit. p. 290. "Fulah" means "white:" Cf. Beecham, Ashantee, or the Gold Coast; p. 161, note.
491 Ibid., op. cit.; and Latham, Varieties of Man, p. 6.
492 Morton, Cr. Æg., pl. xii. fig. 7.
493 Virey, Histoire Naturelle du Genre Humain, i, p. 240; pl. 2: drawn in colors, on a folio scale, by Geoffroy and Cuvier, Mammiferès, 1829: i. pl. 1 and 2; and described in pp. 1-7.
494 Morton, Cr. Æg., p. 16.
495 Prichard, Researches, v. p. 3. Thus amply confirmed by Crawfurd—"There are 15 varieties of Oriental Negroes. * * * * There is no evidence, therefore, to justify the conclusion that the Oriental Negro, wherever found, is of one and the same race." (Edin. New Philos. Jour., 1853. p. 78.—"Negroes of the Ind. Archip.")
496 Churchill's Collection of Voyages, i.; "History of Navigation, supposed to have been written by the celebrated Locke." This information may be relied on, as it was furnished me by Dr. Charles Pickering.—G. R. G.
497 Anthropologie, p. 348.
498 Op. cit.; from "Voyage de l'Uranie."
499 Morton, Catalogue, 1849, No. 1327.
500 Prichard, Researches, i. p. 298, fig. 7.
501 Dumoutier, Atlas, pl. 35, fig. 6.
502 Ibid., pl. 37, fig. 2.
503 Martin, Man and Monkeys, p. 310, fig. 227.
504 Dumoutier, Atlas, pl. 36, fig. 4—"Van Diemen."
505 Prichard, Researches, i. p. 297, fig. 6.
506 Dumoutier, Atlas, pl. 36, fig. 2—"Van Diemen."
507 Op. cit., pl. 34.
508 Martin, Man and Monkeys, p. 312, fig. 229. There is nothing herein stated about the almost inconceivable animality of Papuans, Ahetas (Ajetas) or Negritos, Arruans, Al Foers, which the reader cannot find in a new work—"Ethnographical Library, Conducted by Edwin Norris, Esq., Vol. I. The Native Races of the Indian Archipelago, by George Windsor Earl," London, 1853.
509 Observations faites pendant le 2me voyage de Cook, p. 208.
510 Mœrenhout, ———, ii. p. 248; cited by D'Eichthal, "Races Océaniennes et Américaines," 1845.
511 Polynesian Researches, ii. p. 13.
512 Dumoutier, pl. 26, fig. 6—"Cavernes sepulchrales - Teneriffe."
513 Ibid., pl. 29, fig. 4—"Marquesas."
514 Ibid., pl. 30, fig. 4—"Caverne ossuaire—Taiti."
515 Ibid., pl. 31, fig. 4—"Sepultures abandonnées—Isle Vavao."

No. (of Notes, &c.)

516 Martin, Man and Monkeys, p. 310.
517 Dumoutier, pl. 32, fig. 2—"Isle Mawi."
518 Philadelphia, 2d ed., 1844; pp. 4, 5.
519 Mr. Strain's letter to Dr. Morton, "Rio Janeiro, 7th Decem., 1843"—Proceed. Acad. Nat. Sciences., Phila., Dec., 1844.
520 Putnam's American edition, New York, 1853, p. 36.
521 Ethnography and Archæology; American Journ. of Science and Art, ii. 2d series; New Haven, 1846; tirage à part, pp. 67, 117–9.
522 Crania Americana, p. 145.
523 Rivero and Tschudi (pp. 39–40) doubt the possession by Dr. Morton of crania of the royal Inca family: but the note of the translator (p. 41) may be passed over as inconsequent.
524 The Creole Negro; supra, No. 491.
525 Cr. Americana, p. 130; pl. xi. C.
526 Op. cit., p. 131; xi. D.
527 Peruvian Antiquities, pp. 39–40.
528 Cr. Americana, p. 152; pl. xvi.
529 Op. cit., p. 155; pl. xviii.
530 Op. cit., p. 166; pl. xxii.
531 Op. cit., p. 198; pl. xxxix.
532 Op. cit., p. 220; pl. lii.
533 Op. cit., p. 224; pl. lv.
534 Op. cit., p. 259.
535 Op. cit., p. 257.
536 Anthropologie, pp. 229–30, 232.
537 Martin, Man and Monkeys, p. 273.
538 Ibid., p. 273.
539 Chine, d'après les documents Chinois, p.1.
540 Wood-cut, fig. 329—Paravey, Documents, &c., sur le Deluge de Noé, Paris, 1838, pp. 11, 56:—Pauthier, Chou-king, Part II., chap. i. p. 62; Part IV., chap. xxvii., p. 131:—Ibid., Chine, pp. 55–7.
541 Pauthier, Chine, pl. 22; pp. 120–1.
542 Ibid., pl. 51, fig. 4; pp. 246–8.
543 Ibid., pl. 12; pp. 57–8.
544 Ibid., pp. 472–4.
545 Révolutions des Peuples de l'Asie Moyenne, Paris, 1839; ii. p. 432.
546 Catalogue, 3d ed., 1849; Intro., pp. 1–2.
547 Nat. Hist. of Human Species; Edinb., 1848, p. 157.
548 Bremer, Homes of the New World, Am. ed., 1853, ii. pp. 162–3. [Note, 24 Jan., 1854. Let me confirm my colleague's accuracy by two additional extracts—1st, as regards crosses between American Indians and white men. All readers are aware with what gusto a superior civilization has been attributed to the Mandans; and how sundry instances of fair complexion, light hair and blue eyes, among individuals of that tribe, have also led to surmises that they might even be of Welsh descent! Major John Le Conte pointed out to me a solution in the fact that Lewis and Clark wintered among them with a party of 43 able-bodied men. As a specimen, read the following account of one orgie, on Saturday night, Jan. 5, 1805—"Unus nostrum sodalium multum alacrior et potentior juventute, hac nocte honorem quatuor maritorum custodivit." (Lewis and Clark, Travels to the source of the Missouri river; 1804–6; London ed., 1814; ch. vi., pp. 109–111.)—2d, As respects crosses between Negroes, Indians, and white persons, on the Panama Isthmus; a passage which was indicated to me by Mr. Conrad:—

"The character of the half-castes is, if possible, worse than that of the Negroes. These people have all the vices and none of the virtues of their parents. They are weak in body, and are more liable to disease than either the whites or other races. It seems that as long as pure blood is added to the half-castes proper, when they intermarry only with their own colour, they have many children, but these do not live to grow up; while in families of unmixed blood the offspring are fewer, but of longer lives. As the physical circumstances under which both are placed are the same, there must really be a specific distinction between the races, and their intermixture be considered as an infringement of the law of Nature."—Berthold Seemann, F. L.S.—Narrative of the Voyage of H. M. S. Herald, 1845–51: London, 1853, I., p. 302.—G. R. G.]
549 Martin, Man and Monkeys, p. 210, fig. 180.
550 Ibid.—fig. 181.
551 Ibid.—fig. 182.
552 Savage and Wyman, Troglodytes Gorillæ; Boston, Jour. of Nat. Hist., 1847, p. 27.
553 Martin, op. cit., p. 228.
554 Ibid., p. 280.
555 Ibid., p. 384.
556 Ibid., p. 223.
557 Prichard, Researches, i. p. 290, fig. 3.
558 Martin, op. cit., p. 367.
559 Virey, Hist. Nat., ii. p. 42.
560 Martin, op. cit., p. 254.
561–562 Wood-cuts, figs. 346, 348—Illustrated London News, 1851—"drawn by an English officer at the Cape."
563 Amaryllidaceæ, pp. 338, 339.
564 Races of Men, p. 12.
565 American Jour. of Science and Art, Vol. xxxviii., No. 2.
566 Anatomie comparée, tome ii.

PART II.

567 Geographiæ Sacræ Pars prior; Cadomi, fol., 1651—(Loganian Library, Phila.)
568 Spicilegium Geographiæ Hebræor. exteræ, post Bochart., vol. ii., 1769–80.
569 Gliddon, Otia, London, 1849, pp. 16, 124.
570 Rev. Dr. Eadie, Early Oriental History—Encyclopædia Metropolitana, London, 1852, p. 2.
571 Rev. Dr. Hales, Analysis of Chronology; 2d ed., 1830; Preface, p. 21, and i. p. 352.
572 Pauthier, Livres Sacrés de l'Orient, Paris, Intro., p. 1.
573 Cahen, La Bible, Traduction Nouvelle, Paris, 1831; i. pp. 26–8.
574 Avec un Atlas géographique, pittoresque, archéologique, géologique, &c.—"Ouvrage qui a remporté le prix de la Société de Géographie de Paris, en 1838;" Paris, 6 vol. Text, 8vo., 1839–43.

No. (of Notes, &c.)

575 Bulletins de l'Académie royale de Bruxelles, vi.; and Notions élémentaires de Statistique, Paris, 1840.

576 Voyage dans les steps d'Astrakhan et du Caucase; and Histoire Primitive des Peuples qui ont habité anciennement ces contrées.

577 GoMeR. Bochart, pp. 194–6. — Homer, Odyss. xi. 14.—Diodor., v. 32.—Herod., iv. 100.—Josephus, Antiq. i. 6.—Rawlinson, Commentary, 1850, p. 68.—Dubois; i. 61, iv. 321, 327, 350, 391; v. 22, 35, 44.

578 MaGUG. Bochart, pp. 212–19. — Rev. Moses Stuart, Interpretation of Prophecy, Andover, 1842, p. 123. — De Wette, transl., Parker, i. p. 95–7, &c. — Kur'àn, Ch. xviii., v. 93, 96; xxi. 95, &c. — Pauthier. Liv. Sac. de l'Orient, p. 495: Lane, Selections, p. 140.—Barthelemy, Anciennes Religions des Gaules; Rev. Archéol., 1851, p. 338, note.—Dubois, iv. 321, 345; 363–407.—Josephus, Ant., i. 6. — Hieronymus, Comm. in Ezek. xxxviii, 2. — Lenormant, Cours d'Hist. Ancien., Paris, 1837, p. 289.—Emelin, 1774, and Porter (Travels, ii. 520), 1819 — "wall of Gog and Magog at Derbend." — Anthon, Classic. Dict., 1843; voce "Asi," p. 218. "Scythic" is here used in the sense proposed by Rawlinson (Commentary, pp. 68, 75: and Cuneiform Inscriptions, 1847, pp. 20, 34–7,) and adopted by Norris, (Memoir on the Scythic Version of the Behistun inscription; Jour. R. Asiat. Soc., 1853; xv., Part 1, p. 2. — Sir W. Jones, 6th Discourse, on Persians; Asiatic Researches, 1799, ii. p. 64. — Gliddon, Otia, p. 124. — Westergaard, Median Species of Arrowheaded writing: Antiq. du Nord, 1844; pp. 273–8, 289.—Hincks, Persepolitan Writing, 1846, p. 18. — D'Omalius d'Halloy, Races Humaines, ou éléments d'ethnographie, 1845, "Ossètes," p. 79.

579 MeDI. Bochart, pp. 219–25.—Herod., vii. —De Saulcy, Recherches sur l'Écriture cunéiforme Assyrienne; Paris, 1848, p. 26. — Layard, Babylon, p. 628. — De Longpérier, Lettre à M. Löwenstern; Rev. Archéol., 1847, p. 505. — Rawlinson, Tablet of Behistun.—Birch, Tablet of Karnac, pp. 14–5. — Dubois, iv. 321, 339.

580 IUN. Bochart, pp. 174–6.—Aristophanes, In Acharnum; Act i., scene 3.—Homer., Iliad, xiii, 685.—Pausanias, Achaic., p. 397. — Herodotus, viii. 44. — Rosetta Stone, in Lepsius's Auswahl; or in Birch's Gallery, pp. 114–17, pl. 49: — also, Lenormant, Essai sur le Texte Grec, 1840; pp. 10, 11; lines No. 54; and p. 45.—Hincks (True date of the Rosetta Stone, Dublin, 1842, pp. 6, 8,) claims "March, 197, B. C.," as date of this decree; but a Letronne would first have determined the year of "C.:" vide infra, pp. 665–7. — Champollion, Grammaire Égyptienne, pp. 151, 175; Dict., p. 66. — "Ouinin," in conquests of Seti-Meneptha, and of Ramses II.—De Saulcy, Recherches, p. 26; Inscriptions trouvées à Khorsabad, Rev. Archéol., 1850, pp. 769–72. — Rawlinson, Behistùn, pp. 1, xxvii.—Layard, Babylon, p. 628. Pauthier's Manou, lib. x., v. 44. —Wilford, Asiatic Researches, 1799; iii. p. 358.—Sykes, Jour. R. Asiat. Soc., 1841., vol. vi.; Art. xiv. pp. 434–6. — "J. P. S." (in Kitto, Biblical Encyclopædia, ii., p. 393–400) omits any explanation of Tubal, Meshech, and Tiras, in his "sons of Japheth" (p. 397)! There are numerous similar oversights in Kitto, no less than in Robinson's Calmet. — Dubois, iv. 321. 334.

581 T*t*uBaL. Bochart, pp. 204–13. — Munk, Palestine, p. 420. — De Wette, ii. 366. seq. — Strabo, ii. 129. — Herod., vii. 78. Rawlinson, Commentary, pp. 63–4. — Layard, Babylon, p. 628. — Dubois, iv. 321, 388.

582 MeS*h*eK. Bochart, pp. 204–13.—Herod., iii., 94; vii. 78. — Rawlinson, Commentary, pp. 63–4.—Birch, Stat. Tablet of Karnac, pp. 14–5. — Hincks, Report of Syro-Ægyptian Soc., 1846.—Dubois, ii. 17; iv. 321, 336, 347.

583 T*t*IRaS. Bochart, p. 172–3. For hieroglyphical mention of "Thraces," in Egyptian conquests, see Champollion (Lettres) and Rosellini (MS., iv. 288): for classical, the "Inscrip. of Adulis" —Champollion-Figeac, Eg. Anc., p. 67. —Dubois, iv. 321, 324.

584 AS*h*KeNaZ. Bochart, pp. 196–8.—Pliny, iv. 24.—Kitto, ii. p. 397.—Rawlinson, Commentary, p. 46; "Nimroud Obelisk." — Ibid., London Lit. Gazette, Aug., 1851.—Dubois, iv. 321, 330, 391.

585 RIPaT*t*. Bochart, pp. 198–9. — Strabo, vii. 341. — Pliny, iv. 24. — Dubois, iv. 321, 330.

586 T*t*oGaRMaH. Bochart, pp. 200–4. — Moses Choren., Hist. of Arm., p. 24.— St. Martin, Mémoires sur l'Arménie, 1818; i. pp. 205, 271–8.—Strabo, xii. — Josephus, Ant., i. 1, 6. — Löwenstern, Lettre à M. de Saulcy, Rev. Archéol., 1849, p. 494. — Dubois, ii., p. 9; iv. pp. 332–3.—Jardot, Révolutions, ii. p. 6.

587 ALISaH. Bochart, pp. 176–8.—Homer, Il., ii. 617. — Grote, Hist. of Greece, i. p. 487.—Herod, i. § 146, &c.

588 Wood-cut, fig. 355—Layard, folio Monuments; and Babylon, pp. 343, 350.—De Longpérier, Rev. Archéol., 1844, pp. 224–5; 1847, p. 297. — Stuart, Crit. Hist. and Def., pp. 113, 114, 120. — De Wette, ii. pp. 452–6.—Cahen, Notes on Jonah, vol. xii.—"Berosiana," in Bunsen's Eg. Pl., i. pp. 704–19. — Munk, Palestine, pp. 451–2. — On "Sibylline verses" see Letronne, Examen Archéologique, Croix Ansée, 1846, pp. 33–4.

589 T*t*aRSIS. Acts, xxii. 3. — Lanci, Paralipomeni, i. pp. 150–5. — Gesenius, in Parker's De Wette, i. p. 455, note. — Munk, Pal., p. 29. — Gliddon, Otia, p. 50. — Pickering, Races, p. 373. — Pauthier, Sinico-Ægyptiaca, p. 10. — Bochart, pp. 188–94. — London Lit. Gaz., May, 1852.

590 KiT*t*IM. Bochart, pp. 178–83. — Birch, Ivory ornaments found at Nimroud, pp. 174–5; and Annals of Thotmes III., pp.

No. (of Notes, &c.)

157–60. — Boeckh, Corpus Inscrip. Græc., i. p. 523.—Ptolemy, lib. v. 14.—Josephus, Antiq., i. 6. 1.—Rev. Archéol., 1846, pp. 114–15; and 1847, p. 448.

591 DoDaNIM. Bochart, pp. 183–8. —Wiseman, Connection between Sci. and Rev. Rel., 1836; ii. pp. 168–9.—Champollion-Figeac, Dissert. s. l'Étymologie, p. 8. —Herod., ii., § 52.

592 Wood-cut, fig. 356.—Champollion, Grammaire, pp. 150, 151, 195, 407; Dictionnaire, p. 409.—Hincks, Hierog. Alphabet, p. 16; pl. i., figs. 23, 26, 27.

593 Letronne, Opinions cosmographiques des Pères de l'Église; Rev. des deux Mondes, 1837, pp. 601–33: and Récueil des Inscrip., ii. p. 37, seq. — Raoul-Rochette, Archéologie comparée, 1848; Part ii. p. 190, seq.—Lenormant, Cours d'Hist. Anc., p. 228.

594 KUS*h*. Bochart, p. 238, and 241.—Martin, Études sur le Timée de Platon, Paris, 1841; "Atlantide," i. p. 332.—Walton, Bibl. Polygl.; Proleg., xv. pp. 97–9.—De Wette, i. pp. 228–31.—Wells, Hist. Geog. of O. and N. Test., 1804, pp. 103–105. — Lanci, Paralip., ii. p. 45. Nott, Bibl. and Phys. Hist., p. 143.—Forster, Geog. of Arabia, 1844, i. pp. 26–7, 28, 29. — Burckhardt, Travels in Arab., ii. p. 385.—Rosellini, Monumenti Civili, ii. pp, 394–403.—Gliddon, Otia, p. 133.—Forster, op. cit., i. 14–6.—Letronne, Mém. et Docum., Rev. Archéol., 1849, p. 85. — Cahen, Bible, v.; avant propos, p. 13. — Quatremère, Recher., Coptes.—De Wette, i. pp. 202–6.—Peyron, Coptic Lexicon, voce Ethosh.—Parthey, Vocabularium Copticum, p. 549. Wilkinson, Topog. of Thebes, p. 487; Mod. Eg. and Theb., ii. p. 317.—Birch, Stat. Tabl. Karnac, p. 47. — Anthon, Class. Dict.; and Syst. of Anc. Geog.; voce "Asia."—Rémusat, in Pauthier's Chine, p. 259.—Kitto, Bibl. Cyclop., i. p. 238.

595 Volney, Recherches Nouvelles, Paris, 1822, iv. — Lenormant, Cours d'Hist. Anc., 1838, pp. 24, 129. — Jomard, Arabie; in Mengin, 1839, iii. p. 327–9, and passim. — Fresnel, "Histoire des Arabes avant l'Islanisme," in Jour. Asiat., "4me Lettre" Djeddah, Jan., 1838. — Sale's Introd. to the Kur'àn, Liv. Sac. d'Or., p. 467. — Lane, Selections, p. 17. — Forster, Geog., i. p. 20. — Gesenius, in De Wette, i. pp. 433–4. — Hyde, Hist. rel. veter. Persarum, p. 37.—Kitto, "Cush," i. p. 503.—Assemani, Bibliotheca Orientalis, iii., part 2, p. 568, seq. — Turner, "Himyarite Inscriptions," Trans. Amer. Ethnol. Soc., New York, 1845, art. iv.—Fresnel, Recherches sur les Inscrip. Himyariques, 1845; Jour. Asiatique, No. 11; also, Lettres, Feb., March, April, May, 1845. — Gesenius, Geschichte der Heb. Sprache und Schrift, 1815. — Forster, Geog. of Arabia, i. pp. 24–76, 94–102.

596 Syncellii "Chronographeion," p. 51.—Letronne, in Biot's Recherches sur l'Année vague les Égyptiens, 1831, pp. 25–7.—Biot, Mémoire sur divers points de l'Astron. Anc., 1846, p. 37.—Matter, Hist. de l'École d'Alexandrie, 1844; ii. pp. 190–1.—Barucchi, Discorsi Critici, Torino, 1844; pp. 14, 15. — Böckh, Manetho und die Hundstern-periode, Berlin, 1848; p. 40. — Bunsen, Ægyptens Stelle, 1845; i. pp. 256–63. — Raoul-Rochette, Jour. des Savans, 1846; pp. 141, 241–2. — Lepsius, Chron. der Ægypter, i. p. 446. — Kenrick, Egypt under the Pharaohs, 1851. — Maury, in Rev. Archéol., Juin, 1851; pp. 160–3.

597 MiT*s*RIM. Grotefend's "Analyse de Sanconiathon," trad. Lebas, Paris, 1839; Introduction, pp. 79–85.—Champollion, L'Égypte sous les Pharaons, 1814; i. Chap. 2. — Parthey, Vocab. Copt., pp. 511–2.—Rawlinson, Behistun, 1846, pp. 1, 27.—Commentary, 1850, pp. 60–7.—De Saulcy, Rev. Archéol., 1850, pp. 768–9, 771; pl. 133, No. 19; and Recherches, Inscrip. de Van, 1848, p. 27. Nash, on the term Copt, and the name of Egypt; Burke's Ethnol. Jour., No. 11, 1849, p. 496. — Hincks, Hierog. Alph.; p. 28, pl. i. fig. 78. — Gliddon, Chapters, p. 41.—Rosellini, Mon. Stor., i. p. 58. — Portal, Symboles des Égyptiens, pp. 51, 73. — Lanci, Lettre à M. Prisse, 1847, pp. 99–103. — Lenormant, Cours, p. 233. — Birch, "Merter," in Annals of Thotmes III., p. 138; Eg. Inscrip. in Bibliothèque Nat., p. 12; also, on "Kam, the black country," as found in the Ritual, in Chæremon on Hieroglyphics, p. 11.—Bochart, p. 292.

598 P*h*UT. Bochart, pp. 333–9. — Gliddon, Otia, p. 127.—D'Eichthal, Foulahs, pp. 1, 8, 150. — Jerome, Commentary on Isaiah, lxvi. 19.—Ptolemy, lib. iii. 1.-Pliny, Hist. Nat., v.—Josephus, Antiq., i. 6, 2. — Gräberg de Hemso, Specchio, p. 291, seq. — Cervantes de Marmol, Descripcion general dé Africa, Grenada, 1573; i. fol. 31, seq. — Champollion, Dict., pp. 339–40. — D'Avezac, Afrique Anc., p. 31. — Lenormant, Cours, pp. 233–6. — Hengstenberg, Eg. and Books of Moses; transl. Robbins, p 211.—De Saulcy, Rev. Archéol., 1850, pp. 769, 772.—Birch, Eg. Inscrip., p. 13.

599 KNA*ă*N. Cahen, Genèse, i. p. 25. — Procopius, De bello Vandalico, ii. cap. 20.—St. Augustin, Expos. Epist. Rom.; cited in De Wette, i. p. 431. — Lanci, Bassorilievo Fenicio di Carpentrasso; Roma, 1824, p. 126. — Munk, Inscrip. Phœnicienne de Marseilles; Journal Asiat., 1847, pp. 473, 483, 526; and Palestine, pp. 87–8, 192. — Gesenius, Geschichte der Heb. Sprache, 1815, pp. 8, 9.—De Saulcy, Mém. sur une Inscrip. Phœnicienne, 1847, passim.—Josephus, Cont. Apion., i, 22. — Kitto, i. p. 823, "Hebrew Language."—Eusebius, Præpar. Evang., i. cap. 10. — Lenormant, Cours. p. 236.—Bochart, pp. 339–42.

600 ScBA. Volney, Recherches, iv. p. 232. —Josephus, Antiq. viii. 6. 5.—Ludolph. Hist. Æthiopica, ii. cap. 3. — Forster, Geog., i. p. 157, seq. — Wathen, Arts, Antiq. and Chron. of Egypt, 1842, pp. 69–70. — Hoskins, Ethiopia, p. 339 [not directly, I find, but inferentially. — G. R. G.]. — Fresnel, 4me Lettre, Jan.,

No. (of Notes, &c.)

1838, pp. 71–7; and Inscriptions Himyariques, pp. 34, 67–9. — Pauthier, Chine, pp. 94–100, notes.—D'Herbelôt, Bibliothèque Orientale, voce "Salomon," and "Thahamurath." — De Wette, ii. pp. 248–65.—Forster, Geog., i. pp. 33–8, and Maps. — Bochart, pp. 146–56.

601 K*h*UILaH. Bochart, pp. 161–3.—Forster, i. pp. 9, 38, 54.

602 SaBT*t*aH. Lenormant, Cours, pp. 237–8. — Strabo, xvi. p. 771, Fr. Transl. — Jomard, Arabie, pp. 373, 389–90. — Pliny, vi. 32. — Volney, iv. p. 232. — Fresnel, Inscrip. Himyar., pp. 51–2. — Forster, Geog., i. pp. 57–8. — Bochart, pp. 252–4.

603 RA*â*MaH. Volney, iv. p. 235.—Forster, i. pp. 59–76; ii. 223–7. — Fresnel, 4me et 5me Lettres, 1838.—Wellsted, Trav. in Arabia, 1838, ii. p. 430. — Burckhardt, Arabia, ii. p. 385. — Bochart, p. 247.

604 SaBT*t*eKA. References as above, No. 603.

605 S*s*eBA. Munk, Palestine, p. 438, on "Ezra." — De Wette, ii. pp. 47–8. — Forster, ii. pp. 323–4; and i. pp. 71–3. — Bochart, pp. 249–51.

606 DeDaN. Bochart, p. 248.—Forster, i. 38; and Maps. — Letronne, "Vénus Angérone," Mém. et Doc., Rev. Archéol., 1849, p. 277.—Glaire, Les Livres Saints vengés, Paris, 1845, passim. — Rev. Sidney Smith, Elementary Sketches of Moral Philos., New York ed., 1850; p. 254.—Strauss, Vie de Jesus, trad. Littré, Paris, 1839; Preface, p. 8.

607 NiMRoD. Vide W. W.'s profound articles "Scripture," and "Verse," in Kitto, ii. pp. 717, 910. — [For hallucinations on "Nimrod," see Anc. Univ. Hist., i. p. 275, seq.; Faber, Origin of Pagan Idolatry, and Bryant, Anc. Mythology, passim; Hales, Analysis of Chron., i. pp. 358–9, and ii.] "Nimrod, a Discourse on certain passages of History and Fable," London, 1829, printed for Richard Priestley. — Higgins, Anacalypsis, London, 1836, i. p. 6.—Wiseman, Lectures, i. p. 37.—Birch, Two Egypt. Cartouches, 1846, pp. 168–70.—Lepsius, Chron. der Ægyp., i. p. 223. — Bunsen, Ægyptens Stelle, iii. p. 133. — Sharpe, in Bonomi's Nineveh, 1852, pp. 69–78. —Rawlinson, Commentary, pp. 4, 6, 7, 22.—Layard, Babylon, pp. 33, 123.—De Saulcy, Dead Sea, ii. p. 544.—D'Herbelôt, voce "Nimrod;" and Ouseley, Oriental Collections, ii. p. 375. — Josephus, Antiq. i. 4, 21.

608–609 De Sola, Lindenthal, and Raphall, Scriptures in Heb. and English; London, 1846; p. 40, notes. — Glaire, Liv. Sts. vengés, i. pp. 313–20.—Rawlinson, Commentary, p. 14. — Lanci, Paralipomeni, ii. parte 8va. — Gesenius, in De Wette, i. p. 435.—Meyer, Hebraïsches Wurzel-Wörterbuch; cited by Bunsen, Disc. on Ethnol., 1847, p. 273.—D'Olivet, Langue Hebraïque restituée, 1815; pp. 281, 343. — Bochart, 256–60.

610 Gliddon, MS. "Remarks on the Introduction of Camels and Dromedaries, for Army-Transportation, Carriage of Mails, and Military Field-service, into the States and Territories lying south and west of the Mississippi, between the Atlantic and Pacific coasts — presented to the War-department, Washington, Oct. 1851." As I intend to publish an entire account of this affair for public edification ere long, it is sufficient now to determine the very recent introduction of the Arabian camel into Africa by quoting Humboldt (Aspects of Nature, p. 71); Ritter (Das Kameel, in Asien, viii. pp. 755–9); Procopius (Bello Vandalico, i. 8; ii. 11); Corippus (iv. 598–9); and Bodichon, Études sur l'Algérie, pp. 62–3.—G. R. G.

611 LUDIM. Bochart, pp. 299–310. — Gräberg de Hemso, Marocco, pp. 69, 246, 251, seq.—Castiglione, Recherches sur les Berbères Atlantiques, Milan, 1846; pp. 89, 100–1.—Lacroix, Numidie, p. 4. — D'Avezac, Afrique Anc., p. 28. — Yanoski, L'Afrique Byzantine, pp. 93, 99. — Ebn-Khaledoon, "Fee ahbâr el-Berber," 3d book; transl. Schulz, in Jour. Asiat., 1828; pp. 140–1.—Asiatic Miscellany, p. 148. — Marmol, op. cit., trad. Perrot, 1667, i. p. 68. — Leo Africanus (Hassan ebn Mohammed el Gharnàtee) Africæ Descriptione, 1556, p. 5. — Bertholet, Guanches, Mém. Soc. Ethnol., Paris, 1841; Part i., pp. 130–46. Agassiz, Diversity of Origin of Human Races; Christian Examiner, Boston, July, 1850, p. 16.—Dureau de la Malle, Carthage, pp. 1–3, 13. — Gibbon, Milman's, viii., pp. 227–8. — Bodichon, Études, pp. 32, 64, 103, 109. — Quatremère, 1st art. on Hitzig's Philistaer; Jour. des Savans, 1846, May; pp. 260, 266: — [That these views upon the "Ludìm" are new, the reader can perceive by opening Munk (Palestine, p. 432); Lenormant (Cours, p. 244); Cahen (Genèse i. pp. 27, 184); Kitto (Cyclop., pp. 397–8); and all English commentators.]

612 A*â*NaMIM. Forster, i. pp. 56–9. — De Saulcy, Dead Sea, 1853; i. p. 64; ii. p. 837.—Birch, Hieratic Canon of Turin, p. 6. — Anthon, Class. Dict., p. 872. — Bochart, p. 322.

613 LeHaBIM. Bochart, p. 316. — Anthon, Anc. and Mod. Geog., pp. 708, 749.— D'Avezac, Afrique, pp. 4, 28, 64–9. — Champollion, Ég. s. l. Phar., ii. p. 363. — Parthey, Vocab. Copt., pp. 497, 530. — Gliddon, Otia, p. 131.

614 NiP*h*aiaT*t*uK*h*IM. Bochart, pp. 317–21. Otia, pp. 9, 16, 133, 136.—Nott, Bibl. and Phys. Hist., pp. 144–5.— Champollion, op. cit., i. p. 55, ii. pp. 5, 31, 144 seq. — Parthey, pp. 110, 506, 530. — Herod., ii., § 18.—Champollion, Lettres, p. 124; and the hieroglyphics in Gram., pp. 169, 363, 406; Dict., pp. 339, 341. — Peyron, Gram. Ling. Copticæ, pp. 30, 36–8. — Hengstenberg, p. 211; and Gliddon, Chapters, p. 41.—Lenormant. Cours, pp. 235, 244–5.—Brugsch, Scriptura Ægyptiorum Demotica, p 25.—De Saulcy, Lettre à M. Guigniaut, p. 18.— Lepsius, Lettre à M. Rosellini, p. 66.

No. (of Notes, &c.)

Bunsen, Eg. Pl., i. pp. 285, 471.—Schulz's Ebn Khaledoon, p. 122.—Castiglione, Berbères, p. 104.—Quatremère, Mém. Geog. sur l'Égypte, i. p. 37; and in Jour. des Savans, 1846, p. 266.

615 P*h*eT*t*uRiSIM. Most of the above references here apply. These are special—Peyron, Papyr. Græc., Part ii. p. 27.—Parthey, pp. 56, 291, 500, 539.—Wilkinson, Mod. Eg. and Theb., ii. p. 137. D'Avezac, Afrique, p. 27.—Champ., Gram. pp. 98, 169, 327; Dict., p. 81.—De Hemso, p. 296, seq.—Lacroix, Numidie, p. 6.—Anthon, Anc. Geog., p. 749.—Quatremère, loc. cit., p. 266.

616 KS*h*iLuK*h*IM. Bochart, pp. 323-9.—De Sola, Genesis, p. 42.—Cahen, i. p. 27. Glaire and Franck's Bible, i. p. 50.—Munk, Palestine, pp. 82, 432.—Kitto, i. pp. 399, 388; ii. 398.—Hales, Analysis, i. p. 355.—Ritter, Vorhalle, p. 35, seq. —Morton, Cr. Æg., pp. 23-27, on "Herodotus."—Eadie, Early Orient. Hist.—Mignot, "3me Mém. sur les Phœniciens;" Acad. R. d'Inscrip., Paris, xxxiv. 1770, p. 146.—Marmol, 1ra parte, fol. 31.—Lepsius, Lettre, pp. 14, 18, 44: Pl. A, No. I, 12.—Birch, in Otia, p. 115.—De Longpérier, Rev. Archéol., 1850, p. 450.—Botta, Écrit. cunéiforme Assyr., pp. 6, 93, 192.—Rawlinson, Commentary, pp. 10-14.—De Hemso, p. 246.—Hitzig, Urgeschichte und Mythologie der Philistaer, 1845; reviewed by Quatremère, loc. cit, p. 266.—Koenig, apud Jomard, Récueil des Voyages, 1829; iv. p. 130, seq.—Hodgson, Sahara, pp. 33-5:—and, for "Oases," Wilkinson, Mod. Eg., ii. pp. 353-79.

617 P*h*iLiST*t*IM. Wilford, Asiat. Res.; iii. 1799, pp. 317-20, 322.—Hales, i. pp. 368, 380; after a disclaimer, p. 198.—[On "Col. Wilford," who is the cause of all those Hindostanic stupidities still current among English hagiographers, conf. Klaproth; in the Journal Asiat., Paris, xxv. p. 13, note; and Vans Kennedy, Hindu Mythology, London, 1831; Appendix A, pp. 406-22.] Champollion, Gram., p. 180.—Osburn, Testimony, pp. 137-41, 155.—Mignot, op. cit, p. 148, seq.—Quatremère (op. cit., pp. 258-69, 411-24, 497-510,) dispenses with more than reference to Kitto, ii. pp. 521-4.—Raoul-Rochette, Archéologie comparée, i. pp. 190-2, 373-4.—De Saulcy, Dead Sea, i. pp. 27-9, 55-6.

618 KaP*h*T*t*oRIM. Bochart, pp. 329-33.—Volney, iv. p. 229.—Quatremère, loc. cit.

619 T*s*IDoN. Bochart, p. 342.—Homer, Il. xxiii. 743; Odys., xv. 425.—Justin, lxviii. 3.—De Saulcy, Dead Sea, i. 52, 57-9.—Quatremère, on Mover's "Phönizier," op. cit., p. 503.—Gliddon, Otia, p. 136.—Eadie, Early Or. Hist., pp. 425-6.—Layard, Babylon, p. 627.

620 K*h*eT*t*. Bochart, p. 344-8, for this and the following names.—Lanci, Paralipomeni, i. pp. 13, 144.—Munk, Palestine, p. 78.—Birch, Archæologia, xxxv. 1853. —Layard, Babylon, pp. 142, 354, 633.

621 IBUSI. Osburn, Testimony, pp. 37-43, 123-5, 154.—Champollion, Lettres, pp. 76-7.—De Saulcy, Inscriptions de Van, p. 26.

622 AMoRI. On "Nephilìm," cf. the Paralipomeni.—Talmud, apud Rabbi Ben-Ouziel; Cahen, iv. p. 107, note.—Gliddon, Otia, p. 137.—Rosellini, Mon. Stor., iii. part 1, pp. 368-70; iv. pp. 94, 237-9.—Birch, Gallery, part i. p. 86.—Hincks, Hierog. Alph., p. 13; pl. i. fig. 17.—Osburn, Test., 65, 128-9, 154.—Birch, Stat. Tab. Kar., pp. 20-3.—De Saulcy, Dead Sea, i. p. 347.

623 GiRGaSI. Munk, Palestine, pp. 69, 79.

624 K*h*UI. Hieronymus, Epist. ad Dardanum, 129.—Kitto, Cyclop., voce "Hivite."—Vico, Scienza Nuova, transl. Paris, 1844, p. 288.

625 A*â*RKI. Vaux, Nineveh, pp. 459, 468, 478.—Gliddon, Otia, pp. 137-8.—Anthon, Class. Dict., pp. 1049-53.

626 SINI. Otia, p. 130.—Munk, p. 78.—Osburn's error of "Sinim" for SINKAR (Test., p. 158, No. 30), was corrected by Birch, Stat. Tab. Kar., p. 37.

627 ARUaDI. Osburn, pp. 52, 58, 69, 80, 118, 156.—Vaux, Nineveh, pp. 459, 468, 478.—Layard, Babylon, p. 627.

628 T*s*iMRI. Otia, p. 137.—Bochart, p. 347.

629 K*h*aMaT*t*I. Rawlinson, in Vaux, p. 462, seq.—De Saulcy, Rev. Archéol., 1850, pp. 767-8.—Layard, Babylon, p. 627.—Osburn, pp. 98, 101, 142, 155.—"Vico, et ses Œuvres," Introd., p. 1.

630 A*â*ILaM. Ainsworth, Assyria, &c., pp. 108, 196-216.—Rawlinson, March from Zohàb to Khusistàn, 1836; R. Geog. Soc., ix. p. 47.—Dubeux, Perse, pp. 1, 9, 13, 31.—Frazer, Mesopotamia, p. 22. —Polybius, v. 44.—Strabo, xvi. p. 744. —Layard, Khuzistàn; R. Geog. Soc., xvi. pp. 61-84.—Tychsen, De Cuneatis Inscrip., 1798, pp. 10, 13.—Ouseley, Travels, 1819, p. 325.—Löwenstern, Rémarques; Rev. Archéol., 1850, pp. 687-723.—De Saulcy, Inscrip. trouvées à Khorsabad; Rev. Archéol., 1850, pp. 767-70.—Layard, Babylon, pp. 212, 353, 628.

631-632 ASUR. De Sola, Genesis, note, p. 41.—De Longpérier, Rev. Archéol., 1850, pp. 429-32.—Rich's Narrative of a Journey to Nineveh; London, 1839; Introd., note, p. xvii.—The Friend of Moses, New York, 1852; pp. 181, 185, 200, 215-6, 220.—Rawlinson, Commentary, pp. 26-7.—Birch, in Layard's Nineveh and its Remains, ii., p. 340, note.—Layard, Babylon, pp. 212, 530, 629.

633 ARP*h*a-KaSD. Kitto, Cyclop., i. p. 229; but see ii, p. 398.—Volney, iv. pp. 249-50.—Lenormant, Cours, p. 203.—Bochart, p. 83.—Michaelis, Spicileg. Geog. Heb., ii., p. 75.—Dubois, Caucase, iii. pp. 421, 434, 488; iv. p. 342-3.—St. Martin, Mémoires, i. p. 205.—Ritter, Asien, vii. p. 320, seq.—Ainsworth, Assyria, pp. 152-156; and "An Evening at Diarbekir," Ainsworth's Mag., 1843, iv. pp. 221-6.—Loftus, in Rev. Archéol., 1850, p. 126.—Layard, Babylon, p. 628.

634 LUD. Herod., i. 7; vii. 74. Grote,

No. (of Notes, &c.)

Greece, i. pp. 127–30, 206, 320, 462, 618. — Raoul-Rochette, Archéologie Comparée, i. pp. 38, 206–227, 271–277, 284.—Champollion, Dict., p. 80.—Prisse, Salle des Ancêtres de Thotmes III., pp. 11–12. — Osburn, Test., pp. 27, 30, 44. — Tacitus, Annal. ii. 60, 4. — Birch, Annals of Thotmes III., pp. 158–60.

635 ARaM. Quatremère, Jour. des Sav., 1846, pp. 503–4. — Bochart, pp. 83–5.— Volney, iv. pp. 246–8. — Munk, Palestine, p. 435.—Champollion, Gram., pp. 500–1.—De Rougé, on Statue of Out'a-horsoun, Rev. Archéol., 7me Année, p. 15. — Judas, in op. cit., 1847, p. 622. — Layard, Babylon, p. 628.

636 *ā*UT*s*. De Wette, ii. pp. 554–70.—Bochart, pp. 90, 91. — Forster, "Sinaic Inscriptions," 1851, pp. 12–68; compared with Kircher, Œdipus Ægyptiacus, Amsterdam, 1652; ii. pp. 103–13.—Hunt, Himyaric Inscriptions, 1848; pp. 46.—Fresnel, Recherches, p. 23. — See also the "Asmonean," New York, 1852, March and April.

637 K*h*UL. Bochart, pp. 91–2. — Grotius, Annot., lib. i. de V. R. C.

638 GeT*te*R. Bochart, pp. 92–3.—Pauthier, Liv. Sac. de l'Orient. p. 465; and Kasimirski's "Koran," xxv. 40, 41.— Lane, Selections, p. 12–15. — Volney, iv. pp. 235, 249.—Pliny, iv. 36.—Solinus, c. 23.

639 MaS*h*. De Wette, ii. pp. 253–316.—Bochart, pp. 93–4. Forster, Geog., i. p. 284–5.

640 SaLaK*h*. Bochart, pp. 100–4.

641 *ĕi*BeR. Gliddon, Chapters, pp. 18, 19.— Lane, Modern Egyptians, Pref.—Gesenius, in De Wette, i. pp. 433–4.—Munk, Palestine, p. 102. — Lenormant, Cours, p. 203.—Fresnel, "Lettre à M. Mohl," Jour. Asiat., 1845, pp. 63–65.

642 PeLeG. "Hebrew Language;" see Gesenius, in De Wette, i. p. 459; and Bunsen, Eg. Pl., i. p. 270.—Athenæum Français, No. 1; Juillet, 1852, p. 7. — Lenormant, Cours, p. 214.

643 IoKTaN. Bochart, 109–12. — Fresnel, Arabes avant l'Islamisme, 1836, 1838.— Jomard, Arabie, in Mengin, iii. pp. 330, 346, 389–91. Forster, Geog., i. pp. 77–107.

644 ALMUDaD. Bochart, p. 112. — Volney, iv. p. 252. — Forster, i. pp. 107–11.

645 SeLeP*h*. Same references.

646 K*h*aT*s*RaMUT*t*. Add to the above,— Plate, Province of Hadramaut, Syro-Eg. Soc., 1845, pp. 112–23; and Jomard, op. cit., p. 349.

647 IeRaK*h*. Bochart, 124–7. — Forster, i. p. 115, 137–43. — Fresnel, 4me Lettre, "Djeddah, Jan. 1838."

648 HaDURaM. Bochart, pp. 128–30.—Sale's Introd. to Koran, Liv. Sac. d'Or., pp. 465–8.—Pococke, Specimen Hist. Arabum, p. 41. —Volney, iv. p. 252.

649 AUZaL. Bochart, p. 130–4. — Rosenmüller, Bibl. Geog., iii. p. 171. — Lane, Selections, p. 3. — Volney, iv. p. 253. — Forster, i. p. 145–7.

650 DiKLeH. Bochart, pp. 134–9.—Forster, i. pp. 147–8.

651 *ā*UBaL. References as above.

652 ABIMAL. Idem.

653 SeBA. Bochart, pp. 146–56.—Forster, i. pp. 154–7.

654 AUP*h*iR. Munk, Palestine, p. 294. — Volney, iv. pp. 255–76. — Bochart, pp. 156–61. — Michaelis, Quæstiones, No. 39. — Forster, i. pp. 165–71.

655 K*h*UILaH. } Same authorities.
656 IUBaB. }

---

657 Prichard, Researches, iii. p. 348.

658 Die Deutschen und die Nachbarstamme; Ibid.

659 Strauss, Vie de Jesus; Littré's transl., Paris, 1839; i. pp. 434, 436–7.

660 Oxlee, Letters to Archbishop of Cant., 2d series, London, 1845, p. 37.

661 Hennell, Origin of Christ., p. 299.

662 Vide Fresnel (Arabes avant l'Islamisme, 1836, p. 61), for a marvellous effort in Arabic by the Shèykh Abbas-el-Yamanèetee.

663 So read De Sola, Lindenthal, and Raphall, Genesis, p. 44.

664 Birch, Stat. Tabl. of Karnac, pp. 36–7. — Gliddon, Otia, p. 75.

665 Layard, Babylon, pp. 496, 506, 529, 543.

666 Lacour, ÆLOÏM, i. pp. 115, 129, 144–6.

667 De Sola's Bible, Genesis, p. 44.

668 Josephus, Antiq. Jud., lib. x. 11, 7

669, 670 N. B. These numbers are inadvertently omitted.

671 Cahen, Genèse, i. p. 188.

672 Ethnological Journal, London, 1848, pp. 197–226.

673 Introd. to the Canon. Scrip. of the Old Test.; Parker's transl., Boston, 1843; ii. pp. 78–82.

674 Account of the worship of Priapus, at Isermia, Naples; London, 1786.

675 Stromata, v. § 42.

676 Apuleius, Metamorph.; apud R. P. Knight Symbolical Language of Anc. Art, &c. Soc. of Dilettanti, 1835.

677 Humboldt, Cosmos, III., pp. 122–6.

678 See remains of Orpheus, Hesiod, Aristophanes, Damascius, &c., in Cory's Ancient Fragments, pp. 291–300; and Gliddon, Otia, pp. 55–6, on "Ereb."

679 Civilisation Primitive, 1845, p. 45—"Quia non supplices humi Mutino procumbimus atque Tutuno, ad interitum res lapsas, atque ipsum dicitis mundum leges suas et constituta mutasse?" (Arnobius, lib. iv. p. 133.)

680 Sama Veda, Kena-Oupanishad; Pauthier, Liv. Sac., Introd. p. 18.

681 Academical Lectures, Boston, 1840; ii. pp. 18–30.

682 Cahen, Genèse, i. p. 5, note. — Munk, Palestine, pp. 423, 445.

683 Peri-Archon, lib. iv. c. 2; Huet, Origeniana, p. 167.

684 Homil. vii. in Levit. — Franck, Kabbale, p. 166.

685 Strom., iii. 42; Righellini, Franc-Maçonnérie, i. p. 33.

686 Recognit., x. 30; Ibid., Mosaïsme et Christianisme, iii. p. 499.

687 Ibid., i. p. 29.

688 De Gen. contr. Manichéos, i. 1; Ibid., Maçonnérie, i. p. 33.

689 Epist. ad Helvet., iii.; Lenormant, Cours, p. 122.

690 Cf. Mosheim, i. p. 186.

No. (of Notes, &c.)

691 Hist. of Egypt, p. 574.

692 Hist. de l'École d'Alexandrie, ii. p. 69, seq.; and Biot, Astronomie Ancienne, p. 87, seq.

693 Chron. der Ægypter, i. pp. 125–48.—De Rougé, Rev. Archéol., 1853, pp. 671–86.

694 Cosmas-Ægyptius, Alexandrinus, Indicopleustes, wrote under Justinian, about 535, A. D. His "Topographia Christiana" was printed from MSS. by Montfaucon, in the "Collectio Nova Patrum et Scriptorum Græcorum;" Paris, 1706; fol., Tom. II.—Montfaucon's Latin version, pp. 190–1; Pl. ii. fig. 2.

695 Præfatio in Cosmæ, p. 4: with extracts from St. Augustine, Lactantius, Chrysostom, Severianus, "Beda; multique alii, quos recensere supervacaneum esset."

696, 697, and 698 Franck, Kabbale, pp. 102, 136–7.

699 Montfaucon's translation.

700 Cahen, xv. p. 172.—Noyes's Job, pp. 71, 194, note 18.

701 Harwood, German Anti-Supernaturalism, London, 1841.

No. (of Notes, &c.)

702 Mankind in Europe during the XIIIth century.

703 Lanci, La Sagra Scrittura Illustrata; Roma, 1827; cap. ix. 5; xi. 7.—Ibid. Paralipomeni all'illustrazione della Sagra Scrittura; Parigi, 1845; "Aleph-tau," parts ii. iii. and viii.

P. S. 1st Feb., 1854. To-day's mail has brought me the first number (Jan. 1,) of a "New Series" of the *Ethnological Journal*, edited by Luke Burke, Esq. (John Chapman, publisher, London). I have only space to express my hearty satisfaction at the re-appearance of this much-needed vehicle for free and manly thought; and to state that my colleagues, Dr. J. C. Nott, Dr. Henry S. Patterson, and the Hon. E. Geo. Squier, while vouching with myself for the great erudition, clear intellect, and high moral worth of its editor, have no hesitation in recommending it as an exponent of, as well as an admirable medium for, the most advanced views in Ethnology. —G. R. G.

APPENDIX II.

ALPHABETICAL LIST OF SUBSCRIBERS TO TYPES OF MANKIND.

E. S. Adrich, M. D., San Francisco, Cala.
Prof. Louis Agassiz, Cambridge, Mass.
John G. Aikin, Esq., Mobile, Ala.
J. H. Alexander, Esq., Baltimore, Md.
Thomas S. Alexander, Esq., "
Chilton Allan, Esq., Lexington, Ky.
Mrs. S. G. Allan, Richmond, Va.
Hon. Philip Allen, Providence, R. I.
Philip Allen, Jr., Esq., "
S. Austin Allibone, Esq., Philadelphia, Pa.
Col. J. S. Allison, Lexington, Ky.
S. Ames, M. D., Montgomery, Ala.
Thomas C. Amory, Jr., Esq., Boston, Mass.
C. G. Anderson, Esq., New Orleans, La.
L. H. Anderson, M. D., Mobile, Ala.
S. H. Anderson, M. D., Sumterville, Ala.
Alfred A. Andrews, Esq., Boston, Mass.
C. G. Andrews, Esq., New Orleans, La.
Rich'd Angell, M. D., Huntsville, Ala.
Hon. H. B. Anthony, Providence, R. I.
Nathan Appleton, Esq., Boston, Mass.
Samuel Appleton, Esq., " (2 copies.)
Rob't B. Armistead, Esq., Mobile, Ala.
Capt. Jos. J. Armstrong, "
Hon. Samuel G. Arnold, Providence, R. I.
Richard D. Arnold, M. D., Savannah, Ga.
J. H. Ashbridge, Esq., New Orleans, La.
Athenæum Library, Philadelphia, Pa.
Washington L. Atlee, M. D. "
W. P. Aubrey, Esq., Mobile, Ala.
C. Auzé, Esq., himself and friends, Mobile, Ala. (22.)

Franklin Bache, M. D., Philadelphia, Pa.
G. Bailey, Esq., Charleston, S. C.
Munro Banister, M. D., Richmond, Va.
Geo. C. Barber, Esq., Philadelphia, Pa.
Milton Barlow, Esq., Lexington, Ky.
Edward Barnett, Esq., New Orleans, La.
Henry Barnewall, Esq., Mobile, Ala.
Godfrey Barnsley, Esq., New Orleans, La., (2 copies.)
Dr. Barry, U. S. N., Washington, D. C.
Hon. J. R. Bartlett, Providence, R. I.
E. H. Barton, M. D., New Orleans, La.
Judge Bates, San Francisco, Cala.
Hon. James A. Bayard, Wilmington, Del.
R. Bean, M. D., New Orleans, La.
C. Beard, M. D., "
E. Begouen, Esq., Mobile, Ala.
Isaac Bell, Esq., "
N. B. Benedict, M. D., New Orleans, La.
Henry C. Berrie, M. D., Philadelphia, Pa.
Thos. F. Betton, M. D., Germantown, Pa.
J. G. Bibby, Esq., New Orleans, La.
Clement C. Biddle, Esq., Philadelphia, Pa.
Henry J. Bigelow, M. D., Boston, Mass.
Samuel Birch, Esq., British Museum, London.
James Birney, Esq., Mobile, Ala.
Geo. S. Blanchard, Esq., for Merc. Lib., Boston, Mass.
Col. W. W. S. Bliss, U. S. A., New Orleans, La.
G. W. Blunt, Esq., New York.
Henry S. Boardman, Esq., Philadelphia, Pa.
Geo. Boldin, Esq., "
S. M. Bond, Esq., Savannah, Ga.
James Bordley, M. D., Baltimore, Md.
Henry I. Bowditch, M. D., Boston, Mass.
W. B. Bowman, Esq., Mansfield, O.
M. Boullemet, Bookseller, Mobile, Ala., (10 copies.)
Thos. J. Bouve, Esq., Boston, Mass.
Burwell Boykin, Esq., Mobile, Ala.
E. M. Boykin, M. D., Camden, S. C.
J. F. Boynton, Esq., Syracuse, N. Y.
A. P. Bradbury, Esq., Bangor, Me.
Charles F. Bradford, Esq., Roxbury, Mass.
Dr. Brierly, San Francisco, Cala.
M. Bright, Jr., Esq., Mobile, Ala.
Geo. Brinley, Esq., Hartford, Conn.
Jno. M. Broomal, Jr., Esq., Chester, Pa.
A. Brother, Esq., New Orleans, La.
Geo. L. Brown, Esq., Mobile, Ala.
N. H. Brown, Esq. "
Jno. Brown, Esq., "
Peter A. Browne, Esq., Philadelphia, Pa.
Jos. Bryan, Esq., Savannah, Ga.
George S. Bryant, M. D., Aberdeen, Mi.
G. S. Bryant, Newbern, Ala.
Jos. Brummel, Esq., Richmond, Va.
Sam. D. Buck, Bookseller, Hopkinsville, Ky., (10 cop.)
Thos. C. Buckley, Esq., N. Y.
W. Gaston Bullock, Esq., Savannah, Ga.
Capt. Owen Burns, Wilmington, N. C.
M. Burton, Esq., Richmond, Va.
W. M. Burwell, Esq., Lynchburg, Va.
Dr. Geo. Bush, New York.
W. A. Butters, Esq., Richmond, Va.
H. L. Byrd, M. D., Savannah, Ga.

D. J. Cain, M. D., Charleston, S. C.
James Campbell, Esq., Mobile, Ala.
Edwin Canter, M. D., New Orleans, La.
Geo. W. Carpenter, Esq., Germantown, Pa.
Jesse Carter, M. D., Mobile, Ala.
A. H. Cenas, M. D., New Orleans, La.
Paul Chaudron, Esq., Mobile, Ala., (15 copies.)
Chas. M. Cheves, Esq., Charleston, S. C.

Langdon Cheves, Jr., Esq., Charleston, S. C.
Julian J. Chisolm, M. D., "
Samuel Choppin, M. D., New Orleans, La.
N. T. Christian, Esq., Georgetown, Ga.
Rev. Dr. J. D. Choules, Newport, R. I.
Jno. C. Claiborne, Esq., New Orleans, La.
A. Clapp, M. D., New Albany, Ia.
W. R. Clapp, Esq., Philadelphia, Pa., (2 copies.)
Jas. M. Clark, Esq., Providence, R. I,
Major M. Lewis Clark, St. Louis, Mo., (2 copies.)
C. Cleaveland, Esq., Yazoo City, Miss.
J. Breckenridge Clemens, M. D., Easton, Pa.
G. B. B. Clitherall, Esq., Mobile, Ala.
Stephen Colwell, Esq., Philadelphia, Pa.
Col. M. I. Cohen, Baltimore, Md.
Octavus Cohen, Esq., Savannah, Ga.
Henry A. Coit, Esq., New York.
A. Comstock, Esq., "
A. Comstock, M. D., Philadelphia, Pa.
Timothy Conrad, Esq., "
Miss Anna S. Coolidge, Boston, Mass.
W. C. Cooper, Esq., Savannah, Ga.
—— Corbet, Esq., Brit. Legation, Washington, D. C.
W. W. Corcoran, Esq., Washington, D. C.
Chas. S. Coxe, Esq., Philadelphia, Pa.
Jno. C. Cresson, Esq., "
John Crickard, Esq., New Orleans, La.
Charles P. Curtis, Esq., Boston, Mass., (2 copies).
Thos. B. Curtis, Esq., "
Hermann Curtius, Esq., New Orleans, La.
Theod. Cuyler, Esq., Philadelphia, Pa.

Mrs. R. P. Dana, New York.
W. H. Dandridge, Esq., Gainesville, Ala.
Hon. John M. Daniel, Richmond, Va.
W. C. Daniell, M. D., Savannah, Ga.
John Darrington, Esq., Mobile, Ala.
Isaac Davenport, Esq., Richmond, Va.
Chas. Davis, Esq., New York.
Jos Barnard Davis, F. S. A., Shelton, England.
Major Geo. Deas, U. S. A., Mobile, Ala.
Henry Deas, Esq., "
W. C. Deas, Esq., "
Zach. Deas, Esq., "
G P. Delaplaine, Esq., Madison, Wis.
A. B. Deloach, M. D., Livingston, Ala.
John Devereux, Esq., Raleigh, N. C.
Joseph Devilin, Esq., Mobile, Ala.
Rev. Henry M. Dexter, Boston, Mass.
Thos. Dexter, Esq., Mobile, Ala., (4 copies.)
Chas D. Dickey, Esq., "
Prof. S. Henry Dickson, Charleston, S. C.
L. Poulson Dobson, Esq., Philadelphia, Pa.
Geo. W. Dorr, Esq., New York.
Jas. Augustus Dorr, Esq., "
Geo. Douglass, Esq., Goshen Hill, S. C.
Sam'l R. Dubbs, Esq., Philadelphia, Pa.
B. F. Duncan, Esq., Jackson, Ala.
W. B. Duncan, Esq., New York.
Hon. James Dunlop, Pittburg, Pa.
E. Durand, Esq., Philadelphia, Pa.

A. M. Eastman, Esq., New York
Chas. J. M. Eaton, Esq., Baltimore, Md.
Geo. N. Eaton, Esq. "
Jno. H. Ecky, Esq. Philadelphia, Pa.
Dr. Ege, San Francisco, Cala.
Jno. A. Elkinton, M.D., Philadelphia, Pa.
Albert T. Elliott, Esq., Providence, R. I.
W. N. Ellis, P. M., Lippican, Mass.
David F. Emery, Esq., West Newbury, Mass.
Moses H. Emery, Esq., Philadelphia, Pa.
Robert D. England, M. D., Mobile, Ala.
T. C. English, Esq., Mobile, Ala.
Richard Esterbrook, Esq., New Orleans, La., (2 cop.)
F. A. Eustis, Esq., Milton, Conn.
Alexander Everett, Esq., Mobile, Ala.
C. C. Everett, Esq., Brunswick, Me.
Hon. E. Everett, for Lib. State Dep., Washington.
Hon. Edward Everett, Boston, Mass.

John Fagan, Esq., Philadelphia, Pa.
Prof. J. E. Farman, Georgetown, Ky.
C. C. S. Farrar, Esq., New Orleans, La.
J. Farrell, M. D., "
Daniel Fearing, Esq., New York.
E. D. Fenner, M. D., New Orleans, La.
Chas. W. Fisher, Esq., Philadelphia, Pa.
Redwood Fisher, Esq., "
Dr. Fonerden, for Md. Hospital, Baltimore, Md.
E. G. Forshey, Esq., New Orleans, La.
Geo. Fort, M. D., Milledgeville, Ga.
B. W. Fosdick, Esq., Savannah, Ga.
Wm. B. Fosdick, Esq., Boston, Mass.
Hillary Foster, Esq., Mobile, Ala.
W. Parker Foulke, Esq., Philadelphia, Pa.
Prof. Jno. F. Frazer, "
J. B. Futch, Esq., New Orleans, La.

Charles Ganahl, M. D., Savannah, Ga.
P. C. Gaillard, M. D., Charleston, S. C.
A. Gaines, M. D., Mobile, Ala.
E. B. Gardette, M. D., Philadelphia, Pa.
James Gardiner, Esq,, San Francisco, Cala.
John L. Gardiner, Esq., Boston, Mass.
J. R. Gardner, Esq., New Orleans, La.
L. M. Gaylord, M. D., Sodus, N. Y.
David Geiger, M. D., Charleston, S. C.
R. W. Gibbes, M. D., Columbia, S. C.
Mrs. M. A. E. Gibson, Richmond, Va.
Jno. Gibson, Esq., Mobile, Ala.
David Gilbert, M. D., Philadelphia, Pa.
Hon. Henry D. Gilpin, " (2 copies.)
Thomas Gilpin, Esq. "
F. E. Gordon, Esq., Mobile, Ala.
Theo. Gordon, Esq., "
W. M. Guilford, M. D., Lebanon,Pa.
Wm. Graddy, Esq., Georgetown, Ga.
Calvin Graddy, Esq., "
Edmund A. Grattan, Esq., H. B. M. Cons., Boston.
Jno. Gravely, Esq., Charleston, S. C.
Hon. John C. Gray, Boston, Mass.
Charles Green, Esq., Savannah, Ga.
A. J. Green, M. D., Columbia, S. C.
J. Green, M. D., Washington, D. C.
J. Green, Esq., for Merc. Lib. Co., Baltimore, Md.
D. S. Greenough, Esq., Boston, Mass.
W. W. Greenough, Esq., "
John Grigg, Esq., Philadelphia, Pa.
James Grignon, Esq., H. B. M. Cons., Portland, Me.
Edmund Grundy, Esq., Philadelphia, Pa.

John Haig, Esq., Mobile, Ala.
R. K. Haight, Esq., New York, (5 copies.)
Jno. S. Haines, Esq., Germantown, Pa.
C. S. Hale, Esq., Burlington, N. J.
Rev. A. O. Halsey, Richborough, Pa.
John Halsey, Esq., New York, (5 copies.)
Hon. J. H. Hammond, Charleston, S. C.
M. C. M. Hammond, Esq., "
P. T. Hammond, Lancaster, S. C.

C. F. Hampton, Esq., Columbia, S. C.
W. Hampton, Esq, "
W. Hampton, Jr., Esq., "
Geo. S. Harding, Esq., Savannah, Ga.
General Jos. Harlan, Philadelphia, Pa.
S. N. Harris, M. D., Savannah, Ga.
Jas. B. Harrison, Esq., Georgetown, Ga.
Samuel T. Harrison, Esq., Washington, D. C.
Thos. Willis Hartley, Esq., Philadelphia, Pa.
John Hastings, M. D., San Francisco, Cala.
Judge Hastings, "
Elias S. Hawley, Esq., Buffalo, N. Y.
W. G. Hay, Esq., Philadelphia, Pa.
Geo. Hayward, M. D., Boston, Mass.
E. H. Hazard, Esq., Providence, R. I.
Isaac P. Hazard, Esq, "
Thos. R. Hazard, Esq., "
Rev. G. W. Heacock, Buffalo, N. Y.
Dr. G. C. Hebbe, Washington, D. C.
Alfred Hennen, Esq., New Orleans, La.
Geo. M. Heroman, Bookseller, Baton Rouge, La. (4)
W. C. Henzey, Esq., Philadelphia, Pa.
T. Higham, Jr., Esq., S. C.
C. W. Hill, M. D., Mobile, Ala.
Geo. S. Hilliard, Esq., Boston, Mass.
W. L. Hodge, Esq., for Lib. Trs. Dep., Washington.
W. B. Hodgson, Esq., Savannah, Ga.
Judge Ogden Hoffman, San Francisco, Cala.
J. E. Holbrook, M. D., Charleston, S. C.
Geo. Holly, Esq., Mobile, Ala.
Prof. Francis S. Holmes, Charleston, S. C.
O. W. Holmes, M. D., Boston, Mass.
Thos. F. Hoppin, Esq., Providence, R. I., (2 copies.)
Daniel Horlbeck, Esq., Charleston, S. C.
Henry Horlbeck, Esq., "
Mrs. Lavinia E. A. Howard, Daphne, Mobile Bay, Ala.
Rev. Geo. Howe, D. D., Columbia, S. C.
Dudley Hubbard, Esq., Mobile, Ala.
Benj. F. Huddy, Esq., Philadelphia, Pa.
J. A. Huger, Esq. Charleston, S. C.
R. W. Hughes, Esq. Richmond, Va.
Thos. Hunt, M. D., New Orleans, La.
A. J. Huntington, Esq., "
Albert Hurd, Esq., Galesburg, Ill.
Henry J. Hyams, Esq., New Orleans, La.

Col. Irving, San Francisco, Cala.

Sam'l Jackson, M. D., Philadelphia, Pa.
Henry Jacobs, Esq., Providence, R. I.
Robert James, Esq., Mobile, Ala.
N. R. Jennings, Esq., New Orleans, La.
W. E. Jennings, Esq., Mobile, Ala.
Dr. J. C. Jennings, Bonn, Prussia.
Jas. P. Jervey, M. D., Charleston, S. C.
Gen. Thos. S. Jesup, U. S. A., Washington, D. C.
Gov. David Johnson, Limestone Springs, S. C.
W. E. Johnson, Esq., Camden, S. C.
T. A. Johnston, Esq., Livingston, Ala.
R. F. Johnstone, Esq., Detroit, Mich.
Allen C. Jones, Esq., Mobile, Ala., (2 copies.)
Edw'd E. Jones, Esq., Philadelphia, Pa.
James Jones, Esq., "
James Jones, M. D., New Orleans, La.
W. Cary Jones, Esq., San Francisco, Cala.
Wm. Jones, M. D., Mobile, Ala.
Wm. Jones, Jr., Esq., "
Messrs. Jordan & Bro., Philadelphia, Pa.
W. J. Joynes, Esq., Petersburg, Va.

Hon. J. P. Kennedy, for Lib. Navy Dep., Washington.
Hon. John P. Kennedy, Baltimore, Md.
James Kennedy, M. D., New York.
L. C. Kennedy, Esq., Charleston, S. C.
P. M. Kent, Esq., New Albany, Ind.
Edward M. Kern, Esq., Philadelphia, Pa.
Geo. Kern, Esq., "
Jno. Kern, Jr., Esq., "
Richard H. Kern, Esq., "
Elisha W. Keyes, Esq., Madison, Wis.
E. H. Kimbark, M. D., New York.
A. C. Kingsland, Esq., "
Robert L. Kirk, Esq., Mobile, Ala.
S. D. Kirk, Esq., Charleston, S. C.
James Kitchen, M. D., Philadelphia, Pa.
Jno. Kitchenmann, Esq., "
W. H. Klapp, M. D., "
Sam'l Kneeland, M. D., for Boston Soc. Nat. Hist.
S. Kneeland, Jr., M. D., Boston, Mass.
E. G. Knight & Co., Booksellers, Cleveland, O., (10 c.)
G. Kursheedt, Esq., New Orleans, La.

John De Lacey, M. D., Philadelphia, Pa.
John Lambert, Esq., "
I. A. Lapham, Esq., Milwaukie, Wis.
Prof. C. W. Lane, Milledgeville, Ga.
W. Langermann, Esq., San Francisco, Cala.
Henry Laurence, Esq., Yazoo City, Miss.
Hon. Abbott Lawrence, Boston, Mass.
James Lawrence, Esq., "
Wm. Beach Lawrence, Esq., Newport, R. I.
Jno. Laurence, Esq., Mt. Upton, Chenango Co., N. Y.
Edw'd Lawton, M. D., Boonville, Mo.
D. Leadbetter, Esq., Mobile, Ala.
Vr. Lecesne, Esq., "
Robert Lebby, M. D., Charleston, S. C.
Joseph Leidy, M. D., Philadelphia, Pa.
Col. Opt. H. Leja, Mobile, Ala.
J. C. Levy, Esq., Savannah, Ga.
S. Yates Levy, Esq., Savannah, Ga.
K. H. Lewis, Esq., Tarboro, N. C.
Levi Lewis, Spread Eagle, Pa.
Mifflin Lewis, Esq., Spread Eagle, Pa.
Richard H. Lewis, Esq., Mobile, Ala.
Saunders Lewis, Esq., Philadelphia, Pa.
Winslow Lewis, M. D. Boston, Mass.
Library of South Carolina College, Columbia, S. C.
Library Company of Easton, Pa.
Library of Young Men's Association, Buffalo, N. Y.
Jacob Little, Esq., New York.
Jack Littlejohn, Esq., Spartanburg, S. C.
Wm. Littlejohn, Esq., "
Chas. A. Locke, Esq., Boston, Mass.
J. L. Locke, Esq., Savannah, Ga.
Rev. S. K. Lothrop, D. D., Boston, Mass.
Robert Lovett, Jr., Esq., Philadelphia, Pa.
Andrew Low, Esq., Savannah, Ga.
Henry A. Lowe, Esq., Mobile, Ala., (2 copies.)
Francis C. Lowell, Esq., Boston, Mass., (5 copies.)
John A. Lowell, Esq., "
E. H. Ludlow, Esq., New York.
R. M. Lusher, Esq., New Orleans, La.

Rev. Geo. Macaulay, Milledgeville, Ga.
Wm. Mackay, Esq., Savannah, Ga.
Charles Magarge, Esq., Germantown, Pa.
Jas. Magee, Esq. New Orleans, La,
C. T. Mann, Esq., Yazoo City, Miss.
Peter Marcy, Esq., Mobile, Ala.
James B. Markham, Esq., Mobile, Ala.
J. H. Markland, Esq., Philadelphia, Pa.

Francis Markoe, Esq., Washington, D. C.
B. F. Marshall, Esq.,Mobile, Ala.
Chas. H. Marshall, Esq., New York.
E. Mason, M. D., Wetumka, Ala.
C. H. Mastin, M. D., Mobile, Ala.
H. B. Mattison, Esq.,Washington, D. C.
Joseph Mauran, M. D., Providence, R. I.
B. Mayer, Esq., for Md. Hist. Soc., Baltimore, Md.
W. E. Mayhew, Esq., Baltimore, Md.
Hon. Theo. H. McCaleb, New Orleans, La.
Jas. McClean, Esq., Philadelphia, Pa.
J. H. B. McClellan, M. D., "
Thos. McConnell, Esq., Mobile, Ala.
J. H. McCulloh, Esq., Baltimore, Md.
E. H. McDonald, Esq., Philadelphia, Pa.
T. F. McDow, M.D., Liberty Hill, S. C.
Wm. McGuigan, Esq., Philadelphia, Pa.
Messrs. McKee & Robertson, Hagerstown, Md.
P. B. McKelvey, M. D., New Orleans, La.
Andrew McLaughlin, Esq., Baltimore, Md.
Mrs. McPherson, Baltimore, Md.
M. Megonegal, Esq., Philadelphia, Pa., (2 copies.)
Charles D. Meigs, M. D., Philadelphia, Pa.
J. Aitken Meigs, M. D., "
J. Forsyth Meigs, M. D., "
Thos. Mellon, Esq., "
N. L. Merriweather, Esq., Montgomery, Ala., (5 cop.)
M. H. Messchert, Esq., Philadelphia, Pa.
John G. Michener, Esq., "
Francis T. Miles, M. D., Charleston, S. C.
Clark Mills, Esq., Washington, D. C.
Charles Millspaugh, M. D., Richmond, Va.
J. F. G. Mittag, Esq., Lancaster, S. C.
E. J. Mollet, Esq., New York,
James Moncreif, Esq., New York.
Cyrus C. Moore, M. D., Philadelphia, Pa.
Comm. E. W. Moore, Texan N., Washington, D. C.
S. Mordecai, Esq., Richmond, Va.
James W. Morgan, Esq., Lynchburg, Va.
Israel Morris, Esq., Philadelphia, Pa.
Jacob G. Morris, Esq., "
John S. Morris, Esq., Phœnixville, Pa.
T. H. Morris, Esq., Baltimore, Md.
B. M. Moss, M. D., New Orleans, La.
E. L. Moss, Esq., Philadelphia, Pa.
Valentine Mott, M. D., New York.
James Moultrie, M. D., Charleston, S. C.
John Munro, Esq., San Francisco, Cala.
Wm. M. Murray, Esq., Charleston, S. C.
G. A. Myers, Richmond, Va.

M. H. Nace, Esq., Richmond, Va.
T. C. Newbold, Esq., Philadelphia, Pa.
Thos. A. Newhall, Esq., Germantown, Pa.
H. Newman, Esq., Boston, Mass.
J. B. Newman, Esq., Washington, D. C.
Jos. Newton, Esq., Philadelphia, Pa.
New York Society Library, N. Y.
W M. Nicholls, Esq., Chesterville, S. C.
B. M. Norman, Bookseller, New Orleans, La., (25 cop.)
Gustavus A. Nott, M. D., New Orleans, La.
James Nott, M. D. San Francisco, Cala.
Jno. R. Nunemacher, Esq., New Albany, Ind. (2 cop.)

Rob't W. Ogden, Esq., New Orleans, La.
J. W. Osgood, Esq., Saxonville, Mass.
J. W. Orr, Esq., New York, (5 copies.)
Rev. S. Oswald, York, Pa.

Edward Padelford, Esq., Savannah, Ga.
B. R. Palmer, M. D., Pittsburg, Pa.
John S. Palmer, M. D., Charleston, S. C.
Alexander Pantoleon, A. M. Smyrna, Turkey.
Comm. F. A. Parker, U. S. N., Philadelphia, Pa.
Henry T. Parker, Esq., Boston, Mass.
Capt. James Parker, Mobile, Ala.
Socrates Parker, Esq., Livingston, Ala,
S. Parkman, M. D., Boston, Mass.
Henry S. Patterson, M. D., Philadelphia, Pa.
Morris Patterson, Esq., "
Joseph Patterson, Esq., " (5 copies.)
Louis L. Pauly, Esq., "
Abraham Payne, Esq., Providence, R. I.
W. I. Peale, Esq., Philadelphia, Pa.
Miss Mary Pearsall, "
Davis Pearson, Esq., "
John Penington, Esq. "
Amos Pennebaker, M. D., "
J. A. Pennypacker, M. D., "
Granville J. Penn, Esq., Penn Castle, England.
I. Pennington, Esq., Baltimore,Md., (2 copies.)
Mrs. C. W. Pennock, Philadelphia, Pa.
J. W. Perard, Jr., Esq., New York.
Chas. T. Percival, M. D., Mobile, Ala.
O. H. Perry, Esq., for Vig. Lib. Assoc., Baltimore, Md.
Rob't E. Peterson, Esq., Philadelphia, Pa.
Jesse E. Peyton, Esq., "
Philadelphia Library Company, Philadelphia, Pa.
Jona. Phillips, Esq., Boston, Mass.
John Phillips, M. D., Bristol, Pa.
Hon. P. Phillips, Mobile, Ala.
Charles Pickering, M. D., Boston, Mass.
J. C. Pickett, Esq., Washington, D. C.
E. B. Pierson, M. D., Salem, Mass.
Henry L. Pierson, Esq., New York.
Hon. Albert Pike, Little Rock, Arkansas.
Wm. M. Pippen, Esq., Tarboro, N. C.
J. N. Platt, Esq., New York.
George Poe, Esq., Washington, D. C.
J. G. Poindexter, Esq., New Orleans, La.
Prof. F. A. Porcher, Charleston, S. C.
George Porteus, Esq., Mobile, Ala.
John Potts, Esq., Chihuahua, Mexico.
I. Pratt, M. D., Philadelphia, Pa.
Wm. Pratt, Esq., Baltimore, Md.
Wm. H. Pratt, Esq., Mobile, Ala.
J. H. Prentice, Esq., New York.
J. S. Preston, Columbia, S. C.
H. C. Price, Esq., Chester, Pa.
Isaac Pugh, Esq., Philadelphia, Pa.
Jno. M. Pugh, M. D., West Philadelphia, Pa.
G. P. Putnam & Co., Publishers, New York, (10 cop.)

B. Howard Rand, M. D., Philadelphia, Pa.
Jno. Randall, Esq., New York.
R. C. Randolph, M. D., Greensboro, Ala.
Edmund Ravenal, M. D., Charleston, S. C.
Edward Rawle, Esq., New Orleans, La.
Daniel T. Rea, Esq., Mobile, Ala.
J. B. Read, Esq., Savannah, Ga.
Wm. Reed, Esq., New Orleans, La.
J. J. Reese, M. D., Philadelphia, Pa.
John R. Reid, Esq., New Orleans, La.
D. Elliott Reynolds, M. D., New Orleans, La.
Col. James Rice, San Francisco, Cala.
W. Bordman Richards, Esq., Boston, Mass.
W. W. Richards, Esq., Philadelphia, Pa.
Maurice Richardson, Esq., Great Valley, Pa.
J. L. Riddell, M. D., New Orleans, La.
Mrs. G. W. Riggs, Baltimore, Md.
J. H. Riley & Co., Booksellers, Columbus, O., (5 cop.)
Thomas Ritchie, Esq., Washington, D. C.

Col. George Rivers, Providence, R. I
J. A. Roberts, Greensville, Pa.
W. Lea Roberts, Esq., New York.
F. M. Robertson, M. D., Charleston, S. C
John Blount Robertson, Esq., New Orleans, La.
Col. W. S. Rockwell, Milledgeville, Ga.
Prof. Henry D. Rogers, Boston, Mass.
Chas. H. Rogers, Valley Forge, Pa.
Hon. Molton J. Rogers, Philadelphia, Pa.
Jno. S. Rohrer, M. D., "
O. A. Roorback, Bookseller, New York, (16 copies.)
Wm. Ropes, Esq., Boston, Mass.
A. H. Rosenheim, Esq., Philadelphia, Pa.
James S. Rowe, Esq., Bangor, Me.
Samuel Ruffin, Esq., Mobile, Ala.
E. H. Rugbee, Esq., Providence, R. I.
James Rush, M. D., Philadelphia, Pa.
Mrs. Rush, "
John Russell, Bookseller, Charleston, S. C., (3 copies.)
Charles Ryan, Esq., Philadelphia, Pa.
Rev. Dr. Ryerson, Toronto, Canada, (2 copies.)

B. J. Sage, Esq., New Orleans, La.
Richard G. Sager, Esq., Mobile, Ala.
Hon. James Savage, Boston, Mass.
W. H. De Saussure, Charleston, S. C.
J. P. Scriven, M. D., Savannah, Ga.
Chas. Scott, Esq., Philadelphia, Pa.
John Scoville, Esq., Salisbury, Conn.
E. M. Seabrook, Esq., Charleston, S. C.
Hon. Benjamin Seaver, Boston, Mass.
P. T. Seibel, M. D., Savannah, Ga.
S. E. Sewall, Esq., Boston, Mass.
George C. Shattuck, Esq., Boston, Mass., (2 copies.)
Lemuel Shattuck, Esq., "
Quincy A. Shaw, Esq., "
Robert G. Shaw, Esq., " (2 copies.)
R. O. Shaw, M. D., Mobile, Ala.
W. W. Shearer, Esq., Livingston, Ala.
—— Shepherd, Esq., Cairo, Egypt.
John H. Sherard, Esq., Livingston, Ala.
W. Sherman, Esq., New York.
Nath. B. Shurtleff, M. D., Boston, Mass.
Origen Sibley, Esq., Mobile, Ala.
Hon. Chas. Sitgreaves, New Jersey.
H. N. Skinner, Esq., New York.
J. R. Slack & Co., Booksellers, Steubenville, O., (3 c.)
Jno. Sloan, M. D., New Albany, Ind.
A. A. Smets, Esq., Savannah, Ga.
F. Gurney Smith, Jr., M. D., Philadelphia, Pa.
Howard Smith, M. D., New Orleans, La.
Jacob Smith, Esq., Georgetown, Ga.
J. Broom Smith, Esq., San Francisco, Cala.
Jno. Jay Smith, Esq., Germantown Pa.
Joseph P. Smith, Esq., Philadelphia, Pa.
J. E. J. Smith, Esq., Georgetown, Ga.
John T. Smith, Esq., Livingston, Ala.
Samuel Smith, Esq., New York.
J. A. Spencer, Esq., Philadelphia, Pa.
Truman & Spofford, Booksellers, Cincinnati, O., (5 c.)
Hon. E. Geo. Squier, New York.
Wm. H. Squire, M. D., Germantown, Pa.
W. E. Stacke, Esq., New Orleans, La.
W. H. Stark, Esq., Mobile, Ala.
Albert Stein, Esq., Mobile, Ala.
Jacob Steiner, Esq., Philadelphia, Pa.
J. P. Steiner, Esq., "
Claudius C. Stewart, Esq., Florida.
Wm. Stevenson, Esq., Baltimore, Md.
D. D. Stewart, M. D., "
F. Stewart, Esq., Mobile, Ala.

Scott Stewart, M. D., Philadelphia, Pa.
Wm. Stewart, Esq., Hagerstown, Md., (2 copies.)
John Stoddard, Esq., Savannah, Ga.
Prof. I. M. Stone, Hanover, Ind.
Warren Stone, M. D., New Orleans, La.
Lt. Isaac G. Strain, U. S. N., Philadelphia, Pa.
Wm. Strickland, Bookseller, Mobile, Ala., (10 copies.)
Col. C. B. Strode, San Francisco, Cala., (10 copies.)
Hon. A. H. H. Stuart, for Lib. Dep. Int., Washington.
Albert Sumner, Esq., Newport, R. I.
Hon. Charles Sumner, Washington, D. C.
Chas. G. Swartz, Esq., Philadelphia, Pa.
Jos. Swift, Esq., "
Samuel Swett, Esq., Boston, Mass.
Mrs. T. A. Swett, "

T. A. Tankusley, Esq., Mobile, Ala.
Benjamin Tanner, Esq., Baltimore, Md.
Rev. S. K. Talmage, LL. D., Milledgeville, Ga.
Henry W. Taylor, Esq., Mobile, Ala.
Wm. Taylor, Esq., Richmond, Va.
J. K. Tefft, Esq., Savannah, Ga.
J. S. Teft, Bookseller, Houston, Texas, (10 copies.)
Carlisle Terry, M. D., Georgetown, Ga.
Charles L. Tew, Esq., New Orleans, La.
Richard H. Thomas, M. D., Baltimore, Md.
Edwin Thompson, Esq., Philadelphia, Pa.
John Thorne, Esq., Baltimore, Md.
Col. James J. Thornton, Mobile, Ala.
B. C. Ticknor, Esq., Mansfield, O.
Osmond Tiffany, Esq., Baltimore, Md.
Howard Tilden, Esq., Philadelphia, Pa.
J. Tisdale, Esq., Boston, Mass.
Dr. Toland, San Francisco, Cala.
Gen. Joseph Totten, U. S. A., Washington, D. C.
Henry Toulmin, Esq., Mobile, Ala.
Morton Toulmin, Esq., "
Elisha Townsend, Esq., Philadelphia, Pa.
Robert Trueman, Esq., Boston, Mass.
David H. Tucker, M. D., Richmond, Va.
J. W. Tucker, Esq., Spartanburg, S. C.
Wm. E. Tucker, Esq., Philadelphia, Pa.
Fred'k Tudor, Esq., Boston, Mass.
Alexander Turnbull, Esq., Baltimore, Md.
T. I. Turner, M. D., U. S. N., Philadelphia, Pa.
Prof. M. Tuomey, Tuscaloosa, Ala.
J. W. Tuthill, Esq., New Orleans, La.
J. A. Tyler, Esq., Boston, Mass.

J. [illegible]. Uhlhorn, Esq., New Orleans, La.

Aaron Vail, Esq., New York.
Jacob B. Vandever, Esq., Wilmington, Del.
Col. Henry Vaughan, Yazoo City, Mi.
W. S. Vaux, Esq., Philadelphia, Pa.
A. L. Vegus, Mobile, Ala.
Henry Vollmer, Esq., Philadelphia, Pa.

Henry Wadsworth, M. D., Philadelphia, Pa.
George H. Walker, Esq., New Orleans, La.
Isaac R. Walker, M. D., Spread Eagle, Pa.
Rev. J. B. Walker, Mansfield, O.
J. J. Walker, Esq., Philadelphia, Pa.
S. J. Walker, Esq., Mobile, Ala.
James P. Walker, Esq., Lowell, Mass.
John N. Walthall, Esq., Mobile, Ala.
J. J. V. Wanroy, Esq., "
J. C. Warren, M. D., Boston, Mass.
J. Mason Warren, M. D., "
Jas. S. Waters, Esq., Baltimore, Md.

Col. John G. Watmough, Germantown, Pa.
Thomas H. Webb, M. D., Providence, R. I.
Nicholas Weeks, Esq., Mobile, Ala.
A. J. Wedderburn, M. D., New Orleans, La.
Plowden C. J. Weston, Esq., Hagley, S. C.
T. M. Wetherill, Esq., Laurel Hill, La.
Wm. Wetherill, M. D., Philadelphia, Pa.
W. West, Esq., Philadelphia, Pa.
Chas. M. Wheatley, Esq., Phœnixville, Pa., (4 copies.)
Wm. Augustus White, Esq., N. York.
Benjamin A. White, M. D., Milledgeville, Ga.
Eli White, Esq., New York.
Hon. W. H. Witte, Philadelphia, Pa., (2 copies.)
Rev. R. S. Whitehall, New Orleans, La.
E. D. Whitehead, Esq., Havanna, Green Co., Ala.
W. C. Wilde, Esq., New Orleans, La.
Capt. Charles Wilkes, U. S. N., Washington, D. C
John Williams, Esq., Lancaster, S. C.
W. C. Williams, M. D., Philadelphia, Pa.
Hon. W. Thorne Williams, Savannah, Ga.
W. Williamson, M. D., Philadelphia, Pa.
A. P. Willis, Esq., New Orleans, La.
Chas. Wilson, Esq., Savannah, Ga.
T. McK. Wilson, Esq., Cannonsburg, Pa.
Rev. W. D. Wilson, D. D., Geneva, N. Y.
Hubbard Winslow, Esq., Boston, Mass.
John Wiltbank, M. D., Philadelphia, Pa.
Philip Winfree, Jr., Esq., New Orleans, La.
James W. Winter, Esq., New York.
C. J. Wister, Esq., Germantown, Pa.
James H. Witherspoon, Esq., Lancaster, S. C.
Thomas R. Wolfe, Esq., New Orleans, La.
Wm. B. Wolfe, Esq., Philadelphia, Pa.
A. Wolle, Esq., Bethlehem, Pa.
F. Wolgamuth, Esq., Philadelphia, Pa.
Messrs. Wood & Conner, Carlisle, Pa.
A. T. Wood, Esq., New Orleans, La.
George B. Wood, M. D., Philadelphia, Pa.
Rev. W. D. Wood, D. D., Geneva, N. Y.
Mrs. Woodbury, New York.
H. A. Wright, Esq., Madison, Wis.
Wm. Wright, M. D., Philadelphia, Pa.
Jacob Wyand, Esq., Philadelphia, Pa.
Wm. W. Wyatt, Esq., "
Messrs. Wylie, Mobley & Strait, Lancaster, S. C.
Samuel G. Wyman, Esq., Baltimore, Md.
Thomas K. Wynne, Esq., Richmond, Va.

Gregory Yale, Esq., San Francisco, Cala.
Jno. C. Yeager, Esq., Philadelphia, Pa.
Philip Yeiser, M. D., New Orleans, La.
Harry M. Young, Baltimore, Md.
J. A. Young, Esq., Camden, S. C.
John B. Young, Esq., Richmond, Va.

ADDITIONAL NAMES,

RECEIVED SINCE THE ABOVE LIST WAS MADE OUT.

G. W. Ball, Esq., Philadelphia, Pa.
A. Billings, Esq., Nashville, Tenn.
Beriah Brown, Esq., Madison, Wis.
Wm. H. Van Buren, M. D., New York.
Stacy B. Collins, Esq., "
John Le Conte, Esq., Philadelphia, Pa.
Jno. Le Conte, Jr., Esq., "
T. J. Crowen, Bookseller, New York, (2 copies.)
Gov. Nelson Dewey, Lancaster, Wis.
John Evans, Esq., West Haverford, Pa.
W. Wayne Evans, Esq., Paoli, Pa.
Felix B. Gaudet, Esq., Philadelphia, Pa.
A. T. Gray, Esq., Madison, Wis.
Prof. S. S. Haldeman, Columbia, Pa.
Charles H. Hall, Esq., Philadelphia, Pa.
E. H. Janssen, Esq., Madison, Wis.
Jno. McBride, M. D., Philadelphia, Pa.
B. Meyer, Esq., Baltimore, Md.
Joshua Moss, Esq., Birmingham, England, (2 copies.)
J. West Nevins, Esq., Philadelphia, Pa.
Jo. S. Pender, Esq., Tarboro, N. C.
Library of Pennsylvania Hospital, Philadelphia, Pa.
D. T. Pratt, Esq., Philadelphia, Pa.

THE END.

www.ingramcontent.com/pod-product-compliance
Lightning Source LLC
LaVergne TN
LVHW101439200926
848483LV00004B/410

* 9 7 8 1 4 2 5 5 9 6 6 4 4 *